PION–PION INTERACTIONS IN PARTICLE PHYSICS

PION–PION INTERACTIONS IN PARTICLE PHYSICS

B. R. MARTIN

Department of Physics and Astronomy
University College London

D. MORGAN

Rutherford Laboratory

G. SHAW

Department of Theoretical Physics
University of Manchester
and Daresbury Laboratory

1976

ACADEMIC PRESS

London New York San Francisco
A Subsidiary of Harcourt Brace Jovanovich, Publishers

ACADEMIC PRESS INC. (LONDON) LTD.
24/28 Oval Road
London NW1

United States Edition published by
ACADEMIC PRESS INC.
111 Fifth Avenue
New York, New York 10003

Library of Congress Catalog Card Number: 75–42582
ISBN: 0–12–474740–X

Printed in Great Britain by
J. W. ARROWSMITH LIMITED
Winterstoke Road
Bristol BS3 2NT

Preface

Complete crossing symmetry, absence of spin complications, and the low mass of the pion have made pion–pion scattering a favourite topic of theoretical study for many years. On the other hand, it is only relatively recently that a reliable body of experimental information has become available. This is therefore an opportune time to summarize this information, and to review our theoretical understanding of both this and other processes where the pion–pion interaction plays a key role. That is the aim of this book.

The field as defined above is a very large one and we have not attempted to give a detailed discussion of all topics. On the other hand, we have sought, within the limits of space, to treat the main themes in some depth, in several cases with supporting Appendices. The level of the book is such that it should be of use to all practising physicists in the field of elementary particles, including post-graduate students.

During the course of writing this review we have benefited greatly from discussions with many of our colleagues. We particularly thank: D. Atkinson, J. L. Basdevant, J. P. Baton, D. M. Binnie, W. Blum, P. D. B. Collins, R. Diebold, P. Estabrooks, C. D. Froggatt, C. J. Goebel, L. Gutay, B. Hyams, L. D. Jacobs, C. Jones, G. L. Kane, D. H. Lyth, W. Männer, A. D. Martin, H. Nielsen, G. C. Oades, W. Ochs, M. R. Pennington, J. L. Petersen, A. C. Phillips, M. K. Pidcock, J. Pisút, J. K. Storrow, E. P. Tryon, R. Vinh Mau and F. Wagner. In addition, thanks are due to Miss Una Campbell for her excellent preparation of the diagrams.

B. R. Martin, D. Morgan, G. Shaw
July, 1975

Contents

Preface v

PART I INTRODUCTION

Chapter 1 Introduction 3

Chapter 2 Theoretical Framework and Notation 7

(a) Kinematics. (b) Amplitudes and cross-sections. (c) Isospin and crossing. (d) Unitarity. (e) Mandelstam analyticity

PART II SOURCES OF INFORMATION ON $\pi\pi$ SCATTERING

Chapter 3 Pion Production 23

3.1 Peripheral Di-Meson Production (1): OPE And Its Backgrounds 24

(a) Introduction. (b) Kinematic formalism for the process $PN \to M^*B$. (c) Formulae for OPE. (d) OPE compared to experiment. (e) Models for the background to OPE. (f) Parameterization of the very small t-region. (g) Amplitude analysis. (h) Applications. (i) Summary

3.2 Peripheral Di-Meson Production (2): Extracting $\pi\pi$ and $K\pi$ Phase Shifts 65

(a) Introduction. (b) Preliminary survey of the alternatives. (c) Goebel–Chew–Low extrapolation. (d) Amplitude analysis and extrapolation—Estabrooks and Martin's method. (e) Phase shifts from amplitude analysis of averaged moments in the physical region. (f) Other topics in brief. (g) Conclusions

3.3 Peripheral Di-Meson Production (3): Results On $\pi\pi$ And $K\pi$ Phase Shifts 84

(a) Objectives and Problems. (b) Data and phase shift analyses for $\pi\pi$ scattering. (c) Results on $I=2$ $\pi\pi$ phase shifts. (d) Results on $I=0$ and 1 $\pi\pi$ phase shifts below 950 MeV. (e) $\pi\pi$ phase shifts above 950 MeV. (f) $K\bar{K}$ threshold effect—the S^* resonance. (g) $K\pi$ scattering. (h) Other related channels—$K\bar{K}$ and $\pi\eta$

3.4 Nucleon–Antinucleon Annihilations 121

(a) Formation channels. (b) $N\bar{N}\to 3\pi$ and related processes

Chapter 4 Processes Involving Dipion Exchange 137

4.1 Elastic Pion–Nucleon Scattering 137

(a) Helicity amplitudes for $\pi\pi\to N\bar{N}$. (b) Backward dispersion relations. (c) Partial-wave dispersion relations. (d) $\pi\pi$ phase shifts. (e) Consistency tests.

4.2 Other Processes 150

(a) Nucleon–nucleon scattering. (b) Kaon–nucleon scattering

Chapter 5 Weak and Electromagnetic Interactions 155

5.1 K_{e4} Decay 155

5.2 Other Weak and Electromagnetic Decays 163

(a) Two-pion decays of the neutral kaon. (b) $K\to 3\pi$ and $\eta\to 3\pi$ decays

5.3 Electron–Positron Colliding Beams 173

(a) $e^+e^-\to\pi^+\pi^-$: One-photon annihilation. (b) Hadronic production in the GeV region. (c) Two-photon annihilation

5.4 Other Electromagnetic Interactions 187

(a) Form factors in the spacelike region. (b) Photoproduction of $I=1$ vector mesons

Chapter 6 Summary 196

(a) Status of the principal methods. (b) Information obtained

PART III THEORY AND MODELS OF $\pi\pi$ SCATTERING

Chapter 7 Rigorous Results 205

(a) Rigorous analyticity. (b) High-energy behaviour. (c) Froissart–Gribov representation. (d) Low-energy bounds. (e) Crossing and positivity. (f) Atkinson's method

Chapter 8 Analyticity at Fixed Momentum Transfer 219

8.1 Inverse Amplitudes and Scattering Length Bounds 219
(a) Sum rule bounds. (b) Bounds on λ. (c) Phenomenological results. (d) Extensions of the method

8.2 Partial Waves from Fixed-t Dispersion Relations 226
(a) Fixed-t relations and their derivatives at $t = 0$. (b) Partial wave projections and Roy's equations

Chapter 9 Analytic Properties of Partial Wave Amplitudes . . . 232

9.1 Partial Wave Dispersion Relations 233
(a) Derivation and crossing. (b) Solution of the partial wave equations: the N/D method

9.2 Attempts At A Partial Wave Dynamics 241
(a) Chew and Mandelstam's S-wave dominant approximation. (b) P-wave models: the rho bootstrap

Chapter 10 Uses of Analyticity—Models and Phenomenology . 248

10.1 Introduction 248

10.2 Some Analytic Models 254
(a) Models using crossing symmetric expansions. (b) Partial wave models using the BNR and Martin conditions. (c) Models using partial wave dispersion relations. (d) Inverse partial wave amplitudes

10.3 Dispersion Relation Phenomenology 261
(a) Forward sum rules. (b) Forward and derivative dispersion relations. (c) Application of the Roy relations. (d) Remarks on relations among the threshold parameters

10.4 Analysis of πK Scattering 275

10.5 Concluding Remarks 278

Chapter 11 Dynamical and Field Theoretic Models 280

11.1 The Strip Approximation 280
(a) Introduction. (b) Basic formalism. (c) Practical calculations. (d) Conclusions

11.2 Lagrangian Field Theory Models—Padé Approximants . 291
(a) General remarks on Lagrangian models. (b) Perturbation expansions and the Padé approximant method. (c) Calculations based on specific $\pi\pi$ Lagrangians. (d) Multichannel calculations

Chapter 12 FESR's, Duality and the Veneziano Model 306

12.1 Finite-Energy Sum Rules 306
(a) High-energy scattering and FESR. (b) Application to $\pi\pi$ scattering

12.2 Duality and FESR Bootstraps 311
(a) Two-component duality. (b) Schmid circles and FESR bootstrap

12.3 Dual Models 322
(a) Veneziano model. (b) $\pi\pi \to \pi\omega$: singly or doubly spaced trajectories? (c) Lovelace–Veneziano model. (d) Zero trajectories. (e) Unitarization and phenomenological applications. (f) Extension to other 0^-0^- scattering processes

12.4 Concluding Remarks 345

Chapter 13 Resonance Spectra 348

13.1 Meson States in the Quark Model 348
(a) The L-excitation quark model: general features. (b) SU(3) and its breaking: ideal mixing. (c) Phenomenology of the low-lying states. (d) The Quark model and SU(6). (e) Higher symmetries and decay widths

13.2 Regge Poles and Duality 366

Chapter 14 Predictions from Current Algebra and PCAC . . . 369

14.1 Weinberg's Calculation of Low-Energy $\pi\pi$ Scattering . . 370
(a) Predictions for off-shell $\pi\pi$ scattering. (b) On-shell results: the linear form

14.2 Relationship to Other Approaches 377
(a) Sum rules and the KSFR relation. (b) Veneziano models

14.3 Corrections to the Weinberg Form 381
(a) Higher Order Terms. (b) Unitarity corrections. (c) Comparison with dispersion relation results

14.4 Applications to Related Processes 389
(a) 0^-0^- scattering lengths. (b) Dipion production. (c) Decay processes

PART IV APPENDICES

Appendix A Notations and Conventions 397

(a) Units. (b) The symbols $0 \sim \approx \to$ and $\equiv$. (c) Four-vectors. (d) Gamma matrices and the Dirac equation. (e) Field theory. (f) Isospin conventions

Appendix B Spin Analysis of Di-Meson Production 402

(a) Kinematics and amplitudes. (b) Crossing. (c) Intensity formulae and the spin density matrix for di-meson production. (d) Experiments on polarized targets. (e) Examples of specific final states and exchanges

Appendix C High Energy Scattering and Regge Poles 422

Appendix D Current Algebra and PCAC 427

(a) Weak interaction currents. (b) Current commutators. (c) PCAC hypothesis

References 433

Subject Index 451

Part I

Introduction

Chapter 1

Introduction

The pion–pion interaction permeates and exemplifies medium energy hadron physics. The attraction for theorists lies in the formal simplicity of the basic $\pi\pi$ scattering process, but experimental investigation is hampered by the non-existence of real pion targets. The consequent need to acquire information by indirect means has prevented the rapid accumulation of data. During the last few years, however, a very large and successful experimental effort has been mounted, and the enlarged precision and scope of the data has enabled theoretical questions to be posed much more sharply. In this book, we give a critical account of this experimental progress, and review the theoretical situation in the light of it. We have interpreted the scope broadly, and, wherever it is appropriate, we extend the discussion to include related processes like πK scattering.

The central role of the pion–pion interaction stems from its ubiquity and its simplicity. These, together with its inaccessibility to direct experiment, are the three governing characteristics which have shaped the development of the subject. The first attribute refers to the wide range of phenomena whose detailed form is strongly influenced by the interaction. A celebrated early example was the prediction of the existence of the rho meson from analysis of the nucleon electromagnetic form factors. Similarly, it has long been realized that dipion exchange plays an important role in the nucleon–nucleon interaction. In this case, appreciable quantitative progress has been made quite recently, based, among other things, upon improved knowledge of the $\pi\pi$ scattering amplitudes. These are just two among the many examples which form one important component of our subject matter. It is, of course, just this pervasive nature of the $\pi\pi$ interaction which allows its study by indirect methods to be so successful, thereby surmounting the inaccessibility to direct measurement. We return to this below.

The formal simplicity of the basic $\pi\pi$ scattering process has made it a prime testing ground for many theoretical ideas. The important ingredients of this, absence of spin, complete crossing symmetry, and the lightness of the pion, although obvious, have far-reaching consequences in making the system peculiarly amenable to detailed analysis. This is apparent if one considers the development of ideas leading from finite energy sum rules (FESR) to specific dual models, in which the study of reactions like $\pi\pi \to \pi\pi$ and $\pi\pi \to \pi\omega$ played a prominent role. It is even more apparent if one considers applications of the general principles of analyticity, crossing and unitarity (ACU), a theme which has run through almost the whole history of the subject. The early phase of this development culminated in the so-called "bootstrap hypothesis". This was an attempt to extend these general properties into a full-scale hadron dynamics by combining them with the idea that all particles are bound states of each other ("nuclear democracy"), and the assumed dominance of long-range forces. Within this approach, a central role is guaranteed for the study of pion–pion scattering. Subsequent discoveries—SU(3) and its extensions, together with indications of indefinitely rising Regge trajectories—revealed a structure which such a scheme was not able to accommodate. Modern applications of the ACU conditions have therefore concentrated on their role as constraints, especially to supplement data in phenomenological analysis. The $\pi\pi$ system, with its complete crossing symmetry, offers the greatest scope for such a programme: much work has been done, culminating in the Roy equations and their applications. In particular, this work has enabled the uncertainty in the low mass $\pi\pi$ amplitudes to be narrowed down to essentially one parameter, despite the shortage of direct data. The threshold region, and the associated parameters, are of especial interest for comparisons with the predictions of current algebra, for which $\pi\pi$ scattering is a crucial problem.

The third factor which has governed developments—that experimental knowledge of $\pi\pi$ scattering must be obtained indirectly—implies that any process which involves pion–pion interactions is also a potential source of information. By far the most important are production processes like $\pi N \to \pi\pi N$, $\pi N \to \pi\pi\Delta$ (and $KN \to K\pi N$ etc.) at high energies and small momentum transfers. It was pointed out quite early how information could be obtained on $\pi\pi$ (or πK) scattering by isolating the one-pion exchange contribution to these reactions. Carrying out this programme in a reasonably model independent way requires very large amounts of data, which only the present generation of experiments has supplied. These have enabled the ambiguities in the $I = 0$ S-wave phase shift, the former bane of studies of the low-mass region, to be resolved. Along with the improved precision has come a very large extension in the range of masses explored, which has enabled phase shift solutions to be continued, albeit with residual uncertainties, up to 2 GeV. Pion–pion scattering has thereby come increasingly within the scope of detailed resonance phenomenology and classification schemes. It is important in this respect to investigate related channels, e.g. $\pi K \to \pi K$ and $\pi\pi \to K\bar{K}$; the latter should also be of assistance in eliminating the remaining ambiguities in the $\pi\pi$ phase shifts. The annihilation reactions $\bar{p}p \to \pi^-\pi^+$ etc., although difficult to analyse, afford a new source of detailed information above their thresholds.

Additional information on the $\pi\pi$ interaction, which both checks and supplements that obtained from dipion production, can be obtained in various other ways. Of these, the most important are K_{e4} decay, which in principle affords an essentially model-independent measurement of low-mass $\pi\pi$ phase shifts, and $e^+e^- \to \pi^+\pi^-$, which isolates the $\pi\pi$ P-wave. In both cases, the experimental data are at present somewhat imprecise, but improvement is expected in the near future. This is particularly important in the K_{e4} case, since the threshold region has proved difficult to establish from dipion production experiments. In contrast to these two, most other potential sources of information are, with a few exceptions, too model-dependent or too imprecise to be of practical use. Thus, almost all processes involving three hadrons in the final state fail on the first count; $K_{S,L} \to 2\pi$, and the analysis of NN scattering fail on the second. The modern trend is rather to insert $\pi\pi$ phase shifts as input and to concentrate on understanding other aspects of these reactions. We have mentioned this earlier in connection with the problem of NN forces; other interesting examples are furnished by three-hadron processes, whether in formation (e.g. $\bar{p}p \to 3\pi$, $\pi N \to \pi\pi N$ at low energies) or production (e.g. $\pi N \to (3\pi)N$). These reactions are of increasing phenomenological interest, especially with regard to resonance physics, and it is usual to relate the three-hadron configurations involved, $\pi\pi N$ or 3π, to the subsidiary two-body interactions, $\pi\pi$ or πN, using some variant of the isobar model. There are questions as to how adequate an approximation this is, and with the status of three-hadron dynamics generally; establishing the parameters of the subsidiary two-body scatterings hopefully clears the way for their resolution.

To sum up: the ubiquity of the $\pi\pi$ interaction, owes to the lightness of the pion, its quantum numbers, and its strong coupling to other hadrons. Almost all the subject matter of $\pi\pi$ studies stems from this, from the application of current algebra and the PCAC condition, to the very method whereby most of our experimental information comes.

The arrangement of the book reflects these considerations. Following this introduction, and a short chapter on the basic theoretical framework and notation (which together form Part I), the material is grouped into two main sections: Part II, "Sources of Information on $\pi\pi$ Scattering", and Part III, "Theories and Models of $\pi\pi$ Scattering". In all cases the scope is slightly extended to cover closely related meson processes, especially πK scattering. The two main parts are largely independent. The reader who is primarily interested in theoretical questions, and who is concerned only with the results of Part II, will find them very briefly summarized in Chapter 6 which concludes that part. As is appropriate, Part II is dominated by discussion (Chapter 3) of peripheral production processes, $\pi N \to \pi\pi B$, and its analogues. In order to do justice to the involved chains of inference upon which essentially all $\pi\pi$ phase shift results depend, this chapter is necessarily more technical than the rest of the book. For the reader interested only in the results, a reading of Section 3.3 should suffice, whereas to understand thoroughly their source, a study of Sections 3.1 and 3.2 will also be necessary. The other two chapters of Part II are concerned with dipion exchange processes, $\pi N \to \pi N$ etc. (Chapter 4), and weak and electromagnetic processes (Chapter 5).

Part III has three main themes—the exploitation of general constraints from analyticity, crossing, and unitarity (Chapters 7–10), consequences of duality and the quark model (Chapters 12 and 13), and current algebra and PCAC (Chapter 14). In addition to these, Chapter 11 contains a brief account of two specific dynamical approaches; the strip approximation, and models based on Lagrangians.

Because of the wide scope of this review, we have not attempted to give detailed discussions of all topics. On the other hand, we have sought, within the limits of space, to treat the main themes in some depth, in several cases with supporting appendices. Even so, the very large number of publications has necessitated considerable selection in the work discussed and cited, and we apologize to any author whose work has been unjustly neglected. Finally, because our subject involves a "horizontal" section through particle physics, it has been impossible to avoid a certain amount of forward referencing, although we have tried to keep this to a minimum.

Chapter 2

Theoretical Framework and Notation

In this chapter we will summarize such topics as: the kinematics and isospin formalism used to describe pion–pion and related interactions; unitarity; crossing symmetry; and Mandelstam analyticity. This will provide much of the notation and general theoretical framework for use in later sections. Formulas will in general refer to $\pi\pi$ scattering, but some analogous results for πK scattering will also be given. Some further details, such as the Lorentz index conventions and field definitions, can be found in Appendix A.

(a) Kinematics

The various $\pi\pi$ scattering processes are represented by Fig. 2.1, where $p_i (i = 1, 2, 3, 4)$ are the particles' four-momenta, and α, β, γ and δ are their isospin indices. This diagram represents the three reactions

$$\begin{aligned} \pi_\alpha(p_1) + \pi_\beta(p_2) &\to \pi_\gamma(p_3) + \pi_\delta(p_4), && (s) \\ \pi_\alpha(p_1) + \pi_\gamma(-p_3) &\to \pi_\beta(-p_2) + \pi_\delta(p_4), && (t) \\ \pi_\alpha(p_1) + \pi_\delta(-p_4) &\to \pi_\beta(-p_2) + \pi_\gamma(p_3). && (u) \end{aligned} \tag{2.1}$$

The symbols s, t, u will be used to label the different reactions as above, and also to denote the usual kinematic invariants

$$\begin{aligned} s &= -(p_1 + p_2)^2 = -(p_3 + p_4)^2, \\ t &= -(p_1 - p_3)^2 = -(p_2 - p_4)^2, \\ u &= -(p_1 - p_4)^2 = -(p_2 - p_3)^2. \end{aligned} \tag{2.2}$$

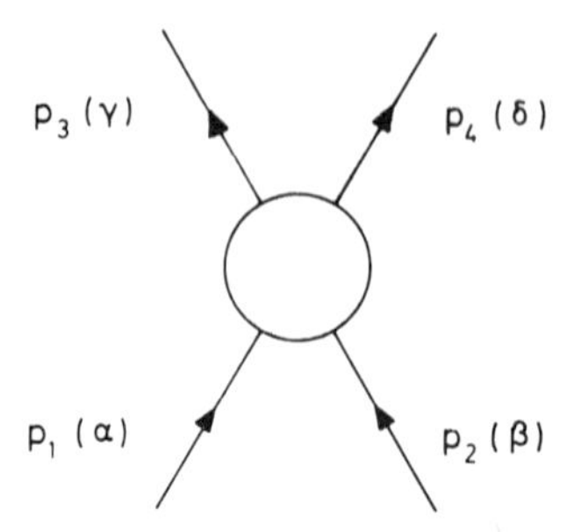

FIG. 2.1. Kinematics and isospin notation for $\pi\pi$ scattering.

Because of momentum conservation, these are subject to the constraint

$$s+t+u=4\mu^2, \tag{2.3}$$

where $\mu \equiv m_\pi$. *Throughout this book the pion mass will usually be set to unity and not explicitly written, except where this could lead to confusion.*

In the s-channel, s is the square of the total centre-of-mass energy W_s, and t and u are given by

$$\begin{aligned} t &= -2q_s^2(1-\cos\theta_s), \\ u &= -2q_s^2(1+\cos\theta_s), \end{aligned} \tag{2.4}$$

where q_s and $\cos\theta_s$ are the magnitude of the c.m. three-momentum in the s-channel, and the cosine of the s-channel c.m. scattering angle, respectively. In the t-channel, in an obvious notation

$$\begin{aligned} t &= W_t^2, \\ s &= -2q_t^2(1-\cos\theta_t), \\ u &= -2q_t^2(1+\cos\theta_t). \end{aligned} \tag{2.5}$$

The physical regions for the various channels follow directly from these equations and are shown in Fig. 2.2.

Another variable which we will frequently use is

$$\nu \equiv q_s^2/\mu^2, \tag{2.6}$$

in terms of which

$$s=4\mu^2(\nu+1); \qquad \omega \equiv W_s/2=\mu(\nu+1)^{\frac{1}{2}}, \tag{2.7}$$

and the $s \leftrightarrow u$ anti-symmetric variable

$$z \equiv \frac{(s-u)}{4\mu^2} = \left(\frac{t-4\mu^2}{4\mu^2}\right)\cos\theta_t. \tag{2.8}$$

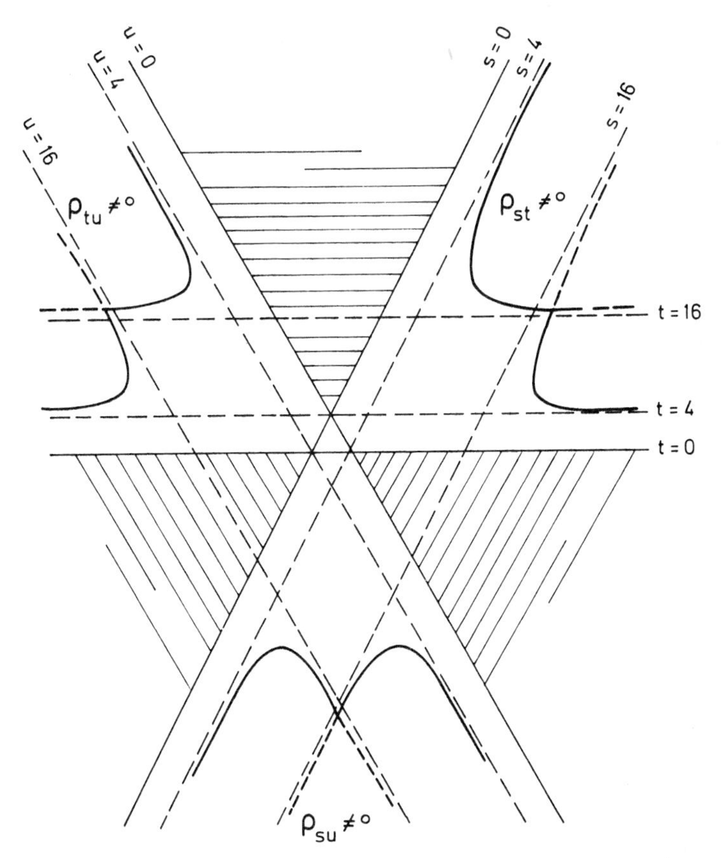

FIG. 2.2. Physical (hatched) and double spectral regions for $\pi\pi$ scattering, (not to scale).

(b) Amplitudes and Cross-sections

In our normalization the S-matrix for the general $2 \rightarrow n$ reaction takes the form

$$S_{fi} = I_{fi} + \frac{i32\pi(2\pi)^4\delta^{(4)}(p_1 + p_2 - p_3 - \cdots - p_{n+2})}{(2\pi)^{3n/2+3}\prod_{i=1}^{n+2}(2p_i^0)^{\frac{1}{2}}} T_{fi},$$

which for the processes of Fig. 2.1 reduces to

$$S_{fi} = I_{fi} + \frac{8i\delta^{(4)}(p_1 + p_2 - p_3 - p_4)}{\pi(16p_1^0p_2^0p_3^0p_4^0)^{\frac{1}{2}}} T_{fi}, \tag{2.9a}$$

where $I_{fi} = \delta^3(\mathbf{p}_1 - \mathbf{p}_3)\delta^3(\mathbf{p}_2 - \mathbf{p}_4)$. The T-matrix, T_{fi}, is Lorentz invariant, the factors proportional to $(2p_i^0)^{-\frac{1}{2}}$ being necessary because the state vectors are non-convariantly normalized, i.e.,

$$\langle p'|p\rangle = \delta^3(\mathbf{p}'-\mathbf{p}).$$

The other factors are such that T_{fi} coincides with the amplitude used by Chew and Mandelstam (1960), whose notation has been followed by most subsequent authors. It also corresponds to the amplitude N_A used in the review of Petersen (1971) if the value $N=2$ is chosen for the arbitrary constant in his formulas. (For further details of our notation and conventions, see Appendix A.) It is also useful to specialize to the CM system and define the reduced S-matrix $\hat{S}_{fi}$ by

$$\hat{S}_{fi} \equiv \frac{4\sqrt{(s)}\delta^4(p_1+p_2-p_3-p_4)}{\pi|\mathbf{p}_1-\mathbf{p}_2|(16p_1^0p_2^0p_3^0p_4^0)^{\frac{1}{2}}} S_{fi}, \tag{2.9b}$$

which enables equation (2.9a) to be rewritten in the form

$$\hat{S}_{fi} = 2\pi\delta(\cos\theta-\cos\theta')\delta(\phi-\phi')+2i\rho(s)T_{fi},$$

where

$$\rho(s) \equiv |\mathbf{p}_1-\mathbf{p}_2|/\sqrt{s} = 2q_s/W_s.$$

The polar angles θ, ϕ refer to the vector $\mathbf{p}_1-\mathbf{p}_2$, and θ', ϕ' to the vector $\mathbf{p}_3-\mathbf{p}_4$.

Cross-section formulas follow from equation (2.9a). With these conventions, the amplitude T_s^I for specific s-channel isospin I (0, 1 or 2) is normalized such that the elastic differential cross-section in the CM of the s-channel is

$$\frac{d\sigma_{el}^I}{d\Omega} = \frac{16\hbar^2}{s}|T_s^I(s, \cos\theta_s)|^2 = \frac{4(\hbar^2/\mu^2)}{\nu+1}|T_s^I|^2, \tag{2.10}$$

and the total cross-section is, by the Optical Theorem,

$$\sigma_{tot}^I = \frac{16\pi\hbar^2}{q_sW_s}\,\text{Im}\,T^I(s, t=0) = \frac{8\pi(\hbar^2/\mu^2)}{[\nu(\nu+1)]^{\frac{1}{2}}}\,\text{Im}\,T_s^I(s, t=0), \tag{2.11}$$

where $\hbar^2 = 0{\cdot}389\ \text{GeV}^2\text{mb}$. We shall usually use units such that $\hbar=1$. Note that

$$\sigma_{el}^I = \varepsilon\int_{4\pi}\frac{d\sigma_{el}^I}{d\Omega}\,d\Omega,$$

where $\varepsilon = \frac{1}{2}(1)$ for the scattering of identical (non-identical) particles. The scattering of isospin eigenstates is treated as the scattering of identical particles. The conventional formulas for the scattering of non-identical particles (e.g. $\pi^+\pi^-$) are obtained by using the isospin decompositions given below.

Most of the time in this book we will be referring to s-channel amplitudes and variables, and the subscripts will often be omitted, except where there is room for ambiguity. Thus, we shall use

$$T^I(s, t, u) \equiv T^I_s(s, t, u); \qquad q \equiv q_s \quad \text{etc.}$$

For the analogous t- and u-channel variables the subscripts will always be retained.

Returning to the s-channel, and suppressing the index s, it is often convenient to use, instead of the invariant amplitudes $T^I(s, t, u)$, the partial wave amplitudes $T^I_l(s)$ defined by

$$T^I(s, t, u) = \sum_l (2l+1) P_l(\cos\theta) T^I_l(s), \tag{2.12}$$

where, for $\pi\pi$ scattering, the sum is over even (odd) values of l for even (odd) values of I because of the restrictions of Bose statistics. The inverse expression is

$$T^I_l(s) = \tfrac{1}{2} \int_{-1}^{+1} d\cos\theta P_l(\cos\theta) T^I(s, t, u), \tag{2.13}$$

and analogous expansions can be made in the t- and u-channels, e.g.

$$T^I_t(s, t, u) = \sum_l (2l+1) P_l(\cos\theta_t) T^I_l(t). \tag{2.14}$$

(The partial wave amplitudes need no channel index as they are functions of a single variable which denotes the channel.)

(c) Isospin and Crossing

(i) $\pi\pi$ *Scattering*

Crossing expresses the principle that the three reactions of equation (2.1) can all be described by a single analytic function $T(s, t, u)$. More precisely, the scattering amplitude (e.g. for s-channel scattering) is given by

$$T_s(s, t, u) \equiv \lim_{\varepsilon \to 0} T(s + i\varepsilon, t, u), \tag{2.15}$$

in the physical region of the s-channel, i.e.

$$s \geqslant 4, \qquad -s \leqslant t \leqslant 0, \qquad u = 4 - s - t, \tag{2.16}$$

and similarly for the other channels. If we exhibit the isospin structure, T may be written in terms of three invariant amplitudes A, B and C by

$$T(s, t, u) = A(s, t, u)\delta_{\alpha\beta}\delta_{\gamma\delta} + B(s, t, u)\delta_{\alpha\gamma}\delta_{\beta\delta} + C(s, t, u)\delta_{\alpha\delta}\delta_{\beta\gamma}. \tag{2.17}$$

These are related to s-channel isospin amplitudes $T^I(s, t, u)$ by using the standard techniques of projection operators. The result is,

$$\begin{aligned} T_s^0(s, t, u) &= 3A(s, t, u) + B(s, t, u) + C(s, t, u), \\ T_s^1(s, t, u) &= B(s, t, u) - C(s, t, u), \\ T_s^2(s, t, u) &= B(s, t, u) + C(s, t, u). \end{aligned} \tag{2.18}$$

Scattering amplitudes for physical charge states in the s-channel are related to the above isospin amplitudes by the relations

$$\begin{aligned} T_s^{++} &\equiv T_s(++, ++) = T_s^2, \\ T_s^{+-} &\equiv T_s(+-, +-) = \tfrac{1}{3}T_s^0 + \tfrac{1}{2}T_s^1 + \tfrac{1}{6}T_s^2, \\ T_s^{+0} &\equiv T_s(+0, +0) = \tfrac{1}{2}T_s^1 + \tfrac{1}{2}T_s^2, \\ T_s^{00} &\equiv T_s(00, 00) = \tfrac{1}{3}T_s^0 + \tfrac{2}{3}T_s^2, \\ T_s^E &\equiv T_s(00, +-) = \tfrac{1}{3}T_s^2 - \tfrac{1}{3}T_s^0. \end{aligned} \tag{2.19}$$

Similarly, amplitudes $T_t^I(T_u^I)$ of specific isospin in the t- (u-) channel can be constructed in terms of A, B and C, and hence in terms of T_s^I using equation (2.17). The resulting equations are

$$T_t^I(s, t, u) = \sum_{I'=0}^{2} C_{II'}^{(ts)} T_s^{I'}(s, t, u), \tag{2.20a}$$

$$T_u^I(s, t, u) = \sum_{I'=0}^{2} C_{II'}^{(us)} T_s^{I'}(s, t, u), \tag{2.20b}$$

where the *crossing matrices* (which are unitary) are given by

$$C_{II'}^{(ts)} = C_{II'}^{(st)} = \begin{bmatrix} 1/3 & 1 & 5/3 \\ 1/3 & 1/2 & -5/6 \\ 1/3 & -1/2 & 1/6 \end{bmatrix} \tag{2.21}$$

$$C_{II'}^{(us)} = C_{II'}^{(su)} = (-1)^{I+I'} C_{II'}^{(st)}. \tag{2.22}$$

Finally, in addition to generalized Bose statistics in the s-channel

$$T_s^I(s, t, u) = (-1)^I T_s^I(s, u, t), \tag{2.23a}$$

the amplitudes must satisfy it in the t- and u-channels also, i.e.

$$T_t^I(s, t, u) = (-1)^I T_t^I(u, t, s), \tag{2.23b}$$

$$T_u^I(s, t, u) = (-1)^I T_u^I(t, s, u). \tag{2.23c}$$

Only the first two equations are independent, and they are equivalent to the relations

$$\left.\begin{array}{l} A(s,t,u)=A(s,u,t) \\ B(s,t,u)=C(s,u,t) \end{array}\right\} \quad \text{under} \quad \left\{\begin{array}{l} s\leftrightarrow s \\ t\leftrightarrow u \end{array}\right. \tag{2.24a}$$

$$\left.\begin{array}{l} A(s,t,u)=C(u,t,s) \\ B(s,t,u)=B(u,t,s) \end{array}\right\} \quad \text{under} \quad \left\{\begin{array}{l} s\leftrightarrow u \\ t\leftrightarrow t \end{array}\right. \tag{2.24b}$$

$$\left.\begin{array}{l} A(s,t,u)=B(t,s,u) \\ C(s,t,u)=C(t,s,u) \end{array}\right\} \quad \text{under} \quad \left\{\begin{array}{l} s\leftrightarrow t \\ u\leftrightarrow u \end{array}\right. \tag{2.24c}$$

between the A, B, C amplitudes of equation (2.17).

In view of these equations, a point of particular interest in the s, t, u plane is the symmetry point

$$s=t=u=4/3.$$

Using equations (2.24), an expansion of these amplitudes can be made about this point (Chew and Mandelstam, 1960) which is usually written in the form

$$\begin{aligned} A(s,t,u) &= B(t,s,u)=C(u,t,s) \\ &= -\lambda+(\tfrac{1}{4}\lambda_1)(s-4/3)+0[(s-4/3)^2,(t-u)^2]. \end{aligned} \tag{2.25}$$

In particular, at the symmetry point itself

$$2T^0=5T^2; \qquad T^1=0. \tag{2.26}$$

(ii) *πK and KK Scattering*

For πK scattering (s-channel) and the crossed process $\pi\pi\rightarrow K\bar{K}$ (t-channel), the decomposition analogous to equation (2.17) is

$$T_{\beta\alpha}=\delta_{\beta\alpha}T^{+}+\tfrac{1}{2}[\tau_\beta,\tau_\alpha]T^{-}, \tag{2.27}$$

where α, β are the isospin indices of the pions, and τ_β, τ_α are the usual Pauli matrices acting on the kaon isospinors. Again, using isospin projection operators, gives in the s-channel

$$\begin{aligned} 3T^{+} &= T^{\frac{1}{2}}(s,t,u)+2T^{\frac{3}{2}}(s,t,u), \\ 3T^{-} &= T^{\frac{1}{2}}(s,t,u)-T^{\frac{3}{2}}(s,t,u). \end{aligned} \tag{2.28}$$

For the t-channel ($\pi\pi\rightarrow K\bar{K}$) we have

$$\begin{aligned} \sqrt{6}T^{+}(s,t,u) &= T_t^0(s,t,u), \\ 2T^{-}(s,t,u) &= T_t^1(s,t,u). \end{aligned} \tag{2.29}$$

The isospin crossing matrices are

$$C_{II'}^{(st)} = \begin{pmatrix} 1/\sqrt{6} & 1 \\ 1/\sqrt{6} & -1/2 \end{pmatrix}; \qquad C_{II'}^{(su)} = \begin{pmatrix} -1/3 & 4/3 \\ 2/3 & 1/3 \end{pmatrix}, \tag{2.30}$$

and generalized Bose statistics for the t-channel gives the crossing relations

$$T^{\pm}(s, t, u) = \pm T^{\pm}(u, t, s). \tag{2.31}$$

For KK scattering, the corresponding relations between the $s(KK \to KK)$, $t(\bar{K}K \to K\bar{K})$ and $u(K\bar{K} \to K\bar{K})$ channels (note particularly the particle ordering) are

$$C_{II'}^{(st)} = \begin{pmatrix} -1/2 & -3/2 \\ -1/2 & 1/2 \end{pmatrix}; \qquad C_{II'}^{(su)} = \begin{pmatrix} -1/2 & 3/2 \\ 1/2 & 1/2 \end{pmatrix}, \tag{2.32}$$

and Bose statistics in the s-channel gives

$$T_s^I(s, t, u) = (-1)^{I+1} T_s^I(s, u, t) \qquad (I = 0, 1). \tag{2.33}$$

A detailed account of the calculation of these, and other, isospin crossing matrices is given, for example, by Petersen (1971).

(d) Unitarity

For the reduced S-matrix, unitarity takes the form

$$\hat{\mathbf{S}}\hat{\mathbf{S}}^{\dagger} = \mathbf{I}.$$

Together with T-invariance, this leads to the expression

$$\delta^4(p_1 + p_2 - p_3 - p_4)\,\mathrm{Im}\,T_{fi}$$
$$= (16\pi)(2\pi)^{4-3n} \sum_n \prod_{i=1}^{n} \frac{d^3 p_i'}{2p_i'^0}\, \delta^4(\Sigma p_i' - p_3 - p_4) T_{fn} T_{ni}^{*}\, \delta^4(p_1 + p_2 - \Sigma p_i'),$$

for the elastic scattering amplitude T_{fi}, where the sum on the right-hand side extends over all available intermediate states, $|p_i'\rangle$. Setting $i \equiv f$ leads directly to the Optical Theorem equation (2.11).

Let us now turn to non-forward elastic scattering and exhibit the angular dependence in the centre-of-mass system. For scattering from an initial direction specified by polar angles (θ_1, ϕ_1) to a final direction (θ_2, ϕ_2), rotational invariance gives

$$\langle \theta_2, \phi_2 | T | \theta_1, \phi_1 \rangle = T(s, \cos\theta),$$

where

$$\cos\theta = \cos\theta_1 \cos\theta_2 + \sin\theta_1 \sin\theta_2 \cos(\phi_1 - \phi_2).$$

Substituting into the unitarity equation, and restricting ourselves to the region of elastic unitarity,

$$\text{Im } T(s, \cos\theta) = \frac{q}{2\pi W\varepsilon} \int d\Omega' \langle 00|T|\theta'\phi'\rangle^* \langle\theta'\phi'|T|\theta 0\rangle, \qquad 4 \leq s \leq 16. \tag{2.34}$$

Finally, if we set

$$1 + t'/2q^2 = \cos\theta',$$

$$1 + t''/2q^2 = \cos\theta'' = \cos\theta' \cos\theta + \sin\theta' \sin\theta \cos\phi',$$

then the above unitarity convolution can be re-expressed in the form

$$\text{Im } T(s, t) = \frac{1}{4\pi Wq\varepsilon} \iint dt'\, dt''\Theta(J) J^{-\frac{1}{2}} T^*(s, t') T(s, t''), \tag{2.35a}$$

where the Jacobian of the transformation, $J^{-\frac{1}{2}}/q^2$, is given in terms of the function

$$J(s, t, t', t'') \equiv [tt't''/q^2 + 2(tt' + tt'' + t't'') - t^2 - t'^2 - t''^2]. \tag{2.35b}$$

In practice, elastic unitarity is often more conveniently imposed in terms of the partial wave amplitudes $T_l^I(\nu)$. Substituting the expansion equation (2.12) into equation (2.34), and its generalization above the inelastic threshold, the result

$$\text{Im } T_l^I(\nu) \geq \left(\frac{\nu}{\nu+1}\right)^{\frac{1}{2}} |T_l^I(\nu)|^2 \frac{1}{2\varepsilon}, \tag{2.36}$$

is obtained. The equality applies to the elastic region $0 \leq \nu \leq 3 (4 \leq s \leq 16)$ and we have exhibited the appropriate inequality for higher energies. The weaker condition

$$\text{Im } T_l^I(\nu) > 0, \qquad \nu \geq 0, \tag{2.37}$$

is often referred to as *positivity*. Equation (2.36) allows us to write the partial wave amplitudes in the familiar form

$$T_l^I(\nu) = \frac{1}{4\varepsilon i}\left(\frac{\nu+1}{\nu}\right)^{\frac{1}{2}} \{\eta_l^I(\nu) \exp[2i\delta_l^I(\nu)] - 1\}, \tag{2.38}$$

where δ, η are real and

$$0 \leq \eta \leq 1, \tag{2.39}$$

the case $\eta = 1$ again applying in the elastic region $4 \leq s \leq 16$. The scattering lengths a_l^I and slope parameters b_l^I are defined by the threshold expansion

$$2\varepsilon \text{ Re } T_l^I(\nu) = \nu^l (a_l^I + b_l^I \nu + \cdots). \tag{2.40}$$

(e) Mandelstam Analyticity

The Mandelstam representation (Mandelstam, 1958a), expressed by the equation

$$
\begin{aligned}
T^I(s, t, u) = & \frac{1}{\pi^2}\int_4^\infty ds' \int_4^\infty dt' \frac{\rho_{st}^I(s', t')}{(s'-s)(t'-t)} \\
& + \frac{1}{\pi^2}\int_4^\infty dt' \int_4^\infty du' \frac{\rho_{tu}^I(t', u')}{(t'-t)(u'-u)} \\
& + \frac{1}{\pi^2}\int_4^\infty ds' \int_4^\infty du' \frac{\rho_{su}^I(s', u')}{(s'-s)(u'-u)},
\end{aligned} \tag{2.41}
$$

or a subtracted form of it, follows from the assumption of simultaneous polynomial boundedness in s, t and u together with the singularity structure suggested by a study of the Feynman diagrams shown in Fig. 2.3, the so-called box diagrams.‡ Since these diagrams involve the lightest mass

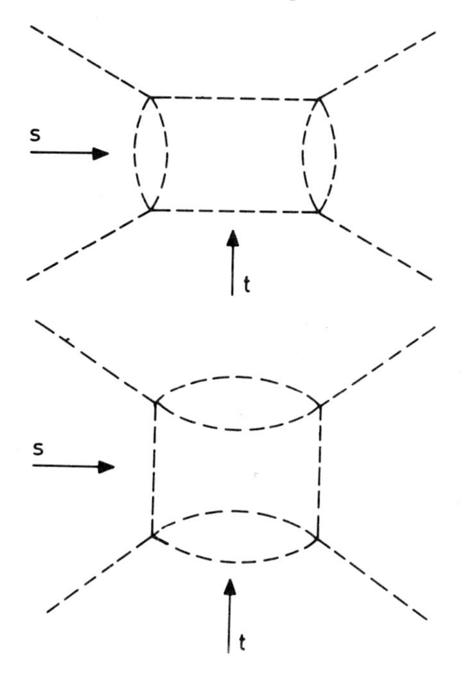

FIG. 2.3. Box graphs giving the boundaries of the double spectral function $\rho_{st}(s, t)$ for $\pi\pi$ scattering.

‡ The simplest box graphs, with a single pion line along each side, are of course forbidden by G-parity.

intermediate states, they should exhibit the discontinuities demanded by unitarity for low s, t, u values, and also the requirements of crossing symmetry. They might thus be expected to give a reasonable guide, at least to the nearby singularities. These arise from the double spectral functions ρ_{st}, ρ_{su}, ρ_{tu}, which are real analytic functions of their arguments, non-zero in the regions shown in Fig. 2.2. The precise boundaries of these double spectral regions can be found by applying the elastic unitarity condition (Mandelstam, 1958b,c). Thus, using the identity

$$\frac{1}{(t'-t)(u'-u)} \equiv \frac{1}{(t'+u'-4+s)}\left(\frac{1}{t'-t}+\frac{1}{u'-u}\right),$$

in equation (2.41), and making the change of variable $s'=4-t'-u'$, one obtains‡

$$T^I(s,t,u)=\frac{1}{\pi}\int_4^\infty dt'\mathscr{D}_t^I(s,t')\left(\frac{1}{t'-t}+\frac{(-1)^I}{t'-u}\right), \tag{2.42a}$$

where $\mathscr{D}_t^I$ denotes the t-discontinuity

$$\mathscr{D}_t^I(s,t)=\sum_{I_t} C_{II_t}^{(st)}\frac{1}{\pi}\int_4^\infty ds'\,\rho_{st}^{I_t}(s',t)\left(\frac{1}{s'-s}+\frac{(-1)^{I_t}}{s'-u}\right), \tag{2.42b}$$

and crossing has been exploited to relate the different cuts. Similarly, the analogous s-discontinuity $\mathscr{D}_s^I$ defined by

$$T^I(s,t,u)=\sum_{I'}\frac{1}{\pi}\int_4^\infty ds'\mathscr{D}_s^{I'}(s',t)\left(\frac{\delta_{II'}}{s'-s}+\frac{C_{II'}^{(su)}}{s'-u}\right), \tag{2.43a}$$

has the spectral representation

$$\mathscr{D}_s^I(s,t)=\frac{1}{\pi}\int_4^\infty dt'\,\rho_{st}^I(s,t')\left(\frac{1}{t'-t}+\frac{(-1)^I}{t'-u}\right). \tag{2.43b}$$

In the physical region $\mathscr{D}_s^I$ can be identified with Im T_s^I (see below) and, substituting equations (2.42a) and (2.43b) into the convolution integral (2.35) expressing elastic unitarity, one obtains

$$\rho_{st}^I(s,t)=\frac{2}{\pi q_s W_s}\int\int\frac{dt'\,dt''\,\mathscr{D}_t^{I^*}(s,t')\mathscr{D}_t^I(s,t'')}{K^{\frac{1}{2}}(s,t,t',t'')}, \qquad s\leq 16. \tag{2.44}$$

The function K (Kibble, 1960), is related to J of equation (2.35) by

$$K\equiv -J=[t^2+t'^2+t''^2-2(tt'+tt''+t't'')-tt't''/q_s^2],$$

‡ In equations (2.42), (2.43) and (2.44), I refers to s-channel isospin, I_t to t-channel isospin.

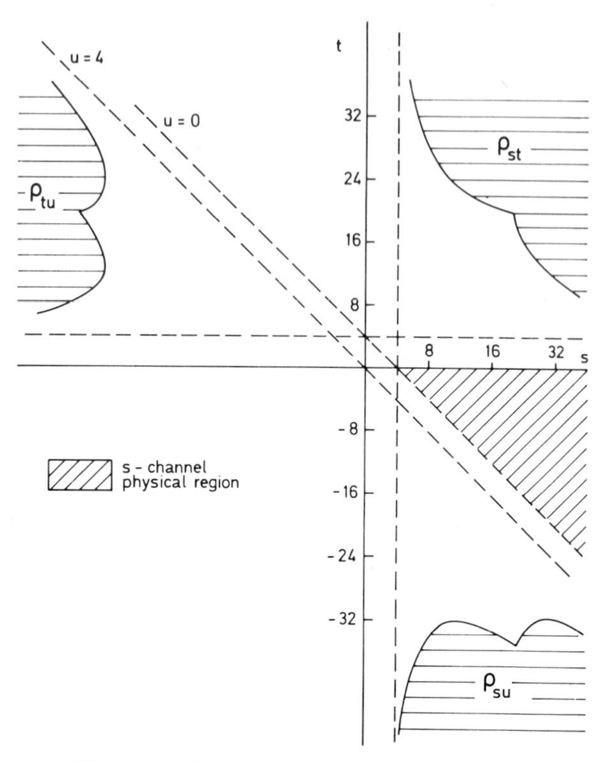

FIG. 2.4. Diagram to illustrate regions of convergence of the partial-wave expansions for $\pi\pi$ scattering.

and the integration in (2.44) extends over the region where K is positive. From this it follows that ρ_{st} is non-zero within the boundaries

$$(s-4)(t-16)-32=0,$$

$$(t-4)(s-16)-32=0,$$

and these curves, and the corresponding curves bounding ρ_{tu}, ρ_{su} are shown in Fig. 2.2.

Having obtained these boundaries, we can now obtain the familiar results on the analytic properties of $T^I(s, t, u)$ when one variable is held fixed. If we hold s fixed, then from equation (2.43b) it follows that there are no singularities in t (and hence $\cos\theta$) provided t lies outside the regions where ρ_{st} and ρ_{su}‡ are non-zero. These can be read off from Fig. 2.2 or better from Fig. 2.4. Thus for $s=16$, there are no singularities in t (or $\cos\theta$) for $-33 \leqslant t \leqslant 21$ which embraces the physical region $0 > t > -s = -16$.

‡ From crossing $\rho^I_{su}(s, u'=t') = (-1)^I \rho^I_{st}(s, t')$.

Hence the partial wave expansion equation (2.12), which by definition converges in the physical region, will also converge in this wider range $-33 \leqslant t \leqslant 21$ ($s = 16$). The corresponding ranges for other s-values can similarly be read off from Fig. 2.4. In particular, it can be seen that $\mathcal{D}_s^I$ (of equation (2.43b)) has no singularities in the range $-32 < t < 4$ for any value of s, so that, for t in this range, the only singularities are the s- and u-channel cuts exhibited explicitly in equation (2.43a). Using this, and the reality of $T^I(s, t, u)$ below threshold, which gives

$$T^I(s^*, t) = (T^I(s, t))^*,$$

and so allows the discontinuity to be expressed in terms of the imaginary part, we obtain the usual form of the fixed-t dispersion relation

$$T^I(s, t, u) = \frac{1}{\pi} \sum_{I'} \int_4^\infty ds' \operatorname{Im} T^{I'}(s', t) \left(\frac{\delta_{II'}}{s' - s} + \frac{C_{II'}^{(su)}}{s' - u} \right), \quad -32 < t < 4. \qquad (2.45)$$

These equations take a simpler form, and the question of subtractions is easier to discuss, if we write them for the t-channel amplitudes $T_t^I(s, t, u)$ expressed as functions of the z variable defined in equation (2.8). Thus for $I = 1$ exchange in the t-channel we have

$$T_t^1(z, t) = \frac{2z}{\pi} \int_{z_0}^\infty \frac{dz' \operatorname{Im} T_t^1(z', t)}{z'^2 - z^2}, \qquad (2.46)$$

where $z_0 = 1 + t/4\mu^2$. According to conventional ideas on high-energy scattering (see Appendix C) this relation is convergent as written. However, the corresponding relation for $I = 0$ exchange, and possibly also that for $I = 2$ exchange would not be expected to converge, so that a subtraction is required. Making this at $z = 0$, we obtain

$$T_t^{0,2}(z, t) = T_t^{0,2}(z = 0, t) + \frac{2}{\pi} \int_{z_0}^\infty \frac{dz' \operatorname{Im} T_t^{0,2}(z', t)}{z'(z'^2 - z^2)}, \qquad (2.46b)$$

which, both according to Regge pole ideas on high-energy scattering (see Appendix C) and results derived from general field theory (see Chapter 7) is now convergent as written.

A final remark concerns the theoretical status of the preceding results on Mandelstam analyticity. Aside from the assumption of simultaneous polynomial boundedness in s and t (which is used in writing equation (2.41) or subtracted variants of it), which is sometimes questioned, the foregoing results are generally accepted as a working hypothesis. However, they do go somewhat beyond what has been rigorously established from general field theory. For further discussion, we refer to Chapter 7.

Part II

Sources of Information on $\pi\pi$ Scattering

Chapter 3

Pion Production

The major source of experimental information on $\pi\pi$ and $K\pi$ scattering, generically $P\pi$ scattering (P for pseudoscalar meson), is from the study of production processes

$$PN \to M^*B,\ B \equiv N, \Delta \text{ etc.} \begin{cases} P \equiv \pi,\ M^* \equiv \pi\pi,\ K\bar{K},\ \pi\pi\pi\pi, \ldots \\ P \equiv K,\ M^* \equiv K\pi,\ K\pi\pi, \ldots. \end{cases} \tag{3.1}$$

For non-strange mesons of mass greater than twice the nucleon mass, one has the additional possibility of studying them in formation using anti-proton beams:

$$\left.\begin{aligned} \bar{N}N \to M^*,\ M^* \equiv {} & \pi\pi \\ & \bar{K}K \\ & 3\pi,\ \pi\pi\omega \text{ etc.} \end{aligned}\right\} \tag{3.2}$$

The former production processes, equation (3.1), afford information on $\pi\pi$ or $K\pi$ elastic and inelastic scattering, principally through the one-pion exchange (OPE) component of production. This poses the problem of separating out the OPE contribution, and thus leads to the study of the production processes, their angular dependences and spin structure, in order to disentangle the backgrounds to OPE from the desired signal. All this is an added complication and source of uncertainty in determining $\pi\pi$ and $K\pi$ phase shifts and inelasticities from experiment, as compared to conventional hadron scattering studies in formation, e.g. $\pi N \to \pi N$. For this reason, we separate the discussion of the production processes into three parts, "OPE and its Backgrounds" (Section 3.1), "Methods for Extracting $\pi\pi$ and $K\pi$ Phase Shifts" (Section 3.2), and "Results" (Section 3.3). For ease of reading, spin formalism for discussing the production processes is collected in Appendix B. In this way, the reader can join the discussion at whatever point he wishes—perhaps working back from the results (Section 3.3) to the methods (Section 3.2), and then back to the generalities of Section 3.1.

Following the discussion of results on $\pi\pi$ and $K\pi$ scattering from the study of production processes in Section 3.3, we turn in Section 3.4 to the study of annihilation processes, $\bar{N}N \to M^*$. The even G-parity sector of this, especially the $\pi\pi$ channel, affords useful direct information on the two-body channels in which we are interested. Angular momentum barriers should make this process an especially sensitive indicator of low-J high-mass resonances. We conclude this last section of the chapter by discussing the odd G-parity three-meson channels, especially 3π. This shows a rich structure which has provoked a lot of theoretical interest; but, as elsewhere, the complexities of a full three-hadron situation have so far defeated attempts to extract definite conclusions on the dynamics of the two-body sub-systems.

3.1. Peripheral Di-Meson Production (1): OPE And Its Backgrounds

(a) Introduction

The di-meson production processes to be considered in this and the following two sections (3.2 and 3.3) constitute the most prolific source of information on meson–meson scattering. The importance of these processes owes primarily to the fact that, over a significant range of beam momenta, they have a conspicuous one-pion exchange (OPE) contribution, schematically indicated in Fig. 3.1.1(a). If this contribution can be isolated, the associated $\pi\pi$ (or $K\pi$) elastic amplitude follows immediately. In practice, there are problems as to how the background to OPE should be treated. Notwithstanding this, the scope for experimental investigation is very large. One has

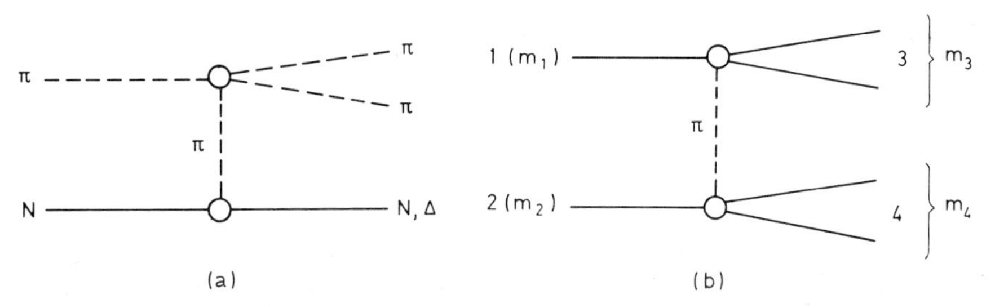

FIG. 3.1.1. One-pion exchange (OPE) graphs: (a) for the specific process $\pi N \to \pi\pi B$, (b) for the general process $1+2 \to 3+4$.

access via different experimental arrangements to all isospin states of the di-meson, and an unlimited range of di-meson masses and angular momenta. Inelastic channels for $\pi\pi$ or $K\pi$ scattering can be investigated by analogous reactions.

The best studied groups of reactions, and the ones to which we shall devote most attention, are $\pi N \to \pi\pi N$ and $\pi N \to \pi\pi\Delta$. For the former "N-reactions" the experimentally accessible processes are

$$\begin{aligned} \pi^+ p &\to \pi^+ \pi^0 p \\ &\to \pi^+ \pi^+ n \\ \pi^- p &\to \pi^- \pi^0 p \\ &\to \pi^+ \pi^- n \\ &\to \pi^0 \pi^0 n, \end{aligned}$$

together with the related processes on deuteron targets. Likewise, for the "Δ-processes" (two-pion production with Δ (1236) formation) there have been important studies of

$$\begin{aligned} \pi^+ p &\to \pi^+ \pi^- \Delta^{++} \\ \pi^- p &\to \pi^- \pi^- \Delta^{++}. \end{aligned}$$

For both types of reaction there also exist good data on $K\bar{K}$ production, e.g.

$$\pi^- p \to K^+ K^- n,\ K_1^0 K_1^0 n,$$

and on 4π production, both of which are of interest in assessing inelasticity in $\pi\pi$ scattering. For KN production processes, there is even more scope for inelastic effects, since there is no constraint analogous to G-parity for inelastic states coupling to $K\pi$.

In seeking to identify the OPE contribution to any of these processes, we are only interested in a portion of the production cross-section; hence one looks at configurations where contributions from competing mechanisms are minimized. Thus, low beam momenta should be avoided because of confusion with isobar formation. Above about 1·5 GeV/c, a distinct exchange-dominated portion of phase space can be identified at small momentum transfers, and this shows a strong OPE signal. Increasing the beam momentum reinforces the separation from isobar formation, and gives access to smaller momentum transfers and larger di-meson masses. On the other hand, contributions from the exchange of higher lying Regge trajectories (e.g. ω, A_2) to the production of a given di-meson mass are relatively enhanced as the beam momentum increases, and will ultimately swamp the OPE signal. There is however good reason, supported by experiment, to expect that this diminishing signal-to-background ratio at fixed di-meson mass will be compensated by a relative improvement as a function of increasing di-meson mass. In any case, there *are* backgrounds to OPE, even in the most favoured situation, and we need to understand something about them in order to proceed to our main objective—the separation of the OPE term and the determination of $\pi\pi$ or $K\pi$ scattering amplitudes.

The subject matter of Section 3.1 somewhat exceeds the strict confines of $\pi\pi$ and $K\pi$ scattering, but this is justified both on account of intrinsic interest, for example, the recent findings concerning the pion Regge trajectory; also because so much of our purported $P\pi$ scattering information‡ rests on rather intricate analyses whose justification is only to be found in our overall understanding of the production mechanisms.

The remainder of this section is organised as follows: (b) Kinematic Formalism for the Process $PN \to M^*B$, (c) Formulae for OPE, (d) OPE Compared to Experiment, (e) Models for the Background to OPE, (f) Parameterization of the Very Small t-region, (g) Amplitude Analysis, (h) Applications, and (i) Summary. Section (b) is somewhat longer, and more technical than the others and could perhaps be omitted on a first reading by those not interested in the details.

(b) Kinematic Formalism for the Process $PN \to M^*B$

The general kinematic formalism for discussing the processes $PN \to M^*B$,§ especially the intricacies of their spin analysis, is outlined in Appendix B on which we draw throughout this section. (Equations from Appendix B will be cited by number, (B.20), etc.) Here, we emphasize the main points and give particular relevant examples, referring to Appendix B for detailed results in the general case.

For the general process $1+2 \to 3+4$ at a fixed beam momentum, where 3 and 4 are groups of particles, a minimum set of kinematic variables comprises the sub-energies m_3 and m_4 and the momentum transfer, t. In terms of these, one can define a doubly inclusive‖ cross section $\partial^3\sigma/(\partial t\, \partial m_3\, \partial m_4)$. In general, we shall be interested in exclusive‖ cross-sections and will require further variables to describe the decay of systems 3 and 4.

We shall concentrate on the class of reactions $PN \to P\pi B$ for the case where the target nucleon is unpolarized and the recoil baryon polarization is undetected. Four independent variables are needed to describe such an experiment at a given beam momentum. These are conveniently taken to be

‡ "$P\pi$ scattering information" means not only $P\pi$ elastic scattering but also inelastic processes like $\pi\pi \to K\bar{K}$.

§ $P = \pi$ or K, $B = N$ or Δ.

‖ An inclusive (exclusive) cross-section is one for which the internal variables describing the systems 3 and 4 are (are not) averaged over.

two of those already referred to—M, the di-meson mass and t, the (squared) momentum transfer to the baryon vertex—and a pair of polar angles θ and ϕ to describe the di-meson decay. The two standard frames of reference for discussing the decay are respectively the s-channel and t-channel helicity frames. In both cases, the angles θ, ϕ relate to the decay viewed in the di-meson rest frame with OY perpendicular to the plane of production; OZ is taken along the di-meson direction of motion in the overall CM, respectively in the s-channel ($N_1P_2 \rightarrow B_3M_4^*$) and the t-channel ($\bar{M}_4^*P_2 \rightarrow B_3\bar{N}_1$). The two frames are connected by a rotation about the y-axis, (B.13)–B.23)—see also equations (3.1.10)–(3.1.13) below.

The t-channel frame is the most convenient for isolating specific exchange contributions (Gottfried and Jackson, 1964); in particular, OPE viewed in this frame only couples to di-mesons with helicity $m = 0$, and consequently gives no ϕ dependence to the cross-section (Treiman and Yang, 1962). This fact, and the sharp t-dependence from the pion-pole factor (see equations (3.1.15) and (3.1.16) below), are the basis of all attempts to isolate the OPE contribution. If there were no background, everything would be straightforward; in particular, the θ-dependence of the cross-section would relate directly to the $P\pi$ differential cross-section. Unfortunately, there are backgrounds, and we have to penetrate deeper into the spin structure of $P\pi$ production. For many purposes, the s-channel frame leads to a simpler description, e.g. for discussing absorption and vector dominance comparisons; the helicity amplitudes for a general set of exchanges have a particularly simple small-t structure in this frame.

The experimentally observed intensity distribution

$$I(s, t, M, \theta, \phi) \equiv \frac{\partial^4 \sigma}{\partial t\, \partial M\, \partial(\cos\theta)\, \partial\phi},$$

is usually presented in terms of the integrated intensity

$$N \equiv 4\pi\langle I\rangle \equiv \frac{\partial^2 \sigma}{\partial t\, \partial M},$$

and Legendre moments of the normalized angular distribution

$$W(\theta, \phi) \equiv \frac{1}{N} I(\theta, \phi) = \sum_{l=0}^{\infty} \sum_{m=0}^{l} a_{lm} \operatorname{Re} Y_l^m(\theta, \phi) \qquad (a_{00} = (4\pi)^{-\frac{1}{2}}). \tag{3.1.1}$$

The latter expansion is often re-written in terms of expectation values as

$$W(\theta, \phi) = \sum_l \{\langle \operatorname{Re} Y_l^0\rangle \operatorname{Re} Y_l^0(\theta, \phi) + 2 \sum_{m=1}^{l} \langle \operatorname{Re} Y_l^m\rangle \operatorname{Re} Y_l^m(\theta, \phi)\}, \tag{3.1.2}$$

($\langle \operatorname{Re} Y_l^m\rangle$ is usually written as $\langle Y_l^m\rangle$ for short).

As discussed in Appendix B, the above intensity distribution may be expressed in terms of the helicity amplitudes, $H^{(c)}_{\lambda_3\lambda_1}(s, t, M, \theta, \phi)$, where $c = t$ or s indicates the helicity reference frame being used (this will sometimes be omitted) and λ_1 and λ_3 denote the initial and final baryon helicity components. In a suitable normalization, the intensity formula is (cf. (B.28))

$$I(s, t, M, \theta, \phi) \equiv \frac{\partial^4 \sigma}{\partial t\, \partial M\, \partial(\cos\theta)\, \partial\phi} = \frac{1}{2} \sum_{\lambda_3\lambda_1} |H^{(c)}_{\lambda_3\lambda_1}|^2. \tag{3.1.3}$$

For discussing the production of relatively low mass di-meson systems such as are currently explored, it is convenient to decompose $H_{\lambda_3\lambda_1}$ into a sum of contributions, $H^{(c)j}_{\lambda_3\lambda_1;m}$ corresponding to the production of di-meson states of spin j and helicity m (B.29). On substituting into the above intensity formula (3.1.3), we obtain, (B.31),

$$I(\theta, \phi) = N \sum_{j_1 m_1 j_2 m_2} (2j_1+1)^{\frac{1}{2}}(2j_2+1)^{\frac{1}{2}} \rho^{(c)j_1 j_2}_{m_1 m_2}\, d^{j_1}_{m_1 0}(\theta)\, d^{j_2}_{m_2 0}(\theta)\, e^{i(m_1 - m_2)\phi}, \tag{3.1.4}$$

where the spin *density matrix* elements for di-meson production, $\rho^{(c)j_1 j_2}_{m_1 m_2}$, are given by the appropriate cross-terms between the helicity amplitudes, averaged over initial and final baryon helicity (B.32):

$$\rho^{(c)j_1 j_2}_{m_1 m_2} = (1/2N) \sum_{\lambda_3\lambda_1} H^{*(c)j_1}_{\lambda_3\lambda_1;m_1} H^{(c)j_2}_{\lambda_3\lambda_1;m_2}. \tag{3.1.5}$$

In all the above formulae, dependences on the other kinematic variables s, t and M are not shown.

The Legendre coefficients, a_{lm} of equation (3.1.1) may be expressed as sums of density matrix elements (B.37); for example, the requirement $a_{00} = (4\pi)^{-\frac{1}{2}}$ is secured by the trace condition

$$\sum_{jm} \rho^{jj}_{mm} = 1.$$

The density matrix elements are not all independent, being constrained by hermiticity and parity conservation (B.34). Furthermore, the independent elements are not all observable in experiments where the baryon polarization is unobserved. For example, if only S- and P-wave di-mesons contribute, the decay distribution may be written in the form

$$\begin{aligned} W(\theta, \phi) = \frac{1}{4\pi}\{1 &+ (\rho^{11}_{00} - \rho^{11}_{11})(3\cos^2\theta - 1) + 2\sqrt{3}\,\mathrm{Re}\,\rho^{10}_{00}\cos\theta \\ &- 3\sqrt{2}\,\mathrm{Re}\,\rho^{11}_{10}\sin 2\theta\cos\phi - 2\sqrt{6}\,\mathrm{Re}\,\rho^{10}_{10}\sin\theta\cos\phi \\ &- 3\rho^{11}_{1-1}\sin^2\theta\cos 2\phi\}. \end{aligned} \tag{3.1.6}$$

The formulae for the Legendre coefficients, a_{lm}, in terms of the ρ's can be read off from this equation. The analogous formulae where S-, P- and D-wave di-mesons are allowed are listed in Table B.1.

For the purposes of this discussion, we will focus on the above S- and P-wave example, which in fact gives a good first approximation to $P\pi$ production up to and through the first vector resonance (ρ or K^* (890)). The trace condition for this case reads

$$\rho_{00}^{00}+\rho_{00}^{11}+2\rho_{11}^{11}=1,$$

and the six independent intensity components may be expressed in terms of the participating helicity amplitudes thus:‡

$$\frac{\partial^2\sigma}{\partial t\,\partial M}=\|H_s\|^2+\|H_0\|^2+\|H_+\|^2+\|H_-\|^2 \qquad (\equiv N)$$

$$(\rho_{00}^{11}-\rho_{11}^{11})\frac{\partial^2\sigma}{\partial t\,\partial M}=\|H_0\|^2-\tfrac{1}{2}(\|H_+\|^2+\|H_-\|^2) \qquad (\equiv\sqrt{5\pi}N\langle Y_2^0\rangle)$$

$$\rho_{1-1}^{11}\frac{\partial^2\sigma}{\partial t\,\partial M}=\tfrac{1}{2}(\|H_+\|^2-\|H_-\|^2) \qquad (\equiv-\sqrt{10\pi/3}N\langle Y_2^2\rangle)$$

$$\mathrm{Re}\,\rho_{10}^{11}\frac{\partial^2\sigma}{\partial t\,\partial M}=\frac{1}{\sqrt{2}}\mathrm{Re}\,\|H_-^*\,.\,H_0\| \qquad (\equiv\sqrt{5\pi/3}N\langle Y_2^1\rangle)$$

$$\mathrm{Re}\,\rho_{00}^{10}\frac{\partial^2\sigma}{\partial t\,\partial M}=\mathrm{Re}\,\|H_0^*\,.\,H_S\| \qquad (\equiv\sqrt{\pi}N\langle Y_1^0\rangle)$$

$$\mathrm{Re}\,\rho_{10}^{10}\frac{\partial^2\sigma}{\partial t\,\partial M}=\frac{1}{\sqrt{2}}\mathrm{Re}\,\|H_-^*\,.\,H_S\| \qquad (\equiv\sqrt{\pi}N\langle Y_1^1\rangle). \qquad (3.1.7)$$

In equation (3.1.7), summation over the independent baryon helicity components is to be understood, for example

$$\mathrm{Re}\,\|H_-^*\,.\,H_0\|\equiv\sum_{\lambda_1(\lambda_3>0)}H_{\lambda_3\lambda_1;1}^{*1-}H_{\lambda_3\lambda_1;0}^{0-}.$$

(Only half the helicity amplitudes need to be referred to, because of the relations from parity conservation (B.11). We choose to work with λ_3, the final baryon helicity, positive.) Also in (3.1.7), we have adopted a short-hand notation for the di-meson spin and helicity

$$(H_S,H_0,H_+,H_-)\equiv(H_{\lambda_3\lambda_1;0}^{0},H_{\lambda_3\lambda_1;0}^{1-},H_{\lambda_3\lambda_1;1}^{1+},H_{\lambda_3\lambda_1;1}^{1-}).$$

‡ Where the final baryon B is unstable, i.e. for $B\equiv\Delta$ (1236), additional information is available from its decay angular distribution. The present discussion applies if this is averaged over.

For convenience, we have used the "$\pm$" combination of the $+m$ and $-m$ di-meson helicity amplitudes which correspond to the exchange of given dominant parity in the t-channel (cf. (B.16)),

$$H^{j\pm}_{\lambda_3\lambda_1;m} = \frac{1}{\sqrt{2}}(H^j_{\lambda_3\lambda_1;m} \mp (-)^m H^j_{\lambda_3\lambda_1;-m}) \qquad (m>0)$$
$$H^{j-}_{\lambda_3\lambda_1;0} = H^j_{\lambda_3\lambda_1;0}. \tag{3.1.8}$$

These have the special virtue that, even for finite s, there are no H^+H^- cross terms in the unpolarized intensity. The "$\pm$" separation commutes with the spin crossing matrix (B.17), so that, for example, OPE, which couples to $m=0$ in the t-channel, only couples to dominantly unnatural exchange in the s-channel.

Equation (3.1.7) and its generalizations to include higher spin contributions clearly expose the problem of isolating OPE effects. If OPE were the sole production mechanism, only the $m=0$ nucleon non-flip amplitudes would couple in the t-channel—i.e. $H^{(t)0}_{\frac{1}{2}\frac{1}{2};0}$ and $H^{(t)1}_{\frac{1}{2}\frac{1}{2};0}$ in the above example; both amplitudes would have the same production phase and both would contain the pion pole factor $1/(t-\mu^2)$, so that it would be straightforward to infer the $P\pi$ scattering amplitudes. Once background enters, the problem is in principle under-determined, unless one relies solely on extrapolating to the pion pole. In the physical region, the OPE signal is accompanied by backgrounds from other effects. For example, in the case where just S- and P-waves contribute, the isotropic term in the t-channel frame, $(1-\sqrt{5\pi}\langle Y^0_2\rangle)N$, receives contributions not only from S-wave production but also from transverse P-wave production, the latter unconnected with OPE

$$(1-\sqrt{5\pi}\langle Y^0_2\rangle)N = \|H_S\|^2 + \tfrac{3}{2}\|H_+\|^2 + \tfrac{3}{2}\|H_-\|^2,$$

as follows immediately from (3.1.7). Such backgrounds have to be removed either by extrapolation or through the use of a model. As we shall see below, both the OPE signal and its backgrounds have characteristic and distinctive helicity dependences, suggesting strategies for purifying the OPE signal preparatory to extracting $P\pi$ phase shifts.

Before concluding this sub-section, there are two more kinematic preliminaries to note. The first concerns the crossing transformation connecting the s- and t-channel helicity frames. As already mentioned, it is convenient to focus on the exchange-parity dominant ($\pm$) combinations of the helicity amplitudes (3.1.8), since these transform separately (B.17). Furthermore, the crossing transformation is a direct product of a transformation on the baryon spin indices, λ_1, λ_3 and a transformation on the di-meson spin. The baryon spin transformation is always diagonal in the exact forward direction; however, for the important case of $\frac{1}{2}^+ \to \frac{1}{2}^+$ via (predominantly) unnatural parity exchange, this situation only extends for a tiny region, $|t_{\min}-t|=0(|t_{\min}|)$; thereafter, the transformation is anti-diagonal near the forward direction, (B.23). This has the effect that $PN \to P\pi N$ via OPE, which rigorously is pure non-flip in the t-channel, is essentially pure flip in the s-channel, except very close to $t_{\min}$:

$$\frac{H^{(s)}_{++}}{H^{(s)}_{+-}} \sim -\frac{\sqrt{-t_{\min}}}{\sqrt{t_{\min}-t}}, \; (t \to t_{\min}) \tag{3.1.9}$$

(cf. (B.55)). For all other situations for the baryon vertex, in particular for the other important practical case of $PN \to P\pi\Delta$ via OPE, non-flip in the t-channel transforms into non-flip in the s-channel. It is an important check on procedures for extracting $P\pi$ scattering parameters from data on $PN \to M^*B$, that the two main sources of information have distinct spin couplings.

The transformation between s- and t-channel density matrix elements is governed by the di-meson spin crossing angle, χ_4:

$$\rho^{(t)j_1j_2}_{m_1m_2} = \sum_{m_1'm_2'} d^{j_1}_{m_1m_1'}(\chi_4)\rho^{(s)j_1j_2}_{m_1'm_2'}d^{j_2}_{m_2'm_2}(-\chi_4). \tag{3.1.10}$$

For large s, χ_4 is given by (see (B.19) and (B.20) for the full formulae at finite s)

$$\cos\chi_4 \underset{s\to\infty}{\sim} -\frac{[t+M-\mu_P^2]}{\mathcal{T}_{24}}$$
$$\sin\chi_4 \underset{s\to\infty}{\sim} -\frac{2M\sqrt{-t}}{\mathcal{T}_{24}}, \tag{3.1.11}$$

with

$$\mathcal{T}_{24} = [(t-(M+\mu_P)^2)(t-(M-\mu_P)^2)]^{\frac{1}{2}}, \tag{3.1.12}$$

and $\mu_P = m_\pi$ or m_K, as appropriate. The density matrix transformations follow from those for the helicity amplitudes (B.17). As an example, for the case of P-wave di-meson production in an $N \to N$ reaction, the explicit transformation is ($s \to \infty$, and except for the very near-forward direction cf. (3.1.9)),

$$\begin{bmatrix} H^{(s)1-}_{x;0} \\ H^{(s)1-}_{x;1} \end{bmatrix} = \mp i \begin{bmatrix} \cos\chi_4 & -\sin\chi_4 \\ \sin\chi_4 & \cos\chi_4 \end{bmatrix} \begin{bmatrix} H^{(t)1-}_{y;0} \\ H^{(t)1-}_{y;1} \end{bmatrix}, \tag{3.1.13a}$$

$$x = ++,\ y = +- \text{ or } x = +-,\ y = ++ : (++ \equiv \tfrac{1}{2}\tfrac{1}{2}, \text{etc.})$$

and

$$(|t| \ll 4m_B^2) \qquad (m_B \equiv m_N, m_\Delta, \text{etc.})$$

$$H^{(s)1+}_{x;1} = H^{(t)1+}_{x;1} \qquad x = ++ \text{ or } --. \tag{3.1.13b}$$

The di-meson crossing angle, χ_4, plays an important role in computing effects which are attributed to the modification of particle exchanges such as OPE by absorption. The particle exchanges have a simple spin structure in the t-channel, whilst absorption is thought to be diagonal in s-channel helicity (see discussion of the absorption model, Section 3.1(e) below).

Finally, on the subject of kinematics, we give formulae for $t_{\min}$, the boundary of the physical region corresponding to forward M^* production. This is an important parameter since it determines whether, for given beam momentum and M^* mass, we have access to the small-t region $|t| = 0(\mu^2)$ where a distinct OPE signal should be seen. The general formula, (B.5), implies a large s behaviour (B.7), which for the equal baryon mass case is $0(1/s^2)$ and for the unequal baryon mass case is $0(1/s)$:

$$t_{\min} \underset{s\to\infty}{\sim} -m_N^2 \frac{(M^2-\mu_P^2)^2}{s^2}, \qquad (m_B = m_N)$$

$$t_{\min} \underset{s\to\infty}{\sim} \frac{-(m_B^2-m_N^2)(M^2-\mu_P^2)}{s}, \qquad (m_B \neq m_N) \tag{3.1.14}$$

($\mu_P = m_\pi, m_K$). Some relevant numerical examples (using the full formula (B.5)) are given in Table 3.1.1.

TABLE 3.1.1. Examples of $t_{\min}$ Values

Reaction	p_{lab} (GeV/c)	$\lvert t_{\min}\rvert/\mu^2$			
		$M^* \equiv \rho$	$M^* \equiv f$	$M^* \equiv g$	$M^* \equiv 5$ GeV
$\pi N \to M^* N$	2·7	0·70	8·5	—	—
	6·0	0·13	1·11	4·1	—
	17·0	0·015	0·12	0·4	—
	40·0	0·003	0·02	0·07	7·8
	200·0	0·0001	0·001	0·002	0·2
$\pi N \to M^* \Delta$	7·0	1·75 (0·4)	6·0 (1·3)	13·6 (2·7)	
	200·0	0·05 (0·01)	0·14 (0·04)	0·26 (0·06)	2·6 (0·6)
		$M^* \equiv K^*$ (890)	$M^* \equiv K^*$ (1420)	$M^* \equiv 1·76$ GeV	
$KN \to K^* N$	6·0	0·12	1·4	4·2	

(For the "Δ-reaction", the spread of $t_{\min}$ corresponding to the Δ-width is shown in brackets.)

(c) Formulae for OPE

We now set down some specific formulae for OPE for future reference. The doubly inclusive cross-section for the process $1+2 \to 3+4$ via OPE (Fig. 3.1.1(b)) is given by the formula

$$\left[\frac{\partial^3 \sigma}{\partial t\, \partial m_3\, \partial m_4}\right]_{\text{OPE}} = \frac{1}{4\pi^3 \hbar^2}\left[\frac{1}{m_2 p_{lab}}\right]^2 [m_3^2 q_3 \sigma_3^{\text{off}}] \frac{1}{(t-\mu^2)^2} [m_4^2 q_4 \sigma_4^{\text{off}}], \quad (3.1.15)$$

where q_3 and σ_3^{off} (q_4 and σ_4^{off}) are the *CM* momentum and cross-section for the process $\pi + 1 \to 3$ ($\pi + 2 \to 4$), t is the momentum transfer and the other notation is self explanatory.‡ Note the appearance of the kinematic factors $m_3^2 q_3 (m_4^2 q_4)$ multiplying $\sigma_3^{\text{off}}(\sigma_4^{\text{off}})$, to cancel flux factors in $\sigma(1\pi \to 3)$, and to supply phase space factors. They have the effect of enhancing high mass signals for the sub-systems 3 and 4 over what would be obtained in formation experiments on 3 or 4. In the important case when 1 and 3 (or 2 and 4) are both nucleons,§ one replaces $[m_i^2 q_i \sigma_i^{\text{off}}]$, $i = 3$ or 4, by $4\pi^2 \hbar^2 m_N^2 f^2(-t)/\mu^2$, where f^2 is the $\pi N\bar{N}$ pseudo-vector coupling constant ($f^2 = 0{\cdot}081$ for $\pi^0 N\bar{N}$, $f^2 = 0{\cdot}162$ for $\pi^{\pm} N\bar{N}$).

For the particular processes $PN \to P\pi B$, the OPE contribution to the intensity takes the form

$$I^{\text{OPE}} \equiv \frac{\partial^4 \sigma}{\partial t\, \partial M\, \partial(\cos\theta)\, \partial\phi} = \frac{M^2 q_{P\pi}}{2\pi^2 q_{\text{in}}^2 s} \frac{d\sigma_{P\pi}^{\text{off}}}{d(\cos\theta)} \frac{K_{NB}(t)}{(t-\mu^2)^2} \theta_{NB}(t, t_{\min}), \quad (3.1.16)$$

where q_{in} is the *CM* momentum of the incoming meson, P, and $q_{P\pi}$ denotes the break-up momentum of the di-meson in its *CM*; θ_{NB} denotes the kinematic cut-off. The baryon vertex factor, $K_{NB}(t)$, has a distinctive form according as the final baryon, B, is a nucleon or a Δ (1236). In the former case

$$\begin{aligned} K_{NN}(t)\theta_{NN} &= f^2 m_N^2(-t/\mu^2)\theta(t_{\min}-t) \\ &\equiv (G_{N\bar{N}\pi}^2/4\pi)(-t/4)\theta(t_{\min}-t), \end{aligned} \quad (3.1.17)$$

‡ The "off-shell" cross-section, σ_3^{off} can be defined as the sum of the on-shell partial wave cross-sections suitably scaled to give appropriate threshold behaviour

$$\sigma_3^{\text{off}} = \sum_{l=0}^{\infty} \left(\frac{q_3^{\text{off}}}{q_3^{\text{on}}}\right)^{2l} \sigma_l^{\text{on}}.$$

One may wish to include additional t-dependent factors—for example form factors and signature and vertex factors arising from Reggeisation. Obviously, the OPE contribution is only well defined at the pole, $t = \mu^2$, and the continuation away from there is a matter of convenience.

§ We follow the baryon-first convention for convenience in handling the spin-formalism.

while for the latter case ($B \equiv \Delta^{++}$)‡

$$K_{N\Delta}(t)\theta_{N\Delta} = 2/3(G^2_{\Delta\bar{N}\pi}/4\pi)(q^2_\Delta)\bar{\theta}_\Delta(t, t_{\min}), \tag{3.1.18}$$

with

$$G^2_{\Delta\bar{N}\pi}/4\pi = 3/2(m^2_\Delta\Gamma_\Delta/q^3_\Delta).$$

(The coupling constants are as defined in Pilkuhn *et al.* (1973).) The quantities m_Δ, Γ_Δ and q_Δ are the mass, width and *CM* break-up momentum of the Δ; the factor $\bar{\theta}_\Delta(t, t_{\min})$ denotes the result of averaging the kinematic cut-off $\theta(t_{\min}(m^2_{\pi^+p}) - t)$ over the Δ^{++} band selected. From the appropriate formula for $t_{\min}$ (3.1.14), this is seen to give a blurring out of the kinematic cut-off over a distance in t

$$\langle \Delta t_{\text{edge}} \rangle \underset{s\to\infty}{\sim} \frac{2(m_\Delta\Gamma_\Delta)(M^2 - \mu_P^2)}{s},$$

thus negligible at sufficiently high energies but appreciable at presently well-explored energies.§

This effect, which causes the cross-section for $PN \to M^*\Delta$ to fall towards the kinematic boundary, is common to all exchanges—OPE and its backgrounds. The situation is quite different when the final baryon is a nucleon, $PN \to M^*N$. In that case, the factor $(-t/\mu^2)$ in (3.1.17) is peculiar to OPE and is a consequence of the pseudo-scalar coupling $(\bar{u}\gamma_5 u)$ at the nucleon vertex.‖ As regards its t-dependence, the OPE cross-section has the form

$$I^{\text{OPE}}_{PN\to M^*N} \propto \frac{-t/\mu^2}{(t-\mu^2)^2}, \tag{3.1.19}$$

and thus rises sharply to a peak at $|t| = \mu^2$ and falls off again at large t. For $PN \to M^*\Delta$, the analogous form is (cf. equation (3.1.18))

$$I^{\text{OPE}}_{PN\to M^*\Delta} \propto \frac{\bar{\theta}_\Delta(t, t_{\min})}{(t-\mu^2)^2}. \tag{3.1.20}$$

Inserting numbers into (3.1.17) and (3.1.18), one sees that, aside from factors for the π^+p mass cut and the folded $t_{\min}$ cut-off effect represented by the factor $\bar{\theta}_\Delta$, the $N \to \Delta$ cross-section is larger than the $N \to N$ cross-section out to about $|t| \approx 5\mu^2(K_{N\Delta}(t)/K_{NN}(t) \approx (5\mu^2/(-t))$. The reaction $PN \to M^*\Delta$

‡ This equation is set down for purposes of discussion, especially to make comparisons with (3.1.17). For the actual analysis of data, equation (3.1.15) is used, inserting experimental πp cross-sections integrated over the selected mass-cut.

§ This formula underestimates the spread at finite s, cf. discussion below (3.1.14).

‖ Likewise, of course for pseudo-vector coupling. The general statement is that a zero-mass pseudo-scalar must decouple from any $B \to B$ transition where the mass and parity is unchanged.

is thus very competitive with $PN \to M^*N$ except for effects arising from the larger $t_{\min}$ (see equation (3.1.14)).

The non-vanishing t-channel helicity amplitudes for OPE corresponding to various di-meson spins j take the form

$$H^{(t)j}_{\frac{1}{2}\frac{1}{2};0} = \frac{(16\varepsilon)^{\frac{1}{2}}(q^{\text{on}}_{P\pi})^{\frac{1}{2}}}{q_{\text{in}}\sqrt{s}}(2j+1)^{\frac{1}{2}}\left(\frac{q^{\text{off}}_{P\pi}}{q^{\text{on}}_{P\pi}}\right)^j \frac{\sqrt{K_{NB(t)}}}{t-\mu^2} T^{j\text{on}}_{P\pi}(M), \qquad (3.1.21)$$

$$(PN \to P\pi B \text{ via OPE}) \quad (\varepsilon = 2 \text{ if } P \equiv \pi,\ \varepsilon = 1 \text{ if } P \not\equiv \pi),$$

where most of the notation is as defined in (3.1.16). $T^{j\text{on}}_{P\pi}(M)$ is the appropriate on-shell $P\pi$ partial wave amplitude, and imparts the phase of this amplitude $\bar{\delta}^j_{P\pi}$ to the production helicity amplitude

$$T^{j\text{on}}_{P\pi}(M) = |T^{j\text{on}}_{P\pi}(M)|\, e^{i\bar{\delta}^j_{P\pi}(M)}.$$

(We use the notation $\bar{\delta}^j_{PM}$ to emphasize that it is the phase at T^j, not the phase shift δ^I_j which is referred to. The distinction arises where there is inelasticity.) The quantity $\bar{\delta}^j_{P\pi}(M)$, which we term the *decay-phase*, is entirely a property of on-shell $P\pi$ scattering, and for the case of simple OPE is the only phase appearing. In general, there will be additional production phases corresponding to other exchange mechanisms—for example Regge phases; however, in a kinematic region dominated by exchange, it may be hoped that all rapid M dependences are represented by the factors appearing in (3.1.21). This is the basis of the factorization model to be discussed in Section 3.2.

The factor $(q^{\text{off}}_{P\pi}/q^{\text{on}}_{P\pi})^j$, with $q^{\text{off}}_{P\pi} = \mathcal{T}_{24}(t)/2M$, (cf. equation 3.1.12) gives correct kinematic threshold factors to $H^{(t)j}_{\frac{1}{2}\frac{1}{2};0}$. For the subsequent discussion, it is convenient to establish a short-hand version of (3.1.21) which emphasizes t-dependence. We therefore write

$$H^{(t)j}_{\frac{1}{2}\frac{1}{2};0} = iG^j_{P\pi}\frac{\sqrt{-t}}{(t-\mu^2)}\left(\frac{\mathcal{T}_{24}(t)}{M}\right)^j, \qquad (3.1.22)$$

$$(PN \to M^*N)$$

with

$$G^j_{P\pi} = \frac{(16\varepsilon)^{\frac{1}{2}}}{q_{\text{in}}\sqrt{s}} f(m_N/\mu)\frac{(2j+1)^{\frac{1}{2}}}{2^j}(q^{\text{on}}_{P\pi})^{-j+\frac{1}{2}} T^{j\text{on}}_{P\pi}(M). \qquad (3.1.23)$$

(Again, $\varepsilon = 2$ or 1, according as $P \equiv \pi$ or K, cf. equation (3.1.21).) As an example, we can define an effective G_ρ for the rho-resonance corresponding to a "quasi-particle" description.

$$G_\rho \equiv \left[\int_\rho dM_{\pi\pi}|G^1_{\pi\pi}|^2\right]^{\frac{1}{2}},$$

evaluated in the narrow resonance approximation. In this way, we find

$$G_\rho^2 = \left(\frac{f_{\rho\pi\pi}^2}{4\pi}\right)\frac{G_{N\bar{N}\pi}^2}{4\pi}\frac{\pi}{2sq_{\text{in}}^2}, \tag{3.1.24}$$

with the couplings given by

$$\frac{G_{N\bar{N}\pi}^2}{4\pi} \approx \frac{4m_N^2 f}{\mu^2} \approx 14\cdot6,$$

and

$$\frac{f_{\rho\pi\pi}^2}{4\pi} = \left[\frac{12\Gamma_\rho}{m_\rho}\right]\left[1 - \frac{4\mu^2}{m_\rho^2}\right]^{-\frac{3}{2}} \approx 2\cdot8. \tag{3.1.25}$$

In using such a "quasi-particle" description for the rho-resonance, all the intensity formulae (3.1.7) are understood to be integrated with respect to $M(\equiv M_{\pi\pi})$ over the resonance. Such a formulation is convenient for making vector dominance comparisons (see below).

(d) OPE Compared to Experiment

As discussed above, the pure OPE term of Fig. 3.1.1 is independent of ϕ in the t-channel frame, so that for all processes $PN \to P\pi B$ (via OPE)

$$\rho_{m_1 m_2}^{(t) j_1 j_2} = 0, \quad \text{unless} \quad m_1 = m_2 = 0. \tag{3.1.26}$$

Furthermore, the OPE contribution to the differential cross-section takes the specific form appropriate either to "N-reactions" ($B \equiv N$) or "Δ-reactions" ($B \equiv \Delta$):

$$\left(\frac{d\sigma}{dt}\right)^{\text{OPE}} = \frac{\text{const}}{(t-\mu^2)^2}\left(\frac{-t}{\mu^2}\right) \quad (N)$$
$$= \frac{\text{const}}{(t-\mu^2)^2}\bar{\theta}_\Delta(t) \quad (\Delta), \tag{3.1.27}$$

(cf. equations (3.1.16)–(3.1.20)).

How do the basic predictions for OPE (3.1.26), (3.1.27), compare with the data? As a first example, we consider data on the reaction

$$\pi^- p \to \pi^+ \pi^- n,$$

which is a major source of information on $\pi\pi$ scattering. Figures 3.1.2 and 3.1.3 show the differential cross-section and s- and t-channel intensity components for this reaction at 17·2 GeV/c as functions of $\sqrt{-t}$ for a band of dipion masses in the rho-region ($0\cdot71 \leqslant M_{\pi^+\pi^-} \leqslant 0\cdot83$ GeV) (Grayer *et*

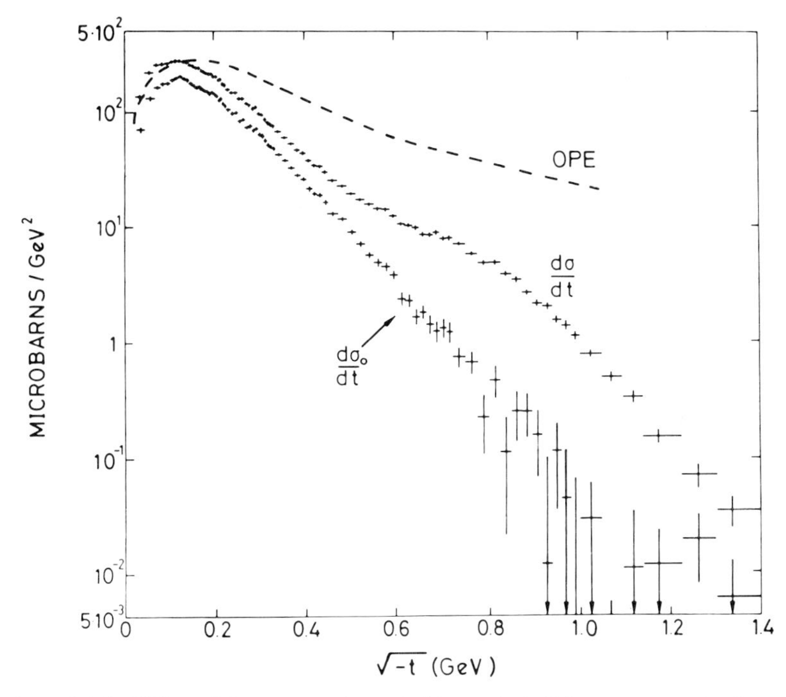

FIG. 3.1.2. Differential cross-section $d\sigma/dt$ and helicity zero projection, $d\sigma_0/dt \equiv (\frac{1}{3}\rho_{00}^{00}+\rho_{11}^{00})\, d\sigma/dt$ in the t-channel helicity frame, for $\pi^- p \to \pi^+\pi^- n$ at 17·2 GeV/c, with $0{\cdot}71 \leqslant M_{\pi^+\pi^-} \leqslant 0{\cdot}83$ GeV (Grayer *et al.*, 1972a). The dashed curve shows the t-dependence const $|t|/(t-\mu^2)^2$ from pure OPE.

al., 1972a,b). The differential cross-section does indeed show a peak at around $|t| = \mu^2$, dropping both towards $t_{\min}$ and larger $|t|$ values, so that a large contribution from OPE is clearly indicated. It is also clear that there are backgrounds to this signal:

1. The data (Fig. 3.1.2), show greater collimation than simple OPE, i.e. $d\sigma/dt$ falls off more rapidly at larger $|t|$ than (3.1.27) would indicate.
2. There is ϕ dependence, and the density matrix elements with $m_1 \neq 0$ or $m_2 \neq 0$ do not vanish in accord with equation (3.1.26) (see Fig. 3.1.3).
3. At least some intensity components $\langle Y_l^m \rangle\, d\sigma/dt$ appear definitely not to vanish at $t = 0$, to which the physical region almost extends at this high beam momentum ($t_{\min} \approx -{\cdot}015\mu^2$).

This is particularly brought out in Fig. 3.1.2, where, besides the full differential cross-section, $d\sigma/dt$, the component $d\sigma_0/dt = (\frac{1}{3}\rho_{00}^{00}+\rho_{11}^{00})\, d\sigma/dt$ is also plotted.‡ The latter quantity should be determined by OPE and does

‡ The quantity $d\sigma_0/dt$ is directly derivable from data, assuming only S- and P-wave dipions (cf. (3.1.7)).

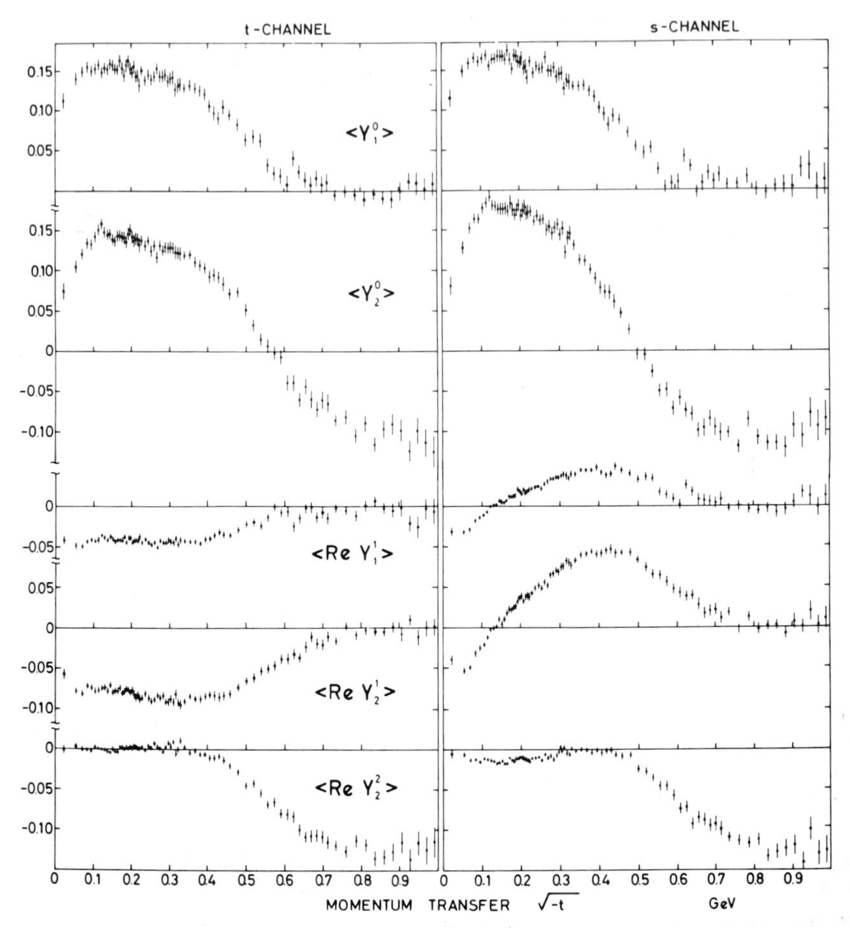

FIG. 3.1.3. s- and t-channel intensity components $\langle Y_l^m \rangle$ for $\pi^- p \to \pi^+\pi^- n$ at 17·2 GeV/c with $0{\cdot}71 \leq M_{\pi^+\pi^-} \leq 0{\cdot}83$ GeV, as functions of $\sqrt{-t}$ (Grayer *et al.*, 1972a).

indeed drop towards $t = 0$, whilst the remainder, $d\sigma_1/dt = (\frac{2}{3}\rho_{00}^{00} + 2\rho_{11}^{11})\, d\sigma/dt$, which is predominantly background, shows a nonvanishing intercept at $t = 0$. The latter term is also much less sharply collimated.

As an example of di-meson production from a "Δ-reaction", Fig. 3.1.4 shows the differential cross-section, $d\sigma/dt$, for the reaction

$$K^+p \to K^+\pi^-\Delta^{++},$$

with $M_{K^+\pi^-}$ in the interval 0·8—1·0 GeV at 12 GeV/c incident K^+ momentum (Barbaro-Galtieri *et al.*, 1973). As for the previous "N-reaction" example, the differential cross-section is sharply peaked towards

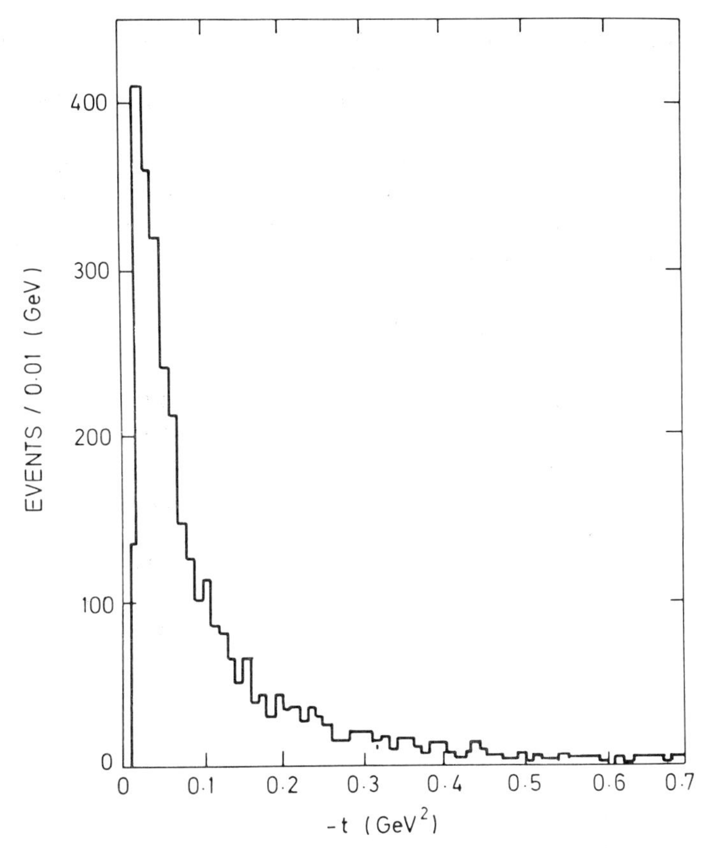

FIG. 3.1.4. Differential cross-section, $d\sigma/dt$, for the reaction $K^+p \rightarrow K^+\pi^-\Delta^{++}$ at 12 GeV/c, with $0{\cdot}8 \leq M_{K\pi} \leq 1{\cdot}0$ GeV (Barbaro-Galtieri *et al.*, 1973).

the forward direction, as expected from OPE, (equation 3.1.18), but shows additional collimation to the simple pole formula towards large $|t|$. The authors do not actually publish density matrix elements; they do however show ϕ-distributions in the t-channel frame (Treiman–Yang test), and these indicate approximate isotropy at small $|t|$ suggesting that OPE dominates there. This seems to be the general picture for "Δ-reactions". Abrams *et al.*, (1970) have published density matrix elements for $\rho^0\Delta^{++}$ production at 3·7 GeV/c and these indicate that the helicity component $H^{(t)}_{\frac{1}{2}\frac{1}{2};0}$ to which OPE contributes is the dominant one at small t.‡ The background problem is

‡ Joint-decay density matrix elements for this experiment have also been published (Barnham *et al.*, 1973), and been fed into a model amplitude analysis (with some rather special assumptions—see Section 3.2(f)) by Irving (1973). The general picture of OPE dominance at small $|t|$ is confirmed.

thus not reckoned to be so serious for "Δ-reactions", especially since the signal-to-background ratio does not vanish at $t_{\min}$. $P\pi$ phase shifts are normally extracted for these reactions by extrapolation to the pion-pole, using form-factors to describe the observed additional collimation over simple OPE. This is also one possible approach for describing "N-reactions", but in view of the more severe background problem a variety of other techniques is also adopted. We shall therefore mainly emphasize "N-reactions" in the following discussion.

(e) Models for the Background to OPE

Concentrating now on the "N-reaction" $PN \to M^*N$, we review likely candidates for the background contributions to simple OPE, and special models which incorporate them. We have already noted three features which ought to be accommodated, i.e. additional collimation, $m \neq 0$ components (in the t-channel), and non-vanishing contributions at $t = 0$. These properties relate to the t-dependence and spin-alignment for a given di-meson mass band dominated by a particular di-meson spin, and we shall also be concerned with what happens when the parameters M and j vary.

According to the currently accepted picture of non-vacuum exchange reactions, we should consider the following modifications to simple OPE:

(i) Form factors and Reggeization,

(ii) Other Regge pole exchanges,

(iii) Cut contributions arising from the absorption of the Regge poles.

We shall discuss some detailed aspects of these possibilities below, but certain general points can be made at once. The first is that form factors can only supply collimation, and cannot explain the departure from the OPE predicted spin alignment.‡ Candidates for additional Regge poles depend on the particular $P \to M^*$ reaction being considered. The most restrictive case, $\pi^- \to M^{*0}$, is also the best studied and the one on which we shall concentrate. Here the obvious candidates, besides the π, are A_2 and A_1 exchange, although the latter may not couple appreciably. For non-charge exchange reactions, and for K-induced reactions, there are additional candidates (e.g. ω exchange in $\pi^\pm \to M^{*\pm}$ for which there is good evidence), but all can be classified as either natural parity exchange, (N), associated with the vector–tensor exchange–degenerate trajectories, or unnatural parity axial–vector exchanges which are predominantly t-channel flip (A).§ Three points follow from this. Firstly as to s-dependence, i.e. dependence on the incoming beam momentum, the (N) exchanges, having higher lying trajectories, must ultimately dominate the π and (A) exchanges as $s \to \infty$ for

‡ Reggeization of the pion does introduce extra spin couplings to $H^{(t)-}_{++;m\neq 0}$ which however must vanish at the pion pole. This is not sufficient to explain the observed effects.

§ For the K-reactions one can have B-meson exchange, which is t-channel non-flip.

given di-meson mass, M. We return below to this point and the related question of M-dependence for alternative trajectories. Secondly, the (N) exchanges, provided they are evasive, i.e. unaccompanied by partners of opposite parity and degenerate mass (or analogous absorptive cut contributions), necessarily decouple at $t=0$ (cf. Appendix B); the non-vanishing background at $t=0$ cannot therefore be attributed to them. Thirdly, (A) exchanges will in general give non-vanishing contributions as $t\to 0$, predominantly (except right in the forward direction $|t_{\min}-t|=0|t_{\min}|$) to the t-channel flip $m=0$ amplitude. A signal for contributions of this type would be a non-vanishing intercept as $t\to 0$ for intensity components such as $\rho_{00}^{(t)jj}\, d\sigma/dt$. For the case of $\pi^-p\to\pi^+\pi^-n$ at 17·2 GeV/c with $M_{\pi\pi}$ in the rho-mass band, the plot, Fig. 3.1.2, of $(\frac{1}{3}\rho_{00}^{00}+\rho_{00}^{11})\, d\sigma/dt$ does not indicate any large contribution of this kind. As we shall see below, many analyses of $\pi\pi$ production assume A_1 exchange to be absent. The most sensitive check on this assumption will come from experiments on polarized targets (see Appendix B).

The final ingredient listed above, for modifying simple OPE, absorption, is the mechanism which comes nearest to explaining all the observed features of the background, and it will consequently occupy the major part of the following discussion. In order to clear the way for this, we start with a brief account of the other possibilities—form factors and additional Regge poles.

(i) *Form Factors and Reggeization*

There exist a number of models in which the production is pictured as proceeding via the exchange of "virtual pions" with modified t-dependence, i.e.

$$H_{[\lambda]}=H_{[\lambda]}^{\mathrm{OPE}}\Phi(t).$$

(The helicity indices are collectively denoted by $[\lambda]$.)

The additional collimations shown by the data (effect (1) in the discussion of the previous sub-section) can be reproduced; however no depolarization is allowed for, so no account is taken of effects (2) and (3). One version, for example the approach of Dürr and Pilkuhn (1965), introduces vertex functions with off-shell angular momentum threshold factors incorporating range parameters, as in nuclear theory. The method, extended to allow for separate vertex functions and propagators, with factorization, has been applied to a systematic discussion of OPE in various processes (Wolf, 1969). Most commonly, form factors of this type are used as an *ad hoc* device, for example in the analysis of $\pi N\to\pi\pi\Delta$, to reproduce the gross features of the t-dependence, prior to extrapolation to the pion pole—see e.g. Protopopescu *et al.* (1973).

Another type of form factor model, inspired by PCAC and the Veneziano model (see Chapters 14 and 12) has been proposed for dipion production by Gutay *et al.* (1969) and by Wagner (1969). Rho and f^0 production are unaltered from OPE except for the introduction of off-shell angular momentum factors as in the Dürr–Pilkuhn model, but very striking effects are predicted for S-wave production. The physically observed production process is understood as proceeding via the exchange of an off-shell pion which conforms to the PCAC condition, i.e. the production amplitude is required to have a zero at $s_{\pi\pi} = t_{\pi\pi} = u_{\pi\pi} = m_\pi^2$. From the general properties of analytic functions of several variables this induces a line of zeros in the $(s_{\pi\pi}, m_\pi^{\text{off}})^2 \equiv t)$ plane; as a result the production amplitudes of S-wave dipions would have zeros at values of momentum transfer which vary with dipion mass and isospin. In particular (in the linear approximation for the off-shell $\pi\pi$ scattering amplitudes, cf. Chapter 14), $d\sigma/dt$ for S-wave $\pi^+\pi^-$ production would have a zero at a point, $t = t_0 \propto (M_{\pi\pi}^2 - m_\pi^2)$ in the physical region. The constant of proportionality would be governed by the ratio of S-wave scattering lengths, a_0^0/a_0^2, and be expected to be of order unity. Evidence in favour of this conjecture was presented by Gutay *et al.* (1969) on the basis of rather low statistics; no further confirmation has at yet appeared and it seems almost certain that the hypothesis is not correct in its simple form. Nonetheless, it may be that the PCAC form factor does play a role, but modified by absorption, which would have a very strong influence on such a structure, especially at small dipion masses where the predicted form factor is very sharp. Despite the lack of other experimental support, there remains the indication that $\pi^+\pi^+$ production is less sharply collimated than S-wave $\pi^+\pi^-$ production, qualitatively as predicted by the model (Wagner, 1969). We shall not discuss PCAC form factors further but, until relevant detailed data are available, one should keep in mind that the last word on this subject may not have been said.

Reggeization of pion exchange both modifies the s-dependence, and, at fixed s, also supplies a form factor. Thus, besides such vertex factors as are inserted (usually exponentials, e^{bt}), the specific Regge factors $(s/s_0)^{\alpha_\pi(t)}$ and $e^{-i\pi\alpha_\pi(t)/2}$ also enter. For an experiment at given energy, the important effect comes from the second factor, the Regge phase. When Reggeization is conjoined with absorption (see below), this can induce differences of production phase between different helicity amplitudes, although probably not very large ones (Froggatt and Morgan, 1972). The phenomenological status of the pion trajectory was for a long time in doubt (in particular the effective trajectories for charged pion photoproduction were persistently flat); the determination of essentially normal trajectory parameters from the systematics of dipion production (Estabrooks and Martin, 1972b) (see Fig. 3.1.10 below) was an important clarification. However, owing to the proximity to the physical region of the pion pole, Reggeization including the extra spin couplings which it admits, does not achieve any important modification at small t. For the extraction of $P\pi$ scattering parameters, it is not something which we need to emphasize.

(ii) *Other Trajectories*

The pion pole, whether Reggeized or not, even, as we shall see below, whether absorbed or not, does not produce substantial differences of production phase or of energy dependence among the different components of production. The simplification that results from this is disrupted by the coupling of other trajectories. As discussed above, the obvious candidates (aside from the possibility of A_1 exchange for which there so far exists no evidence) are the natural parity trajectories belonging to the vector and tensor nonets, e.g. for dipion production ω and A_2 exchange. If these should couple without accompanying absorptive cuts (evasively), then their contributions must vanish in the forward direction as discussed above. For the charge exchange reactions $\pi^- \rightarrow (\pi\pi)^0$ only A_2 exchange ($I_t = 1$) can couple. As we shall see, when discussing results of the recent amplitude analysis, the prevailing opinion at present is that A_2 exchange does couple evasively, so that discrepant production phases are not present close to the forward direction. An opposite viewpoint has however emerged in some discussions of the rather closely related charge pion photoproduction reactions (see e.g. Worden (1973)). For the non-charge exchange reactions $\pi^{\pm} \rightarrow (\pi\pi)^{\pm}$, especially in the rho region, there are important ω exchange contributions ($I_t = 0$) which can be isolated by forming appropriate combinations of density matrix elements for $\rho^{\pm}_{0}$ production (Contogouris *et al.*, 1967). Experiments at 6 GeV/c (Crennel *et al.*, 1971) and 16 GeV/c (Bartsch *et al.*, 1972) establish a rather convincing ω-exchange signal which appears to vanish in the forward direction, peak at around $|t| = 0{\cdot}1$ GeV^2 ($\approx 5m_{\pi}^2$), then fall away to a pronounced dip, compatible with zero at $|t| \approx 0{\cdot}45$ GeV^2. This last feature is commonly taken as evidence for the vanishing of the vertex function at the point where $\alpha_{\omega} = 0$ (a "nonsense wrong signature zero"), and reinforces the idea of the vector–tensor contributions to those reactions being little affected by absorption. Following this line, it is natural to extend the discussion to charged K^* (890) production and consider the general systematics of charged vector meson production (see for example, Michael (1973a)). For our present purpose, the crucial facets of vector and tensor exchanges (N) are, firstly the energy dependence and Regge phase, $(s/s_0)^{\alpha_N(t)}$ and $e^{-i\pi\alpha_N(t)/2}$, with $\alpha_N(0) \approx \frac{1}{2}$ as compared to $\alpha_{\pi}(0) \approx 0$; secondly, the likelihood or possibility that these couplings, including their associated cuts, vanish in the forward direction. A third point to which we return towards the end of this section, is the question of di-meson mass and spin dependence, and the possibility that the (N) exchanges become less important relative to π-exchange, as one moves up the leading $\pi\pi$ trajectory $\rho - f - g$.

(iii) *Absorption Model*

This model, especially as applied to one-pion exchange (OPEA) has quite a long history (Gottfried and Jackson, 1964; Durand and Chiu, 1965; Bander and Shaw, 1965; Ross *et al.*, 1970; Chan *et al.*, 1970). The basic idea is that the primary exchange should be modified, supposedly to take account of absorption effects in the ingoing and outgoing channels (Fig. 3.1.5).‡ In the

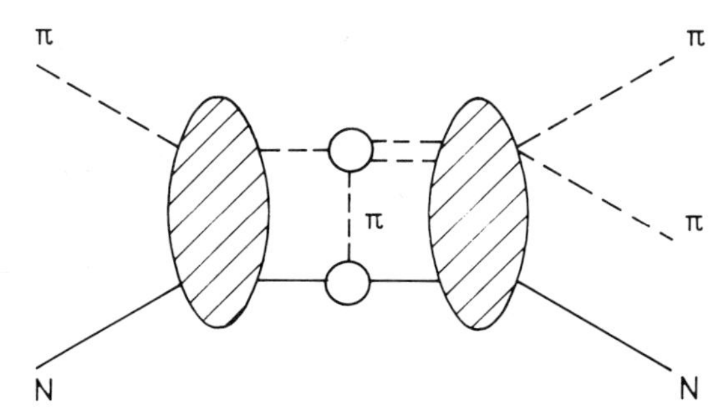

FIG. 3.1.5. Schematic diagram for absorbed one-pion exchange (OPEA).

high-energy limit, the content of the model is often expressed in the impact parameter representation, where the "absorbed" amplitudes in the s-channel take the form

$$H^{(s)}_{[\lambda]}(t) = \tfrac{1}{2}\int_0^\infty db\, bJ_n(b\sqrt{-t})S_{\text{abs}}(b)\int_0^\infty d(-t')H^{(s)\text{OPE}}_{[\lambda]}(t')J_n(b\sqrt{-t'}). \quad (3.1.28)$$

The index of the Bessel function J_n denotes the net helicity flip,§ a concept which plays a key role in the absorption model. For an exchange of given parity, for example OPE, the helicity amplitudes with respectively $(+n_{ac}, n_{nd})$ and $(-n_{ac}, n_{bd})$ are simply related (cf. (3.1.8) and Appendix B). Absorption treats these two amplitudes quite differently because of the necessary kinematic factor, $(-t)^{n/2}$, carried by the Bessel functions in equation (3.1.28), cf. (B.24). No further λ-dependence is ordinarily allowed for in $S_{\text{abs}}(b)$, which is often taken to be of the form $[1 - C\exp(-\gamma b^2)]$. Note

‡ The limitations of the absorption model should be stressed. Qualitatively it accounts for important features of data; but as an instrument of quantitative systematic phenomenology it has grave limitations—specifically, in a given situation, one does not know to what extent it should act, if at all.

§ For the process $ab \to cd$, one distinguishes the flip at the ac vertex, $n_{ac} = \lambda_a - \lambda_c$, that at the bd vertex, $n_{bd} = \lambda_b - \lambda_d$, and the net flip $n = |n_{ac} - n_{bd}|$.

that the discussion has been in terms of s-channel helicity amplitudes. Absorption is, of course, not diagonal in the t-channel frame, and results in the population of states with $m \neq 0$. It also mixes the exchanged parity, and, operating on evasive particle exchanges like π and A_2, achieves non-vanishing forward cross-sections. Provided there are no t-dependent production phases in the particle exchange amplitude being absorbed, the transformation (3.1.28) does not induce any relative production phases (for $S_{\text{abs}}(b)$ purely real). Combined with the assumption that small-t production is dominated by OPE and its associated cuts this is the basis of the conjecture that the production amplitudes may be coherent in production phase, to which we return below.

The William's Model (*Poor Man's Absorption*). A simplified version of the absorption model has been given by Williams (1969, 1970), sometimes referred to as the "Poor Man's Absorption Model" (PMA). This capitalizes on the long-standing observation that one of the prime effects of absorption is to suppress the anomalously large, low j (here $j=\frac{1}{2}$), contributions of OPE to the s-channel helicity amplitudes. (These are the so-called "Kronecker delta" contributions which are not obtained by extrapolating from high j and which do not depend on the range of the exchanged particle.) In Williams' model, spin-dependent effects are represented by the total removal of these anomalous contributions; residual effects of absorption are represented by an overall collimating factor.

It is instructive to see the results of the PMA approach. Consider, for example, P-wave dipion production at high energies (this somewhat simplifies the details). The non-vanishing helicity components arising from OPE, and the modified amplitudes on removing the Kronecker parts, take the form

$$
\begin{array}{llll}
 & \text{OPE} & \text{PMA} & \\
H^{(s)1}_{+-;0} \sim G\dfrac{\sqrt{-t}(t+M^2_{\pi\pi}-\mu^2)}{M_{\pi\pi}(t-\mu^2)} & & G\dfrac{\sqrt{-t}M_{\pi\pi}}{(t-\mu^2)} & (n=1) \\
H^{(s)1}_{+-;1} \sim G\dfrac{\sqrt{2}(-t)}{(t-\mu^2)} & & -G\dfrac{\sqrt{2}\mu^2}{(t-\mu^2)} & (n=0) \\
H^{(s)1}_{+-;-1} \sim -G\dfrac{\sqrt{2}(-t)}{(t-\mu^2)} & & -G\dfrac{\sqrt{2}(-t)}{(t-\mu^2)} & (n=2).
\end{array}
\tag{3.1.29}
$$

The OPE contributions follow for instance from applying the appropriate st crossing transformations (cf. (B.58) and (3.1.10)) to the single OPE t-channel component, $H^{(t)}_{++;0} = iG\sqrt{-t}\mathcal{T}_{24}/(t-\mu^2)$, (cf. (3.1.22) and (B.60)).

The corresponding PMA contributions then follow from the rule of removing the "unnecessary" t-factors in the numerator. The factor G allows for the coupling constants and flux factors $G^j_{P\pi}$ as in (3.1.23); also, in fitting to data, Williams allows a common overall collimation factor, $\Phi(t)$

$$G = G^j_{P\pi}\Phi(t) \qquad (\Phi(\mu^2) = 1). \tag{3.1.30}$$

It is instructive to form the corresponding parity dominant amplitudes, (3.1.8):

$$\begin{aligned} H^{(s)1-}_{+-;0} &= G\frac{\sqrt{-t}M_{\pi\pi}}{(t-\mu^2)} \qquad (\equiv H^{(s)}_{+-;0}) \\ H^{(s)1-}_{+-;1} &= -G\frac{(t+\mu^2)}{(t-\mu^2)} \\ H^{(s)1+}_{+-;1} &= G. \end{aligned} \tag{3.1.31}$$

The corresponding t-channel amplitudes are obtained by applying the rotation inverse to (3.1.13).

The overall t-dependence is of course adjusted in the model via the factor Φ in G (3.1.30). The distinctive prediction concerns the spin alignments, i.e. the ratios of the helicity amplitudes in (3.1.31). For the situation where only S- and P-wave dipions need to be considered, it implies predictions for the ratios of the $\langle Y^m_l \rangle$ moments in the s-channel frame (Ochs, 1972):

$$\frac{\langle Y^1_1 \rangle}{\langle Y^0_1 \rangle} = \frac{H^{(s)1-}_{+-;1}}{H^{(s)1-}_{+-;0}} = \frac{-1}{\sqrt{2}M_{\pi\pi}}\frac{t+\mu^2}{\sqrt{-t}}. \tag{3.1.32}$$

$$\langle Y^0_2 \rangle : \langle Y^1_2 \rangle : \langle Y^2_2 \rangle = t^2 + M^2_{\pi\pi}t + \mu^4 : \sqrt{3/2}M_{\pi\pi}\sqrt{-t}(t+\mu^2) : -\sqrt{6}\mu^2 t, \tag{3.1.33}$$

as can be verified by substituting (3.1.31) into (3.1.7). (To obtain the result it is not actually necessary to assume that S-wave dipion production is all nucleon flip, although this would be expected on the PMA philosophy and on more general grounds.) Figure 3.1.6 shows the comparison of the predictions (3.1.32) and (3.1.33) with data for $\pi^+\pi^-$ production over the rho-mass interval, from the CERN-Munich experiment on $\pi^- p \to \pi^+\pi^- n$ at 17 GeV/c (Grayer *et al.*, 1972a). The predictions are extraordinarily well borne out by the data at small t.

Absorption as a Function of Di-Meson Mass. This quantitative success of the Williams model does not persist beyond the rho-region (Ochs and Wagner, 1973) and the analogous K^* (890) region of $K\pi$ production. At small t, the crucial ingredient of the PMA model is the total absorption of the

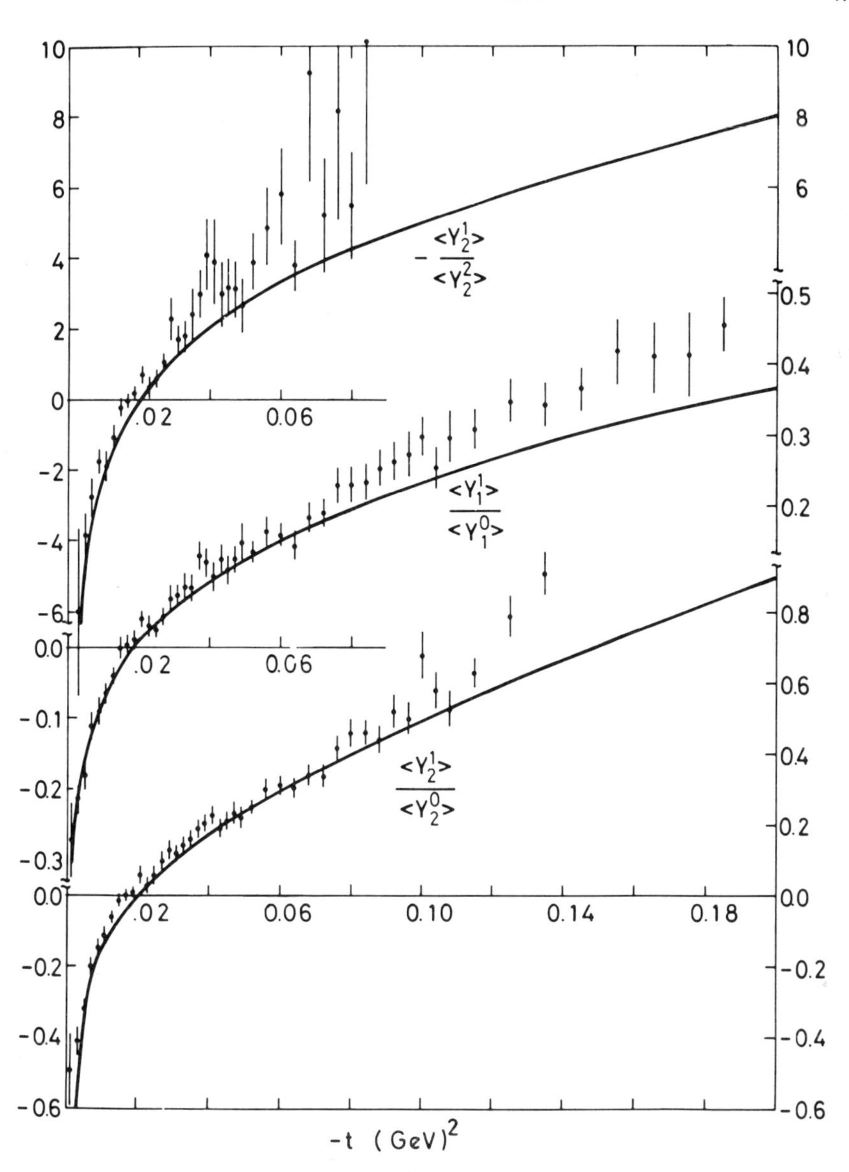

FIG. 3.1.6. Ratios of s-channel Legendre moments, for $\pi^- p \to \pi^+ \pi^- n$ at 17·2 GeV/c with $0{\cdot}71 \leq M_{\pi\pi} \leq 0{\cdot}83$ GeV, compared to the predictions of the Williams model (equations (3.1.32) and (3.1.33)) (Grayer *et al.*, 1972a).

anomalous S-wave in the production amplitude for $P\pi$ scattering. Defining this situation to correspond to an absorptive-cut contribution of strength unity, Ochs and Wagner allow other possibilities for the cut strength (cf.

3.1.29)

$$H^{(s)1}_{+-;1} = G\sqrt{2}\left(\frac{-t}{t-\mu^2}+C\right),$$

and so on for other di-meson partial waves. They do not actually work with individual di-meson partial wave contributions, but instead express the full t-channel amplitude for nucleon flip, $\alpha = \pm 1$, in the form (Wagner, 1973) (valid for small crossing angles, i.e. near $t=0$)

$$H^{(t)}_{\alpha}(\theta, \phi) = G_0\left[\frac{\sqrt{-t}}{t-\mu^2}\Phi_{\pi}(t) - C\frac{M}{M^2-\mu_P^2}\Phi_c(t)\,\mathrm{e}^{i\alpha\phi}\frac{\partial}{\partial\theta}\right]T_{P\pi}(M^2, \theta), \qquad (3.1.34)$$

such that the intensity is given by $\frac{1}{2}\sum_{\alpha}|H^{(t)}_{\alpha}(\theta, \phi)|^2$ (for the derivation of this formula see equation (3.1.39) below). G_0 is the previous G (3.1.23) with the $P\pi$ partial wave factors removed and Φ_{π}, Φ_c are real form factors relating to

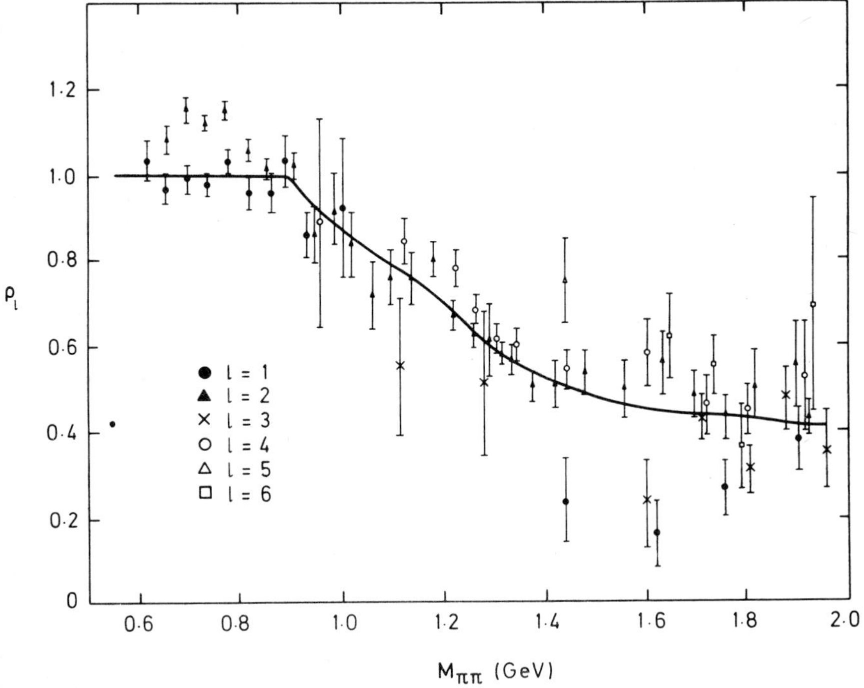

FIG. 3.1.7. Estimate of the relative π-cut contribution at $t=0$ as a function of $\pi\pi$ mass (Ochs and Wagner, 1973). The quantity plotted is the ratio

$$\rho_l = -\frac{\langle Y_l^1\rangle}{\langle Y_l^0\rangle}\frac{M}{\sqrt{l(l+1)}}$$

divided by its value for $l=1$, $M=M_{\rho}$.

the pole and cut contributions. Ochs and Wagners' formula predicts an intensity distribution at small t of the form

$$I(\theta, \phi) = I_0(\theta) + I_1(\theta) \cos \phi,$$

and, under reasonable additional assumptions, allows the real part of the cut-strength parameter, C, to be determined from ratios of the observed Legendre moments $\langle Y_l^1 \rangle$ and $\langle Y_l^0 \rangle$ averaged over the small $|t|$ region. Results for $\pi^+\pi^-$ production are shown in Fig. 3.1.7 (for details see the original reference (Ochs and Wagner, 1973)). This important conclusion that near-forward cut-contributions decrease with increasing di-meson mass is borne out as regards the $\pi^+\pi^-$ system in other analyses, for example that of Estabrooks *et al.* (1973) to be discussed below, and in fact follows rather directly from the data. Similar results to those shown in Fig. 3.1.7 are obtained for $K^+\pi^-$ (Ochs and Wagner, 1973).

(f) Parameterization of the Very Small t-Region

At this point, it is convenient for later application, temporarily to specialize the discussion to the very small t-region. What are the crucial parameters determining the background at $t = t_{\min}$? In the s-channel frame, they will clearly be the values of the non-vanishing no-net-flip amplitudes $H^{(s)j}_{++;0}$ and $H^{(s)j}_{+-;1}$ (cf. (B.24)); the latter terms of course contribute equally to the exchange-parity dominant combinations $H^{(s)j+}_{+-;1}$ and $H^{(s)j-}_{+-;1}$. Near the forward direction, the only important OPE contributions are to states with $|m| = 0$ or 1, as follows from the form of the crossing matrix (B.17). We can thus write the relevant s-channel helicity amplitudes in the form

$$H^{(s)}_{++;0} = G^{(j)}\left[\frac{\sqrt{-t_{\min}}}{t-\mu^2} + A\right] + \cdots$$

$$H^{(s)-}_{+-;1} = G^{(j)}X^{(j)}\left[\frac{t_{\min}-t}{t-\mu^2} + \frac{1}{2}C\right] + \cdots \tag{3.1.35}$$

$$H^{(s)+}_{+-;1} = -\tfrac{1}{2}G^{(j)}X^{(j)}C + \cdots,$$

where "+..." denotes appropriate higher powers of $(t_{\min} - t)$ which we will ignore, and A and C refer to the possible additional no-net-flip contributions. The factor $X^{(j)}$ arises from the first-order expansion of the di-meson st crossing matrix

$$\sqrt{2}\frac{d^j_{10}(\chi_4)}{d^j_{00}(\chi_4)} \approx -\sin \chi_4 \left[\frac{j(j+1)}{2}\right]^{\frac{1}{2}}$$

$$\approx X^{(j)}\sqrt{t_{\min} - t},$$

with

$$X^{(j)} \equiv \frac{2M}{M^2 - \mu_P^2} \left[\frac{j(j+1)}{2} \right]^{\frac{1}{2}}, \tag{3.1.36}$$

(cf. (3.1.11)), replacing the factor $\sqrt{-t}$ by the more accurate approximation $\sqrt{t_{\min} - t}$. Now consider the above form rotated back into the t-channel and consider the *very* small t region such that $|\sin \chi_4| \ll 1$, so that we only need allow for the baryon spin crossing matrix (B.18). In this way, we obtain the small-t parameterization proposed by Froggatt and Morgan (1969):

$$\begin{aligned}
H^{(t)}_{++;0} &= G^{(j)} \left[\frac{\sqrt{-t}}{t - \mu^2} + \frac{\Gamma_0}{\sqrt{-t}} \right] \\
H^{(t)}_{+-;0} &= G^{(j)} \left[\Gamma_0 \frac{(t/t_{\min} - 1)^{\frac{1}{2}}}{\sqrt{-t}} \right] \\
H^{(t)-}_{++;1} &= G^{(j)} \sqrt{2} \Gamma_1 \frac{(t/t_{\min} - 1)^{\frac{1}{2}}}{\sqrt{-t}} \\
H^{(t)-}_{+-;1} &= -G^{(j)} \frac{\sqrt{2}\Gamma_1}{\sqrt{-t}} \\
H^{(t)+}_{++;1} &= 0 \\
H^{(t)+}_{+-;1} &= G^{(j)} \frac{\sqrt{2}\Gamma_1}{\sqrt{-t_{\min}}},
\end{aligned} \tag{3.1.37}$$

with

$$\begin{aligned}
\frac{\Gamma_0}{\sqrt{-t_{\min}}} &= A \\
\frac{\Gamma_1}{\sqrt{-t_{\min}}} &= \frac{CX^{(j)}}{2\sqrt{2}}.
\end{aligned} \tag{3.1.38}$$

Aside from small absorption corrections which vanish as $s \to \infty$ (see Froggatt and Morgan, 1970), the parameter A refers to possible A_1 exchange contributions. If these are absent, and if furthermore we take the high-energy limit $t_{\min} = 0$, then the residual non-vanishing contributions to (3.1.37), written in terms of C, become

$$\begin{aligned}
H^{(t)}_{++;0} &= G^{(j)} \frac{\sqrt{-t}}{t - \mu^2} \\
H^{(t)-}_{++;1} &= H^{(t)+}_{++;1} = \tfrac{1}{2} G^{(j)} C X^{(j)},
\end{aligned}$$

leading to the intensity formula

$$I = \left| \sum_j G^{(j)} \left[\frac{\sqrt{-t}}{t-\mu^2} d^j_{00}(\theta) + \frac{1}{2} C X^{(j)} \frac{(e^{i\phi} + e^{-i\phi})}{\sqrt{2}} d^j_{10}(\theta) \right] \right|^2$$

$$+ \left| \sum_j \frac{1}{2} G^{(j)} C X^{(j)} d^j_{10} \frac{(e^{i\phi} - e^{-i\phi})}{\sqrt{2}} \right|^2$$

$$= \frac{1}{2} \sum_{\alpha=\pm 1} \left| \sum_j \left\{ \frac{\sqrt{-t}}{t-\mu^2} - \frac{C\sqrt{2} X^{(j)}}{2[j(j+1)]^{\frac{1}{2}}} e^{i\alpha\phi} \frac{\partial}{\partial\theta} \right\} G^{(j)} d^j_{00}(\theta) \right|^2.$$

Using equations (3.1.23) and (3.1.36) and re-summing the partial wave series, we obtain

$$I = \frac{1}{2} \sum_{\alpha=\pm 1} \left| G_0 \left\{ \frac{\sqrt{-t}}{t-\mu^2} - \frac{CM}{M^2 - \mu_P^2} e^{i\alpha\phi} \frac{\partial}{\partial\theta} \right\} T_{P\pi}(M^2, \theta) \right|^2. \quad (3.1.39)$$

Aside from the form factors $\Phi_\pi(t)$ and $\Phi_c(t)$ which they employ, this is the intensity formula of Ochs and Wagner, referred to previously (3.1.34). Note that it is assumed that C is independent of j, which of course might not be the case.

In general then, there are two complex parameters A and C characterizing the small-t backgrounds to OPE for each di-meson partial wave. If there is no A_1 exchange, then A is zero apart from finite energy $0(1/s)$ corrections (Froggatt and Morgan, 1970). If C arises from absorption applied to π-exchange, it will be real, and if furthermore absorption is as in the Williams (PMA) model (see above) then C is of order unity. As we have seen (Fig. 3.1.7) this is the case respectively in the ρ and K^* (890) regions, but C decreases at higher masses.

If suitable form factors are supplied, formula (3.1.37) can serve as a model for a somewhat larger t-region. One would multiply the pion pole term by a suitable form factor which can be absorbed into $G^{(j)}$, and generalize Γ_0 and Γ_1 to be functions of t. The resulting model has the distinctive property of having

$$|H^{(t)+}|^2 = |H^{(t)-}|^2,$$

(and consequently $\langle Y_l^2 \rangle = 0$, a property which is exploited by Ochs and Wagner (1973) in their method for extracting $\pi\pi$ phase shifts). For the case of just S- and P-waves, the intensity components take the form (cf. Froggatt and Morgan, 1969):

$$
\begin{aligned}
\rho^{11}_{00} \,.\, \sigma &= \left[\frac{-t}{(t-\mu^2)^2} + \frac{2\,\mathrm{Re}\,\Gamma^P_0}{(t-\mu^2)} + \frac{|\Gamma^P_0|^2}{(-t_{\min})}\right]|P|^2 \\
\rho^{11}_{11} \,.\, \sigma &= \left[\frac{2|\Gamma_1|^2}{(-t_{\min})}\right]|P|^2 \\
\rho^{11}_{10} \,.\, \sigma &= \left[\mathrm{Re}\,\Gamma_1 \frac{(t/t_{\min}-1)^{\frac{1}{2}}}{(t-\mu^2)}\right]|P|^2 \\
\rho^{11}_{1-1} \,.\, \sigma &= 0 \qquad\qquad (3.1.40) \\
\rho^{00}_{00} \,.\, \sigma &= \left[\frac{-t}{(t-\mu^2)^2} + \frac{2\,\mathrm{Re}\,\Gamma^S_0}{(t-\mu^2)} + \frac{|\Gamma^S_0|^2}{(-t_{\min})}\right]|S|^2 \\
\rho^{10}_{00} \,.\, \sigma &= \left[\frac{-t}{(t-\mu^2)^2} + \frac{\Gamma^S_0+\Gamma^{P*}_0}{(t-\mu^2)} + \frac{\Gamma^S_0\Gamma^{P*}_0}{(-t_{\min})}\right](S^* \,.\, P) \\
\rho^{10}_{10} \,.\, \sigma &= \left[\Gamma^*_1 \frac{(t/t_{\min}-1)^{\frac{1}{2}}}{(t-\mu^2)}\right](S^* \,.\, P),
\end{aligned}
$$

where S and P denote the factors $G^{j=0}$ and $G^{j=1}$, and $\Gamma^{S,P}_0$ denote the Γ_0's for $j=0$ and 1. If the small-t background is assumed to emanate from absorption of the pion pole contribution, then $\Gamma^S_0=\Gamma^P_0=0$ (aside from the corrections at finite energies referred to above) and Γ^P_1 is real. At somewhat larger t, additional background terms can enter, for example a $C \sin \chi_4$ contribution to $H^{(t)1}_{++;0}$ (cf. (3.1.42) below).

(g) Amplitude Analysis

The successes of the Williams model and its generalizations suggest the possibility of performing a model *amplitude analysis* of di-meson production, using only the unpolarized intensity information which we have been discussing (Ochs, 1972; Froggatt and Morgan, 1972; Estabrooks and Martin, 1972a). A sufficient condition for being able to do this is that the production amplitudes be coherent in spin and phase, except for relative decay phases for the different di-meson spins. In general, the production amplitude for given di-meson spin and helicity is characterized by two complex numbers ($H^j_{++;m}$, $H^j_{+-;m}$) so that it can be represented as a vector in an abstract space of four dimensions; coherence in spin and phase means that the vectors are all aligned—same ratio of flip to non-flip and same overall production phase. Since the unpolarized intensity components contain no cross-terms between the alternative exchange-parity dominant amplitudes $H^{j\pm}$ (3.1.8), it is only necessary to make the spin-phase coherence (SPC) assumption for the H^+ and H^- sectors separately. Once the SPC assumption is made, there are more than enough intensity components to determine the remaining parameters in the helicity amplitudes, so that testable constraints follow.

As to the *a priori* grounds for expecting approximate SPC, we have already mentioned the main one, which is the helicity structure of OPE and its associated absorptive cuts. For $\pi^- \to (\pi\pi)^0$, the other main contribution is expected to be from Reggeized-A_2 exchange, predominantly nucleon flip from the systematics of $A_2N\bar{N}$ couplings. If there were no absorption, the SPC requirement would certainly be met; all the $H^{(s)-}$ components would be flip and the $H^{(s)+}$ components would have the Regge phase $e^{-i\pi\alpha_{A_2}(t)/2}$ while the $H^{(s)-}$ components would have the phase $e^{-i\pi\alpha_\pi(t)/2}$ (assuming now Reggeized OPE see below). The likely existence of a small non-flip A_2 exchange contribution would not affect the result. Absorption will in general disrupt this pattern, altering Regge phases differently from amplitude to amplitude according to the net flip, and mixing π and A_2 cuts together. A_1 exchange, if present, would in general spoil the spin coherence. These possibilities are schematically summarized in Table 3.1.2. Special mechanisms may operate to neutralize these SPC breaking tendencies in particular

TABLE 3.1.2. Exchange Contributions to Di-Meson Production

	Poles	Cuts		
$H^{(s)-}_{+-} \equiv U_{+-}$:	$\pi \equiv \lvert\pi\rvert\, e^{-i\pi\alpha_\pi(t)/2}$ $m = 0, 1, \ldots$	π-cut $m = 1, (n = 0)$	(A_2^{flip}-cut)	Flip
$H^{(s)+}_{+-} \equiv N_{+-}$:	$A_2^{\text{flip}} \equiv \lvert A_2^{\text{flip}}\rvert\, e^{-i\pi\alpha_{A_2}(t)/2}$ $m = 1, \ldots$	π-cut	(A_2^{flip}-cut)	
$H^{(s)-}_{++} \equiv U_{++}$:	(A_1)			Non-flip
$H^{(s)+}_{++} \equiv N_{++}$:	($A_2^{\text{non-flip}}$)			

(Contributions which are assumed to be small or absent are shown in brackets.)

kinematic regions. For example, the A_2 pole contribution should be rather unimportant at small $|t|$. (It is required to vanish at $t = 0$ and should contain no strong t-dependences since, unlike OPE, there are no near-lying poles in t.) For the same reason, the near-forward A_2-cut should be weak. A_2 exchange is also believed to diminish relative to π exchange with increasing di-meson mass (see below).

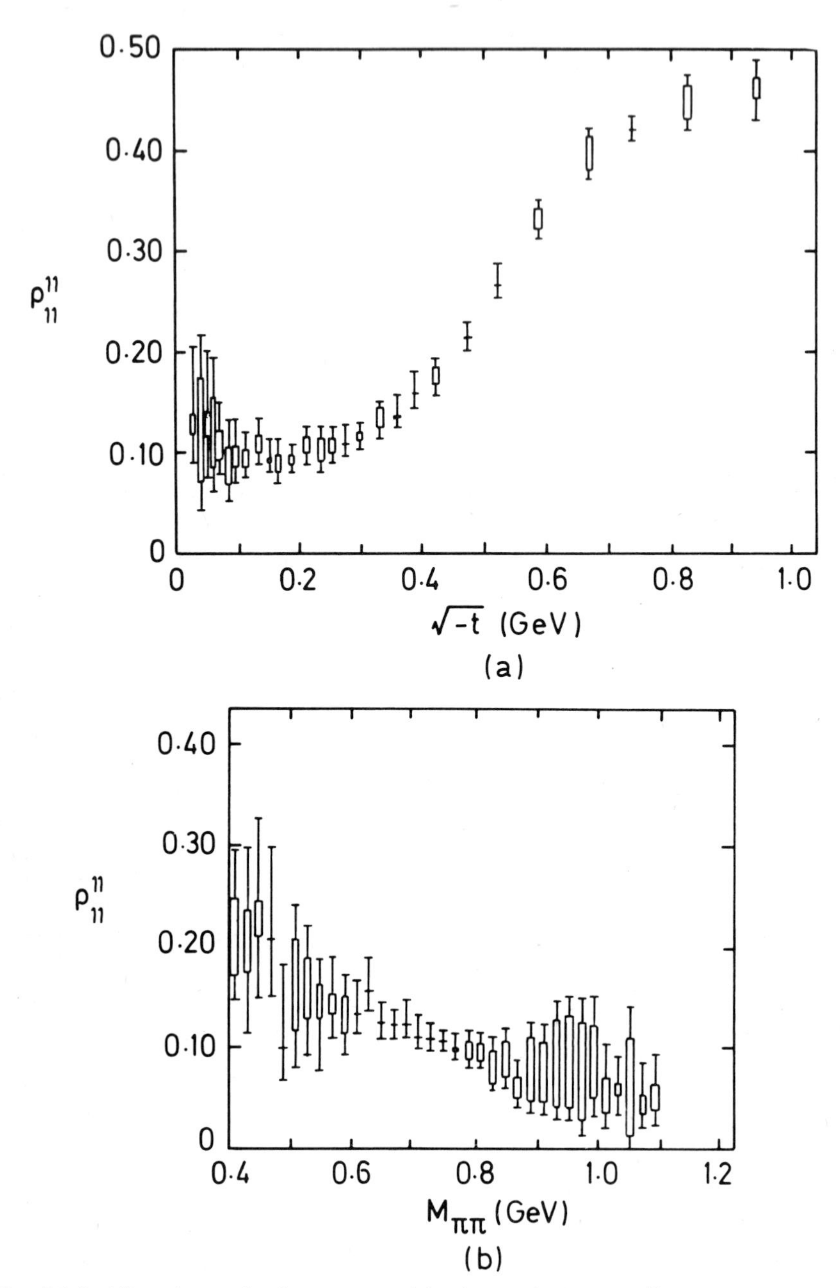

FIG. 3.1.8. Allowed range for the unmeasured density matrix element ρ_{11}^{11} for $\pi^- p \rightarrow \pi^+ \pi^- n$ at 17·2 GeV/c (Grayer *et al.*, 1972b); (a) as a function of $\sqrt{-t}$ for $0{\cdot}70 < M_{\pi\pi} < 0{\cdot}85$ GeV, (b) as a function of $M_{\pi\pi}$ for all $|t|$ between 0·01 and 0·08 GeV^2.

Once the amplitudes have been extracted, the unobserved density matrix elements such as ρ_{11}^{11} can be computed. Predominantly natural and unnatural exchange contributions can be separated; particular intensity components in which the OPE signal should be relatively uncontaminated by background can be isolated. This last is the basis of Estabrook and Martin's method of extracting $\pi\pi$ phase shifts to which we return in Section 3.2.

There is a risk of circularity in making assumptions in order to separate intensity components, then appealing to the results to uphold the assumptions, although consistency in this respect is of course a necessary condition. The constraints which are entailed by assuming SPC do afford some check; for $\pi^- \to (\pi\pi)^0$ it appears that SPC solutions can always be found, within the experimental errors. This of course is not to say that they are unique (see below).

A model independent way to expose the scope allowed by the data to the unobserved density matrix elements is simply to employ general Schwarz-type inequalities (cf. Appendix B). SPC implies special relations among the observed density matrix elements; inequalities in the general case become equalities when amplitudes are aligned in spin and phase, and the unobserved density matrix elements should be very tightly constrained. The CERN-Munich group have tested this approach on their 17·2 GeV/c data on $\pi^- p \to \pi^+\pi^- n$ (Grayer *et al.*, 1972b). Figure 3.1.8(a),(b) shows their results for ρ_{11}^{11} using data with $M_{\pi\pi} < 1{\cdot}1$ GeV and assuming only S- and P-waves. The results are seen to be in very satisfactory agreement with SPC. The authors also compute limits on the unobserved imaginary parts of interference moments and show that they are all compatible with being zero, again in agreement with SPC. Analogous analyses of data in the f^0-region have been given by Charlesworth *et al.* (1973).

For the special case where only S- and P-wave di-mesons are produced the SPC assumptions can be somewhat relaxed (Estabrooks and Martin, 1972a). In the first place, the natural exchange component, H_+, only occurs in the form modulus squared averaged over spin (cf. (3.1.7)), so no special assumption is required. For the H_- terms, an extra relative production phase, ϕ_{10}, can be allowed between the two P-wave components, H_0 and H_-, in the notation of (3.1.7). This, together with the relative decay phase, $\bar{\delta}_{SP}$ between S- and P-wave production, and the four moduli $|H_S|, |H_0|, |H_-|, |H_+|$, gives six parameters to determine from six pieces of experimental information (3.1.7). In terms of this parameterization (3.1.7) takes the form‡ (cf. Estabrooks and Martin, 1972a)

‡ Estabrooks and Martin (1972a) actually allow for the s-channel non-flip contributions associated with OPE at very small t, cf. (3.1.9), whilst continuing to cast their equations in terms of the flip amplitudes.

$$
\begin{aligned}
N &= |H_S|^2 + |H_0|^2 + |H_+|^2 + |H_-|^2 \\
(\rho_{00}^{11} - \rho_{11}^{11})N &= |H_0|^2 - \tfrac{1}{2}[|H_+|^2 + |H_-|^2] \\
\rho_{1-1}^{11}N &= \tfrac{1}{2}[|H_+|^2 - |H_-|^2] \\
\rho_{10}^{11}N &= \frac{1}{\sqrt{2}}|H_-||H_0| \cos \phi_{10} \\
\rho_{00}^{10}N &= |H_0||H_S| \cos \bar{\delta}_{SP} \\
\rho_{10}^{10}N &= |H_-||H_S| \cos (\phi_{10} - \bar{\delta}_{SP}).
\end{aligned}
\tag{3.1.41}
$$

The (SPC) constraint equation corresponds to requiring $|\cos \phi_{10}| = 1$. The above equations (3.1.41) in general admit several discrete alternative solutions. Thus, eliminating the other unknowns, one obtains a cubic equation for $|H_0|^2$. Estabrooks and Martin find in their analysis of 17 GeV/c $\pi^+\pi^-$ production for $M_{\pi\pi} \leq 1$ GeV, that one of the three solutions is unphysical with $|H_0|^2$ negative, and that (within experimental errors) the other two solutions are both physical with rather similar values of $|H_0|^2$. Once $|H_0|^2$ is fixed, the other parameters follow without ambiguity, so that there is just a two-fold ambiguity for the whole set of parameters determining the amplitudes, and this situation persists over the whole dipion mass range up to 1 GeV. In each case, there is a solution with $|\cos \phi_{10}|$ close to unity which the authors term "solution 1", and another solution, "solution 2", with $|\cos \phi_{10}|$ in general deviating somewhat from unity. They prefer "solution 1" because the ensuing $\pi\pi$ phase shifts are in somewhat better agreement with results on $\pi^0\pi^0$ production (see Section 3.3); however, the amplitude analysis as such affords no grounds for preference.

Estabrooks with Martin's analysis of S- and P-wave dipion production is expressed within the framework where spin coherence is assumed, but can be viewed from a more general standpoint (Morgan, 1974a). As we mentioned earlier the experimental observables, as listed in equation (3.1.7), involve $\|H_+\|^2$ and sundry bilinear combinations of the four-component quantities H_0, H_- and H_S (four components from real and imaginary parts of the flip and non-flip amplitudes). The three four-vectors H_0, H_- and H_S in general span a space of three dimensions. Estabrooks and Martin's analysis assumes that they form a plane. This could arise not from spin coherence but from some more general combination of phase and spin alignments. If this "coplanarity" requirement were relaxed, one would be left with just the inequalities for unobserved density matrix elements discussed above. In fact, it can be shown that Estabrooks and Martin's alternative solutions correspond to the extreme cases allowed by the data and that other (non-coplanar) possibilities would lie between them. The full

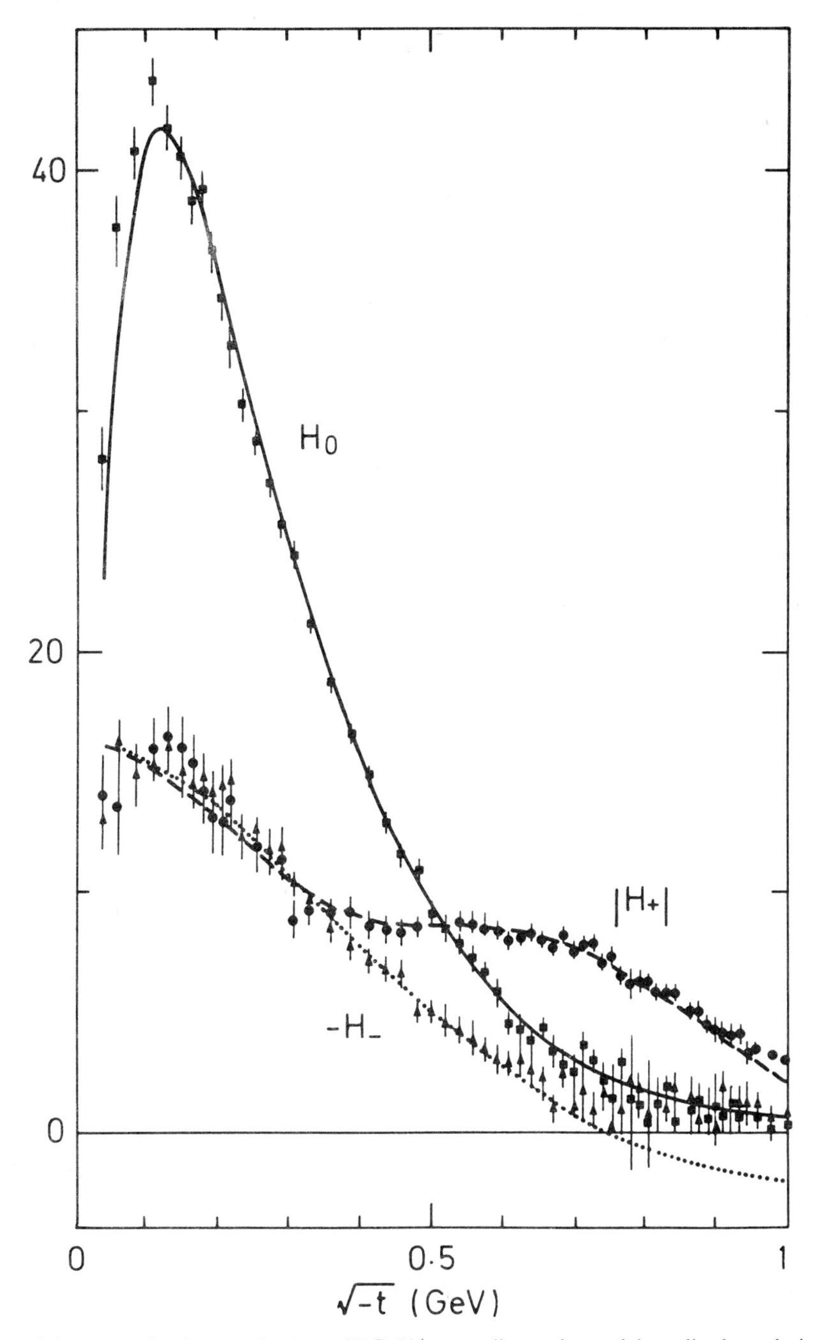

FIG. 3.1.9. ρ production amplitudes at 17 GeV/c according to the model amplitude analysis of Estabrooks and Martin (1972). The quantities shown are the t-channel quantities H_0, H_+, H_- as functions of $\sqrt{-t}$, together with a fit to the simple pole + cut parameterization of equation (3.1.42), from Estabrooks *et al.* (1974a).

SPC assumption entails that H_0 and H_- are not only coplanar with H_S but are parallel (or anti-parallel), $|\cos \phi_{10}| = 1$. Estabrooks and Martin's finding that their "solution 1" is compatible with this requirement reproduces in this respect Grayer *et al's* (1972b) result referred to above, using inequalities on density matrix elements.

An example of the results from the model amplitude analyses is shown in Fig. 3.1.9 from Estabrooks *et al.*, (1974a). This shows the t-channel helicity amplitudes $H_0^{(t)}$, $H_-^{(t)}$ and $H_+^{(t)}$ for 17·2 GeV/c rho production ($0{\cdot}7 < M_{\pi\pi} < 0{\cdot}85$ GeV). One sees the expected OPE signal in $H_0^{(t)}$ and the non-vanishing background contributions to $H_+^{(t)}$ and $H_-^{(t)}$ at $t = 0$. These are the features which are crucial for extracting $\pi\pi$ phase shifts. From the point of view of production mechanism systematics, the predominance at large $|t|$ of the natural parity exchange contribution $H_+^{(t)}$ is also noteworthy. This effect becomes less marked at lower beam momenta (see for example the results at 6 GeV/c from the Argonne Effective Mass Spectrometer (Ayres *et al.*, 1973a)). This ties in with assuming H_+ to have an important contribution from Reggeized A_2 exchange ($d\sigma^{(+)}/dt \propto s^{2\alpha_{A_2}(t)-2}$), and the unnatural exchange to be given primarily by OPE and its associated cut ($d\sigma^{(-)}/dt \propto s^{2\alpha_\pi(t)-2}$). This notion is given quantitative expression by extracting effective trajectory functions from the systematics of dipion production experiments at different beam moments. Results, again from Estabrooks *et al.* (1974a) are shown in Fig. 3.1.10. The unnatural exchange (U) components yield trajectory functions roughly consistent with what one would expect for the π-trajectory. The natural exchange (N) trajectory resembles the (U) trajectories at small-t, and switches to the behaviour one would expect from A_2 exchange at larger t. The authors suggest that this be understood in terms of a model with three exchange components, the π and A_2 poles and a π-cut contribution, C. They therefore propose the following simple parameterization for the P-wave (rho) production amplitudes:

$$\begin{aligned} H_0^{(t)} &= \pi + C \sin \chi_4 \\ H_-^{(t)} &= C \cos \chi_4 \\ H_+^{(t)} &= A_2 + C, \end{aligned} \tag{3.1.42}$$

with the t-dependences of the π and A_2 pole contributions and the π-cut contribution given by

$$\begin{aligned} \pi &= g_\pi \frac{\sqrt{-t}}{\mu^2 - t}\, e^{b_\pi t}\, e^{-i\pi\alpha_\pi/2} \\ C &= g_c\, e^{b_c t}\, e^{-\pi\alpha_c/2} \\ A_2 &= -t g_A\, e^{b_A t}\, e^{-i\pi\alpha_{A_2}/2}, \end{aligned} \tag{3.1.43}$$

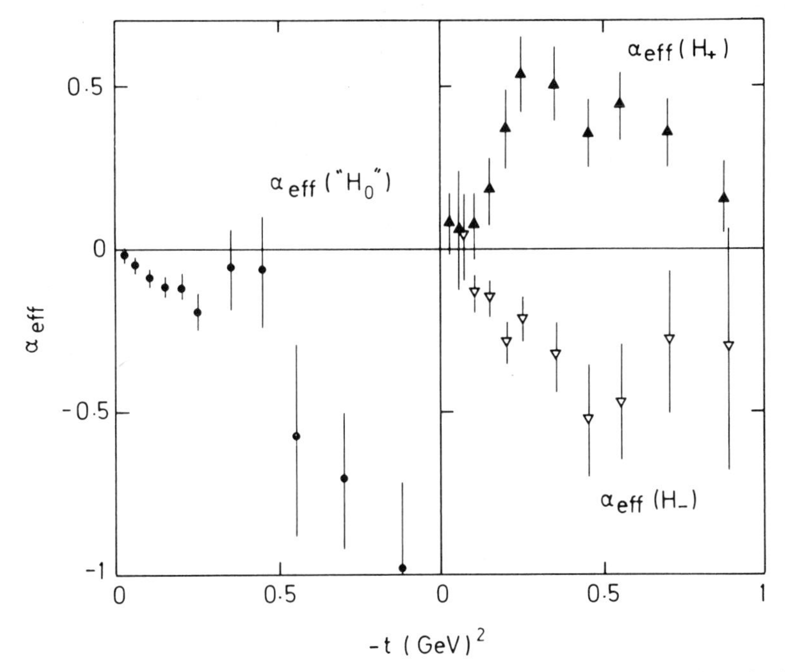

FIG. 3.1.10. Effective trajectory functions for the components $\alpha_{\text{eff}}(H_0)$, $\alpha_{\text{eff}}(H_+)$, $\alpha_{\text{eff}}(H_-)$ of rho-production, from Estabrooks *et al.* (1974a).

where g_π, g_c, g_A and the b's are constants. (Note that this form for the OPE contribution differs from (3.1.21) in not having the factor $\mathscr{T}_{24}$ in the numerator, as well of course as being Reggeized.) The structure of (3.1.42) follows from assuming that C only contributes to the no-net-flip s-channel amplitude (cf. 3.1.13). The evasive non-absorbed form for the A_2 contribution allows this to dominate H_+ only at large t. Assuming $\alpha_c = \alpha_\pi$, the authors proceed to fit the above form to the amplitudes shown in Fig. 3.1.9, and the resulting fit is shown by the curves. They then discuss how the resulting parameters are reasonably consistent with the α_{eff}'s of Fig. 3.1.10; also how their model accommodates quite well the observed patterns of ρ–ω interference, to be discussed below.

(h) Applications

We briefly sketch three areas of application which illuminate the general picture which we have been developing—rho–omega interference, comparison with charge pion photoproduction, and duality predictions for exchange couplings as a function of di-meson mass and spin.

(i) *Rho–Omega Interference*

In much of our discussions, we are driven to making assumptions about production phases in order to derive information on decay phases. In the phenomenon of rho–omega interference,‡ this state of affairs is providentially reversed, affording a special opportunity for probing production amplitudes (Goldhaber, A. S., *et al.*, 1969). The effect stems from the existence of a small G-parity violating 2π decay mode for the ω. As a result, the observed $\pi^+\pi^-$ final state exhibits interference patterns between amplitudes belonging respectively to rho and omega production. The effect is usually analysed in terms of the formula

$$\frac{\partial \sigma^{(m)}}{\partial t\, \partial M} = |H_\rho^{(m)}|^2 \left\{ 1 + \left| \frac{H_\omega^{(m)}}{H_\rho^{(m)}} \right|^2 |B_\omega|^2 + 2\zeta^{(m)} \left| \frac{H_\omega^{(m)}}{H_\rho^{(m)}} \right| \mathrm{Re}\,(B_\omega\, e^{i\phi_{\omega\rho}^{(m)}}) \right\}, \quad (3.1.44)$$

where B_ω describes $\omega \to \pi^+\pi^-$ decay. This is commonly assumed to go predominantly by virtual rho-production so that

$$B_\omega = \text{const} \frac{1}{(m_\omega^2 - M_{\pi\pi}^2 - im_\omega\Gamma_\omega)} \frac{m_\rho^2 - M_{\pi\pi}^2 - im_\rho\Gamma_\rho}{(m_\rho^2 - m_\omega^2 - im_\rho\Gamma_\rho + im_\omega\Gamma_\omega)},$$

(cf. equations (5.3.30) and (5.3.31)) with the constant fixed by the $\omega \to 2\pi$ branching ratio. The remaining parameters in (3.1.44) relate to hadronic production with $\phi_{\omega\rho}^{(m)}$ denoting the relative production phases and $\zeta^{(m)}$ a factor (often set equal to unity) to allow for spin incoherence. The superfix (m) is to allow for different combinations of spin alignment corresponding to natural and unnatural exchange, etc.

It was originally thought (Goldhaber, A. S., *et al.*, 1969) that the predominant effect should be in the unnatural exchange component, (U), but the observations at 3 to 6 GeV/c (Ayres *et al.*, 1973b) show natural (N) exchange also to be important, and it appears that by 17·2 GeV/c the latter effect is the predominant one (Estabrooks *et al.*, 1974b). In terms of Regge pole exchanges, one may think of rho-production as going respectively by π and A_2 exchange for the (U) and (N) components, and ω production as going similarly via B and ρ exchange. This would account for the increasing importance of (N) exchange effects at higher beam momenta. A missing ingredient is the π-cut contribution (cf. discussion in the previous sub-section), and, on including this, a satisfactory overall picture of the (N) exchange component emerges (Estabrooks *et al.*, 1974a). Relative π, B interference phases are likewise inferred from the (U) component.

We conclude this sub-section by showing an example of the observed signal from the data of the Argonne group (Ayres *et al.*, 1973a) (Fig. 3.1.11).

‡ Cf. also Section 5.3(iii); for general reviews see Goldhaber, G. (1970) and Donnachie and Gabathuler (1970).

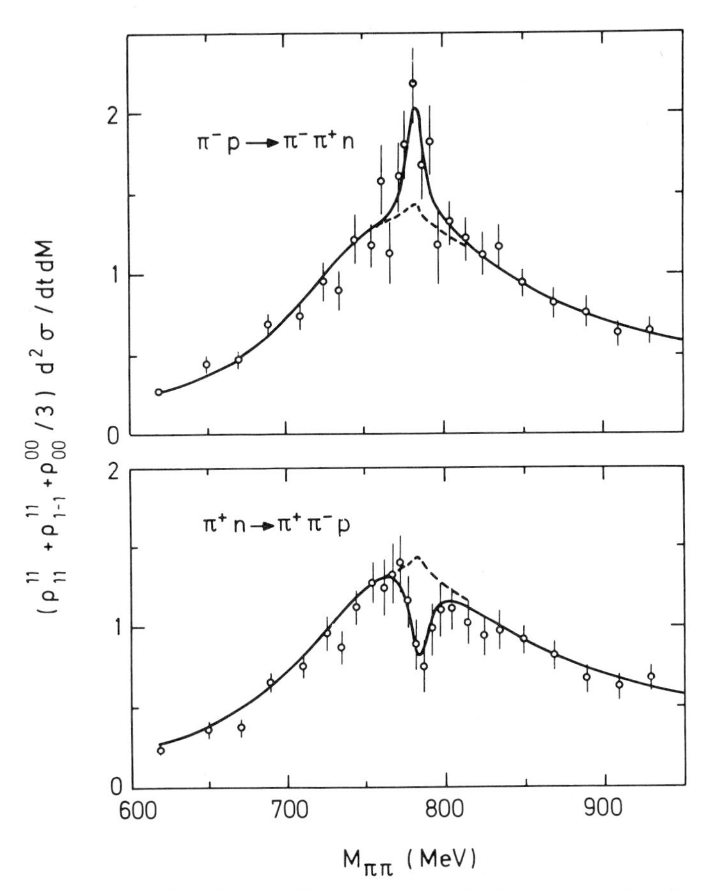

FIG. 3.1.11. Rho–omega interference effects in $\pi^+\pi^-$ production via natural parity exchange at 4 GeV/c (Ayres *et al.*, 1973a).

Besides results on $\pi^-p \to \pi^+\pi^-n$, they also have data on the isotopically reflected reaction $\pi^+n \to \pi^+\pi^-p$ from a π^+d experiment. The expected reversal in sign of the rho–omega interference signal is beautifully manifest.

(ii) *Comparison with Charged Pion Photoproduction*

The existence of quite a large mass of data on the reactions $\gamma p \to \pi^+n$ and $\gamma n \to \pi^-p$, including experiments with polarized beams and with polarized targets, has stimulated a continuing interest in making comparisons with the hadronic di-meson production reactions, $PN \to M^*B$ which we have been discussing. There are two aspects to such comparisons. Firstly, there is the general point that one has the same exchanges as in ρ and ω production (see above): π and B for the (U) exchanges and ρ and A_2 for the (N) exchanges, together with their attendant cuts. One thus has the opportunity to study the same mechanisms in another situation, with the capability of distinguishing odd and even G-parity exchanges from comparisons of π^+ and π^- production, and of distinguishing effects of natural and unnatural exchange by experiments with polarized photons. The pole contributions decouple in the forward direction so that the observed non-vanishing photoproduction cross-section is an unambiguous measure of the strength of the cut components.

The second aspect of such comparisons concerns the very specific hypothesis of vector meson dominance (VMD). This asserts that the appropriate production amplitudes for charged pion photoproduction and for vector meson production are proportional to each other in a suitable frame of reference, generally agreed to be the s-channel helicity frame; in particular, one has a relation between transverse ρ^0 production and the isovector component of photoproduction. On comparing with experiment, the relation is found to be reasonably good for the forward peak and the unnatural exchange component, and bad for the natural exchange component except at $t=0$ (e.g. see Bulos *et al.*, 1971b). The important conclusion is that the relation does not work universally.

In the very small t region, it has long been noted that the essential features of photoproduction can be reproduced by the so-called "gauge invariant electric Born term". Cho and Sakurai (1969, 1970), using vector dominance, have extended this picture to neutral rho production, with results essentially identical in the high-energy limit to those of the William's (PMA) model (cf. (3.1.29)). Here, we discuss the simpler version (Cho and Sakurai, 1969). The scheme is to supplement the OPE Born term by the direct and crossed nucleon Born term with the vector meson universally coupled to the isovector charge. This gives a good description of charged pion photoproduction (the OPE term alone is not gauge invariant so there have to be *some* other terms), and embodies the vector dominance relations between transverse rho-production and photoproduction. Since all the couplings are fixed, longitudinal rho-production is also specified but this is not a consequence of vector dominance. From the helicity amplitudes which are just as in (3.1.29) follow the intensity components

$$\begin{aligned}
\rho_{00}^{11}\left[\frac{d\sigma}{dt}\right]_\rho &= G_\rho\frac{m_\rho^2(-t)}{(t-\mu^2)^2} \\
\rho_{11}^{11}\left[\frac{d\sigma}{dt}\right]_\rho &= G_\rho\frac{(t^2+\mu^4)}{(t-\mu^2)^2} \\
\rho_{1-1}^{11}\left[\frac{d\sigma}{dt}\right]_\rho &= G_\rho\frac{2\mu^2(-t)}{(t-\mu^2)^2},
\end{aligned} \tag{3.1.45}$$

with G_ρ as defined in (3.1.27). The corresponding isovector photoproduction cross-sections result from replacing m_ρ by zero and G_ρ by the appropriate $G_\gamma(f^2_{\rho\pi\pi} \to \frac{1}{2}e^2$, etc. in (3.1.24)).

Equation (3.1.45) neatly brings out the contrasting behaviour of near forward rho production and photoproduction—the latter with a pronounced forward peak, the former with a dip from the vanishing of the predominant longitudinal component. As mentioned above, the model fails at rather small $|t|$ for the natural exchange component of photoproduction.

This model is worth mentioning for its simplicity, but it should be emphasized that it is in no way unique, especially as regards the fulfilment of gauge invariance. As regards backgrounds to OPE, it is essentially a "π-cut model" with the background having the same energy dependence and production phase as the pion pole contribution. There have been alternative descriptions of photoproduction, using the strong-cut absorption model in conjunction with the standard Regge pole exchanges, in which a much more important role is assigned to the A_2-cut in achieving the forward cross-section (Worden, 1972, 1973). Specifically at 16 GeV/c, 40% of the forward no-net-flip amplitude is attributed to the A_2-cut. Many features of 16 GeV/c photo production, including detailed polarization effects, are accommodated. If this were upheld, it would have rather far-reaching consequences, assuming the rho-photon analogy to be at least qualitatively correct. One interesting possibility, to be discussed in the next sub-section, would be to understand, at least in part, the relative decrease of the forward cut contribution to f^0 production as compared to that to ρ production (cf. Fig. 3.1.7). Against all of this interpretation is the energy dependence of forward photoproduction which persistently resembles π-exchange, suggesting that the forward cross-section is predominantly given by the π-cut, as in the simple models discussed above. This would harmonize with the picture already arrived at for forward di-meson production, and support the idea of phase coherence of the small-t backgrounds to OPE which is assumed in several of the methods for extracting meson–meson scattering amplitudes to be discussed in Section 3.2.

(iii) *Duality Predictions for Exchange Couplings as a Function of Di-Meson Mass and Spin*

The isolation of the different exchange components of di-meson production by amplitude analysis, enables contact to be made with new areas of phenomenology. The study of inclusive reactions, $a+b \to c+x$ ($x \equiv$ anything) leads to the concept of "Regge-particle" scattering from the separation of the different t-channel ($a\bar{c} \to \bar{b}x$) exchange components corresponding to different Regge pole exchanges. According to the Regge phenomenology of inclusive reactions (Mueller, 1970) (for a general review see Chan Hong-Mo (1972)), the missing-mass dependence of these different exchange components is itself described at large M^2 by appropriate Regge pole exchange contributions to a forward Regge-particle scattering amplitude. So far everything applies to inclusive experiments, e.g. $\pi^- p \to n +$ anything. However, contact can be made with exclusive experiments of the type we have been studying by appeal to the notion of duality (cf. Chapter 12), using the technique of finite-mass sum rules (FMSR's), which are the analogue in this context of the familiar finite-energy sum rules (FESR's) (again cf. Chapter 12). The upshot is (Hoyer *et al.*, 1973), that the triple Regge coupling formulae appropriate to large missing mass give the average behaviour at small missing mass, where production is dominated by specific two-body scattering. The general prediction for the M^2 dependence of the component of production corresponding to Regge pole i is

$$\frac{\partial \sigma_i}{\partial t\, \partial M^2} \propto (M^2)^{\alpha_M(0)-2\alpha_i(t)}. \tag{3.1.46}$$

The index $\alpha_M(0)$ corresponds to the appropriate component of the forward imaginary Regge-particle scattering amplitude, and, from general duality ideas (cf. Chapter 12), the direct channel resonance contributions to the missing-mass spectrum (the rho-peak, the f^0 peak etc.) should be dual to "meson exchange", so that $\alpha_M(0) \approx \frac{1}{2}$ (for exchange of the usual vector–tensor nonet). "Background" should be dual to Pomeron exchange $\alpha_M(0) \to 1$.

The important idea for our present purpose is the dependence on the trajectory parameter $\alpha_i(t)$ of the original $a\bar{c} \to \bar{b}x$ exchange. There is an immediate prediction that the natural parity A_2 exchange ($\alpha_{A_2}(0) \approx \frac{1}{2}$) should diminish in importance relative to the unnatural parity π-exchange ($\alpha_\pi(0) \approx 0$) in the ratio $1/M$. This accords with the observed systematics of (N) and (U) exchanges in $\pi^+\pi^-$ production, at least qualitatively. A special point concerns the strength of the forward cut contribution. If, as suggested in some analyses of charged pion photoproduction (Worden, 1972, 1973), the forward cut in the rho-mass region is 60% π-cut and 40% A_2-cut, then the decreased value of the cut in the f^0 region would be at least in part explained (Worden, 1973). As mentioned earlier, there are objections to assuming such a large forward A_2-cut, but the above remark tells in favour of the idea.

The whole area of Regge-particle scattering phenomenology is a growing one with very interesting new applications in the field of amplitude analysis (Hoyer *et al.*, 1974). As regards the di-meson mass systematics, there has been a slightly different discussion (Michael, 1973b) based on an explicit dual resonance model.

For our present purpose, the essential qualitative point is the predicted growing predominance of OPE relative to natural parity exchange as one moves to higher masses. This obviously has important bearing on the prospects for experiments at higher energies.

(i) Summary

This concludes our general survey of OPE and its backgrounds in di-meson production. The way is now clear for discussing methods of extracting meson–meson phase shifts (Section 3.2), and the ensuing results (Section 3.3). The bulk of our discussion has been of the "*N*-reactions", where the vanishing of the forward OPE contribution at once offers a challenge and an opportunity to study the background exchanges in detail. With the advent of high-statistics, high-energy $\pi^+\pi^-$ production experiments, very considerable progress has been made in disentangling the various components of production within a reasonable framework of assumptions. These assumptions, which underly the amplitude analysis to which we have referred, await target polarization experiments for their confirmation (see Appendix B). Meantime, accepting the assumptions, we have determinations of the relative amounts of natural and unnatural parity exchange in various situations, and in particular have seen the parameters of the pion Regge trajectory established rather convincingly. As we move on to applying these general ideas to extracting meson–meson phase shifts, we have to remember that although the overall picture is probably substantially correct, we are pressing it very hard in specific regions, especially at small t and away from the prominent meson–meson resonances. All possible cross-checks are therefore to be welcomed.

3.2. Peripheral Di-Meson Production (2): Extracting $\pi\pi$ and $K\pi$ Phase Shifts

(a) Introduction

In the last section, we examined the structure of OPE and its backgrounds in peripheral di-meson production. We now turn to ways of exploiting this understanding to extract information on $\pi\pi$ and $K\pi$ scattering. (We shall again refer to these collectively, including their inelastic channels, as $P\pi$ scattering.) Peripheral production is not the only regime of production reactions for which this can be attempted—at the end of this section we shall briefly refer to isobar analyses of $\pi\pi N$ formation at low energies, $\sqrt{s} \lesssim 1{\cdot}5$ GeV—but it is the most favourable one. There is potential $P\pi$ scattering information both in the OPE signal *and* in its backgrounds. The only top-grade information comes from extrapolating to the pion pole, but the majority of extraction methods accept lower grade information as well in order to boost output statistics.

The available methods fall into three classes:

(i) Extrapolating to the pion pole.
(ii) Enriching the OPE signal by amplitude analysis—then extrapolating.
(iii) Fitting to a model in the physical region.

As we saw in the last section, it is for the "N-reactions" $PN \to M^*N$ ($\pi N \to \pi\pi N$, etc.) that special techniques for distinguishing, and even exploiting, backgrounds are appropriate; in the "Δ-reactions" $PN \to M^*\Delta$ (1236), the OPE signal remains dominant in the forward direction, and extrapolation is therefore particularly suitable. As we shall see, not even extrapolation is an entirely model-independent procedure; there is the question of what and how to extrapolate, whether particular combinations formed from the data are especially smooth in t, and so on.

The subject matter of this section will be discussed in the following order: Firstly, a survey of the main options available; then, a fairly detailed description of the three principle methods referred to above; finally, a brief discussion of other possibilities, for example the potential for extracting $P\pi$ phase shifts from $P\pi$ production experiments at lower energies.

(b) Preliminary Survey of the Alternatives

In order to fix ideas, we return to the example discussed in the previous section, where only S- and P-wave di-mesons need be considered. For convenience of exposition, we repeat equation (3.1.7) in which the six components of the intensity distribution are expressed in terms of the four helicity amplitudes, H_0, H_+ and H_- for P-wave di-mesons and H_S for the S-wave, as follows:

$$
\begin{aligned}
\frac{\partial^2\sigma}{\partial t\,\partial M} &= \|H_S\|^2+\|H_0\|^2+\|H_+\|^2+\|H_-\|^2 && (\equiv N)\\
(\rho^{11}_{00}-\rho^{11}_{11})\frac{\partial^2\sigma}{\partial t\,\partial M} &= \|H_0\|^2-\tfrac{1}{2}(\|H_+\|^2+\|H_-\|^2) && (\equiv \sqrt{5\pi}N\langle Y^0_2\rangle)\\
\rho^{11}_{1-1}\frac{\partial^2\sigma}{\partial t\,\partial M} &= \tfrac{1}{2}(\|H_+\|^2-\|H_-\|^2) && (\equiv -\sqrt{10\pi/3}N\langle Y^2_2\rangle)\\
\operatorname{Re}\rho^{11}_{10}\frac{\partial^2\sigma}{\partial t\,\partial M} &= \frac{1}{\sqrt{2}}\operatorname{Re}\|H^*_-\,.\,H_0\| && (\equiv \sqrt{5\pi/3}N\langle Y^1_2\rangle)\\
\operatorname{Re}\rho^{10}_{00}\frac{\partial^2\sigma}{\partial t\,\partial M} &= \operatorname{Re}\|H^*_0\,.\,H_S\| && (\equiv \sqrt{\pi}N\langle Y^0_1\rangle)\\
\operatorname{Re}\rho^{10}_{10}\frac{\partial^2\sigma}{\partial t\,\partial M} &= \frac{1}{\sqrt{2}}\operatorname{Re}\|H^*_-.H_S\| && (\equiv \sqrt{\pi}N\langle Y^1_1\rangle).
\end{aligned}
\qquad (3.2.1)
$$

As will be recalled from the discussion adjoining equation (3.1.7), the above H's are complex production amplitudes with independent components $H_{\lambda_3\lambda_1}$ corresponding to the different possibilities for the initial and final baryon spin, $\lambda_1 = \pm\frac{1}{2}$, $\lambda_3 > 0$, and the appropriate inner products are to be understood in (3.2.1).

In any situation where the di-meson systems are pictured as being first produced by particular t-channel exchanges, then to evolve through two-body final state interactions, it is natural to write the production helicity amplitudes in the form

$$H_S = h_S \,.\, T^{j=0}_{P\pi}$$
$$H_{\pm}^{0} = h_{\pm}^{0} \,.\, T^{j=1}_{P\pi}, \qquad (3.2.2)$$

where the $T^j_{P\pi}$'s are the appropriate $P\pi$ scattering amplitudes, and the "reduced" helicity amplitudes, h, only depend on M, the di-meson mass, through real slowly varying factors. This is the factorization model (Schlein, 1967) which forms the basis of almost all attempts to extract $P\pi$ scattering information by fitting directly to data in the physical region. Regarding the validity of the factorization assumption, we may remark firstly, that, if production in the kinematic domain under study is really, as we have implied, controlled by various specific t-channel exchanges, then the decay phase factors $e^{i\delta^j_{P\pi}(M)} = T^j_{P\pi}/|T^j_{P\pi}|$ in (3.2.2) are required by unitarity, as in the Watson theorem.‡ It further follows, by analyticity, that resonance poles on the unphysical sheet of $T^j_{P\pi}$ must extend to the relevant H's; this justifies identifying rapid di-meson mass dependences in the H's with those of $|T^j_{P\pi}|$;§ and that is as far as we can go without making special additional assumptions.

For the specific OPE contribution, the factorization assumption of course goes through exactly (equation (3.1.21)), aside from kinematic off-shell corrections and form factors, whether the di-meson mass dependence of $T^j_{P\pi}$ be fast or slow. In the context of the absorption model, the same goes for the π-cut contribution. For other exchanges, the result may be assumed, provided additional relative production phases are allowed in the reduced helicity amplitudes of equation (3.2.2), and provided the possibility of additional slow di-meson dependences is kept in mind. If the latter possibility is neglected, then, for example, resonance shapes may be misinterpreted using physical region data in its raw state.

Effects which would definitely spoil the factorization assumption would be those arising from other pair-wise final state interactions than $P\pi$ scattering—isobar reflections and the like, e.g. $\pi^- p \to \Delta^- \pi^+ \to \pi^+ \pi^- n$. One would hope to avoid these by going to sufficiently high beam momentum and

‡ Watson (1952); cf. the discussion of K_{e4} decay (Section 5.1) and $e^+e^- \to \pi^+\pi^-$ (Section 5.3).
§ It is on this basis that most meson resonances have been discovered as bumps in appropriate mass spectra.

avoiding the dangerous sectors of phase space. However, it is sometimes necessary to make specific corrections for such effects.

Granted the validity of the factorization assumption, we have a number of options, depending on what we are willing to assume concerning the structure of the h's in (3.2.2). The most conservative approach is just to concentrate on the OPE component, equation (3.1.21), and aim to separate it out by extrapolation to the pion pole. In terms of intensity components in the t-channel, this means that we just use the $\langle Y_l^0\rangle$ moments, and seek the contribution to them which behaves as a function of t like $K_{NB}(t)/(t-\mu^2)^2$ (cf. (3.1.16)). For "Δ-reactions", where the backgrounds are relatively mild, this is the most commonly adopted procedure. For "N-reactions", the appropriate vertex factor $K_{NN}(t)$ vanishes at $t=0$, equation (3.1.17), so that the background predominates at the edge of the region from which the extrapolation is being made. If, for instance, one seeks to derive the S-wave production cross-section by extrapolating the isotropic component (as against the $\cos^2\theta$ term) of the di-meson decay distribution,

$$(1-\sqrt{5\pi}\langle Y_2^0\rangle)N = \|H_S\|^2+\tfrac{3}{2}\|H_+\|^2+\tfrac{3}{2}\|H_-\|^2,$$

then a very bothersome background from transverse P-wave production (the $\|H_\pm\|^2$ contributions above) has to be surmounted. The same problem, numerically to a lesser degree, disturbs the determination of longitudinal P-wave production from $\langle Y_2^0\rangle$ (cf. (3.2.1)). All this is not to say that extrapolation is impossible for "N-reactions"; merely that it is difficult.

The alternative to straightforward extrapolation is to assume more about the structure of the background helicity amplitudes, one possibility being to separate out the different amplitude contributions in a model, then to extrapolate suitably. As discussed in Section 3.1(g), it is possible, given not unreasonable assumptions, to perform such an amplitude analysis and separate out the unobserved density matrix elements, e.g. $\rho_{00}^{00}\sigma\equiv\|H_S\|^2$ and $\rho_{00}^{11}\sigma\equiv\|H_0\|^2$ for the case of S- and P-wave production (3.2.1). One can then make suitable extrapolation on these enriched intensity components. This is the basis of the method of Estabrooks and Martin (1972, 1973b, 1974a). They usually choose s-channel helicity amplitudes, since they claim that certain ratios are especially smooth, and hence suitable for extrapolation, in this frame.

Another possibility is to work directly with the physical region intensity components at small t, in particular with these quantities averaged over t. As can be seen from equation (3.2.1) for the example of S- and P-wave production, the resulting intensity components involve averages of the bilinear combinations of helicity amplitudes appearing. The resulting expressions can still be written in the form of the right-hand side of (3.2.1), as a matter of definition, and the factorization assumption (3.2.2) extended to the resulting t-averaged amplitudes. The reduced helicity amplitudes should still be only slowly varying functions of M, so that, provided a sufficient degree of spin and phase coherence survives the t-averaging, a direct determination of the $P\pi$ phase shifts can be made. This is the basis of Schlein's method (Schlein, 1967; Malamud and Schlein, 1967; Grayer *et al.*, 1972c), which we discuss in Section 3.2(e) below.

The above three methods will be the main subject of the remainder of this section. Before going on to discuss them in detail, there are a few further preliminary points to be made. The first is a remark on what the factorization model implies for the observability of density matrix elements. On substituting formula (3.2.2) into the definition (3.1.5), one obtains

$$\rho^{j_1 j_2}_{m_1 m_2} = \frac{1}{2N}\left\{\sum_{\lambda_3\lambda_1} h^{*j_1}_{\lambda_3\lambda_1;m_1} h^{j_2}_{\lambda_3\lambda_1;m_2}\right\} T^{*j_1}_{\cdot}\, T^{j_2}$$

$$\equiv r^{j_1 j_2}_{m_1 m_2}(T^{*j_1}_{\cdot}\, T^{j_2}) \qquad \text{(definition).} \tag{3.2.3}$$

One immediately sees the possibility, in suitable circumstances, and where T^{j_1} and T^{j_2} are known, of measuring the imaginary parts of the reduced density-matrix elements, a favourable circumstance being to have a sharp resonance, say in j_1, interfering with a smooth behaviour in j_2. The phase of the reduced density matrix element is revealed by sweeping through the resonance (Hoyer, 1973). A very similar situation obtains in analysing three-body production by the isobar model (Bowler, 1973) (see Section 3.2(f) below). The methods which assume the factorization model implicitly use the above property.

The other remark concerns backgrounds to OPE at low $M_{\pi\pi}$. In Section 3.1, we have given an extended discussion of such backgrounds, as shown in the data and as predicted in models, especially the absorption model. One important outcome for the present consideration is that the predicted backgrounds increase relative to the OPE signal towards low mass (cf. equations (3.1.29)–(3.1.40)), an effect intimately connected with the crossing angles becoming a sharper function of t (cf. equation (3.1.11)). We may therefore anticipate that extricating the OPE signal will present a harder problem at low masses.

(c) Goebel–Chew–Low Extrapolation

The basic idea of the method (Goebel, 1958; Chew and Low 1959) is simply to form the quantity

$$F(M^2, \cos\theta, t) \equiv (t-\mu^2)^2 \frac{\partial^3\sigma}{\partial t\, \partial M\, \partial(\cos\theta)}, \tag{3.2.4}$$

(where $\cos\theta$ refers to the t-channel helicity frame) and extrapolate it to the pion pole at $t=\mu^2$. Only the pole term survives, and the limit is directly proportional to the on-shell $P\pi$ differential cross-section:

$$F(M^2, \cos\theta, t) \underset{t\to\mu^2}{\to} A\frac{d\sigma_{P\pi}}{d\cos\theta}, \tag{3.2.5}$$

with the constant A given by the explicit formula (equation (3.1.16), (3.1.17), and (3.1.18))

$$A = \frac{M^2 q_{P\pi}}{\pi s q_{\text{in}}^2} K_{NB}(\mu^2). \tag{3.2.6}$$

For the "Δ-processes", the approximate formula (3.1.18) is usually replaced by the appropriate integral of the full OPE formula (3.1.15) over the Δ mass band selected with experimental cross-sections inserted. It is also customary to divide through by Dürr–Pilkuhn form factors (Dürr and Pilkuhn, 1965) prior to extrapolation, in order to have rather slowly varying functions of t to fit (as an example see Protopopescu *et al.* (1973)). For the "N-processes", the factor $K_{NB}(\mu^2)$ is given by the simple expression, $-f^2m_N^2$, with $f^2=0{\cdot}081$ for $\pi^0 N\bar{N}$, 0·162 for $\pi^\pm N\bar{N}$ (cf. equation (3.1.17)).

In order to perform the extrapolation, one can fit the small-t production data for F to a polynomial in t, and use that to extrapolate to $t=\mu^2$. Alternatively, one of the more refined extrapolation techniques available (Cutkosky and Deo, 1968; Ciulli, 1969a,b; Pišút, 1970) can be used. Whichever way the extrapolation is done a problem arises for the "N-processes" in that the OPE signal vanishes at $t=0$ (3.1.17), so that, with limited data, not very good statistical precision is achieved.

This has sometimes been countered by adopting the "pseudo-peripheral" or "evasive" assumption that the background as well as the signal vanishes at $t=0$. One can then extrapolate the quantity $(\mu^2/t)F$. Unfortunately, for the $\pi^+\pi^-$ production case at least, this is strictly belied by the data at 15 GeV/c (Bulos *et al.*, 1971a) and 17 GeV/c (Grayer *et al.*, 1972a,b) (also cf. Fig. 3.1.2). At 15 and 17 GeV/c, the effect can be clearly seen because $t_{\min}$ is very small, but it is reasonable to infer it at lower momenta also, since the form of the density matrix elements does not change very much from 2 to 17 GeV/c (Scharenguivel *et al.*, 1970; Jacobs, 1972). The extrapolated cross-section at $t=0$ is not zero, although it is small, and to assume the contrary is to import unknown systematic errors, although possibly small ones.

An alternative approach is to attempt to extrapolate ratios of intensity components, notably the forward–backward asymmetry (Scharenguivel *et al.*, 1969), chosen because in the absorption model it shows a comparatively smooth t-variation in the physical region. A danger in working with any ratio is that at $t=0$ it is entirely given by the background to OPE, so that a

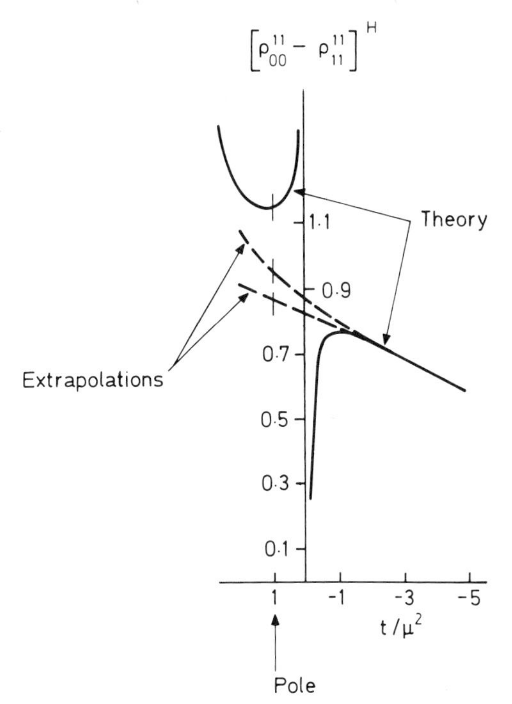

FIG. 3.2.1. Illustration in a model of the dangers of extrapolating ratios (Williams, 1970). The full line shows the quantity $\rho_{00}^{11} - \rho_{11}^{11}$ in the s-channel helicity frame according to Williams' model. The dashed lines show linear and quadratic extrapolations from the form in the physical region to the pion pole.

complicated behaviour may result. It is probably better, whenever the statistics allow it, to work with the original function F of equation (3.2.4). However, it is interesting to examine the smoothness of various ratios as a function of t in a reasonably realistic model in order to see which functions are the smoothest, and also to bring out the dangers of naive extrapolation. A very useful model here is that of Williams described in Section 3.1(e) (equation (3.1.29) and adjoining discussion), as it results in simple analytic forms for the amplitudes. One can take the output of the model in the physical region as "data", and see how well proposed extrapolation procedures fare in predicting the value at the pion pole. An unfavourable example is shown in Fig. 3.2.1 (Williams, 1970; see also Kane, 1970 and Kimel, 1970). The quantity $\rho_{00}^{11} - \rho_{11}^{11}$ in the s-channel helicity frame has been calculated (full line) and linear and quadratic extrapolations (dashed lines) made from the physical region to re-predict the value at the pole.

The conclusion of all this is that one wants large statistics at low t-values, and hence must go to reasonably high beam momenta (e.g. 17 GeV/c) so that these small t-values are accessible. Then it should be possible for the necessary information on the background at $t=0$ to be determined within the extrapolation. However, this is statistically very demanding and one may wish to delimit possibilities within a reasonably general model framework (Kane and Ross, 1969; Froggatt and Morgan, 1969, 1970). For example, the latter model (cf. equations (3.1.37) and (3.1.40)) incorporates the factorization hypothesis, so that t-dependence and M-dependence can be examined separately. The assumed spin structure and t-dependence follows on rather general grounds at very small t (cf. Section 3.1(f)); its extension to somewhat larger t-values, only modified by form factors, constitutes a model. As such it has been used to fit data (Scharenguivel *et al.*, 1970; Jacobs, 1972) with fair success, in particular with parameter values as theoretically expected (Froggatt and Morgan, 1970). Aspects of this parameterization were used in the CERN-Munich group's first analysis of their 17 GeV/c $\pi^+\pi^-n$ experiment (Grayer *et al.*, 1972a), in order to derive additional $\pi\pi$ phase shift information from the t-channel $m=1$ moments. To go further with such parameterizations, takes one into the realm of the amplitude analysis methods, to be discussed below.‡

The extrapolation method is of course applicable to any process of the type $PN \to M_1M_2 \ldots M_nB$ which has an OPE contribution. The equations explicitly given in Section 3.1, apply to two-body processes such as $\pi N \to \pi\pi N$, $K\bar{K}N$, $p\bar{p}N$, etc. The extended class of reactions allow higher multiplicities to be investigated, e.g. $\pi N \to \pi\pi\pi\pi B$. If only the technical experimental problems could be surmounted, measurements on $\pi^{\pm}p \to n+$ anything, with accurate detection of very low-energy neutron recoils, could yield information on $\sigma_{\text{tot}}^{\pi\pi}$ via the component proportional to

$$\sigma_{\text{tot}}^{\pi\pi}[(-t)/(t-\mu^2)^2]$$

in the intensity (Worden, 1971). If practicable, this would be a very attractive technique. Inclusive "Δ-reactions" could similarly be used, and might have practical advantages.

(d) Amplitude Analysis and Extrapolation—Estabrooks and Martin's Method

The method of Estabrooks and Martin (1972, 1973b, 1974a) for extracting $\pi\pi$ phase shifts contains two main steps: Firstly, an amplitude analysis of dipion production (see Section 3.1(g)) to determine the unobserved density

‡ The CERN-Munich data referred to above afford an important test of alternative methods for extracting $\pi\pi$ phase shifts, since several procedures have been tried—straightforward extrapolation and various forms of amplitude analysis. As we shall discuss below, and in the following Section (3.3), the agreement between the various results has not always been satisfactory (cf. Fig. 3.3.6(a)), although a concensus is now emerging (Fig. 3.3.6(b)).

matrix elements, and hence, hopefully, to isolate components of production with an enriched OPE signal. Secondly, the extrapolation to the pion pole of particular components and ratios of components of the resulting signal.

In its original formulation, the method applied to the case where just S- and P-wave di-mesons are produced. In this case, rather mild assumptions suffice to admit the amplitude analysis (see Section 3.1(g) for a discussion of alternative possibilities). Estabrooks and Martin assume that the unnatural exchange component of production is all s-channel flip (i.e. no A_1 exchange), and this enables the six independent observables to be expressed in terms of six parameters (cf. equation (3.1.41)). These comprise the magnitudes of the four relevant helicity amplitudes, $|H_S|$ for S-wave production and $|H_0|$, $|H_-|$ and $|H_+|$ for the three independent spin alignments of P-wave production, and two angles ϕ_{10} and $\bar{\delta}_{SP}$ expressing the non-alignment in spin and phase of the interfering amplitudes H_S, H_0 and H_-.‡

On solving for the parameters, two solutions result, as discussed in Section 3.1(g). One can make the amplitude analysis in either the s or t channel frames. Estabrooks and Martin mostly use amplitudes in the s-channel frame, since it is for these that they find simple extrapolation properties. An example of the resulting amplitudes is shown in Fig. 3.2.2 from Estabrooks and Martin (1973b). This is an application to the 17 GeV/c CERN-Munich experiment referred to previously. This particular analysis refers to production over the rho-mass band, $M_{\pi\pi} = M_\rho \pm 75$ MeV. Analyses have also been made over a very much larger range of masses grouped in small mass bins, leading to $\pi\pi$ phase determinations (Estabrooks *et al.*, 1973; Estabrooks and Martin, 1974a). The former analysis covers the range 440 to 1400 MeV and the latter the range 500 to 1000 MeV. They differ principally in their treatment of D-wave production which plays a role even below 1 GeV. We return to this point below.

For the present, we concentrate on the results for the rho-band, Fig. 3.2.2, and ignore D-wave contributions. The figure shows for the two alternative solutions the quantities $|H_0|$, $|H_+|$, $|H_-|$ and the ratio $|H_S|/|H_0|$ all in the s-channel; also $\bar{\delta}_{SP}$ and $\cos\phi_{10}$. The two solutions are seen to be overall rather similar in appearance, with the main difference concerning the behaviour of $\cos\phi_{10}$ and $\bar{\delta}_{SP}$, and of $|H_-|$ in the region of its minimum. Solution 1 has $|\cos\phi_{10}|$ close to unity, as would be entailed by the assumption of spin and phase coherence (SPC) for H_0 and H_- referred to in Section 3.1(g); solution 2 has $|\cos\phi_{10}|$ definitely not identical to unity. It turns out that for all mass bins, as in the present example, there is always a "solution

‡ That there are just two angles describing the relative alignments of H_S, H_0 and H_- is what the initial assumptions have secured (cf. Section 3.1(g)). Estabrooks and Martin actually allow for the s-channel non-flip contributions associated with OPE at very small t (cf. equation (3.1.9)), whilst continuing to cast their equations in terms of the flip amplitudes. Each appearance of $|H_0|^2$, $|H_S|^2$ and $(|H_0| \,.\, |H_S|)$ is therefore multiplied by the correction factor $R \equiv t/(t - t_{\min})$.

1" approximately fulfilling SPC. It happens that the $\pi\pi$ phase shifts which ensue from solution 1 are in better agreement with data on $\pi^0\pi^0$ production than those from solution 2, and on this ground Estabrooks *et al.* (1973) ultimately select this alternative. However, as regards the amplitude analysis, the solutions are on an equal footing.

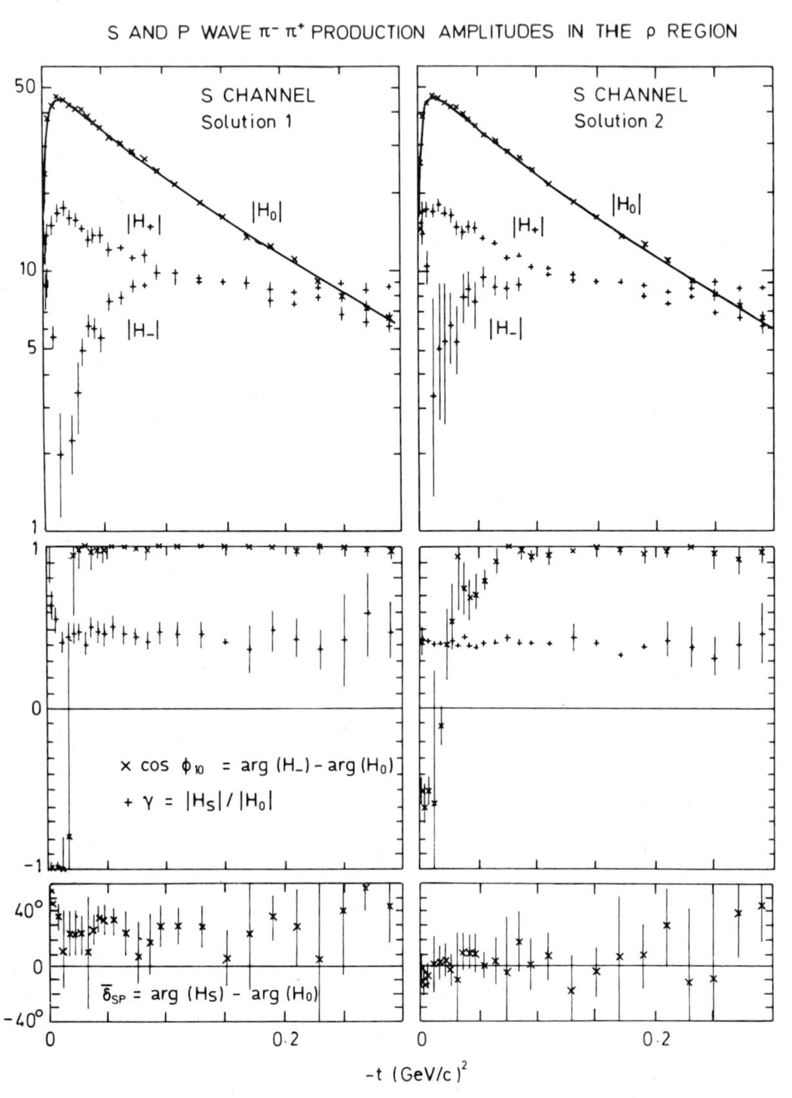

FIG. 3.2.2. Alternative solutions from the *s*-channel amplitude analysis of Estabrooks and Martin (1973b) for $\pi^+\pi^-$ production over the rho-mass band ($M_{\pi\pi} = M_\rho \pm 75$ MeV) in the reaction $\pi^- p \rightarrow \pi^+\pi^- n$ at 17·2 GeV/c. The curves show a fit to H_0 of the form $H_0 = H_0^{\text{OPE}} \times$ exponential form factor.

The next phase of the analysis, in which the actual $\pi\pi$ amplitudes are to be extracted, works entirely with the quantities $|H_0|$, $|H_S|$ and $\bar{\delta}_{SP}$. Estabrooks and Martin's first crucial observation is that the ratio $|H_S|/|H_0|$ and the phase $\bar{\delta}_{SP}$ are very smooth and essentially constant as functions of t (again cf. Fig. 3.2.2), hence excellent candidates for extrapolation. Secondly, they note that the quantity $|H_0|$ (in the s-channel) has a t-dependence which is very well fitted by the formula

$$H_0 \propto \frac{\sqrt{t_{\min}-t}}{t-\mu^2}\, e^{b(t-\mu^2)}.$$

They show that, for the sample of data which they consider, this is a better quantity to extrapolate than the corresponding t-channel quantity. The method for extracting $\pi\pi$ phase shifts follows immediately from transferring the above findings to other mass bins. In fact, the technique adopted by Estabrooks *et al.* (1973) is to fit the s-channel $m=0$ moments to the formula

$$H_0(H_S) \propto \frac{\sqrt{t_{\min}-t}}{t-\mu^2}\, e^{b(t-\mu^2)} \,.\, T_{\pi\pi}^{j=1}(T_{\pi\pi}^{j=0}).$$

The extrapolation of $|H_0|$ determines $|T_{\pi\pi}^{j=1}|$ (allowing of course for the factor $\cos\chi_4\,(t=\mu^2)$ in the dipion crossing angle from extrapolating in the s-channel), and the extrapolations of $|H_S|$ and $\bar{\delta}_{SP}$ determine $|T_{\pi\pi}^{j=0}|$ and $\delta_S-\delta_P$. The latter two pieces of information really comprise one statistically well determined combination $|H_S|\cos\bar{\delta}_{SP}$ and one more poorly determined item, $|H_S|\sin\bar{\delta}_{SP}$. However, given modest information on the $I=2\,\pi\pi$ phase shifts, these are sufficient to determine δ_0^0, assuming δ_1^1 to be known from extrapolation of $|H_0|$. Since the preferred solution has production amplitudes H_0 and H_- which are compatible with the spin-phase coherence (SPC) condition, Estabrooks *et al.* (1973) repeat their analysis assuming this. The output δ_0^0 phase shifts are naturally close to those of the unconstrained solution 1, but tend on the average to fall a few degrees lower below the rho-mass, a point which is of interest when one comes to compare with the results of other analyses.

Even below 1 GeV, allowance has to be made for D-wave production. In order to restrict the number of extra parameters thereby introduced, Estabrooks *et al.* (1973) and Estabrooks and Martin (1974a) assume a special form of spin dependence for D-wave production, analogous to that obtaining in the Williams model (cf. Section 3.1(e)), and such that it can be calculated from the spin-dependence of P-wave production.‡ In the former analysis, the D-wave amplitudes were determined empirically from the experimental $\langle Y_3^{0,1}\rangle$ moments. A first approximation to the S- and P-wave

‡ $H_D^{1\pm}=\sqrt{3}\,H_D^0(H_\pm/H_0)$; $H_D^{2\pm}=0$.

amplitudes was made using $\langle Y_1^0 \rangle - \sqrt{28/27}\langle Y_3^0 \rangle$ in place of $\langle Y_1^0 \rangle$ in equation (3.1.41), to eliminate the predominant P–D interference term (cf. equation (3.2.1) and Table B.1); residual D-wave effects were then allowed for iteratively. Doubt was cast on this procedure, and the reliability of the experimental $\langle Y_3^0 \rangle$'s, by the discrepancy between the resulting empirical D-wave phase shifts and the prediction from fixed-t dispersion relations (cf. Section 3.3(d) and Chapter 10). In their re-analysis of the production data below 1 GeV, Estabrooks and Martin (1974a), therefore impose theoretical D-waves and only use experimental Legendre moments $\langle Y_l^m \rangle$ with $l \leq 2$. Some anomalies of the earlier findings are thereby eliminated; in particular, the $I=0$ S-wave phase shifts, which formerly disagreed with those extracted from the same data either using conventional extrapolation (Grayer *et al.*, 1972a), or model fits in the physical region (Hyams *et al.*, 1973), fell into line with the other determinations (see Section 3.3(d) especially Figs. 3.3.6(a) and (b)). As will be discussed further in the next sub-section, the low-mass P-waves from both the old and the new version of this analysis disagree with the prediction from fixed-t dispersion relations, but this is not a special feature of the method. Overall, the agreement with the results of other methods is very satisfactory and lends support to their use in the higher mass region where an extension of the present methods would be impracticable.

Among the variants of the above procedure discussed by Estabrooks and Martin (1974a) is to perform an amplitude analysis in the t-channel. One should check that there is no OPE signal in the $m \neq 0$ components (Morgan, 1974a). For the special case where amplitude analysis is performed assuming spin and phase coherence, the resulting amplitudes respectively in the s- and t-channel are very simply related by the di-meson spin-crossing transformation. For example, for the P-wave amplitudes

$$\begin{bmatrix} H_0^{(s)} \\ H_-^{(s)} \end{bmatrix} = -i \begin{bmatrix} \cos\chi_4 & -\sin\chi_4 \\ \sin\chi_4 & \cos\chi_4 \end{bmatrix} \begin{bmatrix} H_0^{(t)} \\ H_-^{(t)} \end{bmatrix}, \qquad (3.2.7)$$

(cf. (3.1.13a)), and so on for higher waves.

(e) Phase Shifts from Amplitude Analysis of Averaged Moments in the Physical Region

The extrapolation methods just described do not depend on assuming the content of the factorization model (equation (3.2.2)); if factorization is valid, much information is thereby sacrificed. We now turn to methods which do assume factorization, and, as a result, are able to work directly with the physical region intensity components, indeed, with these components averaged over t. Such an approach is especially suited to the analysis of rapid variations with M, as regards both the dependability of the factorization assumption and the economical use of data.

We are to consider t-averaged intensity components which can be expressed in terms of effective t-averaged helicity amplitudes and reduced amplitudes via suitable definitions. Thus, we define (cf. equation (3.1.5))

$$\langle \rho^{jj}_{mm} \rangle \frac{d\sigma}{dM} \equiv |\hat{H}^j_m|^2, \tag{3.2.8}$$

and (cf. equation (3.2.2))

$$\hat{H}^j_m \equiv \hat{h}^j_m \, . \, T^j, \tag{3.2.9}$$

where the t-averaged reduced amplitude, $\hat{h}^j_m$, is defined to be real. Finally, for the t-averaged interference moments, we write

$$\langle \rho^{j_1 j_2}_{m_1 m_2} \rangle \frac{\partial\sigma}{\partial M} = \hat{H}^{*j_1}_{m_1} \, . \, \hat{H}^{j_2}_{m_2} C^{j_1 j_2}_{m_1 m_2}, \tag{3.2.10}$$

where the coefficient $C^{j_1 j_2}_{m_1 m_2}$ expressing the spin and phase coherence of $\hat{h}^{j_1}_{m_1}$ and $\hat{h}^{j_2}_{m_2}$ has its modulus bounded by unity from Schwarz's inequality. According to the factorization hypothesis, the h^j_m's and the C's are assumed to be slowly varying functions of M. Individual density matrix elements are of course not in general observable, so that further assumptions have to be made concerning spin and phase alignment, i.e. on the C's, in order to separate the various constituents of production. One has the same range of options here as for the amplitude analysis of the intensity components as a function of t, except that the scope for incoherence is greater owing to the averaging over t.

All methods of course focus on the small t region (say $|t| \leqslant 7\mu^2$) where the OPE signal is strong. The method of Schlein (Schlein, 1967; Malamud and Schlein, 1967; Grayer *et al.*, 1972c), which is applicable to the situation where one has just S- and P-wave production, assumes phase coherence and allows spin incoherence: one could just as well make the converse assumptions as in the method of Estabrooks and Martin (1972), discussed above. Either assumption suffices when only S- and P-waves are in question (cf. discussion in Section 3.1(g)).

The approach of Ochs and Wagner (Ochs and Wagner, 1973; Hyams *et al.*, 1973), which we now discuss in detail, caters for an arbitrary number of di-meson partial waves. The assumptions made are, firstly, that one has the factorization condition, equation (3.2.2), with the reduced production amplitudes for given j coherent in spin and phase (up to an unobservable overall phase between natural and unnatural exchange). Secondly, that for the small t-region considered, the t-channel moments $\langle Y^m_l \rangle$ with $m \geqslant 2$ can be assumed to be identically zero, a statement which well approximates the experimental situation. This entails the vanishing of all helicity amplitudes $H_m^{(t)j}$ with $m > 1$, and an exact balance between the unnatural and natural exchange contributions to $m = 1$ production

$$|H_1^{(t)j+}| = |H_1^{(t)j-}|. \tag{3.2.11}$$

As a result, all the H^+ contributions can be eliminated in favour of H^-'s and the differential intensity distribution can be written

$$\frac{\partial^2\sigma}{\partial M\, \partial(\cos\theta)\, \partial\phi} = \tfrac{1}{2} \sum_{\alpha=\pm j} \left| \sum_j (2j+1)^{\frac{1}{2}} (H_0^{(t)j} d^j_{00}(\theta) + H_1^{(t)j-}\, d^j_{10}(\theta)\, e^{i\alpha\phi}) \right|^2, \tag{3.2.12}$$

(cf. equation (3.1.39)), where the t-averaged helicity amplitudes $\hat{H}$ are now

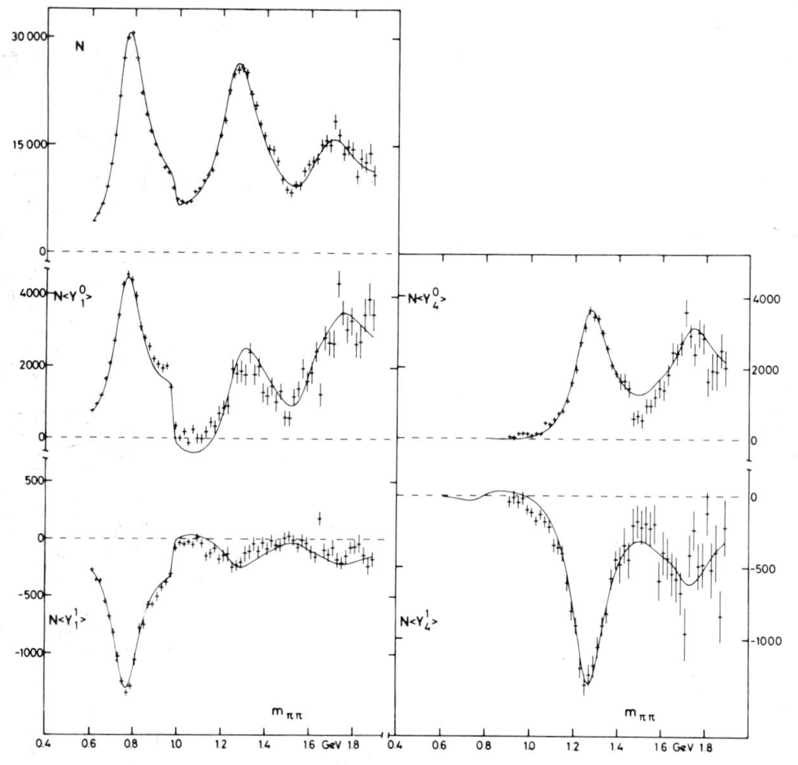

FIG. 3.2.3. Intensity components $N\langle Y_l^m\rangle$ in the t-channel for $\pi^+\pi^-$ production at 17·2 GeV/c, averaged over the t-interval $0{\cdot}01 \leq |t| \leq 0{\cdot}15$ GeV^2 (Hyams *et al.*, 1973). The curves show the fit to these coefficients obtained in the energy-dependent phase-shift analysis by these authors, to be discussed in Section 3.3. The phase-shift analysis covers the mass region 600 to 1900 MeV.

to be understood in all formulae. The method is completed by specifying the form to be assumed for the M- and j-dependence of the reduced helicity amplitudes. For $m=0$ production, Ochs and Wagner assume

$$h_0^j = AM\sqrt{\frac{2j+1}{q}}\,T^j(M), \tag{3.2.13}$$

as for OPE, with A an overall normalization constant, independent of M and j, and q the $\pi\pi$ or $K\pi$ CM momentum. For $m=1$ production, they assume

$$h_1^{j-}/h_0^j = C_j\{1+\alpha_1 M+\alpha_2 M^2+\alpha_3 M^3\}, \tag{3.2.14}$$

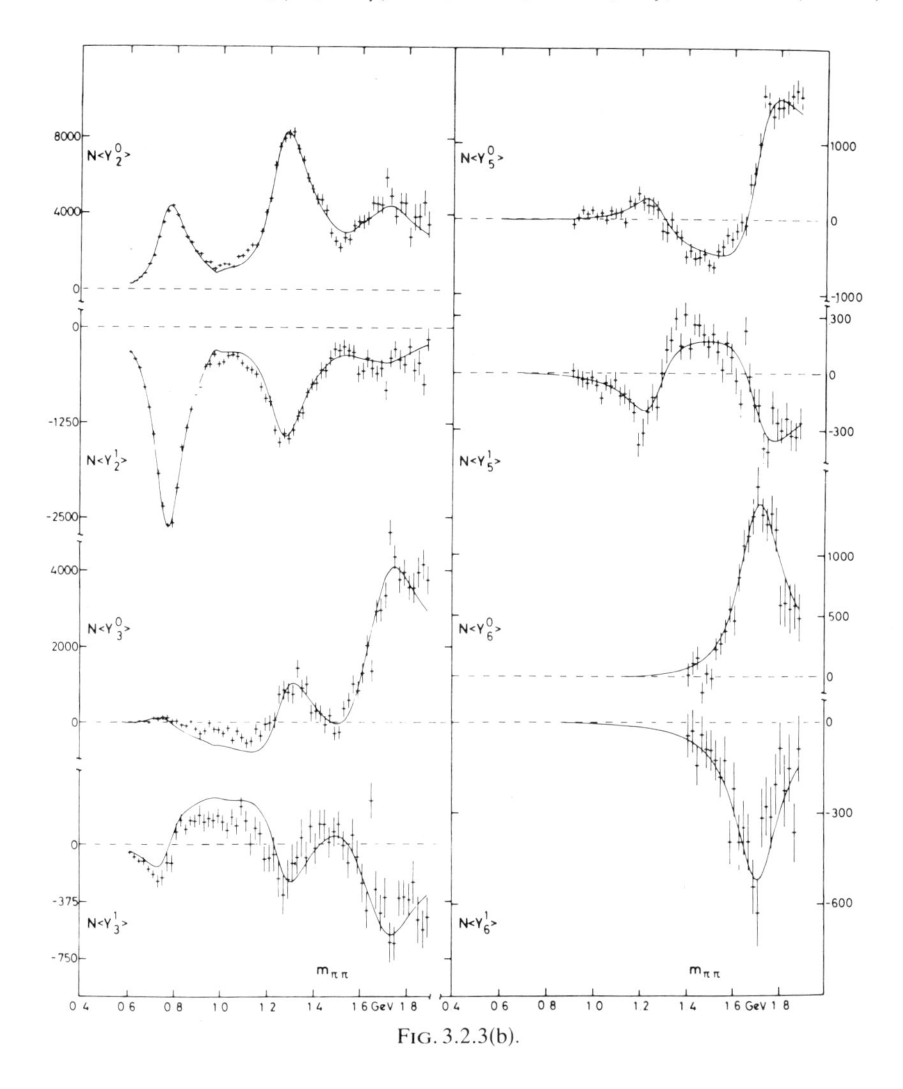

FIG. 3.2.3(b).

with the C_j's and the α's real numbers. These, together with the on-shell $\pi\pi$ phase shifts, constitute the parameters to be fitted. The resulting fit to the t-averaged intensity components of the 17 GeV/c CERN-Munich data (Hyams *et al.*, 1973) is shown in Fig. 3.2.3. We shall be discussing the output $\pi\pi$ phase shifts belonging to this fit in the next section. A general feature of the data apparent in Fig. 3.2.3 which the model explains is the "mirror" symmetry of the $\langle Y_l^0 \rangle$ and $\langle Y_l^1 \rangle$ moments, i.e. the smooth behaviour of $\langle Y_l^1 \rangle / \langle Y_l^0 \rangle$.

Ochs and Wagner justify their assumptions, firstly by relating them to general pictures of OPE and its background such as we discussed in Section 3.1, especially the Williams model and its generalizations (Section 3.1(e) and (f)); in particular, the extent to which spin and phase coherence survives t-averaging is studied in this model, with reassuring results (Hyams *et al.*, 1973). Secondly, they exploit the fact already referred to in Section 3.1(e), that their spin and phase coherence assumptions allow the helicity amplitudes to be determined with information to spare, leading to testable predictions. For example, if just S- and P-waves dominate, the four moments $N\langle Y_l^0 \rangle (l=0, 1, 2)$ and $N\langle Y_2^1 \rangle$ suffice to determine the amplitudes, enabling $N\langle Y_1^1 \rangle$ to be calculated (cf. eq. (3.2.12)). Likewise, as further waves enter, $N\langle Y_3^1 \rangle$ and $N\langle Y_5^1 \rangle$ can be predicted. The resulting comparison (Hyams *et al.*, 1973), is very good (see Fig. 3.2.4).

As we shall see in Section 3.3 the results from applying Ochs and Wagner's method are broadly similar, over the common region of application ($M_{\pi\pi} < 950$ MeV), to those obtained using Estabrooks and Martin's prescription. This is an important cross-check since the latter method makes fewer assumptions and for that very reason is of more restricted scope. It is therefore worth touching briefly on the degree of overlap between the two methods, as applied to data. Firstly, as regards the amplitude analysis, Estabrooks and Martin, whilst not assuming spin and phase coherence, find for the solution which they ultimately choose that they could just as well have done so. In this case, the results of amplitude analysis in the s- and t-channel frames are very simply related (cf. eq. (3.2.7)); in particular, the S–P phase difference, $\bar{\delta}_{SP}$, has the same meaning in either frame. Estabrooks and Martin find this to be a constant as a function of t (cf. Fig. 3.2.2). This implies that its t-average will be equal to the result of extrapolating it. In other respects, the methods are different in detail, for example Ochs and Wagner's determination of the P-wave cross-section depends on their factorization assumption, whereas Estabrooks and Martin extrapolate $H_0^{(s)}$. Even here, both methods are operating on the same separation of S- and P-wave production amplitudes. The principles of the separation are, as we have discussed, reasonably well-founded and tested, but there could be common systematic errors, especially in the low-mass region, where back-

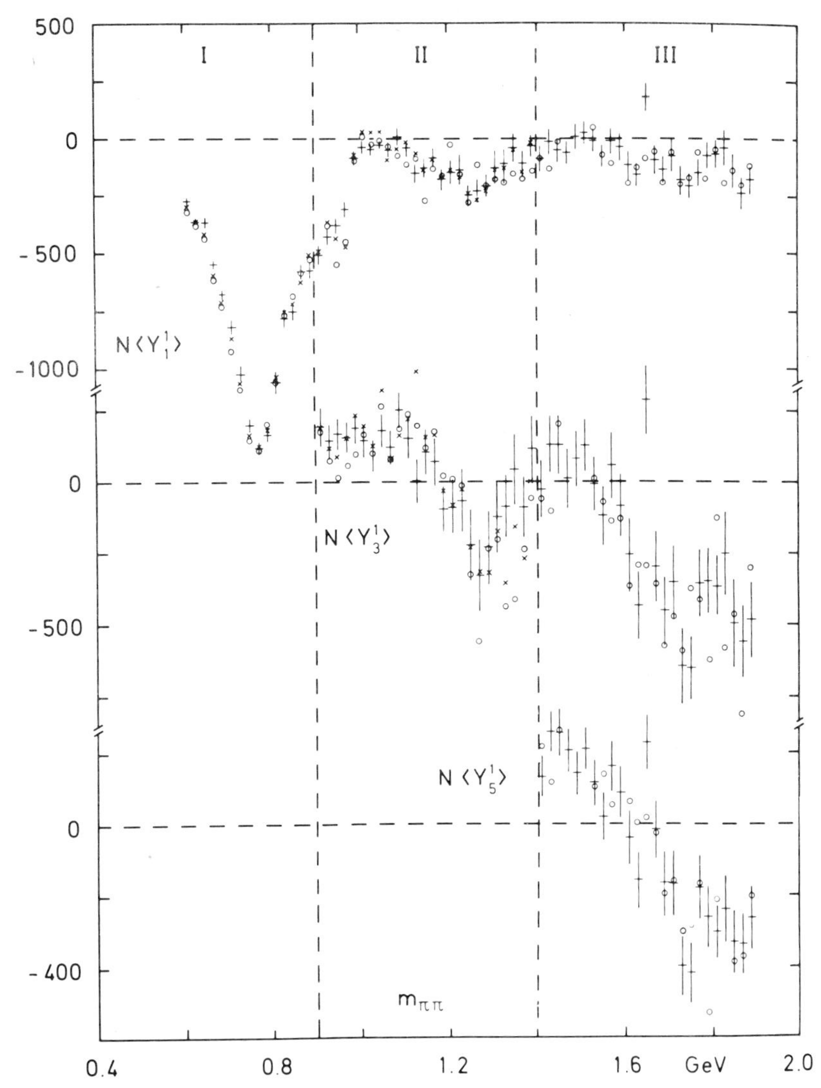

FIG. 3.2.4. Comparison between the moments $N\langle Y_l^1\rangle$, for $\pi^+\pi^-$ production at 17·2 GeV/c, as measured (+) and as predicted (○ and ×) assuming spin and phase coherence for production amplitudes corresponding to the same spin (Hyams *et al.*, 1973). The points (○) refer to the prediction with just the dominant partial waves allowed for—e.g. only *S*- and *P*-waves up to 900 MeV; the (×) points refer to the case where one more partial wave is included.

grounds are proportionately higher and the experimental statistics, and consequently the rigour of testing hypotheses, proportionately lower.

As mentioned earlier, the results (Section 3.3(d)) suggest that there is a problem in that the *P*-wave phase shifts in the low-mass region, $M_{\pi\pi} < 600$ MeV, determined on either method, lie appreciably above the predictions of well-founded analytic models which fit the data in and beyond the rho region. This is already apparent in Estabrooks *et al.*'s analysis and the same trend is revealed much more dramatically in the very low-mass *P*-wave determinations of Männer (1974) (cf. Fig. 3.3.5) by the method of Ochs and Wagner (discussed above). It could be an indication that the procedure for extrapolating to the on-shell *P*-wave begins to go wrong as one goes to lower masses, or, alternatively that the amplitude analysis becomes misleading. (It will be recalled that backgrounds are predicted to become relatively larger as one goes to lower $\pi\pi$ masses (equation (3.1.36)); if the structure is slightly different from that predicted, this could account for the difficulty.) It could be that there is some systematic bias in the data. It is important that each possibility be thoroughly explored since, in all methods of analysis, the *S*-waves are effectively derived from taking ratios to the *P*-wave signal.

(f) Other Topics in Brief

(i) $\pi N \to \pi\pi N$ *at Low Energies*

Pion–pion phase shifts have been fitted to low-energy production data using the isobar model (see e.g. Saxon *et al.*, 1971). The production amplitudes for the various spin-parity states entering are written as sums of contributions from $\pi N \to \pi\Delta$, $\pi N \to \varepsilon N$ and so on. Phase shifts enter via factors $\sin\delta\, e^{i\delta}$ inserted to represent isobar decay, and one has the possibility, within the assumed framework, of seeing informative cross-term effects where different resonance-bands cross. Unfortunately, the whole procedure is very model-dependent so far as determining the final state phase shifts is concerned.‡ It is probably better to concentrate on distinguishing the various J^P contributions and isobar branching ratios, with the final state phases as input (e.g. see Rosenfeld *et al.*, 1975; Cashmore, 1973).

An exception to the model-dependence of $\pi\pi$ information from low-energy experiments is furnished by a result of Tiktopoulos (1970).§ This states that if one could measure the detailed form of the $\pi\pi$ mass spectrum at very low dipion mass for $\pi^+\pi^-$ and $\pi^0\pi^0$ production in the form

‡ Special circumstances could somewhat mitigate the model dependence of isobar model findings. For example, Jones *et al.* (1974) who observe the reaction $\pi^- p \to \pi^+\pi^- n$ close to threshold interpret the pronounced low-mass dip in the $\pi^+\pi^-$ mass distribution as requiring a very small $\pi\pi$ *S*-wave scattering length, $-0{\cdot}06 \leq a_0^0 \leq 0{\cdot}03$. (However, see Morgan and Pennington (1975).)

§ See also the review of Anisovich and Ansel'm (1966) for references to earlier related work.

$$\frac{1}{q_{\pi\pi}^2}\frac{d\sigma_{+-}}{dq_{\pi\pi}} = A + Bq_{\pi\pi} + 0(q_{\pi\pi}^2),$$
$$\frac{1}{q_{\pi\pi}^2}\frac{d\sigma_{00}}{dq_{\pi\pi}} = C - Bq_{\pi\pi} + 0(q_{\pi\pi}^2), \qquad (3.2.15)$$

then the ratio $3|B|/2(AC)^{\frac{1}{2}}$ is a lower bound for the difference $|a_0^0 - a_0^2|$. The result follows from unitarity and is applicable at any beam momentum.

(ii) *Three-pion Production,* $\pi N \to (3\pi)N$

The situation as regards the analysis of these processes (Brockway, 1970; Ascoli *et al*, 1970, 1971, 1973) is closely similar to that occurring in $\pi\pi N$ formation (Cashmore 1973; Morgan, 1971; see also Bowler, 1973). Again the isobar model is used, and again the $\pi\pi$ pair-wise interactions have to be supplied as input.‡

We shall later (in Section 3.4) be discussing 3π formation in $N\bar{N}$ annihilation which has at times inspired greater optimism as to its power to give $\pi\pi$ information. Again we shall conclude, that the available analyses are highly model dependent.

(iii) *Coulomb Interference*

There have been suggestions (Biswas *et al.*, 1968) that there should be a detectable interference between the Coulomb and strong interaction contributions to off-shell $\pi\pi$ scattering, as evidenced in $\pi\pi N$ production. However, it has been pointed out by Michael (1973c) that, when allowance is made for additional terms which are required by gauge invariance, the effect essentially disappears.

(iv) *Amplitude Analysis for "Δ-Processes"*

Despite the greater predominance of the OPE signal relative to its background, one might think to improve the signal in $\pi N \to \pi\pi\Delta$ and its analogues by amplitude analysis. A scheme along these lines has been proposed by Irving (1973) in which relative spin couplings at the Δ vertex are assumed to be as in the Stodolsky–Sakurai model (Stodolsky and Sakurai, 1963), and in the quark model (see e.g. Bialas and Zalewski, 1968). As a scheme for extracting $\pi\pi$ phase shifts, this has the rather serious drawback of misrepresenting the π couplings at the $\pi N\Delta$ vertex, but this could be modified. One would then have an alternative method to straightforward extrapolation.

Another model in which the spin structure is delimited is that of Wagner (1974) whereby the number of independent spin amplitudes for $\pi N \to \rho\Delta$ is

‡ The adequacy of the isobar model in this context has been questioned (Goradia *et al.*, 1974)—cf. discussion of the status of the A_1 effect in Section 13.1(c).

reduced from 12 to 5 enabling a model amplitude analysis of $\pi N \rightarrow \pi\pi\Delta$ to be performed and the different production components to be separated.

(g) Conclusions

As we have emphasized, only the peripheral reactions (high s, low t) offer any reasonable prospect for extracting meson–meson phase shifts in an approximately model-independent fashion. For these reactions, we have a choice of somewhat different ways of doing the analysis, according to whether we do or do not extrapolate, and whether we employ the assumptions of amplitude analysis or not. We have attempted some discussion of the relative *a priori* merits of the alternative schemes. Just as important is the mutual consistency of the final results, and also their compatibility with general theoretical criteria which have not been imposed (e.g. threshold behaviour). This leads on to our next topic.

3.3. Peripheral Di-Meson Production (3): Results on $\pi\pi$ and $K\pi$ Phase Shifts

We now discuss results for $\pi\pi$ and $K\pi$ phase shifts which have been derived by the methods outlined in the preceding section. In looking at these results, one has continually to keep in mind the special problem of systematic errors arising from the necessarily rather intricate analyses. We shall examine the phenomena in domains of increasing di-meson mass, mainly for the $\pi\pi$ system, and then, in less detail, for the $K\pi$ system. There exist reasonably precise data on $\pi\pi$ and $K\pi$ angular distributions up to a di-meson mass of about 2 GeV. The information on the $\pi\pi$ channel has better statistics, and its phase shift analysis is more advanced. Detailed exploration now extends up to and beyond the region of the g-resonance.

In viewing the data and the phase shifts derived from them, we have two objectives. The first and primary one is simply to learn the results. A secondary one is to learn which methods are trustworthy for future applications. The main weight of the $\pi\pi$ evidence comes from experiments on $\pi^+\pi^-$ production, but we shall also refer to data on other final states, $\pi^\pm\pi^\pm$ and $\pi^0\pi^0$, as well as $\pi\pi$ inelastic channels, especially $K\bar{K}$. We now have a fairly consistent detailed picture of the $\pi\pi$ scattering amplitudes up to 1300 MeV. Beyond that, the lack of complete information on the inelastic channels leads to ambiguities—not in the dominant peripheral resonances ρ, f, g and the newly found h (2035) (see p. 112 below) which continue to give clear cut signals in the observed angular moments, but in the subsidiary structures. For example, a ρ' ($I = 1$) P-wave resonance of mass about 1600 MeV, decaying appreciably to $\pi\pi$ is found in some phase shift solutions but not in others.

Regarding the main resonances ρ, f, g, fairly consistent values for the parameters are now found, but there is a trend to appreciably larger widths for the ρ and f than were popular a few years ago. There is satisfactory agreement on at least some of the key parameters of the S^* resonance near the $K\bar{K}$ threshold, and one is able to move on to questions of detail. The S^* is now by far the best established candidate for a 0^{++} nonet member.

Detailed data are now beginning to accumulate on the long-argued $\pi\pi$ threshold region. Some inconsistencies and problems still remain. It is important that these be resolved, not only because we want to know the answers, but also to make sure that we really do understand how to extract $\pi\pi$ phase shifts.

Results on $K\pi$ scattering, which we review rather briefly, appear to resemble the corresponding $\pi\pi$ results very closely, both as regards the pattern of resonances, and the behaviour of the exotic channels. We end with a short comment on the related channels, $K\bar{K}$ and $\eta\pi$, which afford important information for symmetry schemes.

(a) Objectives and Problems

We begin by looking at the general picture of $\pi\pi$ phenomena below 950 MeV. In outline, things are very simple. The inelasticity is very small below the $K\bar{K}$ threshold (see later discussion), and only S-, P- and D-waves seem to be appreciably excited, the latter being small. Thus, the scattering can be described by the five parameters δ_0^0, δ_0^2, δ_1^1, δ_2^0 and δ_2^2 (recall the notation δ_J^I). The most conspicuous effect is the rho resonance in the δ_1^1 phase shift. There are also strong signals of S–P interference in both the $\pi^-\pi^0$ and $\pi^+\pi^-$ final states. The energy structure of the former effect requires a small negative $I=2$ S-wave phase, whilst the latter implies that the $I=0$ S-wave is large in the vicinity of the rho.

Establishing the detailed form of the δ_0^0 phase shift has proved the most troublesome task for low-energy $\pi\pi$ phase shift analysis. In the first place, measurements of S–P interference with a known P-wave lead to a two-fold ambiguity in the resulting S-wave phase shift. This can arise even after extrapolation to the on-shell differential cross-section. In this case, the S–P interference term essentially measures $\sin(\delta_P - 2\delta_S)$, and hence determines $|\delta_S - \frac{1}{2}\delta_P - \pi/4|$. The resulting ambiguity should in principle be resolvable from continuity and from the form of the isotropic term in the $\pi\pi$ decay angular distribution. For the $I=0$ S-wave, this has not in practice always proved decisive, because estimates of the isotropic term are liable to distortion from non-OPE contributions to rho production. Lacking a decision, one has alternative δ_0^0 solutions for each dipion mass band which coalesce somewhere in the neighbourhood of 700 MeV $[(\delta_S - \frac{1}{2}\delta_P - \pi/4) \approx 0]$, the exact position varying from analysis to analysis, but which separate below and above the rho into alternative UP and DOWN branches. This is the classical form of the "UP–DOWN ambiguity" (Schlein, 1967; Malamud and Schlein, 1967; Scharenguivel *et al.*, 1969) which led to the four possibilities UP–UP, UP–DOWN, DOWN–UP and DOWN–DOWN for the overall functional form. Additional and different ambiguities enter when phases are extracted from production intensity components in the physical region ($t < 0$). For example, Estabrooks and Martin's technique of amplitude analysis led to two distinct solutions for all the parameters of the production amplitudes including the relative S–P decay phase (cf. Fig. 3.2.2 and the discussion in the previous Section 3.2). Latterly, most analyses have eliminated the original UP–DOWN ambiguity. Firstly, they find a single solution below 700 MeV (best characterized in relation to earlier alternatives as BETWEEN). The ambiguity potentially remains above that energy, but is, as we shall see, resolved by detailed examination of the structure around the $K\bar{K}$ threshold (the S^* effect). This selects the DOWN branch for δ_0^0 from above the rho up to about 950 MeV, followed by a rapid rise associated with the S^* resonance. The qualitative form of δ_0^0 is thus fairly well established. Earlier phase shift solutions sometimes favoured a narrow ε-resonance of mass about 700 MeV, but this is now essentially excluded.

Once appreciable inelasticity sets in, as it does at about 1 GeV (see Fig. 3.3.10), the problems of phase shift analysis are considerably increased. Nonetheless, the salient features of the dynamics in this region are reasonably well established. The "leading" resonances, $f^0(1270)$, $g(1686)$, h (2035),‡ give strong signals in the mass-spectrum and appropriate Legendre moments. Likewise, the S^* (1000) effect, just through being narrow, reveals itself directly as an $I=0$ S-wave resonance strongly coupled to the $K\bar{K}$ system (see Section 3.3(f)). To establish the details even of these effects, and to uncover additional structure in the lower partial waves, a full analysis is needed.

The difficulty is that, if one just measures the elastic cross-section, the corresponding elastic amplitude is only determined in magnitude, and its phase can in principle be arbitrarily changed (where information on σ_{tot} is lacking):

$$\frac{d\sigma_{\pi\pi}}{d\cos\theta} \equiv |T_{\pi\pi}(M_{\pi\pi}^2, \cos\theta)|^2 = |T_{\pi\pi}\exp\{i\chi(M_{\pi\pi}^2, \cos\theta)\}|^2. \quad (3.3.1)$$

Among other things, this undermines resonance determinations, since structures can be manoeuvred from one partial wave to another. In practical phase shift searches, the scope for uncertainty is curtailed by a number of working assumptions (and also by the requirements of unitarity): firstly, by

‡ See Section 3.3(e) below.

restricting the order of partial waves which are allowed to contribute ($l \leq l_{max}$); secondly, by assigning Breit–Wigner forms to the leading resonances. The l_{max} cut-off ensures that $d\sigma/d\cos\theta$ can be expressed in the form

$$\frac{d\sigma}{d\cos\theta} = C^2 \prod_{i=1}^{l_{max}} (z - z_i)(z - z_i^*) = \left| C\, e^{i\chi(M_{\pi\pi}^2)} \prod_{i=1}^{l_{max}} (z - z_i) \right|^2 \qquad (3.3.2)$$

and transforms the above continuum ambiguity (as a function of z) into the discrete ambiguity ($z_i \leftrightarrow z_i^*$) emphasized by Barrelet (1972).‡ The assumption of Breit–Wigner forms for the leading resonances fixes the overall multiplicative phase $\chi(M_{\pi\pi}^2)$. This is not altogether satisfactory since it prejudges one of the questions one wants to answer.§ Even with these assumptions, one is left with a two-fold ambiguity below 1400 MeV, and a four-fold ambiguity above that point. The latter is particularly troublesome. Until it is resolved,§ one is unable to settle the form of the subsidiary resonance structure, especially in the g-region. In this context, improved information on the $\rho'(1600)$ from other sources (cf. Sections 5.3, 5.4) may well be decisive in selecting the correct solution.

(b) Data and Phase Shift Analyses for $\pi\pi$ Scattering

We now turn to details. The main source of information has been from the study of $\pi^-\pi^0$ production in $\pi^-p \to \pi^-\pi^0 p$ and of $\pi^+\pi^-$ production in $\pi^-p \to \pi^+\pi^-n$ and $\pi^+p \to \Delta^{++}\pi^+\pi^-$. We shall concentrate especially on three examples:

(i) The Saclay bubble chamber experiment on π^-p at 2·77 GeV/c (Baton *et al.*, 1970a,b); (ii) the LBL bubble chamber experiment on $\pi^+p \to \Delta^{++}\pi^+\pi^-$ at 7 GeV/c (Protopopescu *et al.*, 1973)—earlier discussion of this experiment appeared in Alston-Garnjost *et al.* (1971) and Flatté *et al.* (1972); (iii) the spark chamber experiment on $\pi^-p \to \pi^+\pi^-n$ mainly at 17 GeV/c of the CERN-Munich group already referred to in previous sections (Grayer *et al.*, 1972a,c; Hyams *et al.*, 1973; Grayer *et al.*, 1974). These three experiments afford a very fair sample of the available data from the points of view of beam energy, technique, reaction and statistics.

We begin by looking at the results of the Saclay experiment of Baton *et al.* (1970b) which had 11 000 $\pi^-\pi^0$ events and 19 000 $\pi^+\pi^-$ events. Legendre

‡ The Up–Down ambiguity referred to above, although not identical, concerns a similar transformation.

§ Another substitute for the missing information is to impose fixed-t and fixed-u analyticity (Froggatt and Petersen, 1975). Very flexible formulae for $T_{\pi\pi}(s, \cos\theta)$, embodying the required analyticity and certain mild asymptotic requirements, are fitted to the experimentally determined $d\sigma/d\cos\theta$. On applying the technique to $\pi^+\pi^-$ scattering below 1·8 GeV (and imposing the form of $T_{\pi\pi}$ up to the S^*-region) just one solution results. This solution accords with duality much better than the "unassisted" solutions to be discussed below; in particular it does possess a ρ' resonance at about 1600 MeV.

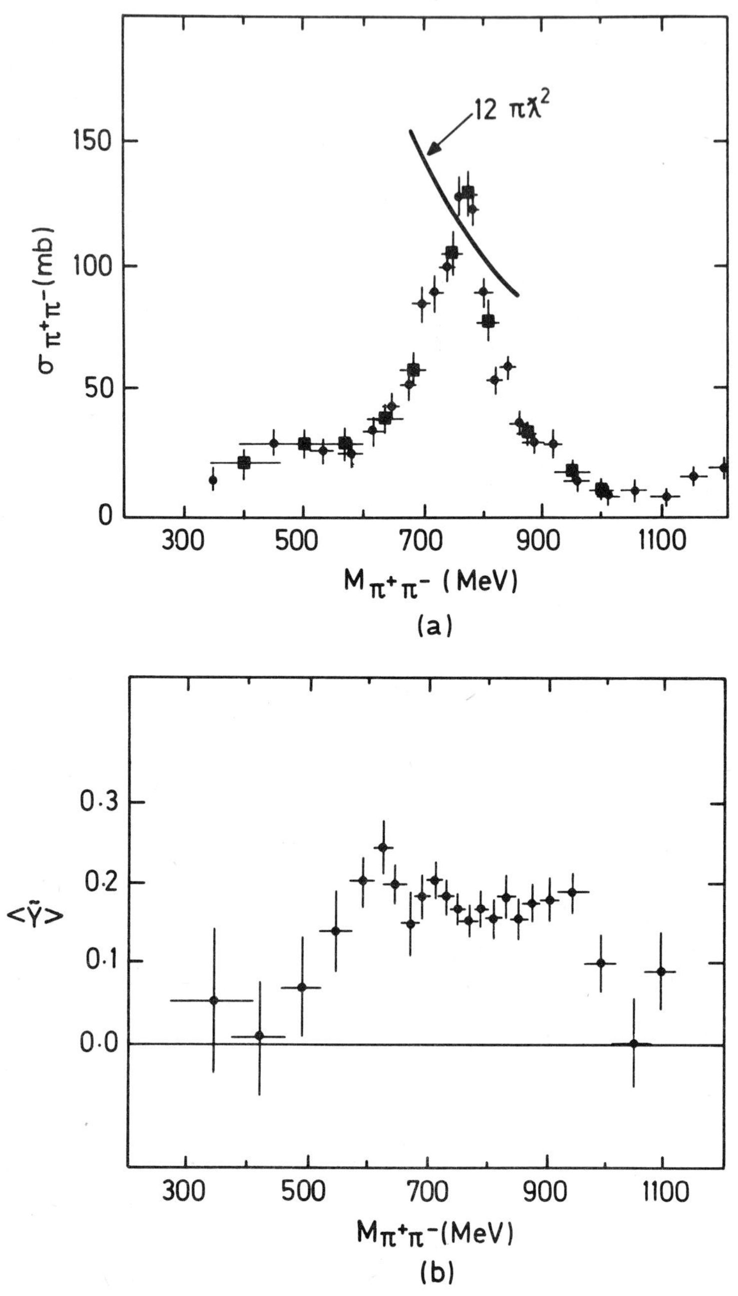

FIG. 3.3.1. Results from the Saclay bubble chamber experiment on $\pi^- p \to \pi^+ \pi^- n$ at 2·77 GeV/c (Baton *et al.*, 1970a,b): (a) Extrapolated cross-section $\sigma_{\pi^+\pi^-}$ in mb; (b) $\langle \tilde{Y} \rangle \equiv (3\pi)^{-\frac{1}{2}} \times$ forward–backward asymmetry $= \langle Y_1^0 \rangle - (\frac{7}{48})^{\frac{1}{2}} \langle Y_3^0 \rangle$.

coefficients for the on-shell $\pi\pi$ differential cross-section are extracted by an evasive extrapolation. As discussed in Section 3.2, the implicit assumption contained in this procedure is not strictly compatible with information on the form of the background (similar inconsistencies are present in the analysis of most medium-statistics experiments). Hopefully, errors from this source are not too large.

Results are presented for the on-shell Legendre coefficients and for the form of the forward–backward asymmetry, $\alpha \equiv (N_F - N_B)/(N_F + N_B)$, where $N_F(N_B)$ denotes the number of events in the forward (backward) hemisphere in the dipion rest system. For purposes of comparison with other results, it is convenient to re-express the latter result in the form

$$\langle \tilde{Y} \rangle \equiv (3\pi)^{-\frac{1}{2}} \alpha_{+-} = \langle Y_1^0 \rangle - (\tfrac{7}{48})^{\frac{1}{2}} \langle Y_3^0 \rangle.$$

The result is shown in Fig. 3.3.1, together with the extrapolated $\pi^+\pi^-$ cross-section.

As a second example, Fig. 3.3.2(a),(b) illustrates the results of the LBL group (Protopopescu *et al.*, 1973) from a study of $\pi^+ p \to \Delta^{++}\pi^+\pi^-$ at 7 GeV/c (27 000 events). These authors show their results both in the physical region for $0 < t_{\min} - t \leq 0{\cdot}1$ GeV^2, and also extrapolated to the pion pole ($t = \mu^2$). The extrapolation to determine the integrated elastic cross-section for on-shell $\pi^+\pi^-$ scattering is performed in quite an elaborate way, using separate Dürr–Pilkuhn form factors for the individual partial waves (for details see the original reference); the on-shell $\langle Y_l^0 \rangle$ $l = 1, \ldots 6$) moments, are obtained from a simple linear extrapolation. Figure 3.3.2(a),(b) shows the results for the $\langle Y_1^0 \rangle$ moment and illustrates the general tendency of the on-shell moments to come out numerically larger than the corresponding off-shell ones (the latter "diluted" by additional non-OPE background contributions). The on-shell moments naturally have larger errors. Comparison of LBL's $\langle Y_1^0 \rangle$ on-shell (Fig. 3.3.2(a)) with Saclay's result for $(3\pi)^{-\frac{1}{2}}\alpha_{+-}$ (Fig. 3.3.1(b)) is generally very satisfactory, especially remembering that they are obtained from different reactions and at different energies. Figure 3.3.2(c) shows the analogous result from the CERN-Munich experiment (Hyams *et al.*, 1973; Grayer *et al.*, 1974). The quantity plotted is $\langle Y_1^0 \rangle$ in the t-channel averaged over the interval $|t_{\min} - t| < 0{\cdot}15$ GeV^2 (see also Fig. 3.2.3 of the preceding section). Comparing with the result of the Δ-reaction, Fig. 3.3.2(b), one sees that, despite the different reaction mechanisms, the small-t off-shell moments are quite similar.

The on-shell $\langle Y_l^0 \rangle$ moments are in a sense the data on which the $\pi\pi$ phase shift analysis has to operate—for the LBL and Saclay analyses literally so. There are small but significant differences in the "data" from the various experiments, and this is one source of disagreement among the results but, as

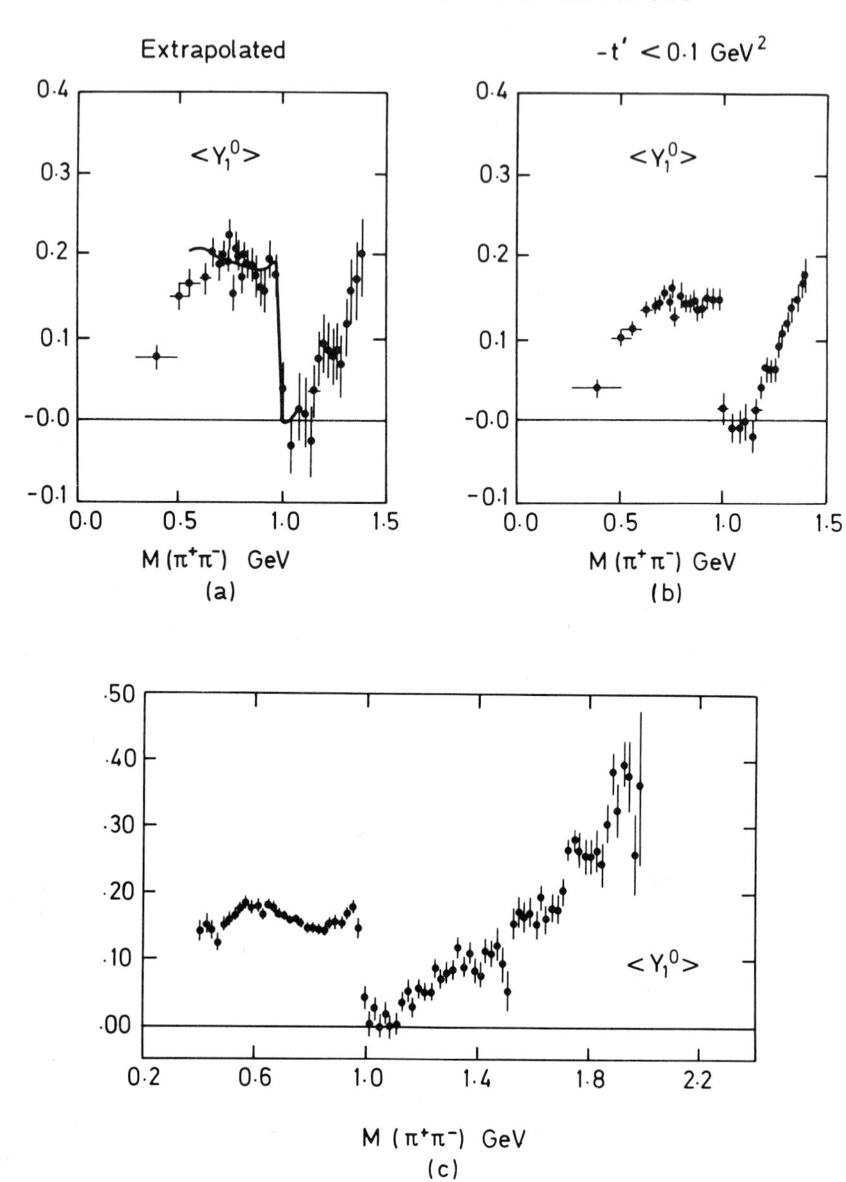

FIG. 3.3.2. Results on the Legendre coefficient $\langle Y_1^0 \rangle$ for $\pi^+\pi^-$ production evaluated in the t-channel frame from the LBL experiment of Protopopescu *et al.* (1973) and the CERN-Munich experiment of Hyams *et al.* (1973): (a) LBL's results extrapolated to the pion pole; (b) LBL's results in the physical region ($t' = |t - t_{\min}| < 0{\cdot}1$ GeV^2); (c) the CERN-Munich results for $|t'| < 0{\cdot}15$ GeV^2.

we shall see, this is not the only source of uncertainty, and a comparable dispersion can arise from applying different methods to the same data.

We now go on to discuss output $\pi\pi$ phase shifts from these and other experiments. Baton *et al.* have fitted $\pi\pi$ phase shifts to their Legendre coefficients. They performed two analyses, the first with no inelasticity allowed, and the second including inelasticity (both are energy-independent fits, i.e. with each mass bin fitted separately). In view of the evidence against appreciable inelasticity below 950 MeV (see later discussion and Fig. 3.3.10) we shall only refer to the former results.

Protopopescu *et al.* (1973) perform an energy-dependent analysis, fitting parametric forms for the on-shell partial wave amplitudes to the extrapolated cross-section and $\langle Y_l^0 \rangle$ moments up to $l = 6$ (e.g. to $\langle Y_1^0 \rangle$ of Fig. 3.3.2(a).

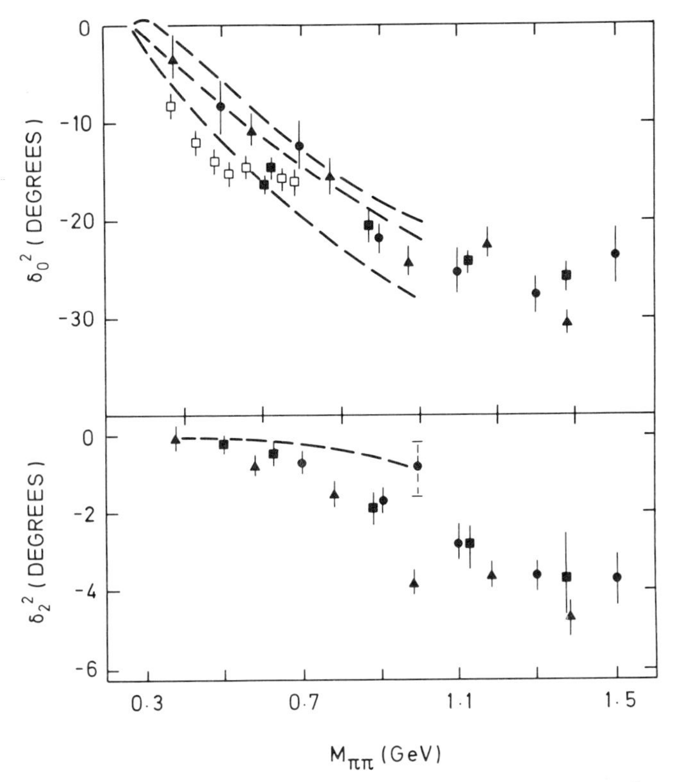

FIG. 3.3.3. Some determinations of the $I = 2$ S- and D-wave $\pi\pi$ phase shifts: (■) Hoogland *et al.* (1974); (▲) Cohen *et al.* (1973); (●) Durusoy *et al.* (1973); (□) Prukop *et al.* (1974). The dashed curves correspond to three specimen theoretical solutions of Basdevant *et al.* (1973, 1974).

For the $I=1$ P-wave and F-wave, and the $I=0$ D-wave, they use standard Breit–Wigner forms plus background, with inelasticity allowed above the $\pi\omega$ threshold. Information taken from other experiments on the $I=2$ S- and D-wave phase shifts, assumed elastic throughout, is supplied in the form of simple (odd) polynomials in q (the presently available data are shown in Fig. 3.3.3). Finally, for the $I=0$ S-wave, they parameterize in terms of a 2×2 inverse K-matrix, to allow for coupling to the $K\bar{K}$ channel. We discuss this aspect (which is directed to analysing the S^* effect) in more detail below. After performing their energy-dependent fit, which as we shall see yields for δ_0^0 a unique Between–Down type of solution up to 900 MeV, followed by a rapid rise, they also make an energy-independent analysis between 500 and 930 MeV in order to probe further their resolution of the Up–Down ambiguity for δ_0^0.

There have been several distinct analyses of the CERN–Munich data (Grayer *et al.*, 1974) to which we assign code-names (Table 3.3.1) for ease of reference. Prior to the 1974 Meson Spectroscopy Conference, one had the three solutions (CCL), (CEM_1) and (COW_1) derived by the alternative methods described in Section 3.2 and all using the 17 GeV/c data. COW_1 is primarily an energy dependent fit, but an energy independent variant of it was also given (the energy dependent results were used as starting values from which the final fit was not allowed to deviate substantially). Subsequent development have been along three lines. Firstly, Estabrooks and Martin (1974a) have performed a careful and extensive re-analysis of the 17 GeV/c production data for $M_{\pi\pi}$ between 500 and 1000 MeV. This analysis (CEM_2) follows the general approach of CEM_1 but imposes theoretical D-waves. Secondly, Männer (1974) has reported phase shifts at much lower masses then heretofore ($M_{\pi\pi}=320$ (80) 560 MeV) (CLM), based on data taken in the CERN-Munich spectrometer with a 7 GeV/c π^- beam. Thirdly, Männer (1974) has reported the results of more extensive phase-shift searches over the $\pi\pi$ mass range 1 to 1·8 GeV (COW_2), and Estabrooks and Martin (1974b) have presented results of a similar investigation (COW_3). Both analyses find the four-fold ambiguity in the phase-shift solutions already alluded to. The new analyses COW_2, COW_3 and CLM are all based on the same assumptions on spin-structure and t-dependence as COW_1, from which however they differ in being properly energy independent. It will be important to see how these assumptions are upheld when CERN-Munich's production experiment using a polarized target is complete.

(c) Results on I = 2 $\pi\pi$ Phase Shifts

We now examine the results from these and other analyses, discussing effects in order of increasing complication, so as progressively to build up a picture

TABLE 3.3.1. Alternative Phase Shift Analyses of CERN-Munich $\pi^+\pi^-$ Data

Reference	Method	Scope (range in MeV, energy dependent (D) or independent (I))	Code-name for present discussion
Grayer *et al.* (1972a)	Non-evasive Chew–Low extrapolation of $m=0$ t-channel moments	500–1500 (I)	CCL
Estabrooks *et al.* (1973)	Estabrooks and Martin's method—model amplitude analysis followed by extrapolation of s-channel moments	440–1400 (I)	CEM_1
Hyams *et al.* (1973) (cf. also Grayer *et al.* (1974)	Ochs and Wagner's method—fitting t-channel moments, averaged over small t, to simple variant of absorption model	600–1900 (D) (also "I" constrained to resemble "D")	COW_1
Estabrooks and Martin (1974a)	As for CEM_1 but imposing "theoretical" D-waves	500–1000 (I)	CEM_2
Männer (1974)	Ochs and Wagner's method applied to 7 GeV data yielding phase shifts at low masses	320 (80) 560 (I)	CLM
Männer (1974) Estabrooks and Martin (1974b)	Re-analysis of 17 GeV/c data by Ochs and Wagners' method	1000–1800 (I)	COW_2 COW_3

of the various phase shifts.‡ We begin with the $I=2$ phase shifts, which can be studied in isolation from the $I=0$ channel. One method is to study the interference moments, $\langle Y_1^0\rangle$, $\langle Y_3^0\rangle$, etc., in $\pi^-\pi^0$ production (Baton *et al.*, 1967a,b, 1970a,b). An alternative is to aim directly at measurements of the $I=2$ cross-section from a study of $\pi^-\pi^-$ or $\pi^+\pi^+$ systems. This has been done using the reactions $\pi^-d \to pp\pi^-\pi^-$ (Cohen *et al.*, 1973; Durusoy *et al.*, 1973), $\pi^-p \to \Delta^{++}\pi^-\pi^-$ (Colton *et al.*, 1971; Losty *et al.*, 1974) and $\pi^+p \to n\pi^+\pi^+$ (Hoogland *et al.*, 1974; Prukop *et al.*, 1974). Where the $\pi^+\pi^+$ or $\pi^-\pi^-$ state is observed, the *sign* of the inferred S-wave phase shift is taken from the $\pi^-\pi^0$ experiments. The main trends, which were already evident in the earlier experiments, are reinforced in the most recent ones. Examples of the resulting δ_0^2 and δ_2^2 values are plotted in Fig. 3.3.3 (a fuller compilation including earlier results can be found in Hoogland *et al.* (1974)). The S-wave phase shift δ_0^2 is seen to fall smoothly from threshold out to at least 1 GeV. Such a behaviour, with essentially one parameter undetermined, is predicted from phenomenological analyses based on analyticity and crossing using information on the $I=0$ and $I=1$ phase shifts over the regions adjoining the rho (see Chapter 10, especially Fig. 10.3.4(b)). This is illustrated by the dashed curves on Fig. 3.3.3 which correspond to alternative specimen solutions from Basdevant *et al.* (1973, 1974). The data shown in Fig. 3.3.3 indicate a value at the rho-mass in the range $\approx -17° \pm 4°$. The recent experiment of Hoogland *et al.* (1974) obtains phase shifts between 625 and 1375 MeV which approximately conform to $\delta_0^2 \approx -0{\cdot}12\ (q/u)$ radians.

Collectively, the data do not point to a precise range of values for the $I=2$ S-wave scattering length a_0^2 ($\equiv \mathrm{Lim}_{q\to 0}\, \delta_0^2/q$), although they all indicate that it is small. The difficulty is sheer lack of statistics. Based on their 5 GeV/c experiment on $\pi^+p \to \pi^+\pi^+n$, the Notre-Dame ANC group (Prukop *et al.*, 1974) quote the result

$$a_0^2 = -0{\cdot}16^{+0{\cdot}02}_{-0{\cdot}03}.$$

However, there is an appreciable spread in output phases according to what form of t-dependence is assumed. The scattering length can be estimated from the phase shifts at higher masses, say above 600 MeV, if δ_0^2 is assumed to extrapolate in the manner of the phenomenological solutions based on the Roy equations (i.e. like the dashed curves of Fig. 3.3.3). In this way, and

‡ The results we discuss are based primarily on the analysis of single reactions. An alternative is to attempt a simultaneous fit to a wide range of $\pi\pi$ data, comprehending the various charge channels. Such an analysis has been performed by a group at Saclay (Villet *et al.*, 1973) using their own data on $\pi^0\pi^0$ production at 2 GeV/c and a representative sample of the available data on other channels. While this is an interesting approach, an objection to doing it is the lack of understanding of how systematic errors vary from channel to channel. Thus there could be a danger that spurious effects may be emphasized.

using the phase shifts of Hoogland *et al.* (1974), Morgan (1974c) made the estimate

$$a_0^2 = -0{\cdot}075 \pm 0{\cdot}05.$$

This, along with the related $I=0$ quantity a_0^0, and the P-wave scattering length a_1^1 to be discussed below, is one of the key parameters characterizing the threshold region.

Experimental findings for δ_2^2, the $I=2$ D-wave, are small and negative (Fig. 3.3.3), qualitatively in agreement with the predictions of the phenomenological analyses (again see Chapter 10) although in magnitude somewhat larger. The D-wave phases of Hoogland *et al.* between 625 and 1375 MeV are well represented by the formula $\delta_2^2 \approx -0{\cdot}003\ (\nu+1)^{3/2}/\nu^{5/2}$ radians ($\sqrt{\nu} \equiv q/\mu$).

At about 1 GeV, inelasticity enters as a complicating factor. Cohen *et al.* (1973) estimate $\sigma_{\text{inel}} \approx 1{\cdot}2$ mb compared to $\sigma_{\text{el}} \approx 3{\cdot}7$ mb at 1400 MeV. Some of the phase-shift solutions which we have quoted use this estimate; Hoogland *et al.*, simply proceed by assuming $\eta_2^2 = 1$ and fit for η_0^2. One would like to have more information, since in all cases $\sigma_{\text{tot}}^{I=2}$ comes out rather small ($\approx$5–7 mb) compared to the "plateau" value one would expect assuming factorization (see Appendix C). As we shall see, a similar effect is found in $I=\frac{3}{2}$ πK scattering.

(d) Results on I = 0 and 1 $\pi\pi$ Phase Shifts below 950 MeV

Having settled the form of the $I=2$ phase shifts, we now proceed to discuss their non-exotic counterparts, firstly, for the region below 950 MeV.

Figure (3.3.4) shows results on the P-wave phase shift, δ_1^1, and the $I=0$ D-wave δ_2^0 below the f-region. The P-wave results are in satisfactory overall agreement, although some spread is revealed on close examination. One aspect of this is brought out by listing resonance parameters which are found for the rho. If we define M_ρ as the 90° point and Γ_ρ in terms of the slope of cot δ_1^1 at resonance (i.e. $\Gamma_\rho = -2((d/dM_{\pi\pi})(\cot\delta_1^1)^{-1})$, then values are found as listed in Table 3.3.2. The values given in the current Particle Data Group tables are entered for comparison. We also list values (where available) arising from alternative definitions for the mass and width. One possibility is to quote the position of the resonance pole,‡ $M_R = M_0 - i\Gamma_0/2$. Alternatively, one can define the upper and lower half-widths from the positions of the 135° and 45° points, $\delta(M+\Gamma_u/2) = 135°$, $\delta(M-\Gamma_l/2) = 45°$, and hence form $\Gamma\,(45°\text{-}135°) = \frac{1}{2}(\Gamma_u + \Gamma_l)$. Other variants are possible, and one has to be careful in comparing results from different analysis that the same definitions are being used.

‡ This choice has been shown to be stable against changes in parameterization for the case of the Δ (1236) (Ball *et al.*, 1973).

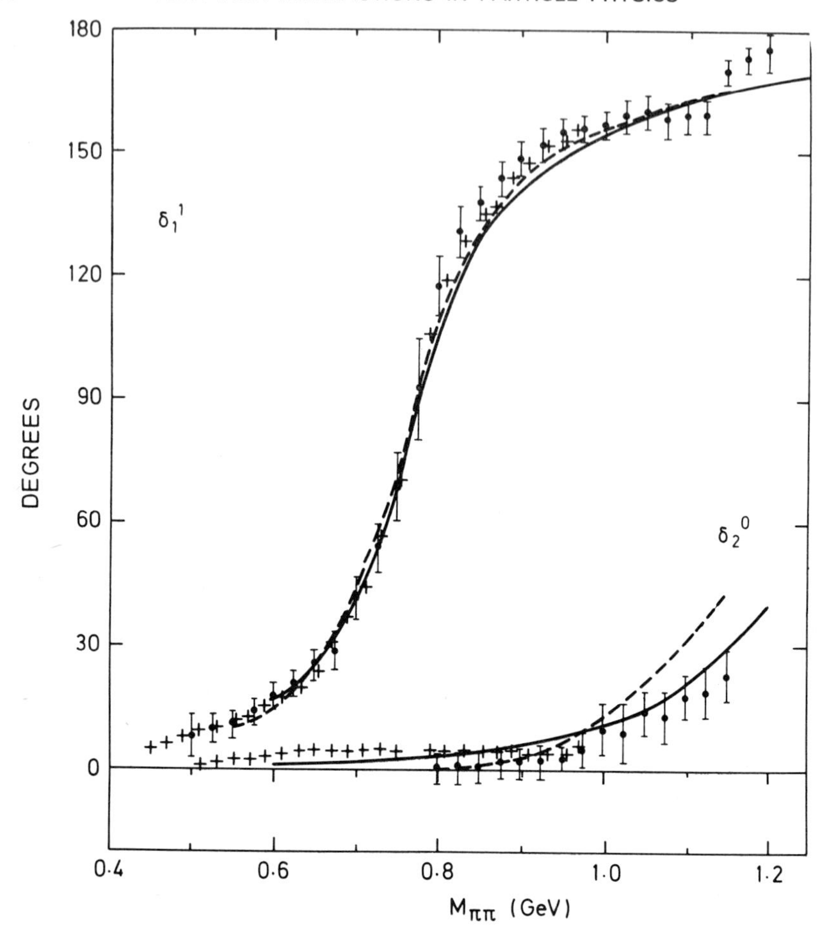

FIG. 3.3.4. Some results on the P-wave and $I=0$ D-wave $\pi\pi$ phase shifts δ_1^1 and δ_2^0: (●) Baton *et al.* (1970b); (+) Estabrooks *et al.* (1973) (CEM$_1$); (– – –) Protopopescu *et al.* (1973) (LBL); (——) Hyams *et al.* (1973) (COW$_1$).

As Table 3.3.2 shows, the spread in the rho-parameters is not inconsiderable; the Particle Data Group values give a fair impression of the recent determinations. Somewhat smaller values for the width, $\Gamma_\rho \approx 120$ MeV, were popular a few years ago. Shifting to present day values involves revising all rho contributions to sum rules upwards by 25%.

A rather sensitive indicator of how the phase shift behaves away from the resonance peak is to form the quantity‡ (cf. Morgan, 1974a)

$$a_{11}(\nu) = \left(\frac{\nu+1}{\nu^3}\right)^{\frac{1}{2}} (1-\nu/\nu_\rho) \tan \delta_1^1$$

‡ The near constancy of $a_{11}(\nu)$ was first emphasized by Olsson (1967).

TABLE 3.3.2. Resonance Parameters for the Rho

Analysis	M_ρ (MeV)	Γ_ρ (MeV)	M_0 (MeV)	Γ_0 (MeV)	Γ(45°-135°) MeV
(a) Saclay (Baton *et al.*, 1970b)	766	133	760	131	127
(b) LBL (Protopopescu *et al.*, 1973)	772	160			
(c) CERN-Munich (CEM_2) (Estabrooks and Martin, 1974a)	770 ± 9	143 ± 13			
(d) CERN-Munich (COW_1) (Hyams *et al.*, 1973)	777	155	778	152	
(e) Particle Data Group (1974)	770 ± 10	150 ± 10			

whose value at ν_ρ is proportional to Γ_ρ and at threshold gives the P-wave scattering length

$$a_{11}(\nu) \underset{\nu \to 0}{\to} a_1^1.$$

Figure (3.3.5) shows plots of $a_{11}(\nu)$ derived from the LBL and CERN-Munich (CEM_2, COW_1 and CLM) analyses compared to the results from the phenomenological analysis using the Roy equations of Basdevant *et al.* (1974). (This incorporates analyticity and crossing and will be discussed fully in Chapter 10.) The theoretical curves correspond to a wide, essentially exhaustive, range of possibilities for the $I=0$ S-wave ($-0{\cdot}05 \leq a_0^0 \leq 0{\cdot}6$ for the corresponding scattering length—see below), and are all seen to give rather a smooth behaviour for $a_{11}(\nu)$. Other analyses confirm this. In contrast, the experimental points from the CEM_2 and CLM analyses, especially the latter which extends down to 320 MeV, are seen to curve upwards very sharply at small ν. It would appear to be very difficult to achieve such an effect within the context of fixed-t dispersion relations and the associated Roy equations. (For a fuller discussion of this point, see Section 10.3.‡) Therefore, one is led to suspect a systematic bias in the δ_1^1 results below say 600 MeV. Although the disputed phase shifts are rather small in magnitude, so that a change of a few degrees is sufficient to remove the disagreement in Fig. 3.3.5, the change is nonetheless important, since δ_0^0 is essentially determined from S–P interference. We shall discuss the implications of this below.

The experimental $I=0$ D-waves (Fig. 3.3.4) qualitatively confirm the predictions of theoretical analysis based on crossing and analyticity in finding δ_2^0 small and positive, as if given by the tail of the f^0. Quantitatively,

‡ See also Basdevant *et al.* (1975).

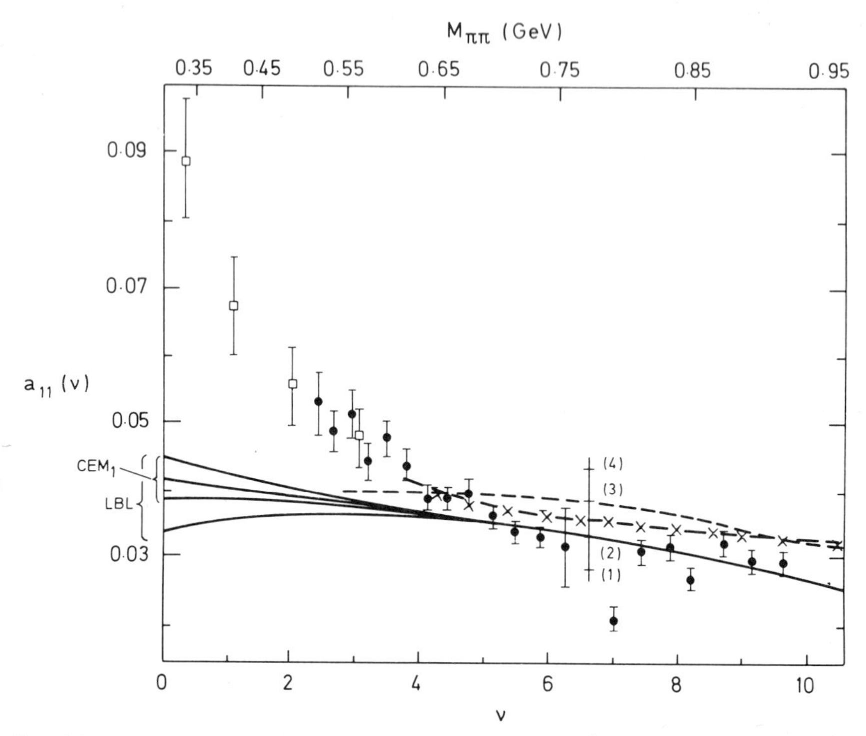

FIG. 3.3.5. Comparison, using the function $a_{11}(\nu) \equiv [(\nu+1)/\nu^3]^{\frac{1}{2}}(1-\nu/\nu_\rho)\tan\delta_1^1$, of empirical P-wave $\pi\pi$ phase-shifts, (□) Männer (1974) (CLM); (●) Estabrooks and Martin (1974a) (CEM_2); (– – –) Protopopescu *et al.* (1973) (LBL) and (–×–×) Hyams *et al.* (1973) (COW_1), with model predictions based on dispersion relations and fitting to empirical $I=0$ S-wave phase-shifts (LBL and CEM_1) above 500 MeV, from Basdevant *et al.* (1974) (——); the corresponding ranges of values for the P-wave scattering length $a_1^1 \equiv a_{11}(0)$ are shown beside the vertical axis. The values of $a_{11}(\nu_\rho)$ corresponding to $\Gamma_\rho = 120$, 140, 160 and 180 MeV are indicated by (1), (2), (3) and (4).

there are similar anomalies to those discussed above for the P-wave in the $I=0$ D-wave phase shifts from the CERN-Munich (CEM_1) analysis. Below say 900 MeV, the experimental values exceed the theoretical estimates by a factor which by 600 MeV is of the order of three, again suggesting some anomaly in the data or in the analyses. Taken overall, the data from the various analyses are in fair agreement with theoretical estimates. Basdevant *et al.* (1974) estimate $\delta_2^0 = 0{\cdot}07°$, $1{\cdot}3°$, $3{\cdot}5°$, $11°$ for $M_{\pi\pi} = 400$, 600, 800, 1000 MeV. In their new analysis (CEM_2), Estabrooks and Martin use the above theoretical estimates for the D-waves and only use $\langle Y_l^m \rangle$ moments with $l \leqslant 2$.

We now turn to the $I = 0$ S-wave phase shift, δ_0^0. As mentioned earlier, this quantity has had a troubled history beset with ambiguities. Figure 3.3.6(a) illustrates the situation prior to the most recent determinations. As is evident, a considerable range of possibilities is encompassed. Saclay (Baton *et al.*, 1970b) and SLAC (Baillon *et al.*, 1972) found "Up" and "Down" branches above 700 MeV—the classical "Up–Down ambiguity"—although at their highest energy, SLAC only found one solution, lying below their "Up" branch. This already suggested that the true behaviour might be to follow the "Down" branch, followed by an upward swing. Likewise, for the Saclay analysis, the falling phase shift above 900 MeV looked unphysical, and again suggested the possibility of a switch of branches between 900 and 1000 MeV.

Such a behaviour was rather conclusively demonstrated in a careful analysis of the behaviour of the experimental $\langle Y_l^0 \rangle$ moments in the neighbourhood of the $K\bar{K}$ threshold as found in the LBL experiment (Alston-Garnjost *et al.*, 1971; Protopopescu *et al.*, 1973), and has been upheld in the analyses of the CERN-Munich experiment. One clear signal of this is the behaviour of the interference moment (Fig. 3.3.2(c)) which *rises* to a peak at around 950 MeV, before falling sharply, indicating that δ_0^0 is rising up through the value of δ_1^1. We defer a detailed discussion of how the rapid changes in this vicinity are associated with the S^*-resonance. For the moment, the crucial point is the selection of the Down branch of δ_0^0 from 700 to 900 MeV.

This leaves the question of how δ_0^0 behaves from threshold up to this region. As Fig. 3.3.6(a) indicates, there was a fair range of possibilities even among the solutions plotted.‡

The new determinations of δ_0^0 (CEM_2, CLM) reported at the 1974 Meson Spectroscopy Conference are shown along with the previous LBL and CERN-Munich (COW_1) energy-dependent phases in Fig. 3.3.6(b). Estabrooks and Martin present their findings (CEM_2) in the form of bands of possibility, having regard to alternative procedures and assumptions which they consider. Provided the Down branch is followed, as seems to be definitely required from the form of the S^* (1000) effect (see below), the new phases are in good agreement with LBL and COW_1. We appear at last to

‡ The CEM_1 analysis, by the method of Estabrooks and Martin, has a second solution for δ_0^0 and δ_1^1 in addition to the one plotted in Figs 3.3.6(a) and (3.3.4). As discussed in Section 3.2, this ambiguity stems from the amplitude analysis, and is not exactly equivalent to the Up–Down ambiguity, although the results are somewhat similar. The alternative δ_0^0 has a Down–Up type of behaviour, and the authors (Estabrooks *et al.*, 1973) reject this on comparison with results on $\pi^0\pi^0$ production (see below). An additional ground for rejecting this "solution 2" is that the associated P-wave phases below the rho are even larger than those of "solution 1", so that the anomaly referred to above is even more accentuated. For these reasons, we omit this solution for δ_0^0 from Fig. 3.3.6(a).

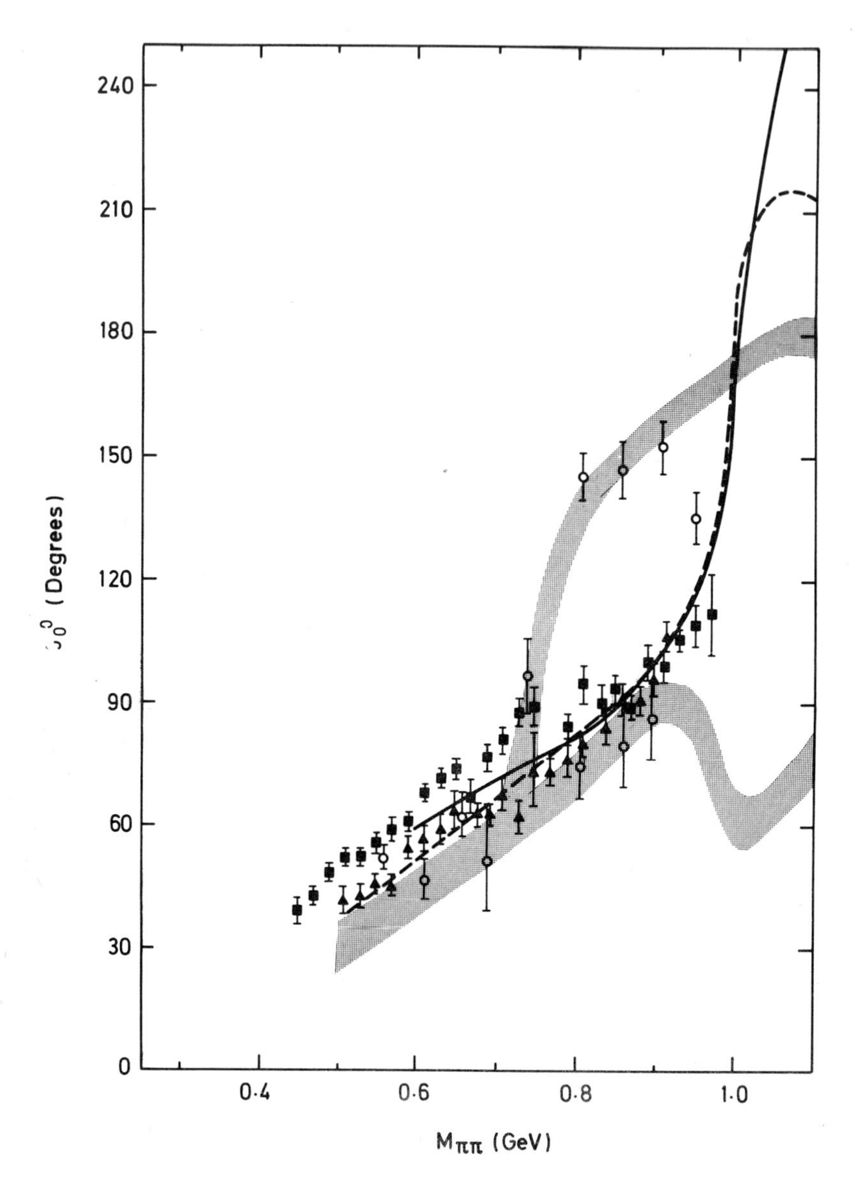

FIG. 3.3.6. Findings for the $I=0$ S-wave $\pi\pi$ phase shift, δ_0^0, below 1 GeV: (a) the situation before 1974: (——) Hyams *et al.* (1973) (COW_1); (– – –) Protopopescu *et al.* (1973) (LBL); (○) Baillon *et al.* (1972) (SLAC); (■) Estabrooks *et al.* (1973) (CEM_1); (▲) Grayer *et al.* (1972a) (CCL); (▒) Baton *et al.* (1970b); (b) 1974—results: Bands of possibility found by Estabrooks and Martin (1974a) (solid curves), and the low-mass points of Männer (1974) (CLM) (●) compared to previous LBL (– – –) and COW_1 (– · –) results.

have an agreed structure for δ_0^0 from 1 GeV down to say 600 MeV. Even below that point, Männer's low-mass results (CLM) appear to join on well to the new phases above 500 MeV; however, the previously noted difficulty with the related P-wave, δ_1^1, raises a doubt to which we return in discussing the threshold region (see below).

Additional information on the $I=0$ phase-shifts comes from the study of $\pi^0\pi^0$ production, via the reaction $\pi^- p \to \pi^0\pi^0 n$. Odd partial waves are forbidden by Bose statistics, so that if D and higher waves can be neglected, the on-shell charge exchange cross-section is given by the simple formula

$$\sigma(\pi^+\pi^- \to \pi^0\pi^0) = \frac{8}{9}\frac{\pi}{q_{\text{in}}^2}\sin^2(\delta_0^0 - \delta_0^2).$$

Given a sufficiently precise and reliable experiment, one would have a direct measure of the phase difference $(\delta_0^0 - \delta_0^2)$, thus determining δ_0^0 up to the uncertainty with which δ_0^2 is known, i.e. to within a few degrees. Despite the practical difficulties in studying this reaction (the need to detect two π^0's

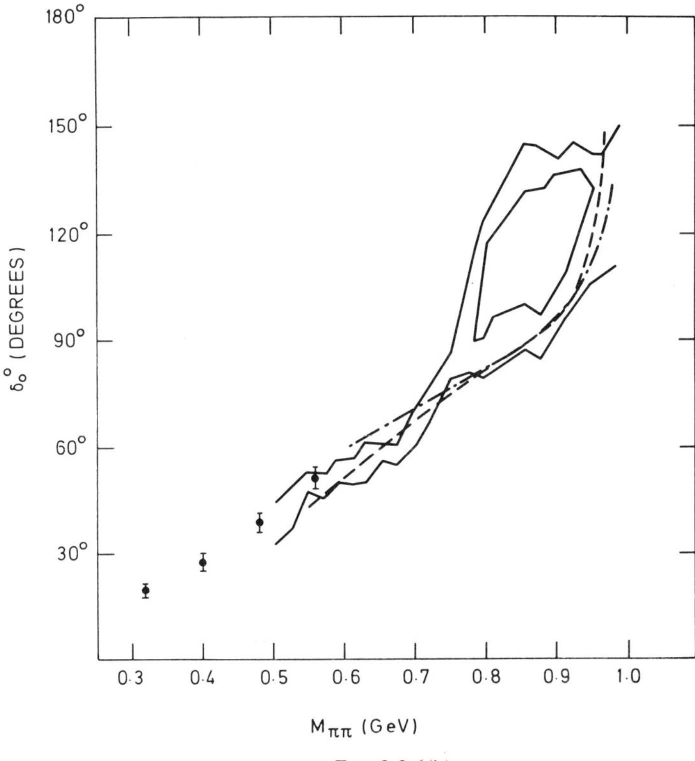

FIG. 3.3.6(b).

reliably), a number of experiments have been performed. No doubt, because of the difficulties, there is a fair measure of disagreement as to the detailed form of the $\pi^0\pi^0$ mass-spectrum, some experiments even finding support for a narrow ε, corresponding to the "Up" solution for δ_0^0. However, of recent experiments, most have supported, at least qualitatively, the Down type of form for δ_0^0. As an example, Fig. 3.3.7 shows the $\pi^0\pi^0$ mass spectrum obtained by the Karlsruhe–Pisa Group (Apel *et al.*, 1972) from a study of $\pi^- p \to \pi^0\pi^0 n$ at 8 GeV/c. Between 600 and 1100 MeV, the spectrum shape agrees well with that predicted from the LBL phase shifts (indicated by the full line). Unfortunately, owing to the incomplete angular acceptance, the relative normalization is uncertain above 1100 MeV so that one cannot calibrate the spectrum lower down from the height of the f^0 peak. Below 600 MeV, the observed spectrum is seen to lie above the continuation of the

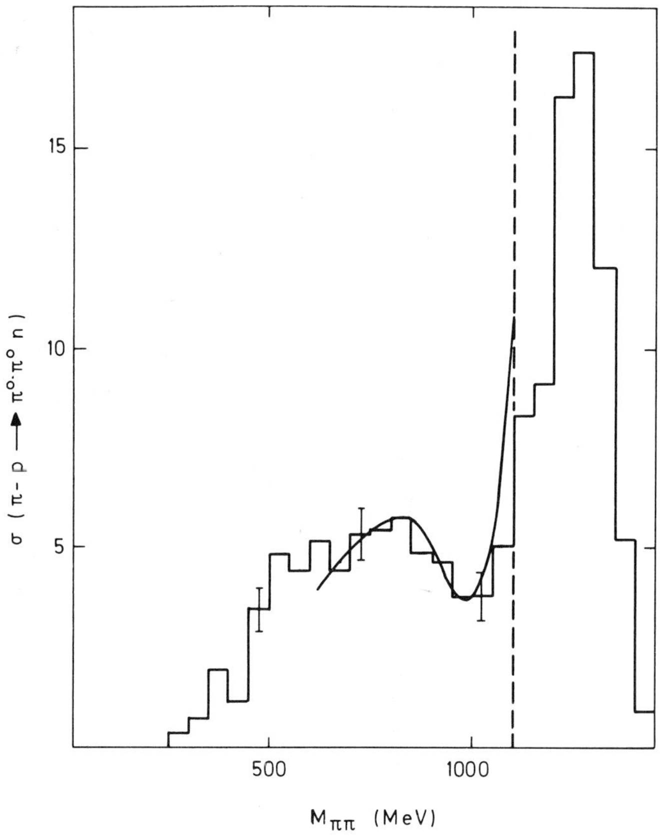

FIG. 3.3.7. Example of the $\pi^0\pi^0$ mass spectrum, from the experiment of Apel *et al.* (1972) on $\pi^- p \to \pi^0\pi^0 n$ at 8 GeV/c. The solid curve corresponds to the LBL phase shifts of Protopopescu *et al.* (1973).

"LBL" curve. This trend is shown by a number of $\pi^0\pi^0$ experiments, all of which would favour a very large value $\gtrsim 0{\cdot}6$ for the S-wave scattering length a_0^0.

(i) *Threshold Parameters*

This brings us back to the question of how the phase shifts behave in the threshold region. Such information is of particular interest in view of the predictions of current algebra, which yield specific values for the threshold parameters, e.g. $a_0^0 \approx 0{\cdot}20 \pm 0{\cdot}04$, $a_0^2 \approx -0{\cdot}045 \pm 0{\cdot}005$, $a_1^1 \approx 0{\cdot}033 \pm 0{\cdot}003$ (see Chapter 14). As we have seen, the new 7 GeV/c CERN-Munich data (CLM) (Männer, 1974) offers rather precise information on δ_1^1 and δ_0^0 down to very low masses, with the new determinations joining on well to the previous ones at higher masses. Against this there is the serious disagreement with the dispersion relation prediction for the P-waves noted earlier (Fig. 3.3.5). This casts doubt not only on the P-waves but also on the low-energy S-waves, which are coupled to the P-waves in the observed experimental quantities. Männer's results imply a P-wave scattering length $a_1^1 \approx 0{\cdot}1$, whilst the theoretical curves indicate a value in the range $a_1^1 \approx 0{\cdot}037 \pm 0{\cdot}005$. For the S-wave scattering length, a_0^0, the new phase shifts (Männer, 1974) imply a rather precise value:

$$a_0^0 = 0{\cdot}44 \pm 0{\cdot}1.$$

As noted earlier, the dispersion relation calculations (Chapter 10) admit a wide range for this parameter, say $-0{\cdot}1 \leq a_0^0 \leq 0{\cdot}6$; they do however predict a strong correlation between a_0^0 and its $I = 2$ counterpart, a_0^2, but, as we have seen (Section 3.3(c)), the latter quantity is not very well determined.

An opportunity for detailed cross-checking will be furnished by a new "Omega" experiment on $\pi^- p \to \pi^+\pi^- n$ at 3·2 GeV/c (Lemoigne *et al.*, 1974). There are preliminary results for $M_{\pi^+\pi^-} = 380$, 340 and 300 MeV from which, assuming a negligible P-wave contribution, the authors of the experiment infer the following value for the appropriate combination of S-wave scattering lengths:

$$|a_0^0 + \tfrac{1}{2} a_0^2| = 0{\cdot}26 \pm 0{\cdot}05.$$

In Chapter 6, we shall compare these two recent determinations with values inferred from analysis of other processes.

(e) $\pi\pi$ Phase Shifts above 950 MeV‡

We have already outlined the problems which beset phase-shift analysis once inelasticity becomes appreciable. We now look at the detailed consequences.

‡ Detailed phenomena adjoining the $K\bar{K}$ threshold, especially the S^* (1000) effect are discussed in the following section (3.3(f)).

The energy region from 950 to 1800 MeV is dominated by the f and g resonances. A prime objective is to discover the full pattern of subsidiary resonances accompanying them. Some dual models (e.g. the Lovelace–Veneziano model—see Chapter 12) suggest that the "leading" sequence of particles ρ, f^0, g should be accompanied by towers of daughter resonances of approximately degenerate mass and descending angular momentum. In particular, a ρ' ($J=1$), and possibly also an ε' ($J=0$) resonance, under the f^0 would be predicted in certain models; likewise, the g would have 2^+ and perhaps also 1^- and 0^+ companions. Over the f-region, the phenomena are, as we shall see, sufficiently simple and well understood for a clear-cut answer to be given. There is no evidence for a ρ' in the vicinity of the f^0 coupling appreciably to $\pi\pi$. There is however considerable evidence for an ε', a fairly broad elastic $I=0$ S-wave resonance situated at about the f-mass.‡

A similar pattern with a daughter two units of angular momentum below its parent was found to obtain in the g-region (Hyams *et al.*, 1973). This result emerged from the energy-dependent analysis of the CERN-Munich data (COW_1 of Table 3.3.1) which found a ρ' ($J=1$) resonance of mass 1590 and width 180 MeV. Despite some inconsistencies of detail (see below), it was natural to associate this with the electromagnetically produced object of mass about 1500–1600 MeV which is seen to decay to four pions (see Sections 5.3 and 5.4). Subsequent energy independent analyses of the same data (COW_2, COW_3) find a four-fold ambiguity in the solutions above 1400 MeV, some with and some without a ρ' $(1600)\rightarrow\pi\pi$ in the P-wave amplitude. The status of this as of all the subsidiary waves, is thus uncertain, a situation which will surely persist until there are good data on the other related channels, i.e. on $\pi^-\pi^0$, $\pi^0\pi^0$ and the inelastic modes.

As we have seen, the existence of alternative phase-shift solutions can be identified with the Barrelet ambiguity, whereby an amplitude zero, z_i, can be replaced by its complex conjugate, z_i^*.§ In practice, assumptions of continuity and the observed dominance by successive peripheral resonances (ρ, f, g) restricts the scope of the discrete ambiguities. As higher waves come into play, successive zero trajectories $\{z_i : T(M^2_{\pi\pi}, \cos\theta = z_i)=0\}$, enter the physical region with the sign of Im z_i required to be negative from the assumed structure.‖ This absence of ambiguity persists until a particular zero is swept close to the appropriate Legendre zero, as the dominant peripheral wave resonates. Im z_i then becomes small and its sign ambiguous. In the energy range of present interest, three zeros play a role ($l_{max}=3$), the

‡ An interpretation of the various resonance-like effects in the $I=J=0$ channel is discussed in Section 3.3(f). According to this, the ε' (1250) and ε (900) effects are viewed as aspects of one and the same elastic resonance interrupted by coupling to the S^* (1000).

§ In practice, this identification is only approximate, in that each solution is a new fit to the observed $\pi N\rightarrow\pi\pi N$ production cross-section at small t, including its errors.

‖ For a general review see Pennington (1973).

third entering the physical region at about 1·5 GeV. (They are conventionally numbered in order of increasing Re z_i.) The ambiguity below 1·4 GeV concerns the first one, and that above 1·4 GeV concerns the first two. As regards the latter, one can classify solutions according as (Im z_1, Im z_2) $=(--)$, $(-+)$, $(+-)$ or $(++)$ at 1·5 GeV (Männer, 1974; Estabrooks and Martin, 1974b). If one assumes that Im z_1 changes sign at 1·2 GeV (Im $z_1 \approx 0$ at 1·2 GeV, a sign change would correspond to a smoother behaviour), the first symbol also specifies Im z_1 say at 1·1 GeV, $(--)$ corresponding to Im z_1 (1·1) >0, etc. (Männer, 1974).‡

The corresponding phase-shift solutions (COW$_2$) are shown in Fig. 3.3.8 and compared with the earlier (COW$_1$) solution which is also plotted along

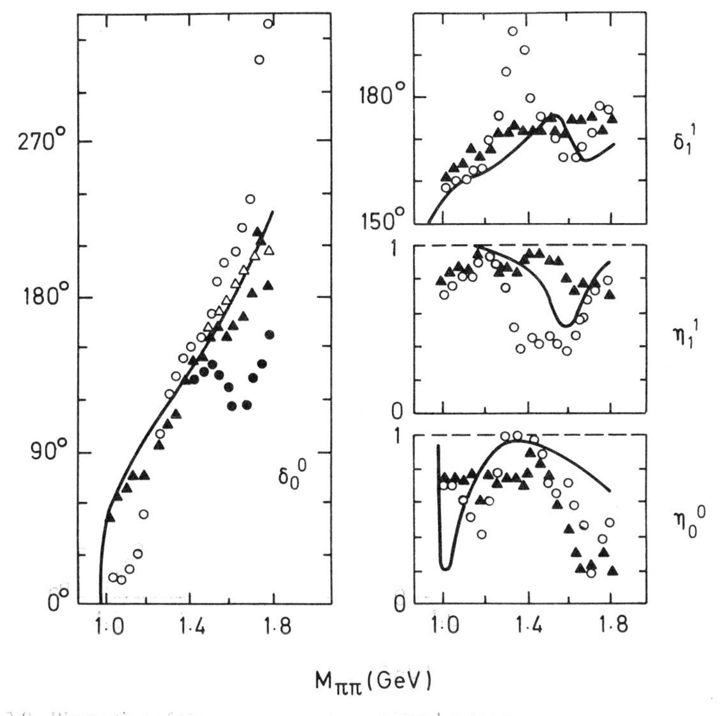

FIG. 3.3.8. Illustration of alternative $\pi\pi$ phase-shift solutions above 1 GeV (▲$(--)$, ●$(-+)$, △ $(+-)$, ○ $(++)$ as described in the text) (Männer, 1974) (COW$_2$). The previous energy-dependent phase-shift results (Hyams *et al.*, 1973) (COW$_1$) (full line) are also shown for comparison (cf. also Fig. 3.3.9).

‡ Alternatively the branches above and below 1·2 GeV could be associated differently. Indeed, Estabrooks and Martin (1974b) (COW$_3$) conclude from the behaviour of $\sigma(\pi\pi \to K\bar{K})$ and the form of the $\pi^0\pi^0$ spectrum that Im z_1 must be positive in the region immediately above 1 GeV, so that their principal alternatives $(--)$ (A) and $(+-)$ (B) arise accordingly as Im z_1 does or does not change sign at 1·2 GeV.

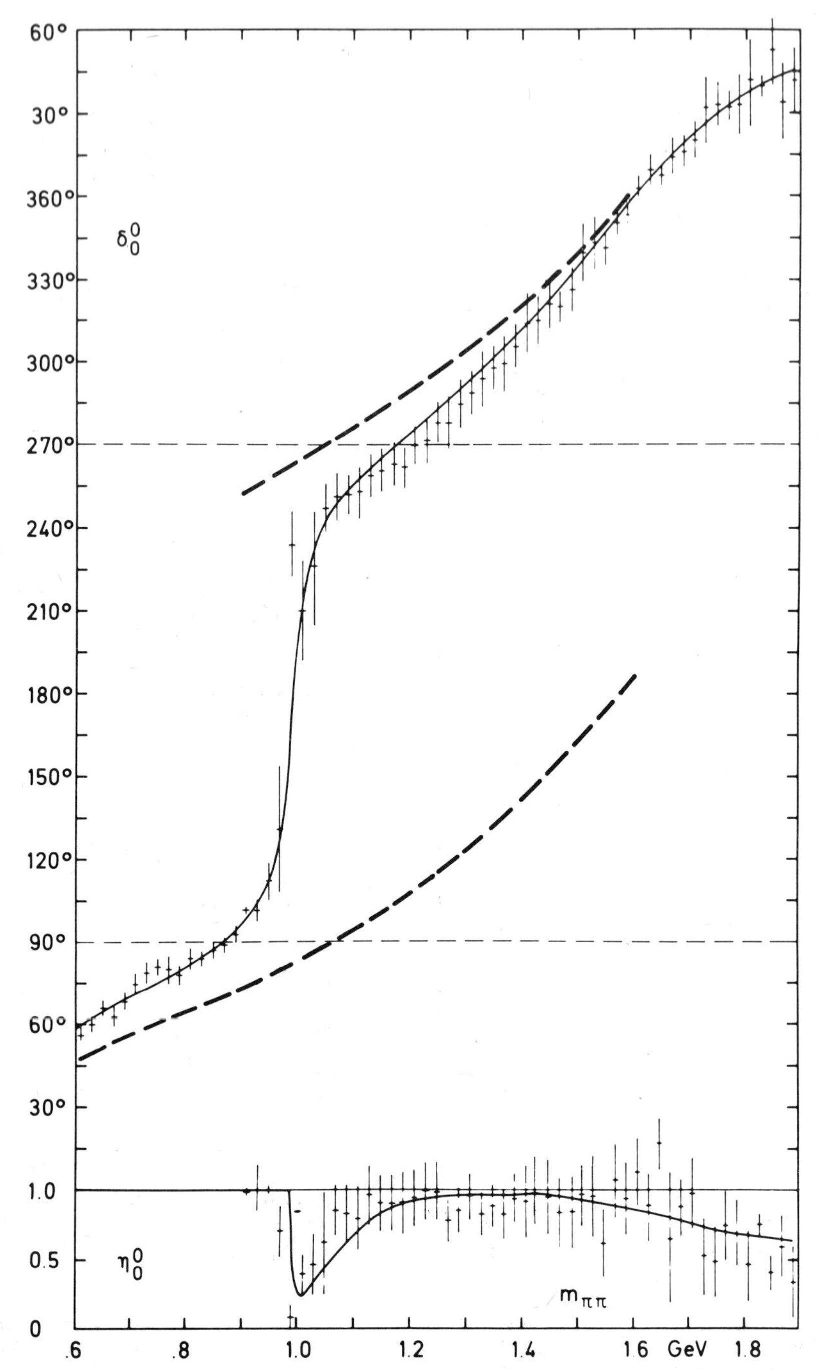

FIG. 3.3.9. $\pi\pi$ phase shifts and inelasticities for $M_{\pi\pi}$ between 600 and 1900 MeV from the energy-dependent analysis (COW_1) of the 17 GeV/c CERN-Munich experiment (Hyams *et al.*, 1973). The points with error bars show the result of the energy-independent variant of this fit: (a) $I=0$ S-wave; (b) $I=1$ P- and F-waves and $I=0$ D-wave. The dashed lines on (a)

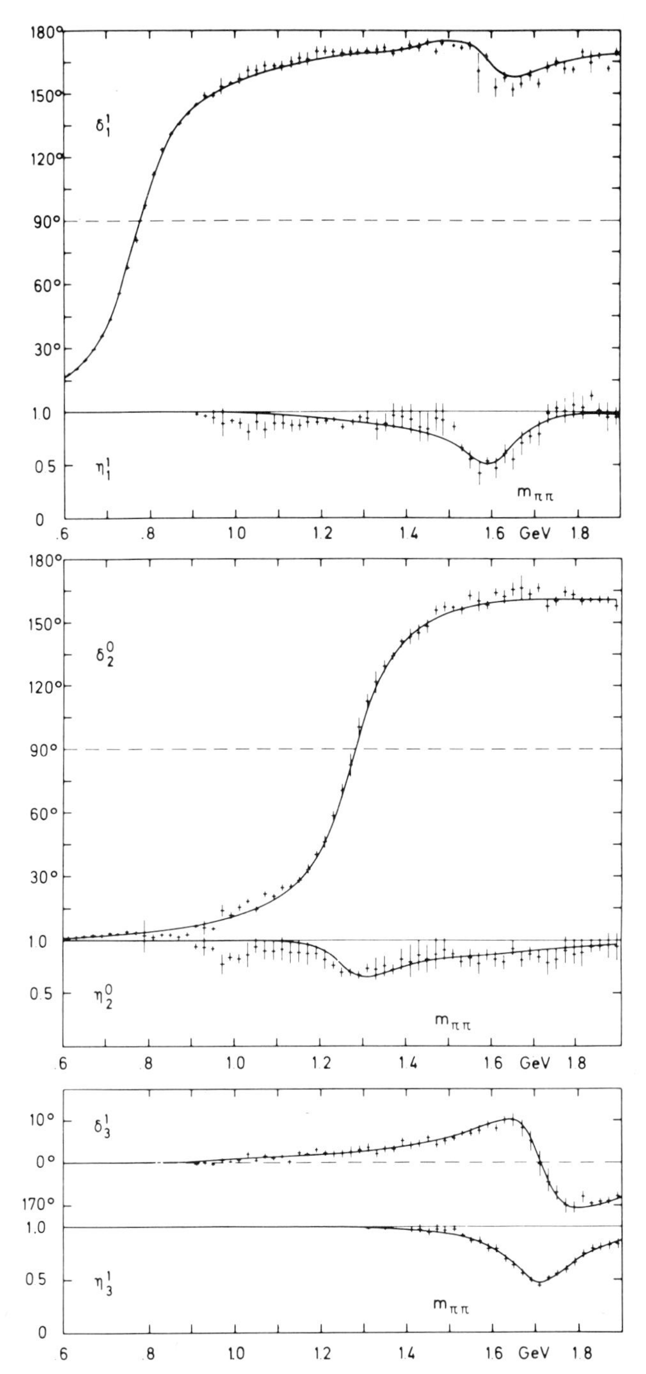

correspond to the residual phase-shift, $\bar{\delta}_0^0 \equiv \delta_0^0 - (\delta_0^0)^{S^*}$, after removal of the phase-shift associated with the S^* (1000), according to the prescription of Morgan (1974b) (cf. Section 3.3(f)), and the same quantity raised by 180°.

with its energy independent variant in Fig. 3.3.9. Although more restrictive as to the class of solutions which it admits, COW_1 extends over essentially the whole available mass range ($600 \leqslant M_{\pi\pi} \leqslant 1900$ MeV) and, through being an energy-dependent fit, readily gives resonance parameters. We shall therefore refer much of the following discussion to COW_1 despite its limitations. It is however to be emphasized that some at least of the COW_2 solutions (e.g. $(--)$) give a markedly better fit to the data.

The gross features of the phase ambiguities are clearly exposed in Fig. 3.3.8. Below the f-region only the first of the $(\pm, \pm)$ alternatives is operative and, as can be seen, the main uncertainty concerns the S-wave, as to whether it has a slow or rapid transit of 90° at about the f-mass. The former possibility associated with the $(-, \pm)$ solutions is the one selected by COW_1 and COW_3; the latter $(+, \pm)$ branch, in its behaviour near 1 GeV, resembles the previously discussed LBL solution. The continuation of these latter solutions above 1·2 GeV have a tendency to show unphysical inelasticities ($\eta_0^0 > 1$) strongly marked in the $(+-)$ branch. (Both COW_2 solutions below 1·2 GeV fail to achieve an η_0^0 corresponding to the observed inelasticity in the S^*-region (cf. Fig. 3.3.10) which undermines the credibility of the associated phase shifts.)

As one proceeds up in energy, the ambiguities proliferate especially in the S- and P-waves which show a variety of resonant behaviours in the g-region (solutions $(\pm +)$ have S-wave resonances and $(+ \pm)$ have P-wave resonances). The only stable feature is the form of the peripheral resonances. To some extent even this is built into the analysis through the assumption of

TABLE 3.3.3. Resonance Parameters for the f^0

Analysis		M_f (MeV)	Γ_f (MeV)	Branching ratios
CERN-Munich (CEM_1) (Estabrooks *et al.*, 1973)		1271 ± 2	182 ± 4	$X = 0{\cdot}81 \pm 0{\cdot}01$ $(\pi\pi)$
CERN-Munich (COW_1)[a] (Hyams *et al.*, 1973)		1279 ± 2	202 ± 6	$X = 0{\cdot}84 \pm 02$ $(\pi\pi)$
LBL (Flatté *et al.*, 1972)		1277 ± 4	183 ± 15	
Anderson *et. al.* (1973)	$(f^0 \to 2\pi^+ 2\pi^-)$	1283 ± 10	185 ± 25	$\dfrac{\pi^+\pi^-\pi^+\pi^-}{\pi^+\pi^-} = 5{\cdot}5 \pm 1\%$
Louie *et al.* (1974)	$(f^0 \to 2\pi^+ 2\pi^-)$	1277 ± 12	194 ± 22	$6{\cdot}5^{+1{\cdot}0\%}_{-1{\cdot}7\%}$
Particle Data Group (1974)		1270 ± 10	170 ± 30	$\pi\pi$ $83 \pm 5\%$; KK $4 \pm 3\%$; $2\pi^+ 2\pi^-$ $4 \pm 1\%$

[a] Unphysical sheet pole.

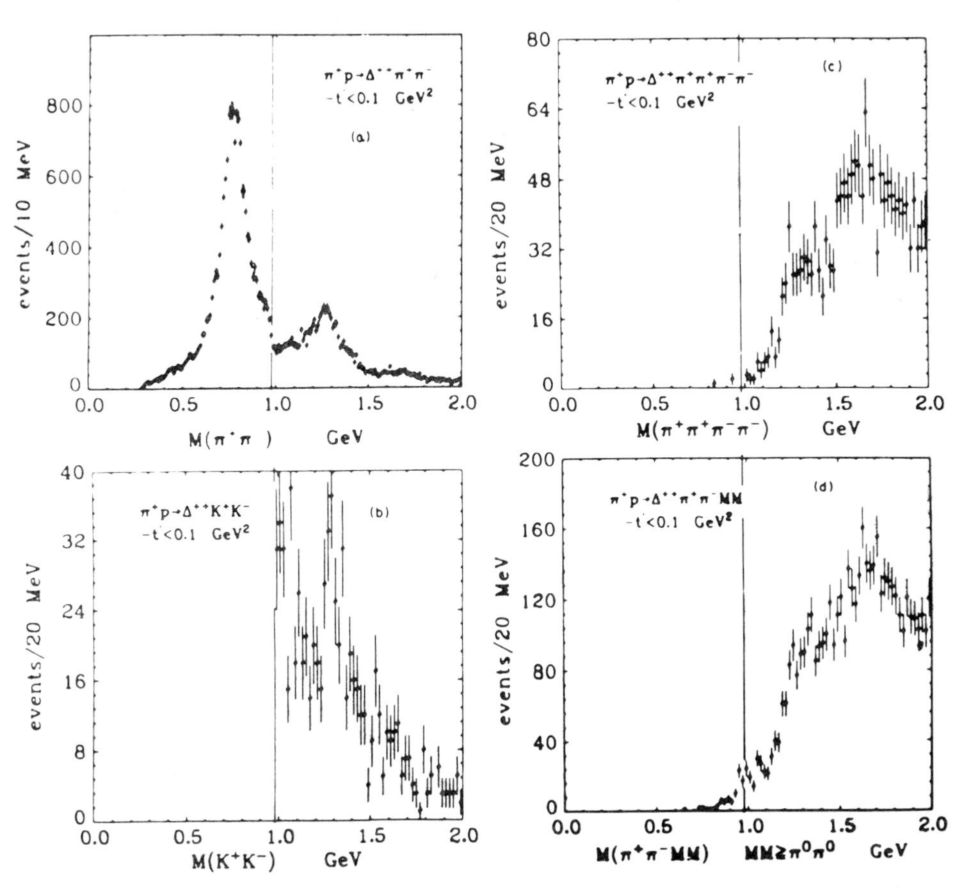

FIG. 3.3.10. Results illustrating the onset of inelasticity from $\pi^+p\to\Delta^{++}$ + mesons at 7 GeV/c (Alston-Garnjost *et al.*, 1971): (a) $\pi^+\pi^-$; (b) K^+K^-; (c) $\pi^+\pi^-\pi^+\pi^-$; (d) $\pi^+\pi^- MM$.

Breit–Wigner forms for the f and g; however, the observed shapes of the appropriate intensity components suggests that this should be a reasonable approximation.

The list of well-substantiated dynamical effects above 950 MeV for which it is appropriate to quote resonance parameters is therefore a short one. Firstly, there is the S^* (1000) phenomenon, the discussion of which we postpone to Section 3.3(f) owing to its special two-channel character. Proceeding upwards in dipion mass, the next dynamical feature is the f^0 resonance which, despite the strong $\pi\pi$ signal, is seen to be definitely inelastic (Fig. 3.3.9). Resonance parameters from COW_1 and other analyses are listed in Table 3.3.3. The various results are seen to be in fairly good mutual agreement with values clustering about $M_f = 1275$, $\Gamma_f = 183$ MeV, $X_{\pi\pi}$

$\approx 0{\cdot}8$. Both the mass and width are somewhat larger than those popular a few years ago.

The resonance-like behaviour of the $I = 0$ S-wave is apparent on Figs. 3.3.8 and 3.3.9. In principle, COW_2 offers two alternative versions, however, as already noted, both fail to achieve the appropriate η_0^0 just above the $K\bar{K}$ threshold. We shall provisionally disregard the possibility of a narrow ε' and follow COW_1 (Fig. 3.3.9) as to details.‡ The inelasticity $1 - \eta_0^0$ is seen to fall from its peak just after the $K\bar{K}$ threshold; the phase shift, after its rapid rise associated with the S^*-resonance, continues upwards at a more steady rate, transiting 270°, at about 1170 MeV and attaining 315° near 1400 MeV. Such an effect has long been suspected from the forms of the $\pi^+\pi^-$ decay angular distribution in the f^0-region. If this were dominated by the decay of helicity $m = 0$ D-waves, one would expect an angular distribution $d\sigma/d\Omega \propto (P_2)^2 \equiv (\frac{3}{2}\cos^2\theta - \frac{1}{2})^2$, with a central hump, and minima at $\cos\theta = \pm 1/\sqrt{3}$. The observed distribution is U-shaped throughout the f^0-band, indicating the presence of a large in-phase S-wave amplitude adding coherently to the D-wave (the S-wave has to be as large as possible, consequently elastic, in order to achieve this). The first quantitative analysis leading to resonance parameters was that of Carroll *et al.* (1972), who proposed the values $M_{\varepsilon'} = 1{\cdot}25 \pm 0{\cdot}04$, $\Gamma_{\varepsilon'} = 0{\cdot}3 \pm 0{\cdot}1$ GeV. These are in good agreement with the values suggested by the phases of Fig. 3.3.9 ($M_{\varepsilon'} \approx 1{\cdot}20$, $\Gamma_{\varepsilon'} \approx 0{\cdot}35$) and also with the values found in CERN-Munich's (CEM_1) analysis ($M_{\varepsilon'} \approx 1{\cdot}24$, $\Gamma_{\varepsilon'} \approx 0{\cdot}2$). There have at various times been reports of additional effects between the S^* and the f, but, since none of them occur in high statistics experiments, we disregard them.

The only other unambiguous resonance signal below 1900 MeV is that corresponding to the g-meson in the $I = 1$ F-wave, for which the COW_1 parameters are:

$$M_g = 1713 \pm 4 \text{ MeV}, \qquad \Gamma_g = 228 \pm 10 \text{ MeV}, \qquad X_g = 0{\cdot}26 \pm 0{\cdot}02.$$

The Particle Data Group (1974), give somewhat smaller values for the mass and width but adopt the same elastic branching ratio with larger errors:

$$M_g = 1686 \pm 20 \text{ MeV}, \qquad \Gamma_g = 180 \pm 30, \qquad X_g = 0{\cdot}26 \pm 0{\cdot}05.$$

They estimate that the total width should be apportioned among the principal decay modes in the ratios

$$2\pi : 4\pi : K\bar{K}\pi : K\bar{K} \approx 0{\cdot}26 : 0{\cdot}7 : 0{\cdot}03 : 0{\cdot}02.$$

‡ Estabrooks and Martin (1974b) (COW_3), after rejecting as unphysical (see above) the variant with a fast-growing phase shift, report a solution of rather similar qualitative appearance to COW_1 but which fails to exhibit resonance behaviour in detail.

The outstanding additional feature of the COW_1 solution (Fig. 3.3.9) is the clear-cut signal of an inelastic P-wave resonance at about 1600 MeV. As we have seen, this only occurs for particular branches of the full range of solutions admitted by the Barrelet ambiguity; until this is resolved, the ρ' (1600)$\rightarrow \pi\pi$ can only be reckoned a possibility, for all its theoretical attractions. The basic difficulty is that daughter resonances do not in general produce a distinct signal in any Legendre coefficient. This is so in the present instance as is shown in Fig. 3.2.3 which gives the $\langle Y_l^0 \rangle$ and $\langle Y_l^1 \rangle$ moments upon which COW_1 is based, and also shows the resulting fit. There is no clear-cut signal at 1600 MeV; furthermore, there is a sharp effect at 1500 MeV in the $\langle Y_2^0 \rangle$ and $\langle Y_4^0 \rangle$ coefficients which the COW_1 fit does not follow.‡ Of the four alternative COW_2 fits, the ones with P-wave resonances $(+-)$ and $(++)$ have poorer χ^2 than the ones that do not, $(--)$ and $(-+)$. In particular, $(--)$ which gives a good overall fit, including the region of the 1500 MeV dip, does not have any subsidiary resonance structure at all over the g-region; in contrast, $(-+)$ and $(++)$ have S-wave structures. Even if the solutions with a ρ' are ultimately preferred,§ the results of COW_2 suggest that the resonance should be broader and more inelastic than the earlier analysis indicated (COW_1 gave $M_{\rho'} = 1590 \pm 20$ MeV, $\Gamma_{\rho'} = 180 \pm 50$ MeV, $X_{\rho'} = 0{\cdot}25 \pm 0{\cdot}05$). This would improve the resemblance to the electromagnetically produced ρ' which is rather broad ($\Gamma \approx 400$ MeV), and for which a very small $\pi^+\pi^-$ signal has been seen (see Sections 5.3 and 5.4).

Of great help in resolving the ambiguities below and through the g-region would be to acquire data on other channels, especially those in which fewer waves can participate, $\pi^-\pi^0$, $\pi^0\pi^0$, $K_1^0K_1^0$, etc. Knowledge of σ_{tot} would also be a helpful constraint.

Detailed high statistics investigations of dipion systems are just beginning to extend above 1900 MeV.‖ Dowell *et al.* (1974) have reported preliminary results of a 12 GeV experiment on $\pi^-p \rightarrow \pi^+\pi^-n$, covering the mass range $0{\cdot}6 \leq M_{\pi\pi} \leq 2{\cdot}4$ GeV. The form of the Legendre moments indicates the possibility of there being a 4^+ meson at about 2 GeV. This is confirmed by data on $\pi^-p \rightarrow \pi^0\pi^0 n$ at 40 GeV/c (Apel *et al.*, 1975) and $\pi^-p \rightarrow K^+K^-n$ at 18 GeV/c (Blum *et al.*, 1975). In both cases a clear peak is seen in the

‡ This might seem to indicate that the f' (1514) does after all couple to $\pi\pi$. A difficulty with this is the seeming absence of a corresponding signal in $\pi\pi \rightarrow K\bar{K}$ (Martin, 1974).

§ The analysis of Froggatt and Petersen (1975) imposing fixed-t and fixed-u analyticity (see footnote following equation (3.3.2)) which apparently avoids the discrete ambiguities does yield a ρ' resonance with parameters $m_{\rho'} = 1600 \pm 50$ MeV, $\Gamma_{\rho'} = 220 \pm 70$ MeV, $X_{\rho'} = 0{\cdot}35 \pm 0{\cdot}10$.

‖ On the basis of rather low statistics and making a number of assumptions, Walker (1973) (see also Robertson *et al.* (1973)) has extracted total and elastic cross-sections for $\pi^+\pi^-$ and $\pi^-\pi^-$ up to 4 GeV, obtaining results of the expected order of magnitude.

appropriate $\langle Y_8^0 \rangle$ moment, the corresponding resonance parameters being

$$M_h = 2020 \pm 30 \text{ MeV} \qquad \Gamma_h = 186 \pm 60 \text{ MeV}$$

and

$$M_h = 2050 \pm 25 \text{ MeV} \qquad \Gamma_h = 225 {}^{+120}_{-\ 70} \text{ MeV}$$

for the $\pi^0\pi^0$, K^+K^- reactions respectively. This state is naturally interpreted as the Regge recurrence of the f^0, and to a good approximation the ρ, f^0, g and h mesons lie on a straight line in the Chew–Frautschi plot.

(f) $K\bar{K}$ Threshold Effect—The S^*-Resonance

The predominant inelastic effect in the region just above 1 GeV is the very strong opening of the channel $\pi\pi \to K\bar{K}$. Appreciable 4π production seems only to occur somewhat above 1 GeV, except possibly for the $\pi\omega$ channel which is pure $I = 1$ (see Fig. 3.3.10 from Alston-Garnjost *et al.* (1971); see also French (1968), Anderson *et al.* (1973)). There exists a reasonable body of data on the reactions $\pi^-p \to K^+K^-n$ and $\pi^-p \to K_1^0K_1^0n$. Below the A_2, say up to 1·26 GeV, and aside from a small amount of ϕ production in K^+K^-, production seems to be entirely consistent with OPE,‡ so that the data lead to inferences on $\sigma(\pi\pi \to K\bar{K})$. The most striking effect is the sharp rise of this cross-section to a peak at around 1·03 GeV, followed by a fall (see Fig. 3.3.11(b)). The sharp onset suggests that one is dealing with an S-wave effect, and the decay angular distribution confirms this. Furthermore, estimates of the magnitude of the peak (Grayer *et al.*, 1973) indicate a value comparable to the S-wave unitarity limit (Fig. 3.3.11(b)). The observed energy dependence strongly suggests the presence of an energy pole on an unphysical sheet, either a resonance or a $K\bar{K}$ bound state coupling to $\pi\pi$. A number of alternative fits along these lines were found even before the advent of the high precision experiments (see the review of Beusch (1970)).

If inelasticity sets in very sharply in the $I = 0$ S-wave $\pi\pi$ channel, one would look to see a distinct signal in the elastic channel, and this is found. A very sharp drop is observed in the $\pi\pi$ interference moment $\langle Y_1^0 \rangle$ just before the $K\bar{K}$ threshold (see Figs 3.3.1(b) and 3.3.2). That the effect is indeed attributable to a rapid movement of the S-wave was carefully argued by the LBL group from the behaviour of the $\langle Y_l^0 \rangle$ moments (Alston-Garnjost *et al.*, 1971). This discussion subsequently received quantitative support in the LBL group's energy dependent phase-shift analysis (Protopopescu *et al.*, 1973), and in various analyses of the CERN-Munich data.

‡ Neutral $K\bar{K}$ states are not eigen-states of G parity ($G(K\bar{K}) = (-1)^{I+J}$), and so the assumption of coupling to the $\pi\pi$ system has to be examined.

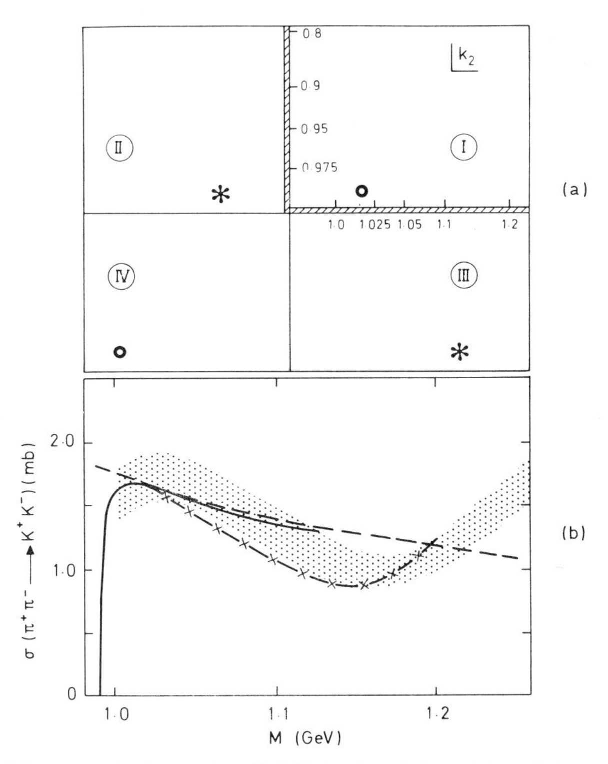

FIG. 3.3.11. Diagrams relating to the S^* (1000) effect (adapted from Morgan (1974b)): (a) Location on the k_2-plane ($k_2 \equiv [E^2/4 - m_K^2]^{\frac{1}{2}}$ of the poles (*) and associated zeros (0) of S_{11} according to Hyams *et al.* (1973) (COW_1). The numbered sectors correspond to the associated sheets of the energy plane (I) (II) (III) and (IV). Physical energies map on to the shaded segments of the axis, representative values being shown in GeV; (b) Comparison of $\sigma(\pi^+\pi^- \to K^+K^-)$ in mb, as inferred from experiment (Grayer *et al.*, 1973) (the shaded band), with alternative one and two-pole parameterizations: (——) Grayer *et al.*'s own constant K-matrix fit; (-×-) form predicted using Hyams *et al.*'s two poles, as in (a) above, for the S-wave contribution; (- - -) S-wave unitarity limit.

Such behaviour has all the hallmarks of a resonance. The question to decide is what kind of resonance and how best to parameterize it. The special situation in which the effect occurs, close to the higher threshold of two, and seemingly only two channels ($\pi\pi$ and $K\bar{K}$, both two-body S-waves), determines the framework within which answers are to be sought (see equation (3.3.3) below). For such a system, one has a four-sheeted energy plane, corresponding to the 2×2 choices for the signs of the channel momenta, $k_i = \pm(\frac{1}{4}E^2 - m_i^2)^{\frac{1}{2}}$ ($m_1 = m_\pi$, $m_2 = m_K$),‡ with all four sheets lying

‡ Sheets I, II, III and IV are defined by ($\mathrm{Im}\, k_1$, $\mathrm{Im}\, k_2$) = (++), (−+), (−−) and (+−) respectively.

close to the domain over which the effect is observed. The structure around the $K\bar{K}$ threshold is conveniently brought out in terms of the k_2-plane (Fig. 3.3.11(a)). Below the threshold, experimental observations correspond to the positive imaginary axis, and above threshold to the positive real axis; rapid changes could correspond to poles on Sheets II, III or IV or some combination of these. The problem is to find out which situation actually obtains.

Up to a point, there is an agreed answer. All recent analyses of the S^*-effect concur in finding a resonance pole, $E_{S^*}^{\mathrm{II}}$, on Sheet II of the energy plane (i.e. the one obtained by passing through the elastic cut). Table 3.3.4 lists a number of recent determinations of the pole position. Besides the

TABLE 3.3.4. Sheet II Pole Parameters for the S^*

Analysis	M_0 (MeV)	Γ_0 (MeV)
Protopopescu *et al.* (1973) (a)	997 ± 6	54 ± 16
(LBL) (b)	982 ± 6	74 ± 16
Grayer *et al.* (1973)	1012 ± 6	34 ± 10
Estabrooks *et al.* (1973) (CEM_1)	997	10
Hyams *et al.* (1973) (COW_1)	1007 ± 20	30 ± 10
Binnie *et al.* (1973)	987 ± 7	48 ± 4

LBL analysis, there are three solutions based on the CERN-Munich results, of which the first listed uses data on K^+K^- production, in addition to that on $\pi^+\pi^-$. The final entry in the table, which refers to the observation of S^* production at threshold (Binnie *et al.*, 1973) agrees well with the high-energy determinations, although the background is observed to be different.‡

Once the existence and approximate location of $E_{S^*}^{\mathrm{II}}$ is agreed on, it remains to decide whether this exhausts the structure.

It turns out to be conceptionally simpler, especially for relating the S^* to other 0^+-resonances, if there is an additional Sheet III pole, such as would arise if the S^* corresponded to a two-channel Breit–Wigner formula (Morgan, 1974b; Fujii and Fukugita, 1974) (see equation (3.3.7) below). Such an additional pole is found in the energy-dependent phase-shift analysis of the CERN-Munich $\pi^+\pi^-$ data (Hyams *et al.*, 1973), its location being $E_{S^*}^{\mathrm{III}} = 1049 - 250i$ MeV, and the energy independent variant of COW_1 exhibits the same tendency.§ In itself, this is not compelling, since only $\pi\pi$ data is fitted, and the extra pole might merely reflect a bias of the parameterization; however, support comes from the observed shape of $\sigma(\pi\pi \to K\bar{K})$ (see

‡ Unitarity of course requires that resonance poles be faithfully reproduced in different situations. This is a nice example.

§ Fujii and Fukugita (1974) find this to be much better reproduced by a two-pole rather than a one-pole analytic form, obtaining for the former case, $E_{S^*}^{\mathrm{II}} = 1001 - 22i$ MeV, $E_{S^*}^{\mathrm{III}} = 960 - 60i$ MeV.

below). The implications of there being an additional Sheet III pole are also discussed by Protopopescu *et al.* (1973) (see especially their Appendix A), who find such extra poles for both their main solutions, that for the favoured solution being at 930–18i MeV. However, this feature is found not to be stable, even as to its existence, against changes of parameterization, and the authors therefore regard its status as being undecided by their data. They look to improved $K\bar{K}$ data to settle the question.

The reaction $\pi\pi \to K\bar{K}$ indeed affords the most sensitive indicator of S-wave structure near its threshold. Presently available data (Grayer *et al.*, 1973) somewhat favours the two-pole picture (Morgan, 1974b); in particular, the shape of $\sigma(\pi\pi \to K\bar{K})$ (Fig. 3.3.11(b)) is reproduced better by a two-pole form corresponding to the COW_1 phase shifts, than by Grayer *et al.*'s own constant K-matrix fit, which by construction only has one pole. (Note that the former is a prediction and the latter a fit.) Given modest assumptions on the $\pi\pi \to K\bar{K}$ P- and D-wave amplitudes, the *sign* of the $\pi\pi \to K\bar{K}$ S-wave amplitude can be determined from the form of the $\langle Y_l^m \rangle$ moments. A full phase-shift analysis of the presently available $\pi\pi \to K\bar{K}$ data should afford a definitive test of the two-pole hypothesis. Pending this, the idea remains conjectural, but with the evidence tipping in its favour. Quantitative estimates of the $E_{S^*}^{\mathrm{III}}$ pole position and coupling constant derived therefrom (see below) are highly provisional. It is however important to identify the pole structure of any given solution in order to separate resonance effects from background. One may hope that the latter, background, phase, which one will seek to relate to variations of the amplitude over a large energy range, will prove to be somewhat stable against changes in the detailed resonance assignment.

The above results come from embedding suitable parameterizations of the 2×2 T-matrix describing the $I=J=0$ channel into $\pi\pi$ phase-shift analyses, with the $\pi\pi$ data sometimes supplemented by information on $\pi\pi \to K\bar{K}$. For interpreting alternative pole structures in the $I=J=0$ sector, it is convenient to cast the parameterization into the following form, in which the diagonal S-matrix elements are expressed as ratios:

$$S_{11} = \frac{\Phi(-k_1, k_2)}{\Phi(k_1, k_2)}; \qquad S_{22} = \frac{\Phi(k_1, -k_2)}{\Phi(k_1, k_2)}, \tag{3.3.3}$$
$$(\Phi(k_1, k_2) \equiv \Phi^*(-k_1^*, -k_2^*)),$$

where, as before, k_1 and k_2 denote the channel momenta.‡ Resonance poles correspond to zeros of Φ, whose properties guarantee the occurrence of

‡ $\Phi(k_1, k_2)$, the so-called "Jost function", is of course a continuous function of just one variable. If one starts by parameterizing the T-matrix in terms of the inverse K-matrix, $\mathbf{M} \equiv \mathbf{K}^{-1}$ it takes the form $\Phi(k_1, k_2) \equiv (M_{11} - i\rho_1)(M_{22} - i\rho_2) - M_{12}^2$ where $\rho_i = 2k_i/E$, or just k_i, according to the normalization adopted.

associated numerator zeros of S_{11} and S_{22} (cf. equation (3.3.6) below). For the case of interest, all these effects are close by; their automatic inclusion is the prime virtue of the Φ-representation. It is natural to maintain this property in separating resonance and background effects, and to express Φ as a product of resonance and background terms each of the required form

$$\Phi = \Phi^R \Phi^B. \tag{3.3.4}$$

As a result, the diagonal S-matrix elements will likewise be factored. This establishes a method for removing a resonance contribution by subtracting the corresponding phase shifts and dividing the associated inelasticities. We shall use this to re-interpret the structure of the $I=0$ S-wave. The separation of resonance and background effects is, as always, somewhat a matter of convention, since one can choose alternative forms for Φ^R with the same zeros. Provided simple forms are chosen, this is not too much of a problem.

Simple examples of one- and two-pole structures are, firstly for a one-pole form‡

$$\Phi = [k_2^R - k_2]\,\mathrm{e}^{-i\delta_B}, \tag{3.3.5}$$

from which

$$S_{11} = \frac{[-(k_2^R)^* - k_2]}{[k_2^R - k_2]}\,\mathrm{e}^{2i\delta_B}; \qquad S_2 = \frac{k_2^R + k_2}{k_2^R - k_2}, \tag{3.3.6}$$

and for a two-pole form

$$\Phi^R = [(M_{S^*}^2 - s)\Lambda(s) - 2i(M_{S^*}^2/E)(g_1^2 k_1 + g_2^2 k_2)], \tag{3.3.7}$$
$$(\Lambda(M_{S^*}^2) = 1).$$

The former is approximately equivalent to assuming constant K-matrix elements, as was done in the CERN-Munich groups' analysis of their own $\pi^+\pi^-$ and K^+K^- data (Grayer *et al.*, 1973). The latter Breit–Wigner form (with a simple linear form for $\Lambda(s)$) was used by Morgan (1974b) to extract reduced widths from the COW_1 results. Requiring the pole positions, $E_{S^*}^{\mathrm{II}}$ and $E_{S^*}^{\mathrm{III}}$, to be reproduced, he obtained $(g_{\pi\pi}^{S^*})^2 \approx 0{\cdot}2$ (corresponding to $\Gamma_{\pi\pi}^{S^*} \approx 180$ MeV), $(g_{K\bar{K}}^{S^*})^2 \approx 0{\cdot}7$ and $M_{S^*}^2 = 0{\cdot}966\ \mathrm{GeV}^2$. These pole positions and the associated zeros of S_{11} (cf. equation (3.3.6)) are illustrated in Fig. 3.3.11(a). Fujii and Fukugita (1974), fitting to Hyams *et al.*'s energy-independent phase shifts, obtain results corresponding to $(g_{\pi\pi}^{S^*})^2$

‡ The one-pole form is not readily assimilated to conventional notions of a coupled channel resonance; for example the parameters E_0 and Γ_0 related to the pole position, $E_0 - i\Gamma_0/2 \equiv E^R \equiv 2|m_2^2 + (k_2^R)^2|^{\frac{1}{2}}$, do not bear the usual relation to the phase and inelasticity changes. This stems from the use of just two parameters, Re k_2^R and Im k_2^R, instead of the conventional three as in (3.3.7), to describe the resonance.

$=0{\cdot}087\pm0{\cdot}014$ ($\Gamma^{S^*}_{\pi\pi}=81\pm13$ MeV), $(g^{S^*}_{K\bar{K}})^2=0{\cdot}20\pm0{\cdot}09$ and $M^2_{S^*}=0{\cdot}983$ GeV^2. The discrepancies between these two determinations reflect the spread in the input "data". The main virtue of presenting the results in this way is to indicate orders of magnitude for comparison with related processes, and also to emphasize $(g^{S^*}_{K\bar{K}})^2>(g^{S^*}_{\pi\pi})^2$.

Once the rapid resonant variations associated with the S^* are parameterized, one can identify the residual background phase shift and relate it to other features of $\delta^0_0(E)$. Adjoining the S^*-region, the phase shift δ^0_0 executes slow transits respectively of 90° ($\varepsilon(900)$ effect) and 270° (ε' (1250) effect), in both cases with $\eta^0_0\approx1$. If one removes the contribution of the S^* according to the recipe of equation (3.3.4), these two effects coalesce to yield a single, slow, essentially elastic, transit of 90° at about 1100 MeV. The qualitative idea is not tied to a particular interpretation of the S^*, although the details obviously are. As an example, the form of $\bar{\delta}^0_0\equiv\delta^0_0-(\delta^0_0)^{S^*}$ deduced from the COW$_1$ phase shifts by Morgan (1974b) is plotted in Fig. 3.3.9(a). (The associated $(\eta^0_0)^{S^*}$ corresponds rather closely to η^0_0, so the remnant amplitude is essentially elastic.) Presented in this way, the data suggest that one is dealing with a single, broad and essentially elastic ε (1100–1300) resonance, accompanying the S^* (1000) in the $I=J=0$ channel. (The range of mass corresponds to uncertainty as to what background phase one should allow.) This is just what one needs to fill the $I=0$ slots in the 0^+ level of the quark model (cf. Chapter 13), with the $I=1$ δ (976) (decaying to $\eta\pi$ and possibly $K\bar{K}$) and $I=\frac{1}{2}$ κ (1250) (see Sections 3.3(g) and 3.3(h) below) providing the other members.

Within the limits set by data and the models used, this idea can be pursued quantitatively by extracting reduced widths for the other 0^+ decays, for comparison with the $(g^{S^*}_{\pi\pi})^2$ and $(g^{S^*}_{K\bar{K}})^2$ referred to above (Morgan, 1974b); a provisional SU(3) solution is found with markedly non-ideal mixing, $\theta_s\approx70°$ (instead of 35° for an ideal nonet, cf. Chapter 13). Quite different solutions have been discussed by Rosner (1974a) and Conforto (1974), the latter demanding very narrow δ and κ widths. Much more experimental work is needed on all the relevant channels to check and select among these ideas. (For a further discussion of 0^+ phenomenology see Morgan (1975).)

(g) $K\pi$ Scattering

$K\pi$ differential cross-sections have been extracted by identifying the OPE signal in various production processes of the form $KN\to K\pi N(\Delta)$ using both K^+ and K^- beams. Despite the larger range of candidates for background to the OPE signal, as compared to the analogous $\pi\pi$ production reactions, a reasonably consistent body of information has emerged. There exist results on the following processes listed by their reaction amplitudes and the associated isotopic decompositions:

$$T(K^{\pm}\pi^{\pm} \to K^{\pm}\pi^{\pm}) = T^{(\frac{3}{2})}$$

$$T(K^{\pm}\pi^{\mp} \to K^{\pm}\pi^{\mp}) = \tfrac{1}{3}T^{(\frac{3}{2})} + \tfrac{2}{3}T^{(\frac{1}{2})}$$

$$T(K^{+}\pi^{-} \to K^{0}\pi^{0}) = T(K^{-}\pi^{+} \to \bar{K}^{0}\pi^{0}) = \frac{\sqrt{2}}{3}(T^{(\frac{3}{2})} - T^{(\frac{1}{2})}).$$

As regards coverage of charge channels, the experimental situation is better than for $\pi\pi$, because the K^0 is easier to detect than the π^0. There are more states because there is no limitation from Bose statistics, and inelastic channels play a larger role since there is no restriction analogous to G-parity. As a result, $K\pi\pi$ states can couple to all except the S-wave $K\pi$ states. The other inelastic channels which could, additionally, couple to the S-wave appear not to be prominent below 1·4 GeV.

Phase-shift searches, less elaborate and with more assumptions, proceed as for the $\pi\pi$-channel and there are close resemblances in the resulting structure. Just as for the $\pi\pi$-channel, there are prominent "leading" resonances, K^* (892) ($J^P = 1^-$) and K^* (1421) ($J^P = 2^+$), with earlier reports (Carmony *et al.*, 1971; Aguilar-Benitez *et al.*, 1973) of a $J^P = 3^-$ successor to these, K^* (1800), now well confirmed (Brandenburg *et al.*, 1974).‡ The non-exotic $I = \frac{1}{2}$ channel thus has an exchange-degenerate leading trajectory analogous to the ρ–f–g sequence. Accompanying the leading partial waves, there are significant effects from states lying lower in J, conspicuous from their interference with the leading states. So far, interest has centred on the behaviour of the non-exotic S-wave, $\delta_0^{\frac{1}{2}}$. The situation regarding this is summarized in Fig. 3.3.12 (adapted from the compilation given by the Particle Data Group (1974)). As the figure indicates, potential Up–Down ambiguities (like those found in $\pi\pi$ scattering near the rho-mass) occur over the K^* (892) and K^* (1421) regions. The first of these has however been essentially eliminated in favour of the Down-branch (excepting the possibility of an extremely narrow S-wave resonance $\Gamma < 7$ MeV) from analysis of a 12 GeV/c $K^+p \to \Delta^{++}K^+\pi^-$ LBL experiment (Barbaro-Galtieri *et al.*, 1973; Matison *et al.*, 1974). Theoretical prejudice would likewise disfavour the higher ambiguity. Certainly the simplest possibility is that the phase just rises steadily upwards transiting 90° between 1200 and 1400 MeV, so that one has a single broad elastic resonance, usually termed κ. If one assumes such a behaviour, resonance parameters can be fitted; for example Lauscher *et al.* (1974) find

$$M_\kappa = 1245 \pm 30 \text{ MeV}, \qquad \Gamma_\kappa = 485 \pm 80 \text{ MeV}.$$

‡ There is some discrepancy in the detailed resonance parameters found in these three experiments. The first two found $M \approx 1760$, $\Gamma \approx 60$–80 MeV, whilst the last mentioned finds $M \approx 1800$–1850, $\Gamma \gtrsim 200$ MeV.

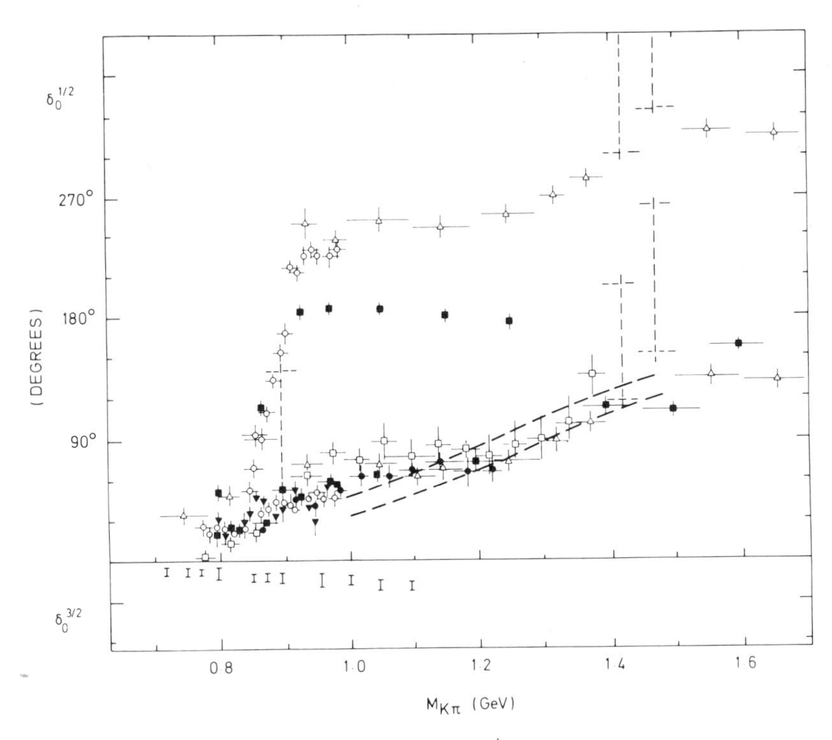

FIG. 3.3.12. $K\pi$ S-wave shifts. Compilation of $\delta_0^{\frac{1}{2}}$ determinations (based on Particle Data Group (1974)) (○) Bingham *et al.* (1972a), (●) Mercer *et al.* (1971), (■) Yuta *et al.* (1971), (△) and (⁝) Firestone *et al.* (1972), (▼) Matison *et al.* (1974), (□) Baker *et al.* (1973). Up solutions of the type shown are excluded by Matison *et al.* (1974) who would require any κ (900) to have width $\Gamma < 7$ MeV. The dashed band (::::) indicates the phase-shift which would correspond to the resonance-fit of Lauscher *et al.* (1974). Also shown is a recent determination of $\delta_0^{\frac{3}{2}}$ (Linglin *et al.*, 1973).

Similar results have been reported by Yuta *et al.* (1973) and Cords *et al.* (1974).

The resemblance to $\pi\pi$ scattering extends to the exotic $I=\frac{3}{2}$ channel, which is best studied in isolation in the doubly charged, $K^{\pm}\pi^{\pm}$, channels, but can also be studied in interference. Despite some earlier conflicting results (see the review of Trippe (1971)), there is now good agreement on the form of the partial waves. The S-wave phase shift is found to be small, negative, and smooth, and the P-wave negligible below 1 GeV. As an example, the S-wave results of Linglin *et al.* (1973), using data on $K^-p \to K^-\pi^-\Delta^{++}$ at 14 GeV/c, are plotted along with $\delta_0^{\frac{1}{2}}$ in Fig. 3.3.12. The same group (Bardadin-Otwinowska *et al.*, 1974) subsequently extracted $K^-\pi^-$ inelastic cross-sections up to $M_{K\pi} = 2{\cdot}8$ GeV. The results interestingly complement and, from the point of view of energy range, surpass the analogous ones on $\pi^-\pi^-$. As for the $\pi\pi$ case, the exotic total cross-section ($\sigma_{tot}^{K^-\pi^-}$ (2·6 GeV) ≈ 6 mb) is small compared to the "plateau" value inferred using factorization ($\sigma_{K\pi}^{\infty} \approx 11$ mb).

Regarding threshold behaviour, both for $I = \frac{1}{2}$ and $\frac{3}{2}$, the qualitative form of the data resembles that predicted from current algebra (Chapter 14). Statistics are low for making quantitative tests. Fitting an effective-range form to their data for $800 \leq M_{\pi K} \leq 1000$ MeV Barbaro-Galtieri *et al.* (1973), extrapolate‡ to a scattering length value $a_0^{\frac{1}{2}} = 0{\cdot}22 \pm 0{\cdot}04$ which is in good agreement with predictions from current algebra§ (e.g. $a_0^{\frac{1}{2}} \approx 0{\cdot}16$ (Griffith, 1968)).

(h) Other Related Channels—$K\bar{K}$ and $\pi\eta$

Although this topic exceeds the theme of peripheral (OPE) production, it is natural to refer to it at this point. Results are as yet limited, but they are important in filling out the picture of 0^-0^- production, especially for comparison with symmetry schemes (Chapter 13).

We have already referred (Section 3.3(f)) to data on pion-induced $K\bar{K}$ production ($\pi N \to K\bar{K}N$, etc.), which has a strong OPE component, leading to inferences on the process $\pi\pi \to K\bar{K}$. Besides the region of the S^*-effect which we stressed, there are now good data extending beyond 2 GeV (Männer, 1974). As well as signals from the well-established peripheral resonances ρ, f^0, A_2, g, there is also clear evidence for a 4^+ resonance at about 2 GeV. This has already been discussed in Section 3.3(e). Information on lower lying resonances must await detailed partial wave analysis.

Data on strangeness changing $K\bar{K}$ production ($\bar{K}N \to K\bar{K}Y$, etc.), although limited in statistics, furnish important information, for example on the f' (1514) resonance ($\Gamma \approx 40 \pm 10$ MeV), of which no signal is visible in data on $\pi\pi \to K\bar{K}$. With the increasing use of triggered detectors, statistics should improve, enabling more elaborate analyses to be made, perhaps allowing $\sigma(K\bar{K} \to K\bar{K})$ to be estimated.

Another important channel which would repay much more study is $\pi\eta$.‖ The only established effects are the $A_2(J^{PC} = 2^{++})$ and $\delta(J^{PC} = 0^{++})$ reso-

‡ Not only is this a long extrapolation but the formula used fails to cater for the sub-threshold zeros which current algebra implies.

§ Fox and Griss (1974), fitting to data on $K^-\pi^+$ and $K^+\pi^-$ production at 2 GeV/c and other momenta, using an effective range formula, obtain $a_0^{\frac{1}{2}} \approx 0{\cdot}25$, compatible with their own estimate $a_0^{\frac{1}{2}} \approx 0{\cdot}22 \pm 0.04$ for the current algebra value, after allowance for unitarity corrections according to a particular prescription.

‖ Any *P*-wave resonance would be an "exotic of the second kind" in the language of the quark model (cf. Chapter 13).

nances for which the Particle Data Group (1974) list the parameters:

$$A_2 : M \approx 1310 \pm 10 \text{ MeV}, \Gamma \approx 100 \pm 10 \text{ MeV} \; (A_2 \rightarrow \eta\pi / A_2 \rightarrow \text{all} \approx 0{\cdot}15),$$

$$\delta : M \approx 976 \pm 10 \text{ MeV}, \Gamma \approx 50 \pm 20 \text{ MeV}.$$

It is believed that the δ-resonance is also connected with an enhancement observed in $I=1$ $K\bar{K}$ production near threshold.

The δ (976) completes the family of 0^{++} scalars, for which we have discussed $I=0$ candidates (S^* (1000) and ε (1100–1300)) in Section 3.3(f), and the $I=\frac{1}{2}$ candidate, κ (1250), in Section 3.3(g). Along with that of the κ (1250), its width is an important ingredient in $SU(3)$ comparisons, fixing the octet coupling constant (cf. Section 3.3(f) and Chapter 13).

3.4. Nucleon–Antinucleon Annihilations

(a) Formation Channels

Nucleon–antinucleon annihilation affords an opportunity of studying meson systems in formation without the complications of baryons in the final state. Resonance signals can be sought in specific channels such as $N\bar{N} \rightarrow \pi\pi$, $K\bar{K}$, and also in the elastic and total cross-sections. A careful study of these reactions was partly stimulated by early results on the process $\pi^- p \rightarrow p +$ missing mass (Chikovani *et al.*, 1966) which indicated a series of narrow structures S, T, U, beyond the g meson, with squared masses continuing the linear sequence m_ρ^2, m_f^2, m_g^2. This is just what would be expected under a regime of linearly rising Regge trajectories, and is also what would be obtained in simple harmonic oscillator quark models (Greenberg, 1964). For this reason it is convenient to speak of the S, T, U mass regions, even though these early results have not been corroborated in subsequent missing-mass experiments (Bowen *et al.*, 1972). This terminology has added point since, although the original evidence for narrow resonance peaks is now largely discredited, a number of broader effects of roughly corresponding mass, have been reported in various $\bar{p}p$ experiments. (For general reviews see: Montanet (1972), Diebold (1972), Bizzarri (1973), and Montanet (1974).)

The simplest measurements indicating such effects are $\bar{p}p$ and $\bar{p}d$ total cross-sections, and small, but statistically significant, bumps have been seen. Experiments at Brookhaven over the momentum range 1·0 to 3·3 GeV/c (Abrams *et al.*, 1967, 1970) indicated bumps in the $I=1$ cross-section at masses of 2190 and 2350 MeV with widths of 85 MeV and 140 MeV, respectively, and also a bump in the $I=0$ cross-section at 2375 MeV with a width of 190 MeV. These effects have been confirmed in the $\bar{p}p$ case by Cohen *et al.* (1972), and preliminary results of an experiment over the lower momentum range 0·38 to 1·05 GeV/c show a further effect at 1930 MeV with width less than 20 MeV (Carroll *et al.*, 1974). These effects need not all be associated with resonance formation, and even if they are they say nothing about the associated J^P values.

The best way to obtain more detailed information is to study specific two-body reactions, in particular $\bar{p}p \to \pi\pi, K\bar{K}$; these couple exclusively to the natural parity channels in which we are primarily interested. At present, the reaction for which there is most information is

$$\bar{p}p \to \pi^+\pi^-.$$

Differential cross-sections for this process have been measured over the momentum range 0·7 to 2·4 GeV/c ($1{\cdot}99 \leq W \leq 2{\cdot}57$ GeV) in a series of spark chamber experiments (Fong, 1968; Nicholson *et al.*, 1969, 1973; Eisenhandler *et al.*, 1973). In addition, there are various bubble chamber experiments extending to lower energies‡ (cf. Bizzarri, 1973), and polarization measurements at a single energy, 1·64 GeV/c (Ehrlich *et al.*, 1972). (Cross-sections for the time-reversed reaction $\pi^+\pi^- \to p\bar{p}$ have also been extracted from data on $\pi^- p \to \bar{p}pn$ by identifying the OPE signal (Grayer *et al.*, 1972d; Ayres *et al.*, 1972; Hyams *et al.*, 1974). Comparison of these results with the directly measured cross-sections is very satisfactory, and lends support to other OPE extrapolations.) The observed differential cross-sections are richly structured and vary considerably with energy. This can be seen in Fig. 3.4.1 where we show a sample of recent results at four of the available energies (Eisenhandler *et al.*, 1973).

Before discussing the detailed interpretation of these cross-sections it is useful to make some comments concerning which partial waves one would expect to be excited from the general systematics of $\pi\pi$ reactions, and allowing for the lower centre-of-mass momentum available to the $p\bar{p}$ pair. Since $p\bar{p} \to \pi^+\pi^-$ is governed by non-vacuum exchanges, it is natural to compare with the $I_t = 1$ part of $\pi\pi$ scattering. This turns out to be predominantly peripheral, the cross-sections being dominated by resonances in partial waves corresponding to an impact parameter $R \approx 0{\cdot}8$ fm.§ One might therefore expect $\bar{p}p \to \pi\pi$ also to be peripheral, and dominated by resonances in partial waves corresponding to a similar impact parameter. The J-values corresponding to $R = 0{\cdot}8$ fm, as calculated from

$$[J(J+1)]^{\frac{1}{2}} = (kq)^{\frac{1}{2}}R, \qquad (3.4.1)$$

where k and q are the centre-of-mass momenta, are shown in Fig. 3.4.2 for

‡ The chief point of interest in these experiments is the decrease from a large positive value through zero of the forward–backward asymmetry as W varies from 1900 to 1950 MeV.

§ Note that even for the $\pi\pi$ system, the dominance by peripheral resonances is only approximate, lower (daughter) states also being excited.

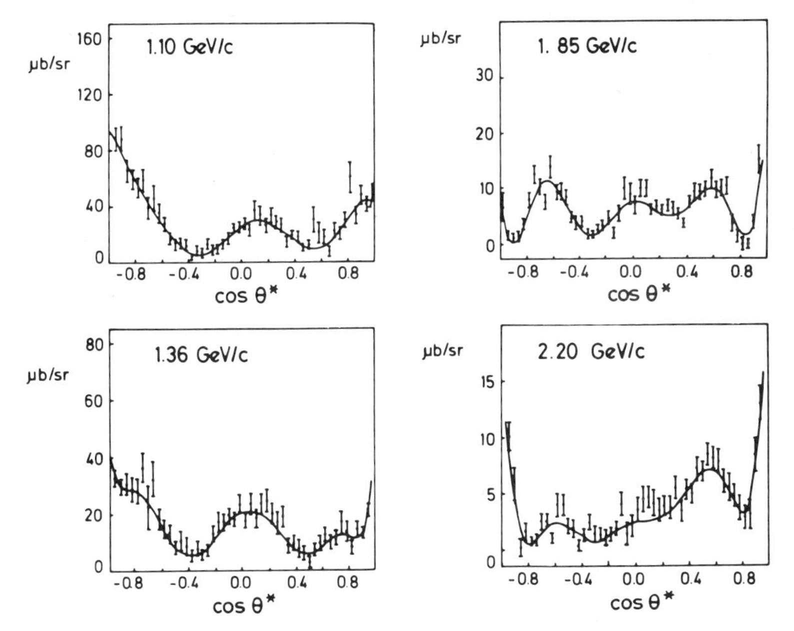

FIG. 3.4.1. Differential cross-sections for $\bar{p}p \to \pi^- \pi^+$ from the experiment of Eisenhandler *et al.* (1973). (This is only a sample of the data from this experiment.)

both reactions. From this figure it is clear that the annihilation process is not expected to be appreciably coupled to the leading meson trajectory. Thus at 1·5 GeV/c ($s \approx 5$ GeV^2) one might expect the cross-section to be dominated by the F-wave implying particular approximate relations between the coefficients A_l in

$$\frac{d\sigma}{d\Omega} = \sum_l A_l P_l(\cos\theta), \tag{3.4.2}$$

for $l \leq 6$. In practice the coefficients up to A_8 have appreciable values, and the simple relations implied by dominance of the F, or any other single partial wave, are not satisfied. It appears that significant contributions are required from several partial waves, including some with low J values; that is, those corresponding to a "central" rather than a "peripheral" interaction.‡ One does, however, expect the predominant J values which are excited to be somewhat lower than for elastic $\pi\pi$ scattering.

Up to the present, all analyses of the above data have been of the "resonance-plus-background" variety. Since many partial waves evidently

‡ A fuller discussion of peripherality, and the way in which the low-energy data are related to higher-energy behaviour via finite-energy sum rules, is given in Chapter 12.

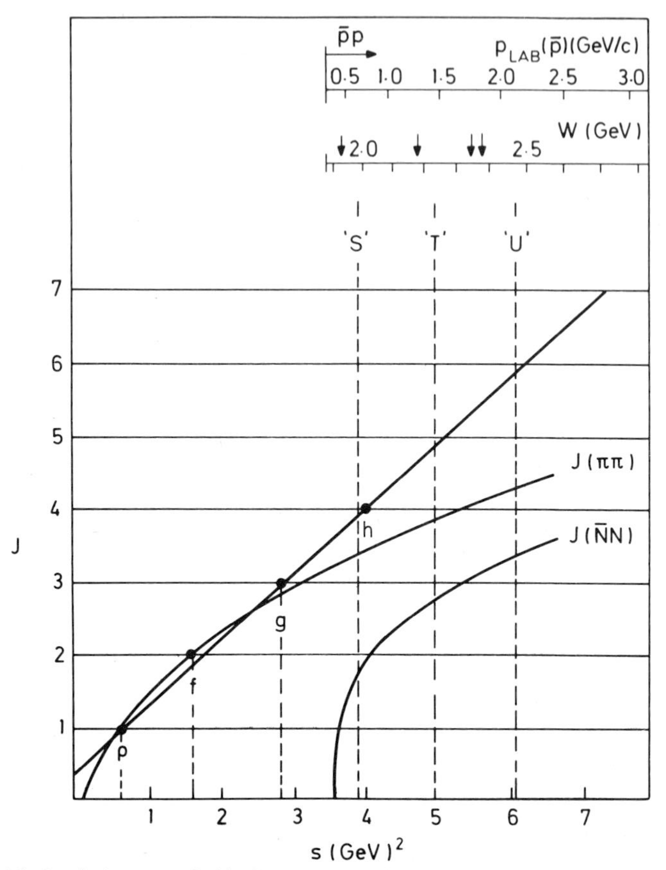

FIG. 3.4.2. Loci of most probable J-values excited in a peripheral interaction with interaction radius $R = 0{\cdot}8$ fm. $J(\pi\pi)$ and $J(N\bar{N})$ correspond to the reactions $\pi\pi \rightarrow \pi\pi$ and $N\bar{N} \rightarrow \pi\pi$, respectively. The arrows indicate bump positions in $\bar{p}p$ and $\bar{p}d$ total cross-sections.

contribute, any resonance parameters extracted from differential cross-section data will be very dependent on assumptions made concerning the background. An early attempt of this type was made by Nicholson *et al.* (1969) who achieved a fit to their folded angular distributions‡ using a simple parameterization involving just two resonances. However, this fit turned out to be in disagreement with the polarization data (Ehrlich *et al.*, 1972), and with the odd Legendre coefficients (Eisenhandler *et al.*, 1973). The authors of the latter paper also attempted to fit their own much more detailed data assuming resonance plus smooth backgrounds. They con-

‡ Positive and negative pions were not distinguished, so that the A_l values of equation (3.4.2) were only available for even l.

cluded that no resonances could be established on the basis of the present data.

An alternative analysis, incorporating a specific physical model for the background, has been made by Donnachie and Thomas (1973). The novelty of their scheme is that they assume a rather radical model for the non-resonant background. This is an extension of the quark rearrangement mechanism for high-t hadronic collisions of Gunion *et al.* (1972) to lower energies than hitherto. Donnachie and Thomas thus interpret the non-peripheral character of $p\bar{p}$ annihilations as being due to the presence of this mechanism. In addition, normal resonance and Regge contributions at low and intermediate energies are included. This is an interesting speculation, and the authors show how it can help in understanding the transition in the character of annihilation distributions above 2·5 GeV/c. As regards the analysis of the T, U region, the main effect is to achieve a strongly energy-dependent background. The resulting fit requires the existence of meson states at 2·13 GeV ($I=1, J^P=3^-$) and 2·34 GeV ($I=0, J^P=4^+$). The existence and parameters of these, of course, depends crucially on the model for the non-resonant backgrounds.

It is clear from the above discussion that the extraction of resonance parameters from present $p\bar{p}\to\pi^+\pi^-$ data is only possible on the basis of special models. A more general analysis must await forthcoming polarization measurements. In the meantime there have been some interesting speculations concerning more global features of the data, in particular the behaviour of the zero trajectories (Odorico, 1969, 1972), which are identified with the loci of dips in the appropriate angular distributions. Eisenhandler *et al.* (1973) have done this for their cross-section data (Fig. 3.4.1), leading to the zero trajectories of Fig. 3.4.3. Also shown are the zero trajectories for $\pi^\pm p$ scattering, and the positions of prominent resonances in the s, t, u channels. As we will discuss in Section 12.3(d), Odorico has suggested that the linear zero trajectories implied by the Veneziano model may well obtain in nature, despite the other failings of this model. In this picture, the zero trajectories interpolate between crossed channel pole intersections, are approximately linear, and interpolate from channel to channel. In the present case, it is natural to try to associate the backward dips in $\pi^\pm p$ scattering with the near forward and backward dips in the annihilation channel.

A noticeable feature of the dip patterns for the annihilation process is the approximate forward–backward asymmetry (cf. Fig. 3.4.1) which may be in keeping with the idea of doubly-spaced Regge trajectories (i.e. coincident resonance of the same parity) (Bugg, 1974). This is not to say that such a zero pattern is necessitated. Within the framework of simple dual models this depends on the nature of the crossed channels and the relations between the

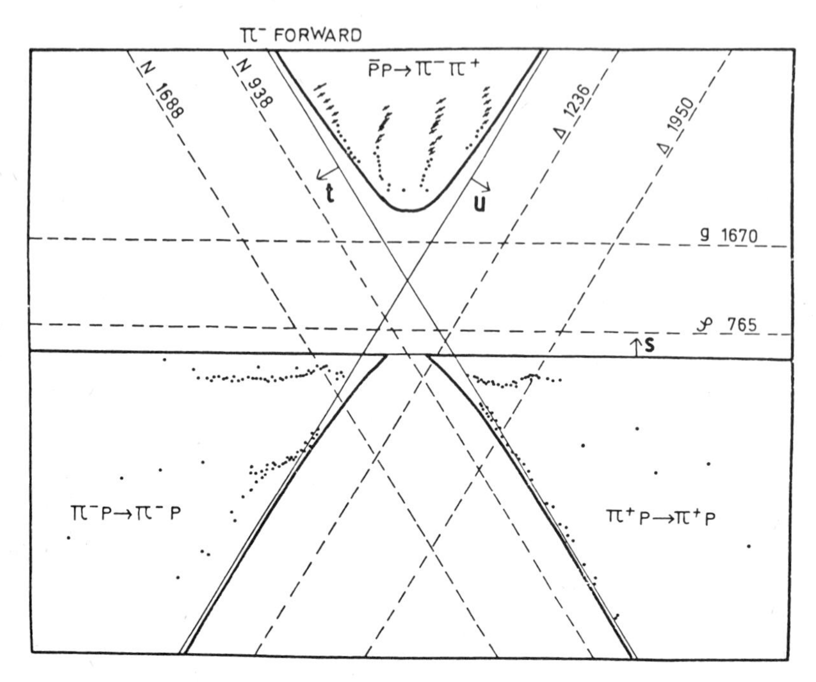

FIG. 3.4.3. The four lines of data points show zero positions as identified by dips in the differential cross sections for $\bar{p}p \to \pi^-\pi^+$ (Eisenhandler *et al.*, 1973). The four sets of dots (two each for $\pi^{\pm}p \to \pi^{\pm}p$) show the zero trajectories for $\pi^{\pm}p$ elastic scattering, and the dashed lines indicate the positions of prominent resonances in each of the three channels.

trajectory functions. For the case of $\pi^+\pi^-$ elastic scattering, Eguchi *et al.* (1974) have emphasized that if in the Veneziano model one requires $\alpha(s)+\alpha(t)+\alpha(u)=1$, then although doubly-spaced trajectories with fixed-u zeros are obtained, the zero patterns are forward–backward symmetric only at the resonance energies (cf. Chapter 12).

Besides the $p\bar{p} \to \pi^+\pi^-$ reaction, there are also data on the analogous reactions $p\bar{p} \to K^+K^-$, $K_s^0K_L^0$, but they are of lower quality. The K^+K^- distributions (see e.g. Eisenhandler *et al.* (1973)) differ appreciably from their $\pi^+\pi^-$ counterparts. Although in principle additional odd G-parity resonances could couple in this case, the Legendre coefficients show less structure; this may be connected with the fact that the u-channel exhibits no resonances (i.e. is "exotic" in the language of the quark model—see Chapter 13). The data on the neutral modes are very sparse and, at present, interest centres on the conflicting evidence as to whether there is or is not a narrow (35 MeV) wide bump in the total cross-section at 1970 MeV. For details on this point, we refer to the review of Bizzarri (1973).

(b) $N\bar{N} \to 3\pi$ and Related Processes

(i) *Introduction*

Studies of nucleon–antinucleon annihilation to three mesons, especially to three pions, have proved stimulating but ambivalent. They do not yield model-independent information on the $\pi\pi$ system, but in the context of models have much to say about the resonance spectrum, especially on the existence of daughter trajectories (cf. Chapter 13). In particular, a P-wave daughter of the f^0, coupled strongly to $\pi\pi$, is persistently demanded. This is in contradiction to all other evidence, and is an undoubted indication that the models used are inadequate. Nevertheless, they have aroused considerable interest, and some of their properties, for example the suggested zero structure, may possibly be of more general validity than the resonance spectra and couplings demanded by particular analyses.

The most readily available data are on capture at rest. Slow $\bar{p}$ beams experience large ionization losses and readily stop (for example, a 500 MeV/c $\bar{p}$ beam has a range of only 1 metre in liquid hydrogen). On being brought to rest, the $\bar{p}$, like other negative particles, undergoes a process of capture into some high $(\bar{p}p)$ atomic orbital, followed by a cascade down through atomic levels, culminating in "nuclear" capture, in this case the annihilation process. It is usually believed, following the original discussion of Day, Snow and Sucher (1960), that the annihilation reaction goes overwhelmingly via S-wave capture,‡ but, as we shall see, there exists evidence to the contrary.

Selection rules follow from parity, isospin, and charge conjugation invariance. For $N\bar{N}$ annihilations the latter two concepts entail the conservation of G-parity.§ All isospin multiplets which are eigenstates of charge conjugation (zero baryon number and strangeness) are assigned a G-parity, plus or minus; for example $G(\pi) = -1$. Since G-parity is a multiplicative quantum number, a system which decays under strong interactions to n pions has $G = (-1)^n$.

For an $N\bar{N}$ system of spin S, isospin I ($I, S = 0, 1$) and orbital angular momentum L, $G = (-1)^{I+L+S}$ (the same rule governs $Q\bar{Q}$ states in the quark model (cf. Chapter 13), $K\bar{K}$ states, and indeed any compound of an isodoublet with its anti-particles).

‡ For a fairly recent appraisal see Bizzarri (1973). The effect of relaxing this assumption will be discussed later.

§ The G-parity operator was originally introduced in just this context (Michel, 1953; Amati and Vitale, 1955; Goebel, 1956; Lee and Yang, 1956).

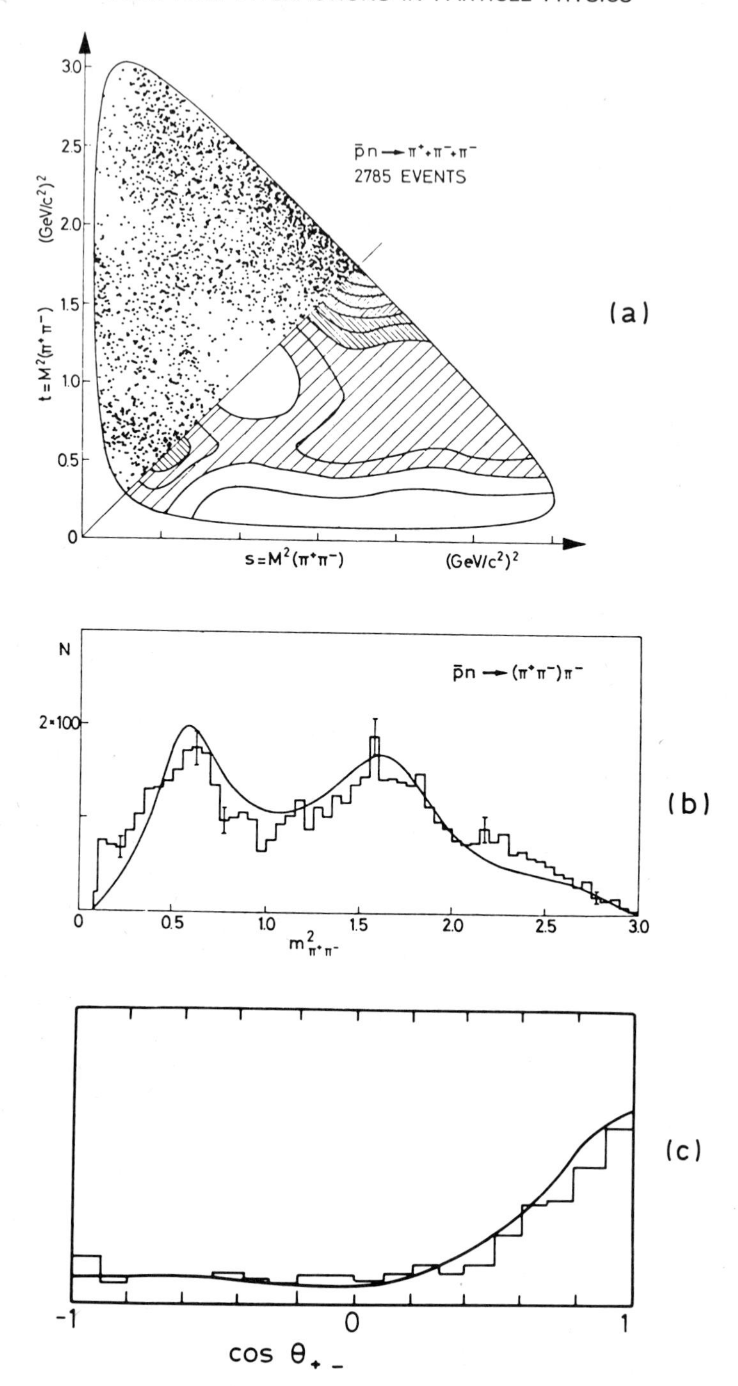

3.0
2.5
2.0
1.5
1.0
0.5
0
$t = M^2(\pi^+\pi^-)$ $(GeV/c^2)^2$
$\bar{p}n \to \pi^+ + \pi^- + \pi^-$
2785 EVENTS
$s = M^2(\pi^+\pi^-)$ $(GeV/c^2)^2$
(a)
N
$\bar{p}n \to (\pi^+\pi^-)\pi^-$
2×100
0
0.5
1.0
1.5
2.0
2.5
3.0
$m^2_{\pi^+\pi^-}$
(b)
-1
0
1
$\cos\theta_{+-}$
(c)

(ii) *Annihilation at Rest*

The most interesting reaction studied is

$$\bar{p}n \to \pi^+\pi^-\pi^-, \tag{3.4.3}$$

observed on deuterium at rest‡. The Dalitz plot distribution for this process is shown in Fig. 3.4.4(a), and the projection on the $M^2_{\pi^+\pi^-}$ axis in Fig. 3.4.4(b) (Anninos *et al.*, 1968). Conspicuous features of the data are: (1) the hole in the centre of the plot, and (2) the pronounced concentration of events for

$$s(\equiv M^2_{\pi_1^-\pi^+}) \approx t(\equiv M^2_{\pi_2^-\pi^+}), \qquad u(\equiv M^2_{\pi_1^-\pi_2^-}) \qquad \text{small.}$$

The latter effect is brought out very clearly in Fig. 3.4.4(c), which shows the distribution across the Dalitz plot§ in the f^0 band $1{\cdot}14 \leq M_{\pi^+\pi^-} \leq 1{\cdot}37$ GeV.

As mentioned above, it has been conventional to discuss annihilation at rest assuming that the reaction goes entirely via S-wave capture. According as the capture proceeds via the singlet or triplet spin state, $S = 0$, 1, and isospin state $I = 0$, 1, the final state is required to have spin-parity $J^P = 0^-$ or 1^-, with G-parity $G = (-1)^{I+S} \equiv (-1)^{I+J}$. In particular, annihilation to three pions is pictured as proceeding either through a state with the quantum numbers of the ω ($I = 0$), or the pion ($I = 1$), only the latter possibility being open for the particularly interesting case of $\bar{p}n$ annihilation. Annihilation at rest to *two* pions would likewise go via a state with the quantum numbers of the rho.

Published S-wave analyses fall into one of two classes. In the first, resonance fits are made with the production amplitudes assumed to be sums of isobar contributions of the form

$$M(s, t) = \{A_0(s) + A_1(s)[t-u] + A_2(s)[3(t-u)^2 - (t+u)^2]\} + (t \leftrightarrow u). \tag{3.4.4}$$

Breit–Wigner forms are assumed for the A_l. This was the approach of Anninos *et al.* (1968), who made fits to their data with the isobar combinations $\pi^-\rho^0$, π^-f^0 and $\pi^-\varepsilon$. Here ε denotes a broad S-wave resonance in the rho-region, which serves to represent the large δ_0^0 phase in this energy range. The authors concluded that $\pi\rho$ did not couple appreciably, and that, within

‡ Annihilation in flight will be considered below.
§ For given s, the distribution across the Dalitz plot, i.e. in $(t-u)$, corresponds linearly to the cosine of the $\pi_1^-\pi^+$ dipion decay angle θ_{+-} in its centre of mass.

FIG. 3.4.4. Data on $\bar{p}n \to \pi^+\pi^-\pi^-$ at rest from Anninos *et al.* (1968) compared to theoretical fits: (a) Dalitz plot distribution; (b) projection on $M^2_{\pi^+\pi^-}$ axis; (c) distribution across the f^0 band $1{\cdot}14 \leq M_{\pi^+\pi^-} \leq 1{\cdot}37$ GeV. The theoretical fits in (a), and (b), are from Lovelace (1968), and in (c) is from Pokorski *et al.* (1972).

the model, a strong $I=2$ S-wave interaction near threshold was needed to account for the enhancement exhibited in Fig. 3.4.4(c). This is in conflict with the behaviour of this wave indicated in the much more reliable dipion production studies. Subsequently, Gleeson *et al.* (1970) considered more elaborate versions of this final-state, or isobar, type of model involving a fourteen parameter fit to the data. They included S- and P-wave daughter states of the f^0, as well as the states considered by Anninos *et al.* (1968), and succeeded in accounting for the experimental distributions without an anomalous $I=2$ S-wave contribution. They concluded that the most important contributions arise from the two S-wave daughters, but a P-wave daughter of the f^0 (cf. Chapter 12) is also required.

The enriched resonance spectrum adopted by Gleeson *et al.* (1970) is that suggested by the Veneziano model for $\pi\pi$ scattering, discussed at length in Section 12.3. This model was first applied to the $\bar{p}n$ annihilation problem by Lovelace (1968) as part of his original application of Veneziano's ideas to $\pi\pi$ dynamics. The annihilation is again pictured as proceeding via a one-pion intermediate state as in Fig. 3.4.5. The production amplitude is then

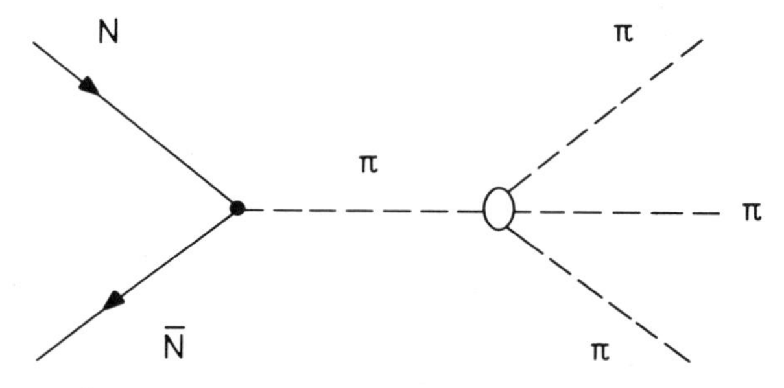

FIG. 3.4.5. $N\bar{N}\to 3\pi$ at rest pictured as proceeding via a one-pion intermediate state.

proportional to the $\pi\pi$ scattering amplitude continued to the appropriate kinematic region; the virtual pion is thus a long way off mass shell. Lovelace (1968), and subsequent followers of his approach (Altarelli and Rubinstein, 1969; Gopal *et al.*, 1971) then represent this amplitude by a sum of "Veneziano" contributions,‡

$$M(s,t)=\sum_{n,m} C_{nm}V_{nm}(s,t). \tag{3.4.5}$$

Here

$$V_{nm}(s,t)\equiv\frac{\Gamma[n-\alpha(s)]\Gamma[m-\alpha(t)]}{\Gamma[m+n-\alpha(s)-\alpha(t)]}, \tag{3.4.6}$$

‡ The properties of such terms, and the resonance spectra associated with them, are discussed in detail in Section 12.3 and 13.2.

and the values of n, m, are restricted to $n \geq 1$, $m \geq 0$. The trajectory function $\alpha(s)$ is assumed to be of the form

$$\alpha(s) = \alpha_0 + \alpha' s + i\alpha''(s - 4m_\pi^2)^{\frac{1}{2}}\theta(s - 4m_\pi^2). \tag{3.4.7}$$

The imaginary part enables finite-width corrections to the narrow-width Veneziano forms ($\alpha'' = 0$) to be taken into account.‡ The parameter α'' and the coefficients C_{nm} are adjusted to fit the distributions over the Dalitz plot. An immediate feature of this model is that provided the condition

$$n + m \leq 3, \tag{3.4.8}$$

is satisfied, the pole in the denominator of equation (3.4.6) at $\alpha(t) + \alpha(s) = 3$ implies a zero for the amplitude at the same point. This zero corresponds to the observed "hole" in the centre of the Dalitz plot distribution Fig. 3.4.4(a), and constitutes the main success of the model.

Lovelace (1968) considered just the contribution of V_{11}. This is quite different from his model for on-shell and nearly on-shell scattering, which retains instead only V_{10}. The effect of changing from V_{10} to V_{11} is to decouple the resonances on the leading trajectory (i.e. the ρ^0 and f^0) so that the important resonance contributions come necessarily from the daughter states. In this way a rather good fit was obtained to the highly structured Dalitz plot with very few parameters, as is shown in Fig. 3.4.4. The choice of V_{11} is, however, quite arbitrary, and subsequent authors have performed fits to the mass distributions (Altarelli and Rubinstein, 1969) and to the whole Dalitz plot (Gopal *et al.*, 1971) allowing contributions from several terms. Subsequently, Pokorski *et al.* (1972) stressed the similarities between this and the final-state resonance approach described above, and made explicit the resonance content of the alternative Veneziano descriptions, listing the couplings to the various states. These differ widely between different fits, but in all cases there is an appreciable contribution from the P-wave daughter of the f^0 which we emphasize is required to couple strongly to $\pi\pi$. This is in direct contradiction to all other evidence, and is an indication of serious failings in the above models.§ In practice, the role of this P-wave daughter is apparently to explain the sharp forward peaking‖ in the f^0 mass range, shown in Fig. 3.4.4(c). Since the domain in question corresponds to the

‡ The problems inherent in this procedure are discussed in Section 12.3(c).

§ As discussed in Sections 12.3 and 13.2, the prediction of a P-wave daughter of the f^0 coupling strongly to $\pi\pi$ is a serious failing of the *single-term* Veneziano model for this process. It is not necessarily required within a general Veneziano model, with satellites allowed. Note that a P-wave state of this mass, which decouples from $\pi\pi$, decaying for instance to $\pi\omega$, is not forbidden by present evidence. Such a possibility has been suggested from electromagnetic form factor considerations (see Shaw (1972), Bramon (1973) and Section 5.4).

‖ The only fit without this resonance is that of Anninos *et al.* (1968) who, as noted above, invoked an equally anomalous $I = 2$ S-wave interaction to give a partial explanation of the same effect.

overlap of the f^0 bands in the s and t channels, this may well be associated with a breakdown of the simple isobar model assumptions, which are common to both types of ansatz considered above.

Another possible source of difficulty with all the above models is the assumption of pure S-wave capture. Experimental evidence to the contrary comes from the reported observation of annihilation at rest to $\pi^0\pi^0$ (Devons *et al.*, 1971), and on the ratio $(\bar{p}d \to \pi^-\pi^0 p)/(\bar{p}d \to \pi^-\pi^+ n)$, for which the result $0{\cdot}68 \pm 0{\cdot}07$ has been reported (Gray *et al.*, 1973); if all capture proceeded via S-waves, in this case via the $^3S_1$ state, the ratio would be 2, assuming isospin invariance. The latter result implies that $(75 \pm 8)\%$ of all annihilations at rest to two pions proceeds via odd L-states; the former result implies $(39 \pm 8)\%$ for the same ratio. These observations specifically relate to annihilation to *two* pions, but they indicate that capture at rest from other orbital states, in particular from the P-state, may also have to be included in discussing annihilation to three pions.

Donnachie and Thomas (1974) have performed an isobar model analysis allowing P-wave as well as S-wave annihilation. In this way they are able to account for the data without invoking anomalous effects; no P-wave daughter of the f^0 and no low-energy $I=2$ enhancement being required. The amount of P-wave annihilation suggested by the fit is about 70% in accord with the large amount of P-wave annihilation suggested by the data on two-pion channels.

In view of these difficulties, what does the Veneziano ansatz accomplish? Consider, for example the contribution V_{11}. This has poles at integer values of $\alpha(s)+\alpha(t)$ (suitably displaced into the complex plane by the $\operatorname{Im}\alpha$ component of equation (3.4.7)), corresponding to the ρ and f^0 bands, and a family of fixed-u zeros corresponding to integer values of $\alpha(s)+\alpha(t)$ from the gamma function in the denominator.‡ As emphasized by Odorico (1970), these lines of zero pass through the network of s and t pole intersections (cf. Fig. 3.4.6) eliminating double (i.e. simultaneous) poles. Away from the pole intersections, the zero trajectories are associated with dips in the intensity distribution. Achieving the observed zero pattern is a success of the Veneziano approach, the hole in the centre of the Dalitz plot corresponding to $\alpha(s)+\alpha(t)=3$. As Odorico has stressed, simple zero patterns may be a feature of more general validity than the Veneziano ansatz itself, since zero trajectories passing through the intersection points are demanded on more general grounds, and their pattern may well be simple. Such a scheme can be extended to other intermediate spin-parity states, "S"$\to 3\pi$ by, invoking the appropriate $S\pi \to \pi\pi$ Veneziano amplitude§

‡ For a fuller discussion of these zeros, see Section 12.3(d).

§ One could also invoke the B_5-model, which is a specific extension of Veneziano's scheme to three-body production (see e.g. Berger (1971)).

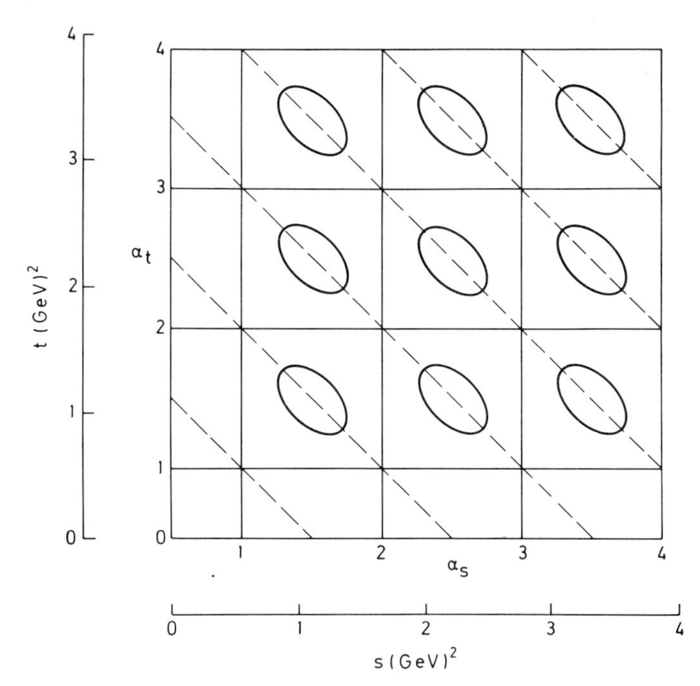

FIG. 3.4.6. Zero pattern on the (s, t)-plane as suggested by the Veneziano model. The pattern is to be compared with the Dalitz plots of Figs 3.4.4(a) and 3.4.7.

(Goebel, Blackmon and Wali, 1969). This results in contributions closely resembling (3.4.6), but with additional factors to take account of the spin coupling to the initial beam direction (which are of course absent for the 0^- contribution):

$$M^s(s, t) = \frac{\Gamma[m-\alpha(s)]\Gamma[n-\alpha(t)]}{\Gamma[m+n-p-\alpha(t)-\alpha(s)]} \times (\text{spin factors}),$$

with $p \geq 0$. It is clear that the same zero pattern *can* prevail as in the situation where just one intermediate partial wave operates, but a high degree of correlation among different resonance contributions is entailed. However, such possibilities have to be kept in mind, particularly when discussing capture in flight.

(iii) *Annihilation in Flight*

Experimental results for capture in flight (Bettini *et al.*, 1971) are shown in Fig. 3.4.7 in the form of a plot of contours of equal density. The features (1) and (2) noted above in the at-rest data are reproduced at approximately the same value of s and t. In addition, there are further features in the extended

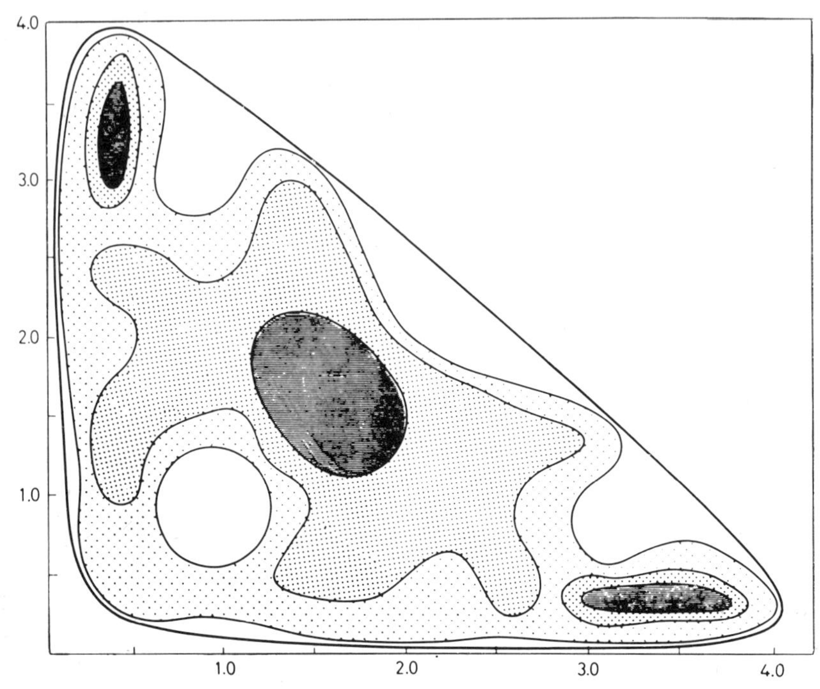

FIG. 3.4.7. Data on $\bar{p}n \to \pi^+\pi^-\pi^-$ in flight (1·2 GeV/c) presented in the form of contours of equal densities (Bettini *et al.*, 1971).

portion of the (s, t) plane, in particular, (3) maxima at large s, small t, and large t, small s.

At these momenta, it makes little sense to attempt analyses based on the $\pi\pi$ Veneziano model and hence predominantly S-wave capture.‡ Furthermore, the Veneziano pattern of zeros, which it might be hoped would extend into the new situation, predicts unobserved dips at (s, t) and (t, s) $\approx (2{\cdot}3,\ 1{\cdot}0)\ \text{GeV}^2$. This latter circumstance prompted Odorico (1970) to propose an alternative form for the production amplitude with a different pattern of zeros. However, this model does not explain the high s, low t, and high t, low s maxima, and entails exotic $I = 2$ meson resonances. Another feature of the scheme is the implication of an appreciable component of P-wave capture at rest. This is in contrast to previous assumptions regarding at-rest data (and, indeed, to many aspects of the annihilation data to various channels (Bizzarri, 1972)), but has some support, at least in the 2π-channel, from the experiments of Devons *et al.* (1971) and Gray *et al.* (1973), and the fit to $\bar{p}n \to 3\pi$ of Donnachie and Thomas (1974) discussed above.

‡ Veneziano type fits from a single partial wave have been made by Bettini *et al.* (1971), but are of questionable significance.

Finally, it is worth emphasizing that more information is available for capture in flight than is contained in a Dalitz plot distribution. The full kinematic description of in-flight capture is given in terms of five variables, the three sub-energies (squared) of the pairs of two-body final states, and two Euler angles relating the initial beam direction to the final configuration (cf. Section 3.1 and Appendix B). Evidence of various J^P contributions should be primarily sought in the angular moments in the latter angles, whereas in practice most attempts to discriminate among alternative J^P contributions have relied on identifying the appropriate kinematic threshold factors on the Dalitz plot in the conventional way.‡

(iv) *Other Processes*

Other charge modes of $\bar{p}$ annihilation at rest have also been studied. In view of the uncertainties associated with the interpretation of all such processes we shall discuss them only briefly. For example, Jengo and Remiddi (1969) have discussed the process

$$\bar{p}p \to \pi^-\pi^+\pi^0.$$

In this case, S-wave capture can proceed from the singlet (π-like) or from the triplet (ω-like) $\bar{p}p$ S-state. Jengo and Remiddi estimate that the singlet state contributes 4.4% of the total $\bar{p}p \to \pi^-\pi^+\pi^0$ annihilation. This component can be expressed in terms of the amplitude for $\bar{p}n \to \pi^-\pi^-\pi^+$. The remaining triplet contribution appears to be mainly dominated by ρ bands in the three alternative pion pairings. The main phenomena are thus explicable in terms of known effects. Regarding the neutral modes, Devons *et al.* (1973) have reported data on annihilation at rest to $3\pi^0$ and to $\pi^0\pi^0\eta$. The Dalitz plot for the $3\pi^0$ channel resembles that for $\pi^-\pi^-\pi^+$ in having a depletion of events at the centre.

An interesting study of the process

$$\bar{p}p \to \pi^+\pi^-\omega,$$

as seen in $\pi^+\pi^-(\pi^+\pi^-\pi^0)$ final states from annihilation at rest has been made (Bizzarri *et al.*, 1969). There are technical problems in assessing the background to true ω production, but assuming this to be satisfactorily surmounted, the resulting $\pi^+\pi^-\omega$ distribution is analysed as arising from a superposition of resonance contributions to the production amplitude. In the $\pi^\pm\omega$ sub-systems, 1^+ and 1^- contributions are required near the mass of the B meson. The 1^- signal could be construed as a P-wave daughter of the

‡ Identifying the dominant isobar-particle orbital angular momentum contributions from the p^{2l} dependence on the recoil momentum is the analogue of selecting dominant partial wave contributions to two-body scattering from the momentum dependence of σ_{tot}.

f^0. However, in view of the presumably spurious signal for such a state in the $\pi\pi$ subsystem of 3π annihilations, this should be regarded with reservation. Another interesting effect is that the $\pi^+\pi^-$ mass spectrum has a pronounced dip at about 940 MeV. The authors attribute this to a zero in the $I = 0$ S-wave $\pi\pi$ amplitude. A similar effect is seen in $\bar{p}p \to \pi^+\pi^+\pi^-\pi^-$ (Diaz *et al.*, 1970). It is tempting to associate these observations with the S^* effect (cf. Section 3.3(f)), despite the shift in mass.

This concludes our survey of three-meson annihilation processes with emphasis on 3π-channels. As we have seen, there are difficulties in existing models which will require a more highly developed three-body dynamics for their resolution. Consequently, despite the great interest aroused by this topic, its residual impact on the study of the dipion system is so far rather small.

Chapter 4

Processes Involving Dipion Exchange

4.1. Elastic Pion–Nucleon Scattering

In this section we will consider what can be learnt about $\pi\pi$ scattering from experimental information on the reactions

$$(s) \quad \pi + N \to \pi' + N',$$

and

$$(u) \quad \pi' + N \to \pi + N',$$

where the usual Mandelstam variables are used to denote the respective channels. The basic idea is to use the data on these reactions to deduce values for the amplitudes for the unphysical process

$$(t) \quad \pi + \pi' \to \bar{N} + N',$$

and then to obtain the $\pi\pi$ phase shifts by use of the generalized unitarity condition. This was in fact one of the first ways used to learn about the $\pi\pi$ interaction, and early work (see e.g. Hamilton and Spearman (1961)) established the need for a strong attractive $\pi\pi$ contribution to πN scattering. However, quantitative information is more difficult to obtain.

The problem is somewhat analogous to the well-known case of the nucleon electromagnetic form factors, where, although the existence of vector mesons was first predicted from just such considerations, attempts to gain detailed information on the parameters of these resonances have been less successful. We might expect that similar difficulties could occur in the

πN case. Some information about the errors involved is given by the fact that, while the method is used mainly to learn about the $I = J = 0$ phase shift δ_0^0, it can also be used for the $I = J = 1$ wave, on which much more reliable information is available from other sources. By considering how the method approximates the known answer in the latter case, one gets a useful clue as to the reliability of the result obtained for the less well-known δ_0^0 phase shift.

How in fact is the information obtained? The relevance to $\pi\pi$ scattering results, as noted above, from applying the generalized unitarity condition in the unphysical region between the 2π and 4π thresholds in the t-channel (Mandelstam, 1960). Thus, for an amplitude leading to a given eigenstate of I, J

$$f_J^I(t) = \pm|f_J^I(t)| \exp[i\delta_J^I(t)], \quad (4.1.1)$$

where $\delta_J^I(t)$ is the corresponding $\pi\pi$ phase shift. Although this relation is only strictly valid for the elastic region $4 \leqslant t < 16$, it is commonly used over the range $4 \leqslant t \leqslant 50$, the justification being that the coupling for $2\pi \to 4\pi$ appears to be weak below ≈ 900 MeV.

Thus, the basic idea is very simple. In the physical region of the s-channel, $s \geqslant (m+\mu)^2, t \leqslant 0$, the πN amplitudes can be obtained from scattering data, either directly or via phase shifts. If these amplitudes can then be analytically continued to the t-channel physical region $4 \leqslant t \leqslant 50$, $s \leqslant 0$, then the $\pi\pi$ phase shifts may be obtained from equation (4.1.1).

In fact, the analytic continuation is not carried out directly using amplitudes which are functions of both s and t, but, in practice, either partial wave dispersion relations, or πN backward dispersion relations, are introduced as intermediate steps in order to reduce the problem to a continuation in a single variable. Although the method using the partial wave dispersion relations was the first to be applied (Hamilton and Spearman, 1961) and yielded interesting results, that based on backward angle dispersion relations (Atkinson, 1962) is much simpler because of the simpler analytic properties of these amplitudes, and we will describe this method first. However, since we will work in terms of t-channel helicity amplitudes, we must firstly introduce some necessary notation.

(a) Helicity Amplitudes for $\pi\pi \to N\bar{N}$

Pion-nucleon scattering is described by two invariant amplitudes $A(s, t)$ and $B(s, t)$ (Chew *et al.*, 1957a), and it is convenient to introduce the isospin combinations $A^{(\pm)}$ and $B^{(\pm)}$, where, in the s-channel

$$A^{(+)} = \tfrac{1}{3}(A^{\frac{1}{2}} + 2A^{\frac{3}{2}}); \qquad A^{(-)} = \tfrac{1}{3}(A^{\frac{1}{2}} - A^{\frac{3}{2}}),$$

and, in the t-channel

$$A^{(+)} = \frac{1}{\sqrt{6}} A^0; \qquad A^{(-)} = \tfrac{1}{2} A^1.$$

In the t-channel the invariant amplitudes may be expanded into helicity amplitudes $f_{\pm}^J(t)$ for the process $\pi\pi \to N\bar{N}$ where $+(-)$ refers to the nucleon non-flip (flip). The expansions are (Frazer and Fulco, 1960a,b)

$$A = \frac{-8\pi}{p^2} \sum_{J=0}^{\infty} (J+\tfrac{1}{2})(pq)^J \left\{ P_J(z) f_+^J(t) - \frac{mzP_J'(z)}{[J(J+1)]^{\frac{1}{2}}} f_-^J(t) \right\}, \tag{4.1.2}$$

and

$$B = 8\pi \sum_{J=1}^{\infty} \frac{(J+\frac{1}{2})}{[J(J+1)]^{\frac{1}{2}}} (pq)^{J-1} P_J'(z) f_-^J(t), \tag{4.1.3}$$

where the pion and nucleon centre-of-mass momenta q and p are given by

$$q^2 = t/4 - \mu^2; \qquad p^2 = t/4 - m^2,$$

and

$$z \equiv \cos\theta_t = (s + p^2 + q^2)/(2pq).$$

Because of Bose statistics the sums run over even (odd) values of J for $A^{(+)}$ and $B^{(+)}$ ($A^{(-)}$ and $B^{(-)}$).

The phase condition (4.1.1) may now be expressed more precisely as

$$f_{\pm}^J(t) = |f_{\pm}^J(t)| \exp[i\delta_J(t)], \tag{4.1.4}$$

where, for J even (odd) the $\pi\pi$ phase shift $\delta_J(t)$ has $I = 0(1)$.

(b) Backward Dispersion Relations

A useful variable to use in this case is $\nu \equiv q^2(s)$. The singularities of $A(\nu, \cos\theta_s = -1)$ as a function of ν are the two cuts $-\infty < \nu \leqslant -\mu^2$ and $0 \leqslant \nu < \infty$. The amplitude $B(\nu, \cos\theta_s = -1)$ also has these cuts, and, in addition, a pole at

$$\nu = \nu_0 \equiv -\mu^2(1 - \mu^2/4m^2),$$

due to the single-nucleon term. It is also convenient to introduce the spin non-flip amplitude $F^{(+)}(t)$ defined by

$$F^{(+)}(t) \equiv \frac{1}{m} A(\nu, \cos\theta_s = -1) + \frac{\omega}{E} B(\nu, \cos\theta_s = -1), \tag{4.1.5}$$

where $E = (m^2 + q^2)^{\frac{1}{2}}$; $\omega = (\mu^2 + q^2)^{\frac{1}{2}}$ and $t = -4\nu$ for $\cos\theta_s = -1$.

The main part of the problem is, as shown in Fig. 4.1.1, to analytically continue from the right-hand cut, where the amplitude can be obtained from πN scattering data, to the left-hand cut, from which one can deduce $\pi\pi$ phase shifts. One way to proceed is as follows: firstly the problem is reduced from one of continuing away from a cut to one of continuing away from a region of holomorphy by introducing a so-called *discrepancy function* (Hamilton *et al.*, 1962a,b,c), defined in this case by

$$\Delta_F(\nu) \equiv F^{(+)}(\nu) - \frac{4\pi f^2}{m(\nu-\nu_0)} - \frac{1}{\pi}\int_0^{\infty} d\nu' \frac{\operatorname{Im} F^{(+)}(\nu')}{\nu'-\nu}, \tag{4.1.6}$$

where f^2 is the πN pseudo-vector coupling constant. From the analytic properties of $F^{(+)}(\nu)$ it follows that $\Delta_F(\nu)$ has only the cut $-\infty < \nu \leqslant -\mu^2$.

In the physical region of the s-channel $\nu \geqslant 0$, the first term on the right-hand side of (4.1.6) can be evaluated from πN phase shifts for $\nu \leqslant \nu_p$ where ν_p corresponds to the highest energy at which such phase shifts are available. The second term is given in terms of the known πN coupling constant, and the integrand of the third term can be evaluated using πN

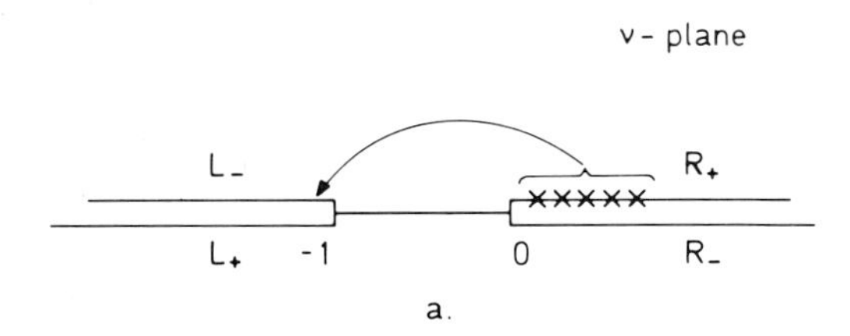

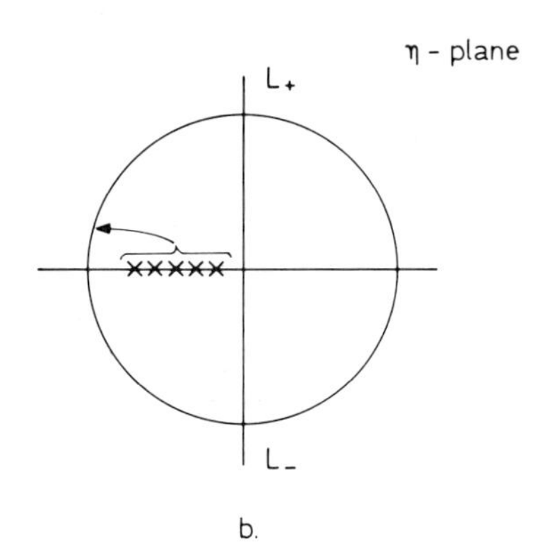

FIG. 4.1.1. Singularities of the backward amplitude $F^{(+)}(\nu)$ as a function of (a) ν, and (b) the conformal variable $\eta(\nu)$.

phase shifts for $\nu' \leqslant \nu_p$ and a Regge pole model at higher energies. Thus, a complete evaluation of $\Delta_F(\nu)$ for $0 \leqslant \nu \leqslant \nu_p$ is possible.

The problem now is to continue $\Delta_F(\nu)$ to values of $\nu \leqslant -\mu^2$. The classical method of analytic continuation from a holomorphic region is by a series expansion. However, an expansion in ν cannot be used as it diverges for all $\nu < -\mu^2$. To overcome this problem Atkinson (1962), following earlier work (Frazer, 1961; Ciulli and Fischer, 1961), conformally mapped the ν-plane into a circle in the $\eta = \eta(\nu)$ plane, where the cut $-\infty < \nu \leqslant -\mu^2$ maps on to the boundary of the unit circle, as shown in Fig. 4.1.1. The discrepancy $\Delta_F(\eta)$ can now be expanded in a power series in η, i.e.

$$\Delta_F(\eta) = \sum_n a_n \eta^n,$$

and the coefficients a_n determined from the fit on the right-hand cut. Since this series can be shown to converge on the whole circle (Atkinson, 1962), it can be used to continue $\Delta_F(\eta)$ to values on the circle, and hence, by mapping back to the ν-plane, to obtain values of $\Delta_F(\nu)$ for $\nu \leqslant -\mu^2$, as required.

The above extrapolation is to the boundary of a region of holomorphy. Recent work on the mathematics of the stability of such extrapolations (Bowcock and John, 1969; Pišút *et al.*, 1969; Prešnajder and Pišút, 1969), has shown that in the absence of *exact* input information the results obtained are unstable if one extrapolates to the value of the function at a point, but *are* stable if, instead, one extrapolates to the *average value over an arc*, the stability increasing with the length of the arc. In recent applications to backward πN dispersion relations (Nielsen *et al.*, 1970), improved mappings have been used (Ciulli, 1969a,b; Cutkosky and Deo, 1968), which, among other features, have the property that a minimum number of coefficients a_n is required. A review of the whole topic of analytic extrapolations as used to determine $\pi\pi$ phase shifts has been given by Pišút (1970), to which we refer the reader for fuller details.

Having obtained values of $\Delta_F(\nu)$ for $\nu \leqslant -\mu^2$, one can calculate values of $F^{(+)}(\nu)$ in this region from the dispersion relation (4.1.6). To obtain $f_+^0(t)$ we have to make a partial wave expansion of $F^{(+)}(t)$. Thus, using equations (4.1.2) and (4.1.3) in (4.1.5), we have

$$F^{(+)}(t) = \frac{-8\pi}{mp^2} \sum_{J,\text{even}} (J+\tfrac{1}{2})(pq)^J f_+^J(t), \tag{4.1.7}$$

and since $\cos\theta_s = -1$ implies $\cos\theta_t = -1$, it follows that this expansion converges for all $t \geqslant 4$. If we now assume that for $4 \leqslant t \leqslant 50$, waves with $J \geqslant 2$ can be neglected, then, from (4.1.7)

$$f_+^0(t) = \frac{-mp^2}{4\pi} F^{(+)}(t). \tag{4.1.8}$$

The values of $\operatorname{Re} f_+^0(t)$ and $\operatorname{Im} f_+^0(t)$ obtained by Nielsen *et al.*, using this assumption, are shown in Fig. 4.1.2. Since $\operatorname{Re} f_+^0(t)$ goes through zero, the $\pi\pi$ phase shift δ_0^0 contains a resonance. This result had been obtained in earlier work (Donnachie *et al.*, 1966), and was considered strong evidence for an S-wave $I=0$ resonance. However, if we use equation (4.1.4) with δ_2^0 very small, then although

$$\operatorname{Im} f_+^2(t) \approx \delta_2^0(t)|f_+^2(t)| \approx 0,$$

for the real part we have

$$\operatorname{Re} f_+^2(t) \approx |f_+^0(t)|,$$

which is not necessarily small.

Nielsen *et al.* showed that the convergence of the expansion (4.1.7) for $\operatorname{Re} F^{(+)}(t)$ is rather poor, and that this is due mainly to the nucleon pole term. By explicitly separating out this term they were able to obtain an expansion which avoided this serious convergence problem. Furthermore, by neglecting the *imaginary* part of the D-wave helicity amplitude they were able to estimate values for the real part, which were then included as corrections in equation (4.1.7). The final values for $\operatorname{Re} f_+^0(t)$ obtained are shown in Fig. 4.1.2, and should be a considerable improvement on the values obtained using equation (4.1.8).

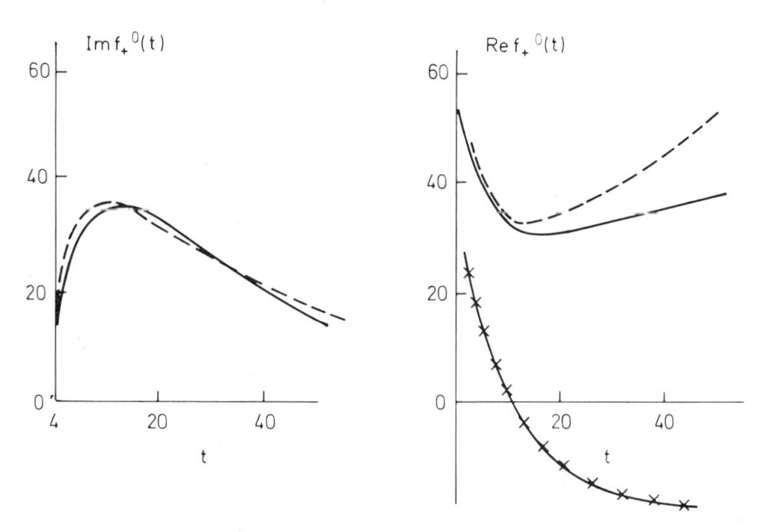

FIG. 4.1.2. Values of $\operatorname{Re} f_+^0(t)$ and $\operatorname{Im} f_+^0(t)$: (——) as calculated from πN partial-wave dispersion relations; (– – –) as calculated from πN backward dispersion relations; (×××) $\operatorname{Re} f_+^0(t)$ as calculated from backward dispersion relations using equation (4.1.8) (neglect of $\pi\pi$ D-waves).

(c) Partial Wave Dispersion Relations

The singularities of the πN partial wave amplitudes $f_{l\pm}(s)$ as a function of s are shown in Fig. 4.1.3, which also shows how each singularity arises (Frautschi and Walecka, 1960; Hamilton and Spearman, 1961). The problem, once again, is to extrapolate from the s-channel and u-channel cuts, where the amplitude can be calculated from physical πN scattering data and the use of crossing symmetry, to the circle cut, which arises from the t-channel $\pi\pi \to N\bar{N}$. Again, a discrepancy function can be defined as an intermediate step, in which the calculable singularities are subtracted off before making the necessary analytic continuation.

In this case, however, the actual calculations, as compared to the backward amplitude method, are of considerable technical complexity. This arises for several reasons. Firstly, the analytic structure is much more complicated. Secondly, the absorptive part of the circle contribution can only be expanded into t-channel helicity amplitudes over a limited region of the circle (Frazer and Fulco, 1960a,b), so that only part of the low-mass $\pi\pi$ exchange is evaluated. Finally, the corresponding expansions for the real parts converge nowhere on the real t-axis, and have to be dealt with separately. These technical problems can all be solved, but the actual calculations are intricate, and we will not describe them here. The details may be found in the paper of Nielsen *et al.*, and the lecture notes of Hamilton (1970).

The results obtained by Nielsen *et al.* for $f_+^0(t)$ are shown in Fig. 4.1.2. They are in good agreement with the results obtained from backward πN

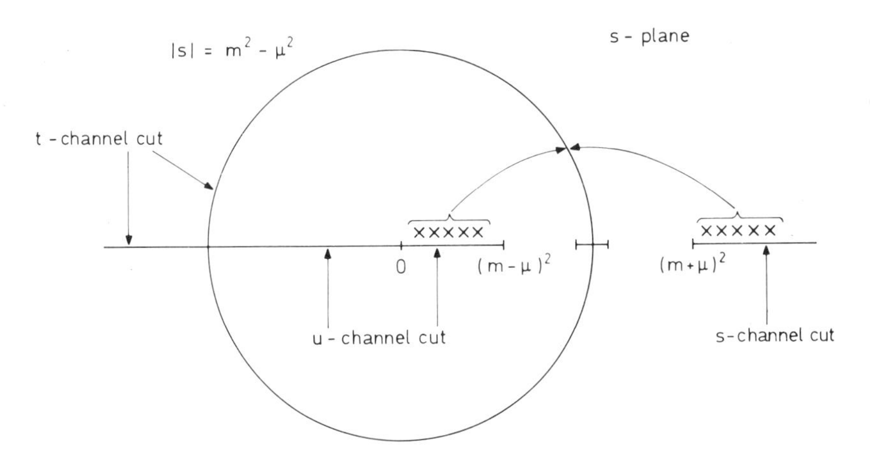

FIG. 4.1.3. Singularities of the πN partial wave amplitudes as a function of s. The origin of each singularity is also shown. (The short cut is due to the nucleon Born term in the u-channel.)

dispersion relations, and it is an important consistency check on the reliability of the methods that both approaches lead to compatible results.

(d) $\pi\pi$ Phase Shifts

The first results using the πN partial wave dispersion relations were obtained by Hamilton and his co-workers (Hamilton *et al.*, 1961, 1962a,b,c). These authors worked mainly with the S-wave amplitude $f_{0+}^{(+)}(s)$ and found that a large attractive δ_0^0 $\pi\pi$ phase shift was required at low energies. Inverting the argument, they were then able to show (Donnachie *et al.*, 1964; Donnachie and Hamilton, 1965) that this resulted in an attractive long-range force, symmetric in spin and isospin, which was required to understand the higher partial waves in πN scattering at low energies. Later, the backward amplitude method (Atkinson, 1962) was applied to the amplitude $F^{(+)}(t)$ by Donnachie *et al.* (1966). However, as mentioned previously, these authors used equation (4.1.8) without the D-wave correction, leading to discrepancies with partial-wave results. This discrepancy was removed by the work of Nielsen *et al.* (1970). These authors have carried out a detailed evaluation and comparison of both methods, making several improvements, including D-wave corrections to equation (4.1.8), and their results for $f_+^0(t)$ are shown in Fig. 4.1.2. The corresponding phase shifts δ_0^0 are shown in Fig. 4.1.4.

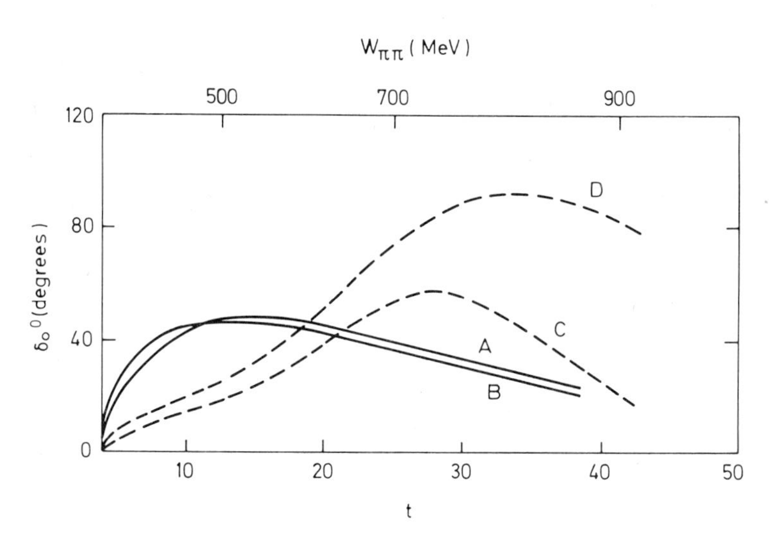

FIG. 4.1.4. Possible values of $\delta_0^0(t)$: (A), as obtained from partial wave dispersion relations; (B), as obtained from backward dispersion relations; (C) and (D), typical results obtained by deforming the shape of Im $f_+^0(t)$ subject to the error estimates discussed in the text.

Although the phase shifts obtained by the two methods are in excellent agreement with each other, as we have pointed out in (b) the results are likely to be much better for averages than for individual values, so that the question of errors needs to be considered very carefully. Nielsen *et al.* have tried to do this by repeating their calculation for the helicity amplitude $f^1_+(t)$, which is related to the rho-meson. A rather complete picture of this amplitude is given by the work of Höhler and his co-workers (Höhler *et al.*, 1968), who have used not only πN data but, in addition, some form for the $\pi\pi$ P-wave, which is comparatively well known. A comparison of the two calculations is shown in Fig. 4.1.5. If one assumes that similar errors exist in both $f^0_+(t)$ and $f^1_+(t)$ then these results suggest that local errors of twice the average value are likely to be present, whereas the errors on the integrated values are probably not more than 50%. Clearly these error estimates allow a very wide class of possible $I = 0$ S-wave $\pi\pi$ phase shifts. In order to try and find the invariant properties of this class, Nielsen *et al.* tried deforming the shape of Im $f^0_+(t)$, subject to the error estimates given above, and for each form recalculated Re $f^0_+(t)$ and the $\pi\pi$ phase shift δ^0_0. Two examples of the resulting phase shifts are shown in Fig. 4.1.4. In addition to the rather well-established result that the phase shift is on average large and positive over the whole low-energy region, Nielsen *et al.* were able to conclude in this way that phase shifts with a clearly resonating behaviour below ≈800 MeV. are improbable, thus providing further evidence against the "Down–Up" type of phase shift (see Section 3.3).

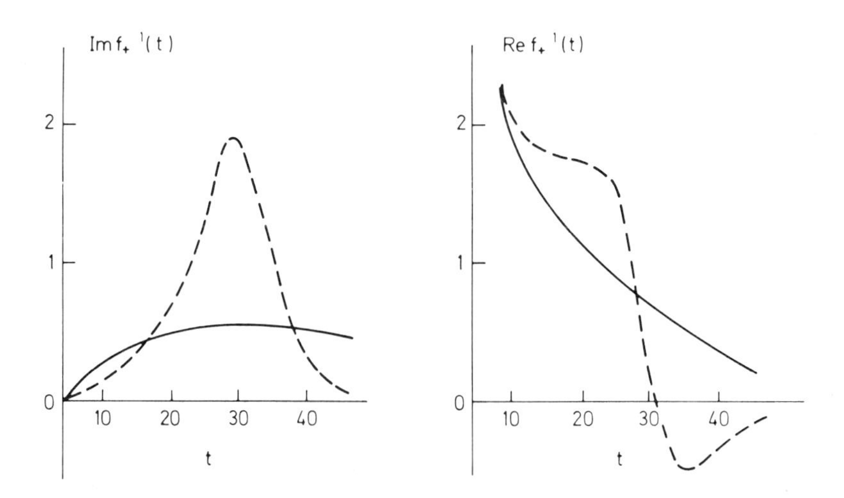

FIG. 4.1.5. Values of Re $f^1_+(t)$ and Im $f^1_+(t)$; (——) as obtained by Nielsen *et al.*; (– – –) as obtained by Höhler *et al.*

(e) Consistency Tests

In the approach discussed above, the idea is to deduce information about the $\pi\pi$ interaction in a model-independent way using only πN data and the usual analyticity assumptions. This is a very ambitious task, and it is clear from the results of Nielsen *et al.* (1970) that, at present, only rather qualitative statements can be made about the $\pi\pi$ phase shifts. A less ambitious, but more fruitful, approach is to use the methods as consistency tests for particular $\pi\pi$ amplitudes. The idea here is to *assume* a form for the $\pi\pi$ amplitude and then, by means of an Omnès equation and a suitable analytic continuation, obtain the helicity amplitudes $f^J_\pm(t)$ for $\pi\pi \to N\bar{N}$. These can then be confronted with πN data via partial wave or backward dispersion relations. We will outline the method for the partial wave case, as that has been the most widely used, and provides the most sensitive consistency tests.

The first stage of the calculation is to obtain the helicity amplitudes $f^J_\pm(t)$. This is done by defining an Omnès function

$$D(t) \equiv \exp\left\{-\frac{1}{\pi}\int_4^T dt' \frac{\delta^I_J(t')}{t'-t}\right\}, \tag{4.1.9}$$

where T is the largest value of t for which the appropriate $\pi\pi$ phase shift is assumed to be known. The function

$$g(t) \equiv D(t) f^J_\pm(t),$$

then has only the cuts $-\infty < t \leq a$ and $T \leq t < \infty$, where $a = \mu^2(4 - \mu^2/m^2)$. Values of $g(t)$ for $4 \leq t < T$ may be found by analytic continuation as follows. Since $\text{Im}\, f^J_\pm(t)$ may be calculated quite accurately in terms of πN data for $-26 \leq t \leq a$ by projection from fixed-t dispersion relations, $\text{Im}\, g(t)$ is known in this interval, and so it is convenient to perform the continuation via the discrepancy method discussed previously. The discrepancy function here is

$$\Delta(t) \equiv g(t) - \frac{1}{\pi}\int_{-26}^{a} dt' \frac{\text{Im}\, g(t')}{t'-t}, \tag{4.1.10}$$

and, from the analytic properties of $g(t)$, $\Delta(t)$ has cuts $-\infty < t \leq -26$, and $T \leq t < \infty$. These singularities can be treated by the conformal mapping methods also discussed previously, and a Legendre expansion in the conformal variable used to extrapolate values of $\Delta(t)$ for $-26 \leq t \leq a$ to the region $4 \leq t \leq T$. Inverting (4.1.10) then leads to values for the helicity amplitudes $f^J_\pm(t)$, i.e.

$$f^J_\pm(t) = \frac{1}{D(t)}\left\{\Delta(t) + \frac{1}{\pi}\int_{-26}^{a} dt' \frac{D(t')\,\text{Im}\, f^J_\pm(t')}{t'-t}\right\}. \tag{4.1.11}$$

The second stage is to use these values of $f_{\pm}^{J}(t)$ to calculate the contribution to πN partial-wave amplitudes. Again the discrepancy method is useful. Thus we define discrepancies $\Delta_{l\pm}(s)$ for the πN reduced partial wave amplitudes $F_{l\pm} \equiv f_{l\pm}(s)/q^{2l}(s)$ by

$$\Delta_{l\pm}(s) \equiv \text{Re } F_{l\pm}(s) - \frac{1}{\pi}\int_{(m+\mu)^2}^{s_p} ds' \frac{\text{Im } F_{l\pm}(s')}{s'-s}$$

$$-\frac{1}{\pi}\int_{R^2/s_p}^{(m-\mu)^2} ds' \frac{\text{Im } F_{l\pm}(s')}{s'-s} - F_{l\pm}^{N}(s), \tag{4.1.12}$$

where $R = m^2 - \mu^2$; s_p is the highest energy at which πN amplitudes are known (in practice $s_p \approx 247\mu^2$); and $F_{l\pm}^{N}(s)$ is the long-range part of the nucleon Born term. The discrepancies may be calculated for $(m+\mu)^2 \leqslant s < s_p$ and $R^2/s_p \leqslant s \leqslant (m-\mu)^2$ from physical πN amplitudes and using crossing symmetry (Nielsen *et al.*, 1970). From the analytic properties of $F_{l\pm}(s)$ (see Fig. 4.1.3), it follows that $\Delta_{l\pm}(s)$ has the real-axis cuts $-\infty < s \leqslant R^2/s_p$, $s_p \leqslant s < \infty$, and the circle cut $|s| = m^2 - \mu^2$. For $l \geqslant 1$ the factor $q^{-2l}(s)$ in the definition of the reduced amplitudes $F_{l\pm}(s)$ suppresses the more distant right- and left-hand cuts, and most of the energy dependence of $\Delta_{l\pm}$ should be due to the front of the circle, the other singularities giving more slowly varying terms. The former contributions are due to low-mass $\pi\pi$ exchange, and are calculable in terms of Im $f_{\pm}^{J}(t)$ (as given by equation (4.1.11)) by convergent expansions for angles on the circle restricted to $|\phi| \leqslant 66°$ (Frazer and Fulco, 1960a,b). Thus

$$\Delta_{l\pm}(s) = \frac{1}{2\pi i}\int_{-66°}^{66°} d\phi \frac{\text{Disc } F_{l\pm}(s')}{s'-s}\left(\frac{ds'}{d\phi}\right) + H_{l\pm}(s), \tag{4.1.13}$$

where $H_{l\pm}(s)$ is a slowly-varying function of s. The consistency test consists in comparing values of $\Delta_{l\pm}(s)$ calculated from (4.1.12) and (4.1.13) for a given input $\pi\pi$ amplitude (and, of course, a fixed set of πN data).

Several attempts have been made to use these consistency conditions. In one (Elvekjaer and Nielsen, 1971) the helicity amplitude $f_{+}^{0}(t)$ has been used to test hypotheses about the form of δ_0^0 by examining the consistency of $\Delta_{l\pm}^{(+)}$ for $l = 1$ and 2. These calculations have been extended to include the amplitudes $f_{\pm}^{1}(t)$, which are related to δ_1^1, and backward dispersion relations (Nielsen and Oades, 1972a). The discrepancies obtained are more precise than those of earlier analyses because the πN input data has improved considerably (Bugg *et al.*, 1971; Almahed and Lovelace, 1972). A technical variation in the calculations of Nielsen and Oades is the use of πN partial wave amplitudes in the region $(m-\mu)^2 \leqslant s \leqslant (m+\mu)^2$, thereby enabling the discrepancies to be calculated very close to the front of the circle, and increasing the sensitivity to the low-energy $\pi\pi$ interactions.

For the amplitudes $F_{l\pm}^{(-)}(s)$ $(l=1, 2)$ good agreement was found with a $I=J=1$ $\pi\pi$ phase shift given by the rho-resonance with parameters $M_\rho = 765$ MeV, $\Gamma_\rho = 125$ MeV, and $a_1^1 = 0{\cdot}035$. A typical result (for $\Delta_{1-}^{(-)}$) is shown in Fig. 4.1.6. For the helicity amplitude $f_+^0(t)$ two basic forms for δ_0^0 were examined. In the notation of Morgan and Shaw (1970) they are DU1 (Down–Up type), and BD1 (Between–Down type). (See Chapter 10, and Fig. 10.3.1, for this terminology.) Both forms give a good description of the s-dependence of the P-wave discrepancies $\Delta_{1\pm}^{(+)}$, but for the D-wave discrepancies $\Delta_{2\pm}^{(+)}$ the DU1 solution gives completely the wrong s-dependence, whereas the BD1 solution, while giving the general features of the s-dependence correctly, produces contributions which are too large close to

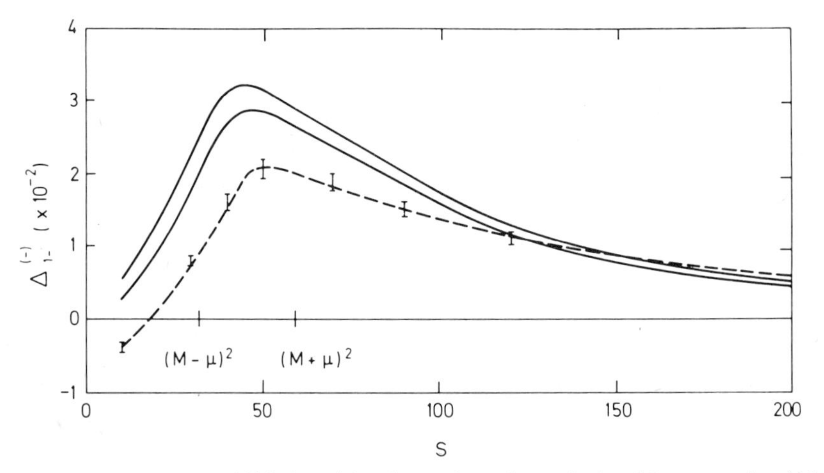

FIG. 4.1.6. Values of $\Delta_{1-}^{(-)}(s)$: (——) band covering values calculated from equation (4.1.12); (– – –) contribution, with associated error, from Im $f_+^1(t)$ integrated over front of circle, i.e. the first term in equation (4.1.13).

$s=(m^2-\mu^2)$, the front of the circle. A typical example (for $\Delta_{2-}^{(+)}$) is shown in Fig. 4.1.7. Because of the weight factor $q^{-2l}(s)$ in the definition of $F_{l\pm}(s)$, the long-range parts of the D-wave discrepancies are given almost entirely by that part of the circle for which $|\phi| \lesssim 20°$, which corresponds to $\pi\pi$ masses less than about 400 MeV. A satisfactory fit to both P- and D-wave discrepancies could only be achieved if the BD1 phase shift was modified below 450 MeV by allowing a small negative scattering length $a_0^0 \approx -0{\cdot}06$ and a zero at $\approx$310 MeV. The resulting improvement in the fit to the discrepancy $\Delta_{2-}^{(+)}$ is shown in Fig. 4.1.7.

These calculations (Nielsen and Oades, 1972a) clearly demonstrate the sensitivity of the discrepancies for the reduced D-wave amplitudes to the

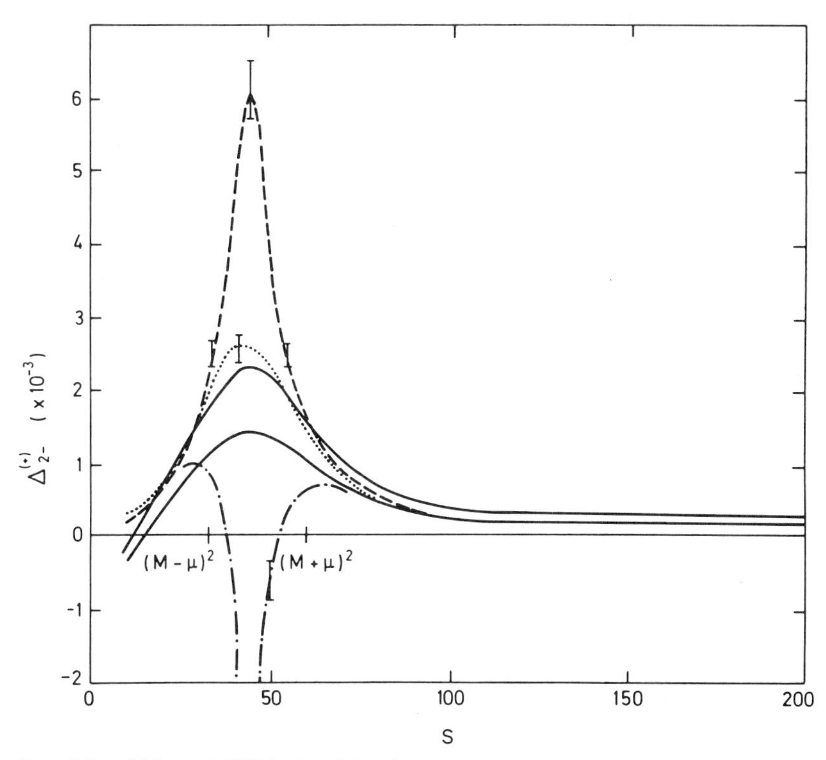

FIG. 4.1.7. Values of $\Delta_{2-}^{(+)}(s)$: (——) band covering values calculated from equation (4.1.12); (- - -) contribution, with associated errors, from Im $f_+^0(t)$, obtained using the BD1 solution for δ_0^0, integrated over front of circle, i.e. the first term in equation (4.1.13); (—·—·—) corresponding DU1 values; (·······) corresponding values obtained using the modified BD1 solution.

very low-energy $\pi\pi$ interaction. They also show that the basic form of the BD1 solution of Morgan and Shaw is preferred over the other forms tested, a conclusion also reached by Elvekjaer (1972) using a suitably weighted long-range part of the S-wave discrepancy. However, the modifications to the BD1 solution close to threshold suggested by Nielsen and Oades (1972a) have to be treated with caution. This is because to obtain the partial waves $F_{l\pm}^{(+)}(s)$ in the gap $(m-\mu)^2 \leqslant s \leqslant (m+\mu)^2$, and hence $\Delta_{l\pm}^{(+)}(s)$ there, fixed-t dispersion relations were used (Nielsen and Oades, 1972b), and this requires the real parts of the invariant amplitudes for $(m+\mu)^2 \leqslant s \leqslant s_p$ and $0 \leqslant t \leqslant 4$. These were reconstructed from partial wave expansions, but as $s \to (m^2-\mu^2)$ (the front of the circle), values of the real parts are required closer to $t=4$, which is the boundary of the ellipse of convergence. Thus, there is a potentially serious source of error in $\Delta_{l\pm}^{(+)}$ in precisely the region where there is an apparent violation of the consistency condition. It is impossible to make a model-independent estimate of the actual error in $\Delta_{l\pm}^{(+)}$ that is introduced in this way, but Nielsen and Oades (1972a) estimated, on the basis of a simple model, that these uncertainties lead to a somewhat larger range of values for a_0^0, and concluded that their results require $-0{\cdot}15 \leqslant a_0^0 \leqslant 0{\cdot}10$.

Subsequently, the same authors (Nielsen and Oades, 1974) have re-examined this interesting point by considering consistency conditions, similar to those above, for the derivatives of Re $F_{2\pm}^{(+)}(s)$ close to the physical threshold $s=(m+\mu)^2$. This avoids the possible dangers of using amplitudes in the region $(m-\mu)^2 \leqslant s \leqslant (m+\mu)^2$, while still retaining the sensitivity of the partial wave dispersion relations for the reduced D-waves. Using forms which are qualitatively similar to the Between–Down form used previously, but which satisfy the Roy equations (see Chapter 10) they obtain values for a_0^0 consistent with their previous estimates, but with larger errors, $a_0^0 = 0{\cdot}06 \pm 0{\cdot}20$.

4.2. Other Processes

In this section we consider briefly two other processes, NN and KN scattering, which also have important forces due to 2π exchange. The aim here is not to deduce the form of the $\pi\pi$ interaction from a study of these reactions, but rather to show that currently accepted $\pi\pi$ amplitudes, along with other input information, lead to a consistent picture of the elastic data in these channels. The calculations are thus similar in spirit to those described in Section 4.1(e) for elastic πN scattering.

(a) Nucleon–Nucleon Scattering

The longest range force in NN scattering is one-pion exchange (OPE), and is unambiguously determined in terms of the πNN coupling constant. The OPE terms gives a reasonable qualitative representation of the basic features of NN phase shifts for $l \geqslant 2$ and $k_{\text{lab}} \leqslant 500$ MeV/c, but for more detailed agreement one must also include other forces, specifically multipion exchange. One approach is via simple Lagrangians, calculated in Born approximation, but this is not very satisfactory, because it is found that to achieve a reasonable fit to the experimental NN phase shifts it is necessary to introduce an isoscalar boson of low mass (for example around 400 MeV). (Early calculations are reviewed by Kramer (1970); see also Partovi and Lomon (1970, 1972).) Since this term (and also ρ-exchange) is found to be important, attention has focussed on detailed calculations of the 2π-exchange term (TPE)—the second longest-range force. Modern calculations use dispersion relation techniques, and although basically not new (the idea was originally proposed in field theoretic language by Charap and Fubini (1959, 1960)), have only recently been revived (Brown, 1970) and shown to be capable of yielding useful results. This is largely due, of course, to the greatly increased information now available on πN and $\pi\pi$ scattering.

The earliest attempt to estimate the TPE force was that of Charap and Tausner (1960) who used a combination of dispersion and fourth-order perturbation theory. Their method was used by Cottingham and Vinh Mau (1963) who extended it following the dispersion relation approach of Amati *et al.* (1960a,b,c, 1963), using the Δ (1232) and ρ-meson as the main input. This was the first dispersion theoretic attempt to correlate TPE with the known properties of the $\pi\pi$ and πN interactions. Different $\pi\pi$ inputs were later investigated by Kapadia (1967). Following these early calculations, several more detailed analyses have been published: (Chemtob and Riska, 1971; Brown and Durso, 1971; Chemtob *et al.*, 1972; Epstein and McKellar, 1972; Cottingham *et al.*, 1973a,b), and we will briefly describe their approach and some of the results obtained.

(i) *Dispersion Theoretic Calculations*

Nucleon–nucleon scattering is described by five invariant amplitudes $P_j(j = 1, 2, \ldots, 5)$, and charge is specified by superscripts $\pm$ such that pure isospin states are given by

$$P_j^0 = 3P_j^+ - 6P_j^-; \qquad P_j^1 = 3P_j^+ + 2P_j^-.$$

The basic assumptions of Cottingham *et al.* (1973*a*) and Chemtob *et al.* (1972), the most complete of the above calculations, is that the amplitudes $P_j^\pm$ satisfy the Mandelstam representation. If the choice of amplitudes of Amati *et al.* is made, then $P_j^\pm$ satisfy the following dispersion relations (written unsubtracted for convenience).

$$P_j^\pm(s, t, u) = \begin{pmatrix} 0 \\ \dfrac{g^2_{NN\pi}\delta_{j5}}{2(t-\mu^2)} \end{pmatrix} + \frac{1}{\pi}\int_{4\mu^2}^{\infty} dt' \frac{\rho_j^\pm(s, t') \mp (-1)^j \rho_j^\pm(u, t')}{t'-t}. \tag{4.2.1}$$

In addition, the weight functions $\rho_j^\pm$ satisfy the dispersion relations

$$\rho_j^\pm(s, t) = \frac{1}{\pi}\int_{4m^2}^{\infty} ds' \frac{d_j^\pm(s', t)}{s'-s}. \tag{4.2.2}$$

The double spectral functions $d_j^\pm(s, t)$ are related to the absorptive parts of elastic πN scattering amplitudes by a set of double dispersion relations, which are obtained by using unitarity for the process $N\bar{N} \to 2\pi \to N\bar{N}$ and fixed-t dispersion relations for πN scattering. (For the details see Chemtob *et al.* (1972).) The results are strictly valid for values of t restricted to the πN physical region, whereas the double spectral functions are required in (4.2.2) for $t \geq 4\mu^2$. Thus an extrapolation from the πN physical region has to be made.‡ Both equations (4.2.1) and (4.2.2) must, in general, be subtracted, and the number of subtractions required can be determined from Regge-pole models. The subtraction constants are given in terms of amplitudes for $N\bar{N} \to \pi\pi$.

In the calculations of Chemtob *et al.* (1972), the πN input was restricted to a few of the lowest lying N^*-resonances, used in the narrow-width approximation. In the more detailed work of Cottingham *et al.* (1973a), the absorptive parts of the πN amplitudes were taken from phase shift analyses of Donnachie *et al.* (1968) and Davies (1970) (the so-called CERN "experimental" and Glasgow "A" solutions, respectively). Both calculations need to extrapolate the absorptive parts to positive values of t.‡ The subtraction functions were evaluated in terms of $\pi\pi \to N\bar{N}$ helicity amplitudes taken from analyses of elastic πN scattering (see Section 4.1). Cottingham *et al.* (1973a) used two solutions for the S-wave amplitudes; that of Nielsen and Oades (1972a) (corresponding to the BD1 δ_0^0 phase-shift solution of Morgan and Shaw (1970)); and that of Nielsen *et al.* (1970) (corresponding to a non-resonant form for δ_0^0). For the P-wave amplitudes they also used two solutions: that of Nielsen and Oades (1972a); and that of Höhler *et al.* (1968).

With these ingredients one can calculate the TPE contribution to the equivalent NN potential. To calculate phase shifts which can be compared with experiment, however, one must add the OPE term and the "fourth-order" contributions, both of which are given in terms of the πNN coupling constant (the latter was calculated taking into account the modifications suggested by Partovi and Lomon (1972) to the method of Charap and Tausner (1960)), and also a term to represent 3π-exchange. For the latter, Cottingham *et al.* (1973a) used ω-exchange in the narrow-width approximation. Finally, since the scattering amplitudes in this model are not necessarily unitary, small unitarity corrections have to be made, although the phase shifts are generally insensitive to the actual method of unitarization.

Phase shifts have been predicted by Cottingham *et al.* (1973a) using standard techniques (see e.g. Furuichi (1967)) for kinetic energies less than 425 MeV. Since the method is a peripheral one it would only be expected to give meaningful results for higher partial waves. The results show that for waves with $2 < l < 6$ the model gives phases in good agreement with those from NN phase shift analysis (MacGregor *et al.*, 1969) with only two exceptions. These are ${}^1H_5$ and ${}^3G_5$, which are only accessible in np scatter-

‡ Arguments are given in Cottingham *et al.* (1973a) to suggest that the extrapolation is reasonable for $t \leq 80\mu^2$.

ing, and therefore possibly not well determined experimentally. This is a significant improvement over OPE alone, and in many cases is appreciably better than OPE + fourth-order terms. The predictions for P- and D-waves are in qualitative agreement with the data. Similar results are obtained in the calculations of Chemtob *et al.* (1972).

Although the calculations involve no free parameters, different input data were used to test the sensitivity of the predictions. The results are relatively insensitive to the πN phase-shift analysis used, except at the higher energies, but *are* sensitive to the form of the $N\bar{N} \to \pi\pi$ S-wave amplitude. The best results are obtained for the model of Nielsen *et al.* (1970), although this is not claimed to be conclusive. The most important result (borne out by both Cottingham *et al.* (1973a) and Chemtob *et al.* (1972)) is that there are important contributions from uncorrelated 2π-exchange, and that current forms for $\pi\pi$ scattering (particularly for δ_0^0) lead, via a chain of dispersion theory calculations, to a consistent picture of $\pi\pi$, πN and NN scattering.

(b) Kaon–Nucleon Scattering

KN and $\bar{K}N$ scattering also provide, at least in principle, consistency checks on $\pi\pi$ amplitudes, although in practice the tests are not as direct as in πN scattering. There are several reasons for this: t-channel unitarity involves, via crossing, πK amplitudes as well as those for $\pi\pi$ and πN; G-parity in this channel does not forbid 3π-exchange; and KN and $\bar{K}N$ amplitudes are not so well known as πN ones. However, there is one advantage that these reactions have over πN scattering. Owing to the small mass of the $\pi\pi$ intermediate state in the t-channel relative to the $K\bar{K}$ initial state, the relevant singularity approaches very close to the physical KN threshold and would therefore be expected to play a significant role in the low-energy dynamics of these reactions. This is indeed found to be the case (Martin and Miller, 1972, 1973).

Calculations proceed in a way analogous to those described in Section 4.1(e) for πN scattering. The nearby parts of the left-hand cut (corresponding to the longest-range forces), which are due to the exchange of low-mass states in the t- and u-channels (ρ, ε, ω, Λ, Σ) are evaluated. For the Λ, Σ these contributions are given directly in terms of the kaon–nucleon coupling constants. For the $\pi\pi$ states, the calculation is done by applying unitarity to the process $K\bar{K} \to \pi\pi \to N\bar{N}$ assuming models for both the $\pi\pi \to K\bar{K}$ and $\pi\pi \to N\bar{N}$ amplitudes. The latter are obtained from fixed variable dispersion relations for πN and πK scattering via a specific assumed form for the $\pi\pi$ phase shifts. The ω-meson is inserted in the narrow-width approximation. These long-range forces can then be used with rescattering (s-channel) corrections to estimate the effective short-range force via partial-wave sum rules. Finally, all the terms are used in partial wave dispersion relations to predict low-energy phase shifts.

In a recent paper (Hedegaard-Jensen *et al.* 1973) this program has been carried out using models for the t-channel amplitudes corresponding to the Between–Down type of solution for both the δ_0^0 $\pi\pi$ phase shift, and the $I=\frac{1}{2}$ πK S-wave amplitude. Predictions are made for the P-wave $I=0$, 1 KN scattering lengths, and three out of four of those quantities are found to be in good agreement with values obtained directly from phase-shift analysis and forward dispersion relation sum rules (Knudsen and Martin, 1973). These results are encouraging, and confirm the more detailed conclusions of Sections 4.1(e), and 4.2(a), although the chain of calculations in the KN case is much longer than in the former examples, and is more dependent on relatively poorly known input information.

Chapter 5

Weak and Electromagnetic Interactions

5.1. K_{e4} Decay

Among all the semileptonic decay processes of the K-meson the four-body modes

$$K \to \pi\pi l \nu_l,$$

where $l \equiv e$ or μ, are uniquely important. This is because, aside from the usual weak interaction issues that arise in these decays, e.g., the validity of the $\Delta I = \frac{1}{2}$ and $\Delta S = \Delta Q$ rules, there exists the possibility of extracting direct information on $\pi\pi$ phase shifts, albeit over a limited range of energies. This was first noted by Shabalin (1963). The theoretical range of centre-of-mass energies is from 280 MeV to 490 MeV, but since the $\pi\pi$ mass spectrum falls rapidly above about 350 MeV, the actual range is considerably smaller (see the $M^2_{\pi\pi}$ distribution of Fig. 5.1.4, p. 162). Nevertheless, the phase shifts are, in principle, obtainable in a model-independent manner, and this is an important point.

The method of analysis depends on observing the interference between S- and P-wave dipions in the decay, and it is therefore important that both waves are present with reasonable strengths. This situation is obtained experimentally, and sizable forward–backward asymmetries are observed in the dipion decay (see the $\cos\theta_\pi$ distribution of Fig. 5.1.4, p. 162). The interpretation is complicated, however, by the various helicity alternatives for the dilepton pair, and analysis of the dilepton decay distribution is necessary to disentangle the allowed contributions.

Experimentally, only the decays

$$K^+ \to \pi^+\pi^- e^+ \nu_e \quad \text{and} \quad K^- \to \pi^+\pi^- e^- \bar{\nu}_e,$$

i.e. $K^{\pm}_{e4}$, have been studied, and so we shall restrict our discussion to these modes, illustrating the kinematics for K^+_{e4}. The four-momenta of the various particles will be denoted as shown in Fig. 5.1.1, and it is convenient to define the following independent four-vector combinations:

$$P \equiv p_+ + p_-; \qquad R \equiv p_+ - p_-; \qquad L \equiv p_l + p_\nu.$$

If the interaction is of the usual local $V-A$ current–current type, then the decay can be kinematically parameterized in terms of five variables. When a

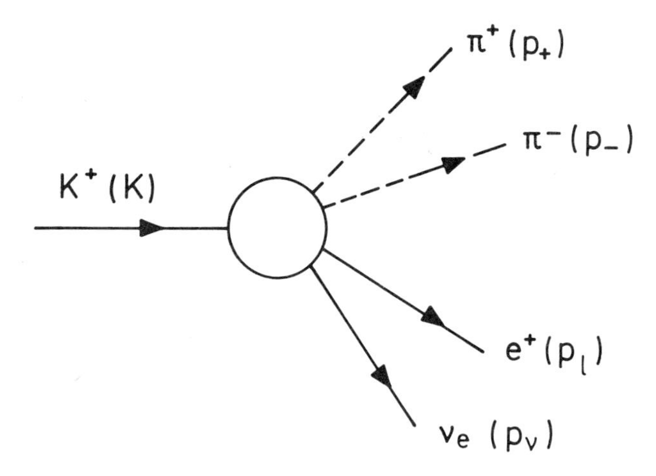

FIG. 5.1.1. Kinematics of K^+_{e4} decay.

suitable choice of these variables is made, the structure with respect to two of them can be made explicit, and independent of hadronic effects. The strong interaction effects are then contained in form factors which can depend at most on the remaining three variables. The following choice of variables achieves this (Cabibbo and Maksymowicz, 1965). Two variables are taken to be the squared invariant masses of the dipion and dilepton pairs, i.e.,

$$s_\pi \equiv -P^2 \quad \text{and} \quad s_l \equiv -L^2,$$

and the other three are chosen to be: θ_π the angle formed by the π^+, in the dipion rest frame, and the line of flight of the dipion, as defined in the K-meson rest frame; θ_l, the similar angle formed by the e^+, in the dilepton rest frame, and the line of flight of the dilepton, as defined in the K-meson rest frame; and ϕ, the angle between the normals to the planes defined in the

K-meson rest frame by the $\pi^+\pi^-$ and $l^+\nu$ pairs. These angles are shown in Fig. 5.1.2.

The matrix element for the decay in first-order perturbation theory is

$$M = \frac{G}{\sqrt{2}} \sin\theta_c [\bar{u}(p_\nu)\gamma_\lambda(1+\gamma_5)\,v\,(p_l)]\langle p_+p_-|A^\lambda + V^\lambda|K\rangle, \qquad (5.1.1)$$

where A^λ and V^λ are the strangeness-changing weak hadronic axial-vector and vector currents, respectively; θ_c is the Cabibbo angle; and G is the usual

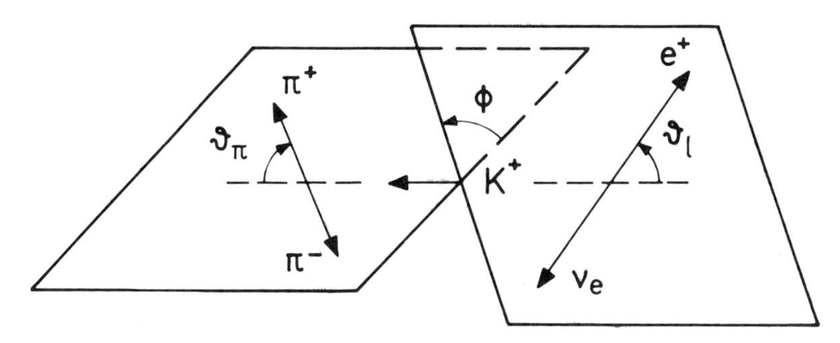

FIG. 5.1.2. Decay angles for K_{e4}^+ as defined in the text.

Fermi coupling constant. From general invariance considerations the hadronic matrix elements have the structure

$$m_K^3\langle p_+p_-|V^\lambda|K\rangle = h\varepsilon^{\lambda\mu\nu\sigma}K_\mu P_\nu R_\sigma, \qquad (5.1.2)$$

$$m_K\langle p_+p_-|A^\lambda|K\rangle = fP^\lambda + gR^\lambda + rL^\lambda, \qquad (5.1.3)$$

where the dimensionless form factors, f, g, h, and r are functions of the three variables s_π, s_l and θ_π. These form factors not only contain information about the weak interactions, but also carry direct strong interaction information. In a partial wave expansion of the form factors with respect to the angular momentum of the dipion system, a partial wave amplitude of definite angular momentum and isospin will have the same phase as the corresponding $\pi\pi$ scattering amplitude, provided time-reversal invariance holds.‡

In practice, two further simplifying assumptions must be made. The first is the validity of the $\Delta I = \frac{1}{2}$ rule. This is necessary because for the even partial waves in K_{e4}^+ decay both $I = 0$ and $I = 2$ $\pi\pi$ final states can, in principle, be present. However, if we assume the $\Delta I = \frac{1}{2}$ rule then only $I = 0$ states are

‡ In the absence of time-reversal invariance it is still possible to obtain the $\pi\pi$ phase shifts from a comparison of the conjugate reactions K_{e4}^+ and K_{e4}^-, provided that CPT invariance holds (Lee and Wu, 1966), but we will not consider that possibility here.

allowed. For the odd partial waves, the $\pi\pi$ system is in a pure $I=1$ state, and no problem arises.‡ Secondly, it is reasonable to assume that for the very low dipion masses involved in the decay only S- and P-waves are important. With these assumptions, and the further one of CP invariance, Watson's final-state interaction theorem§ allows us to write

$$\begin{aligned} f &= f_s\, e^{i\delta_0^0} + f_p\, e^{i\delta_1^1} \cos\theta_\pi, \\ r &= r_s\, e^{i\delta_0^0} + r_p\, e^{i\delta_1^1} \cos\theta_\pi, \\ g &= g_p\, e^{i\delta_1^1}, \\ h &= h_p\, e^{i\delta_1^1}, \end{aligned} \tag{5.1.4}$$

where f_s, f_p, r_s, r_p, g_p and h_p are *real* functions of the invariants s_π and s_l, and δ_0^0 and δ_1^1 are the S-wave $I=0$ and P-wave $I=1$ $\pi\pi$ phase shifts, respectively. A further simplification occurs for K_{e4} decay because the form factor r in equation (5.1.4) occurs in equation (5.1.1) together with a factor $\gamma\cdot L$. Using the Dirac equation, this term in the decay distribution is proportional to the square of the electron mass, and may therefore be neglected.

The decay distribution, summed over lepton spins, is

$$\begin{aligned} d^5\Gamma = {} & \frac{\pi G^2 \sin\theta_c}{(2\pi)^8 16 m_K^5} I(s_\pi, s_l, \theta_\pi, \theta_l, \phi) \\ & \times ds_\pi\, ds_l\, d\cos\theta_\pi\, d\cos\theta_l\, d\phi. \end{aligned} \tag{5.1.5}$$

With our particular choice of variables the distribution function I may be factored into a sum of bilinear forms

$$I = \sum_{n=1}^{9} F_n(\theta_l, \phi) I_n(s_\pi, s_l, \cos\theta_\pi), \tag{5.1.6}$$

where the functions I_n involve the form factors f_s, f_p, g_p and h_p, and the phase-shift difference $\Delta \equiv \delta_0^0 - \delta_1^1$, combined with complicated kinematical factors, but do not depend on θ_l or ϕ. Their full structure may be found in the paper of Pais and Treiman (1968).

If enough decay events existed, they could be binned simultaneously in the five variables and analysed to give Δ as a function of s_π, and the form factors as functions of s_π and s_l. However, with present published data (Birge *et al.*, 1965; Ely *et al.*, 1969; Schweinberger *et al.*, 1971; Bourquin *et al.*, 1971; Basile *et al.*, 1971; Zylbersztejn *et al.*, 1972; Beier *et al.*, 1972, 1973) such an analysis is generally not practicable, and it is usual to work with

‡ A test of the $\Delta I = \frac{1}{2}$ rule is in fact possible in K_{e4} decay if the decay modes $K^+ \to \pi^+\pi^- e^+ \nu_e$, $K^+ \to \pi^0\pi^0 e^+ \nu_e$ and $K^0 \to \pi^-\pi^0 e^+ \nu_e$ are studied.
§ Watson (1952); also cf. Section 3.2(b).

distributions that are integrated over one or more of the variables. Two practical suggestions have been made.

Pais and Treiman (1968) proposed integrating over s_l and $\cos\theta_\pi$ to obtain a distribution in the three variables s_π, ϕ and $\cos\theta_l$. The functions I_n of equation (5.1.6) are then replaced by new functions $\langle I_n\rangle$ which only depend on s_π. These authors point out that information on Δ can be obtained directly by looking at certain ratios of these average coefficients *independent of the values of the form factors and of assumptions regarding their energy dependences.* The phase-shift difference occurs only in the terms $\langle I_4\rangle$, $\langle I_5\rangle$, $\langle I_7\rangle$, and $\langle I_8\rangle$, which are of the form

$$\begin{aligned} \langle I_4\rangle &= A(s_\pi)\cos\Delta; \qquad \langle I_5\rangle = 2B(s_\pi)\cos\Delta; \\ \langle I_7\rangle &= 2A(s_\pi)\sin\Delta; \qquad \langle I_8\rangle = B(s_\pi)\sin\Delta, \end{aligned} \tag{5.1.7}$$

where A and B are functions of s_π alone. Thus the ratios

$$R_1 \equiv \frac{\langle I_7\rangle}{2\langle I_4\rangle}, \quad \text{and} \quad R_2 \equiv \frac{2\langle I_8\rangle}{\langle I_5\rangle}, \tag{5.1.8}$$

yield two independent estimates of $\tan\Delta$.

Should these expressions for $\tan\Delta$ not yield the same answer this would mean that one or more of our assumptions, i.e., time-reversal invariance, the validity of the $\Delta I=\frac{1}{2}$ rule, and the absence of waves with $l\geq 2$, would be incorrect. These assumptions can actually be tested from K_{e4} decay itself. Thus, for example, a necessary condition for the absence of $l\geq 2$ waves is that the $\cos\theta_\pi$ spectrum (all other variables being integrated over) must contain no higher powers than $\cos^2\theta_\pi$. Also, if $\langle I_9\rangle\neq 0$ either $\Delta I\neq\frac{1}{2}$ and/or time-reversal invariance is violated. (This result is true in the absence of waves with $l\geq 3$.) Tests using current K_{e4} decay have revealed no incompatibilities with these various assumptions (Zylbersztejn *et al.*, 1972; Beier *et al.*, 1972).

If necessary, the number of variables can be reduced still further by integrating over s_π. In this case the functions $\langle I_n\rangle$ are replaced by $\langle \bar{I}_n\rangle$ where

$$\langle \bar{I}_n\rangle \equiv \int_{4\mu^2}^{m_K^2} ds_\pi \langle I_n\rangle,$$

and the Pais–Treiman ratios $\bar{R}_1$ and $\bar{R}_2$ then yield values of Δ "averaged" over the $\pi\pi$ mass spectrum.

An alternative method of analysis was proposed by Cabibbo and Maksymowicz (1965), and later extended by Berends *et al.* (1968). In this approach the various angular distributions are considered separately. Within the approximations used by Pais and Treiman, the distributions in

$\cos\theta_\pi$, $\cos\theta_l$ and ϕ may be written

$$\frac{d\Gamma}{ds_\pi\, d\cos\theta_\pi} = \sum_{n=0}^{2} A_n \cos n\theta_\pi,$$
$$\frac{d\Gamma}{ds_\pi\, d\cos\theta_l} = \sum_{n=0}^{2} B_n \cos n\theta_l, \tag{5.1.9}$$
$$\frac{d\Gamma}{ds_\pi\, d\phi} = \sum_{n=0}^{2} (C_n \sin n\phi + D_n \cos n\phi),$$

where the coefficients A_n, B_n, C_n and D_n contain the phase-shift difference Δ together with bilinear functions of the form factors (averaged over s_l), multiplied by kinematical quantities. In the situation of limited data, one can again make assumptions about the dependence of the form factors on s_π, and average over the $\pi\pi$ mass spectrum, to yield a mean value of Δ.

Early experiments were unable to yield useful values for Δ, even after quite severe averaging procedures, due to a lack of data, and only recently have experiments produced samples large enough for reliable results to be obtained. The first of these experiments is that of Zylbersztejn *et al.* (1972) with 1609 K_{e4}^+ events. These were divided into three mass bins, and the dependence of the form factors and Δ on s_π determined in each region (assuming no form factor dependence on s_l). Three different methods were used: a χ^2 fit to the one-dimensional angular distributions of equation (5.1.9); a maximum-likelihood fit to the distribution (5.1.5); and an approximation to the Pais–Treiman method. In all cases the parameters (A_n, Δ, etc.) were assumed to be constant over the range of each bin. The different methods yield consistent results, and show that there is no significant dependence of the form factors on s_π.

The values obtained for $\Delta(\simeq\delta_0^0$, since δ_1^1 is small in this region) in the three mass regions are shown in Fig. 5.1.3. The scattering length a_0^0 was obtained by parameterizing δ_0^0 as

$$\left(\frac{\omega}{q}\right)\tan\delta_0^0 = a_0^0 + \beta q^2, \tag{5.1.10}$$

and using this expression (averaged over each energy bin) to fit the points of Fig. 5.1.3. With β set at the Weinberg current algebra estimate ≈ 2 (see Chapter 14), a_0^0 from the three sets of points was found to be consistent with $0{\cdot}6 \pm 0{\cdot}25$. Using equation (5.1.10) to fit the K_{e4} data directly yielded similar values of a_0^0 (and similar form factors also). Fixing $a_0^0 = 0{\cdot}2$ and varying β gave a best fit for $\beta \approx 10$, but this was poorer than the fit with $a_0^0 = 0{\cdot}6$. These two solutions are shown in Fig. 5.1.3. Finally, allowing both a_0^0 and β to vary gave a best fit for the former value of $a_0^0 \approx 0{\cdot}6$ with β poorly determined.

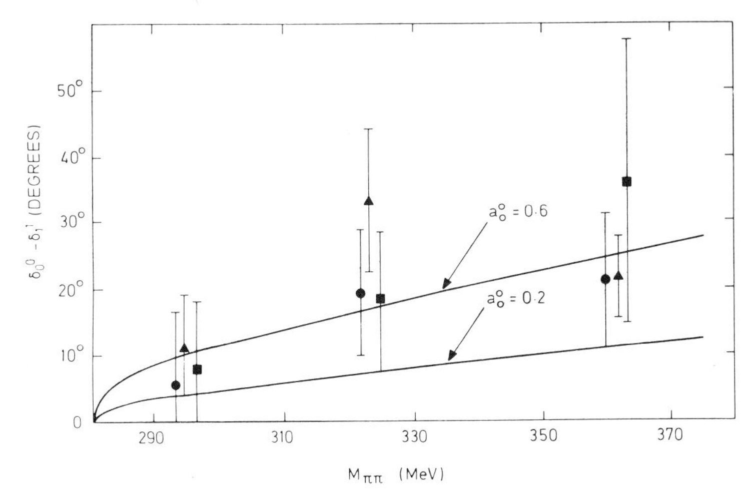

FIG. 5.1.3. Values of $\delta_0^0 - \delta_1^1$ for three mass bins (obtained by three methods of analysis) from the data of Zylbersztejn *et al.* (1972), and fits to the form (5.1.10) referred to in the text.

The second experiment is that of Beier *et al.* (1973) which has 3341 K_{e4}^- and 4800 K_{e4}^+ events. Time-reversal invariance was assumed, thus enabling both types of data to be analysed together (in addition to providing useful checks on calibration). The measured one-dimensional angular distributions are shown in Fig. 5.1.4, and were analysed using equation (5.1.9). (The distribution in cos θ_l was integrated over to avoid possible difficulties with radiative corrections.) Firstly, a fit was made to the remaining four distributions assuming constant form factors f_s, f_p, g_p and h_p (apart from a P-wave barrier factor allowed in f_p), and the result is shown in Fig. 5.1.4. The form factor f_p was found to be poorly determined, and setting $f_p = 0$ and refitting does not appreciably change the other form factors, or the average value of Δ. With $f_p = 0$, the dependence of the other form factors on s_l and s_π can be determined, and the results from four regions of $s_l - s_\pi$ space show that this dependence was very small. The fit was then repeated (with constant form factors) for three regions of s_π, and the resulting values of Δ are shown in Fig. 5.1.5.

To extract the scattering length, δ_0^0 was parameterized by equation (5.1.10) with $\beta = 0$, and the K_{e4} data in the region $280 \leq M_{\pi\pi} \leq 350$ MeV refitted. The best fit gives $a_0^0 = 0{\cdot}17 \pm 0{\cdot}13$, and a similar value of a_0^0 is found by fitting all three values of Δ with β set close to the current algebra value.

An alternative parameterization for extrapolating the $\pi\pi$ phase shift to threshold, and hence extracting the value of a_0^0, has been proposed by Tryon (1974a). He has constructed a model which incorporates the correlations

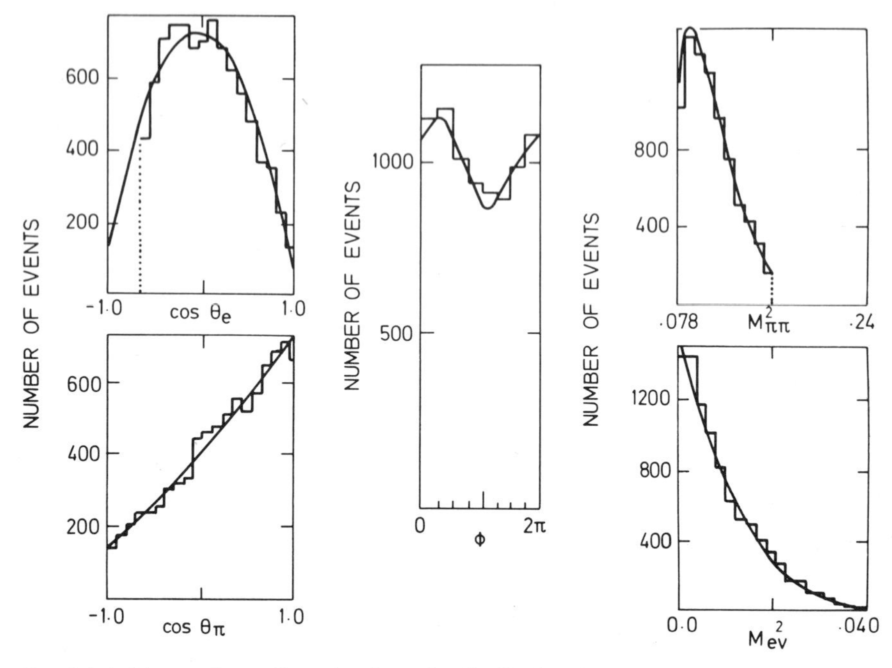

FIG. 5.1.4. Measured one-dimensional angular distributions for $K^{\pm}_{e4}$ decays from the experiment of Beier *et al.* (1973). The solid lines are the fits referred to in the text.

between threshold parameters which result from dispersion relations together with experimental information on the $\pi\pi$ phase shifts at higher energies. (These are discussed in Chapter 10.) Given these assumptions, the low-energy behaviours of both δ_0^0 and δ_1^1 are uniquely specified in terms of the single unknown quantity a_0^0, which is determined by fitting the phases of Zylbersztejn *et al.* (1972) and Beier *et al.* (1973) (all three bins of this latter experiment). The results $a_0^0 = 0{\cdot}50^{+0{\cdot}29}_{-0{\cdot}26}$ and $a_0^0 = 0{\cdot}21 \pm 0{\cdot}08$ are obtained, respectively. These are compatible both with each other, and with the results obtained using the form equation (5.1.10), but are useful confirmations that the scattering length is not dependent on that specific parameterization. A combined fit to both sets of data yields $a_0^0 = 0{\cdot}26 \pm 0{\cdot}08$.

Finally in Fig. 5.1.5 we show the preliminary results of a very recent experiment with 30 200 events (Extermann *et al.*, 1975). Neglecting δ_1^1, these phase shift values were fitted to a two parameter effective range formula‡ to give

$$a_0^0 = 0{\cdot}31 \pm 0{\cdot}10.$$

‡ It should be noted that this form cannot accommodate the subthreshold zero predicted by current algebra. (See the concluding remarks of Section 14.2.)

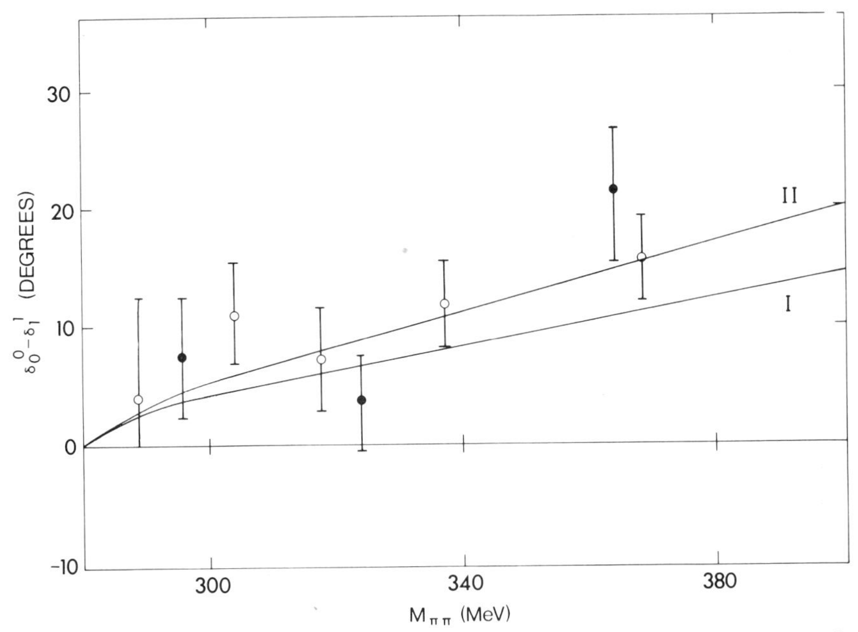

FIG. 5.1.5. Values of $\Delta\equiv\delta_0^0-\delta_1^1\approx\delta_0^0$ from recent K_{e4} data: (●) Beier *et al.* (1973); (○) Extermann *et al.* (1975). The curves are predictions from the current algebra models of (I) Weinberg (1966) and (II) Morgan and Shaw (1972) with scattering lengths of 0·16, 0·21 respectively (see Chapter 14).

The errors on both the phase shift and scattering length values should be significantly smaller when analysis is complete.

In summary, we see that data on K_{e4} decay have improved greatly in the last two or three years, and are now yielding valuable information on δ_0^0 very close to threshold. In particular, the two most recent experiments (Beier *et al.* 1973; Extermann *et al.*, 1975) have many more events than the previous best (Zylbersztejn *et al.*, 1972) with a consequent reduction in the errors. At the same time, the phase shift values are smaller than those suggested by earlier analyses (although strictly compatible within the large errors of those earlier estimates), and are in good agreement with the predictions of models based on current algebra (see Fig. 5.1.5). This result, if maintained in even more accurate experiments, is of considerable theoretical interest.‡

5.2. Other Weak and Electromagnetic Decays

In this section we will consider some other weak and electromagnetic decay processes—$K\to 2\pi$, $K\to 3\pi$ and $\eta\to 3\pi$—which can, in principle, also provide information on $\pi\pi$ scattering. These do not exhaust the possible

‡ See Chapter 14.

decays that have a bearing on the $\pi\pi$ interaction,‡ but they are the ones which have been most extensively studied, and for which abundant data exist.

(a) Two-Pion Decays of the Neutral Kaon

After the discovery that the long-lived component of the $K^0-\bar{K}^0$ system decays into the $\pi^+\pi^-$ mode (Christenson *et al.*, 1964), various phenomenological analyses of neutral kaon decays were made, firstly by Wu and Yang (1964), and later by others (Wolfenstein, 1966; Bell and Steinberger, 1965; Lee and Wu, 1966). These analyses were based on the description of the $K^0-\bar{K}^0$ system originally given by Gell-Mann and Pais (1955) and subsequently developed by Treiman and Sachs (1956), and Good (1957).

If we assume CPT invariance,§ and adopt the convention which relates K^0 and $\bar{K}^0$ in such a way that

$$CP|K^0\rangle = -|\bar{K}^0\rangle,$$

then we may express the short- and long-lived components, (K_S, K_L) of the $K^0-\bar{K}^0$ system as linear combinations

$$|K_S\rangle \equiv p|K^0\rangle - q|\bar{K}^0\rangle,$$

$$|K_L\rangle \equiv p|K^0\rangle + q|\bar{K}^0\rangle,$$

where p and q are complex numbers with

$$|p|^2 + |q|^2 = 1.$$

The various $K\to\pi\pi$ decay amplitudes will be denoted as follows

$$A_{S,L}(+-) \equiv A[K_{S,L}\to\pi^+\pi^-],$$
$$A_{S,L}(0,0) \equiv A[K_{S,L}\to\pi^0\pi^0], \quad (5.2.1)$$

and

$$A_{S,L}(I=0,2) \equiv A[K_{S,L}\to(\pi\pi)_{I=0,2}].$$

In terms of the above amplitudes, two useful sets of parameters to describe neutral K decays into two pions may be defined. They are:

$$\varepsilon \equiv \frac{A_L(I=0)}{A_S(I=0)}; \qquad \varepsilon' \equiv \frac{A_L(I=2)}{A_S(I=0)}; \qquad \omega \equiv \frac{A_S(I=2)}{A_S(I=0)}, \quad (5.2.2)$$

‡ Others are $\eta\to\pi^+\pi^-\gamma$, $\psi'\to\psi\pi\pi$ decays. In the first case Conway (1970) has shown that it is possible to extract the quantity $\delta_1^1-\delta_2^0$ in a rigorous fashion from this decay. However the experiments are very difficult. In the second case, while data exists, it is doubtful whether useful information on the $\pi\pi$ phases can be abstracted from them (Morgan and Pennington, 1975).

§ Discussions of the $K^0-\bar{K}^0$ system in the absence of CPT invariance have been given (see e.g. Lee and Wu (1966)), but we will not consider that possibility here.

and

$$\eta_{+-} \equiv \frac{A_L(+-)}{A_S(+-)}; \qquad \eta_{00} \equiv \frac{A_L(00)}{A_S(00)}; \qquad \eta \equiv \frac{A_S(00)}{A_S(+-)}. \tag{5.2.3}$$

Each set has certain advantages, and both are widely used. They are related as follows:

$$\eta_{+-} = \frac{\sqrt{2}\varepsilon + \varepsilon'}{\sqrt{2} + \omega}; \qquad \eta_{00} = \frac{\varepsilon - \sqrt{2}\varepsilon'}{1 - \sqrt{2}\omega}; \qquad \eta = \frac{\sqrt{2}\omega - 1}{\sqrt{2} + \omega}. \tag{5.2.4}$$

Using CPT invariance, and taking into account final-state interactions, the transition amplitudes for K^0 and $\bar{K}^0$ into $(\pi\pi)_{I=0}$ and $(\pi\pi)_{I=2}$ states may be written

$$\begin{aligned} A[K^0 \to (\pi\pi)_{I=0,2}] &= iA_{0,2}\, e^{i\delta_0^{0,2}}, \\ A[\bar{K}^0 \to (\pi\pi)_{I=0,2}] &= -iA^*_{0,2}\, e^{i\delta_0^{0,2}}, \end{aligned} \tag{5.2.5}$$

where δ_0^0 and δ_0^2 are the S-wave $I = 0,2$ $\pi\pi$ phase shifts at the energy corresponding to the kaon mass.

It is conventional to fix the relative phase of $|K^0\rangle$ and $|\bar{K}^0\rangle$ in such a way that the amplitude A_0 can be taken to be real and positive, i.e.

$$A_0 = A_0^* > 0.$$

If we adopt this convention then the parameters ε, ε' and ω, expressed in terms of A_0 and A_2, become

$$\varepsilon = \left(\frac{p-q}{p+q}\right), \tag{5.2.6}$$

$$\varepsilon' = \frac{pA_2 - qA_2^*}{pA_0 + qA_0} \exp[i(\delta_0^2 - \delta_0^0)], \tag{5.2.7}$$

$$\omega = \frac{pA_2 + qA_2^*}{pA_0 + qA_0} \exp[i(\delta_0^2 - \delta_0^0)]. \tag{5.2.8}$$

Thus, there are two ways to obtain information on the phase-shift difference $\delta_0^2 - \delta_0^0$, from a study of either ω or ε'. Unfortunately, as we shall see below, the former method depends on making a theoretical estimate of the relative importance of $\Delta I = \frac{3}{2}$ and $\Delta I = \frac{5}{2}$ terms in the decays $K^+ \to \pi^+\pi^0$ and $K_S \to \pi\pi$. On the other hand, a study of ε' can obviously give no information on the phase-shift difference if ε' is consistent with zero.

(i) *K_S Decay*

From equations (5.2.4) and (5.2.8) we have

$$|\eta|^{-2}=\left|\frac{\sqrt{2}+\operatorname{Re}A_2/A_0\exp[i(\delta_0^2-\delta_0^0)]}{1-\sqrt{2}\operatorname{Re}A_2/A_0\exp[i(\delta_0^2-\delta_0^0)]}\right|^2 \times\left[1+0\left(\varepsilon\frac{\operatorname{Im}A_2}{A_0}\right)\right]. \qquad (5.2.9)$$

Now the data on $K_L\to\pi\pi$ indicate that $|\varepsilon|=0(10^{-3})$, and that $\operatorname{Im}A_2 \lesssim 0{\cdot}1 \operatorname{Re}A_2$ (see e.g. Lee and Wu (1966)), so that using equations (5.2.3), and neglecting quadratic terms in $\operatorname{Re}A_2/A_0$, equation (5.2.9) may be written

$$\frac{\Gamma(K_S\to\pi^+\pi^-)}{\Gamma(K_S\to\pi^0\pi^0)}-2=6\sqrt{2}\frac{\operatorname{Re}A_2}{A_0}\cos(\delta_0^2-\delta_0^0)+C, \qquad (5.2.10)$$

where the term C represents electromagnetic corrections, e.g., phase-space volume differences due to the mass differences of the charged and neutral pions, and Coulomb interaction effects. The correct expression for C was first given by Belavin and Narodetsky (1968), and confirmed by later calculations (Neveu and Scherk, 1968; Nachtmann and de Rafael, 1969). (An earlier calculation (Abbud *et al.*, 1967) omitted certain terms and gave a result an order of magnitude too large). Numerically, $C=-0{\cdot}006$. (This value is taken from Nachtmann and de Rafael, who point out a numerical error in the evaluation of Belavin and Narodetsky.) In the numerical work that follows we will arbitrarily assign a 100% error to this term, i.e. we will use

$$C=-0{\cdot}006\pm0{\cdot}006. \qquad (5.2.11)$$

The left hand side of equation (5.2.10) may be obtained directly from the experimental K_S decay rates. Using the compiled values for these (Particle Data Group, 1974), we have‡

$$\frac{\Gamma(K_S\to\pi^+\pi^-)}{\Gamma(K_S\to\pi^0\pi^0)}-2=0{\cdot}205\pm0{\cdot}041. \qquad (5.2.12)$$

Thus, in order to obtain a value for $\delta_0^2-\delta_0^0$, it is necessary only to obtain an estimate of $\operatorname{Re}A_2\approx|A_2|$.

To obtain a value of $\operatorname{Re}A_2$ we will compare the rate for K_S decay with the rate for the decay $K^+\to\pi^+\pi^0$. Using CPT invariance, and taking into

‡ In the numerical calculations which follow, values for the various parameters of $K^0-\bar{K}^0$ phenomenology are taken from the compilation of the Particle Data Group (1974). These parameters have fluctuated in value in the past, and the reader is warned that even now they may be subject to change.

account final-state interactions, the amplitude for K^+ decay may be written

$$A^+(+0) = iA_2^+ \exp(i\delta_0^2). \qquad (5.2.13)$$

If we assume that the decay $K_S \to \pi\pi$ proceeds *predominantly* via the $\Delta I = \frac{1}{2}$ rule, the ratio $|A_2^+|/|A_0|$ may be obtained directly from the ratio of the decay rates, the value obtained being

$$\frac{|A_2^+|}{|A_0|} = 0\cdot039 \pm 0\cdot001. \qquad (5.2.14)$$

However, since transitions from an initial kaon to a $(\pi\pi)_{I=2}$ state may be via $\Delta I = \frac{3}{2}$ and/or $\Delta I = \frac{5}{2}$ the amplitudes A_2 and A_2^+ are, in principle, different superpositions of irreducible isospin amplitudes $A^{(\frac{3}{2})}$ and $A^{(\frac{5}{2})}$, i.e.

$$\begin{aligned} A_2 &= \frac{1}{\sqrt{2}} A^{(\frac{3}{2})} + \frac{1}{\sqrt{2}} A^{(\frac{5}{2})}, \\ A_2^+ &= \frac{\sqrt{3}}{2} A^{(\frac{3}{2})} - \frac{1}{\sqrt{3}} A^{(\frac{5}{2})}. \end{aligned} \qquad (5.2.15)$$

If the K_S and K^+ decays are assumed to proceed via a conventional current–current interaction, then $\Delta I = \frac{1}{2}, \frac{3}{2}$ for $K_S \to \pi\pi$ and $\Delta I = \frac{3}{2}$ for $K^+ \to \pi^+\pi^0$. Thus $\Delta I = \frac{5}{2}$ transitions proceed via electromagnetic corrections, so that $|A^{(\frac{5}{2})}|/|A_0| = 0(\alpha)$. Using this result in equation (5.2.14) gives

$$|A^{(\frac{5}{2})}| \lesssim 0\cdot2 |A^{(\frac{3}{2})}|.$$

It follows from equation (5.2.15) that $|A_2|/|A_2^+|$ can lie in the range 0·57 to 1·13. We will therefore use

$$\frac{|A_2|}{|A_2^+|} = 0\cdot85 \pm 0\cdot28$$

which, combined with equation (5.2.14), gives

$$\frac{\text{Re } A_2}{A_0} = \frac{|A_2|}{|A_0|} = 0\cdot033 \pm 0\cdot011. \qquad (5.2.16)$$

Finally, using equations (5.2.16), (5.2.12) and (5.2.11) in equation (5.2.10) gives

$$|\delta_0^2 - \delta_0^0| \lesssim 70°, \qquad (\text{modulo } \pi). \qquad (5.2.17)$$

It is instructive to note that if $\Delta I = \frac{5}{2}$ transitions are neglected, as is frequently the case, then $A_2 = \sqrt{(2/3)} A_2^+$, and a misleadingly precise value $(39 \pm 14)°$ is obtained.‡

‡ Recent analyses of peripheral dipion production (see Section 3.3, especially Figs 3.3.3 and 3.3.6(b)) suggest a value in the range $(\delta_0^0 - \delta_0^2) \approx 47 \pm 9°$.

(ii) K_L *Decay*

From equations (5.2.6) and (5.2.7), recalling that $\varepsilon \ll 1$, we can write ε' in the approximate form

$$\varepsilon' = \frac{i \operatorname{Im} A_2}{A_0} \exp\left[i(\delta_0^2 - \delta_0^0)\right], \tag{5.2.18}$$

and thus $(\delta_0^2 - \delta_0^0)$ may be found if ε' is known, and non-zero. However, if $\varepsilon' = 0$, as is the case in some suggested theories of *CP* violation (e.g. the super-weak theory of Wolfenstein (1966)), then no information can be obtained.

Since $\omega \ll 1$ it may be neglected in equations (5.2.4) for η_{+-} and η_{00}, and so ε' may be found from the approximate expressions

$$\eta_{+-} = \varepsilon + \frac{\varepsilon'}{\sqrt{2}}; \qquad \eta_{00} = \varepsilon - \sqrt{2}\varepsilon', \tag{5.2.19}$$

by eliminating ε. The best current values of η_{+-} and η_{00} are (Particle Data Group, 1974)

$$\begin{aligned} |\eta_{+-}| &= (2{\cdot}17 \pm 0{\cdot}07)10^{-3}; \qquad \phi_{+-} = (46{\cdot}6 \pm 2{\cdot}5)^\circ, \\ |\eta_{00}| &= (2{\cdot}25 \pm 0{\cdot}09)10^{-3}; \qquad \phi_{00} = (49 \pm 13)^\circ, \end{aligned} \tag{5.2.20}$$

where ϕ_{+-} and ϕ_{00} are the phases of η_{+-} and η_{00}. Unfortunately, these lead to values of Re ϵ' and Im ε' both of which are consistent with zero and have large errors, and so no determination of $(\delta_0^2 - \delta_0^0)$ is possible.

As an alternative to the above method, one could avoid using a value of η_{00} (whose phase still has a large error) by obtaining ϵ' directly from the equation for η_{+-} using an input value for the parameter ε. This latter quantity is related to the charge asymmetry in three-body semileptonic K-decays,

$$\Delta \equiv \frac{\Gamma(K_L \to \pi^- l^+ \nu_l) - \Gamma(K_L \to \pi^+ l^- \bar{\nu}_l)}{\Gamma(K_L \to \pi^- l^+ \nu_l) + \Gamma(K_L \to \pi^+ l^- \bar{\nu}_l)}$$

by the following expression (see e.g. Martin and de Rafael (1967))

$$\Delta = 2 \operatorname{Re} \varepsilon \left(\frac{1 - |x|^2}{|1 - x|^2}\right) + 0(\varepsilon^2), \tag{5.2.21}$$

where

$$x \equiv \frac{A(\bar{K}^0 \to \pi^- l^+ \nu_l)}{A(K^0 \to \pi^- l^+ \nu_l)},$$

is a parameter which measures the violation of the $\Delta S = \Delta Q$ rule in these decays.

Combining the compiled best average values of Δ and x (Particle Data Group, 1974) gives

$$\Delta=(3\cdot4\pm0\cdot1)10^{-3}; \qquad \frac{1-|x|^2}{|1-x|^2}=1\cdot00\pm0\cdot08,$$

and

$$\text{Re } \varepsilon=(1\cdot70\pm0\cdot19)10^{-3}. \tag{5.2.22}$$

The phase θ_ε of ε may be obtained from the unitarity condition for K^0 decay (see e.g. Martin and de Rafael (1967)), and to a good approximation

$$\tan\theta_\varepsilon=\frac{2(m_L-m_S)}{\Gamma_S}=0\cdot957\pm0\cdot014,$$

i.e.

$$\theta_\varepsilon=(43\cdot7\pm0\cdot5)°. \tag{5.2.23}$$

Finally, using equations (5.2.22), (5.2.23) and (5.2.20) in equation (5.2.19) leads to a somewhat better determination of ε'. However, it is still consistent with zero, and so again no determination of the $\pi\pi$ phase-shift difference is possible.

(b) $K\to3\pi$ and $\eta\to3\pi$ Decays

(i) *Introduction*

The three-pion decays of the kaon have been studied extensively since Dalitz first used the phase-space plot to analyse 13 events of the type $K^+\to\pi^+\pi^+\pi^-$ (Dalitz, 1953, 1954). The observed decay modes are

$$\begin{aligned} K^\pm &\to \pi^+\pi^-\pi^\pm && \tau^\pm \text{ decay} \\ &\to \pi^0\pi^0\pi^\pm && \tau'^\pm \text{ decay} \\ K_L^0 &\to \pi^+\pi^-\pi^0 && \\ &\to \pi^0\pi^0\pi^0. && \end{aligned} \tag{5.2.24}$$

Because of the small energy release in these decays, the final-state pions would naturally be expected to be predominantly in an overall S-state, and the energy distributions of the pions to be dominated by phase space considerations. To a first approximation this is borne out by experiment, but the spectra are accurate enough to show small, but statistically significant, departures from phase space. If one adopts the view that these deviations are due to $\pi\pi$ scattering effects modifying a constant bare weak interaction matrix element,‡ then one might hope that an analysis of the Dalitz plots of these decays could yield information on low-energy $\pi\pi$ amplitudes.§

Early papers tried to extract the S-wave $\pi\pi$ scattering lengths from these decays. However, severe approximations are needed to reduce the complicated three-body problem to a tractable form, and since the deviations from phase space are small, such attempts are too ambitious. Rather, these decays should be looked upon as providing further consistency tests, within the framework of definite models, for suggested $\pi\pi$ amplitudes. In this sense, they differ from K_{e4} and $K_{2\pi}$ decays, but are more in the spirit of NN and KN scattering which were discussed in Section 4.2.

Several approaches have been suggested.‖ For example, simple pion-pole models (Bég and DeCelles, 1962; Barton and Rosen, 1962; Bég, 1962; Wali, 1962) gave a reasonable description of the data, as did models postulating the existence of a low-lying $I=0$ S-wave $\pi\pi$ resonance (the so-called σ at $\approx$400 MeV) (Brown and Singer, 1964; Mitra and Roy, 1964; Kacser, 1963), but neither yield any useful information about $\pi\pi$ amplitudes. The most widely studied are the specific final-state interaction models, which are also applicable to the data for $\eta\rightarrow 3\pi$ decays by the appropriate change of mass.

Much work has been done on these models. Gribov (1958) has developed a formalism based on expanding the decay amplitudes in power series in the relative momenta of the pions. This was later extended by Anisovich and co-workers (see e.g. Anisovich (1963); Anisovich and Anselm (1966)). A recent application of this model is that of Bunaytov *et al.* (1973) who have analysed the accurate $\tau^{\pm}$ data of Ford *et al.* (1972) assuming that the final-state pions are in a relative S-state, and the validity of the $|\Delta I|=\frac{1}{2}$ rule. They found the S-wave $\pi\pi$ scattering length values

$$|a_0^0|=0{\cdot}72\pm0{\cdot}07; \qquad |a_0^2|=0{\cdot}07\pm0{\cdot}09. \tag{5.2.25}$$

A more specific approach to final-state interactions is that of Barbour and Schult (1967a,b) who have used the Faddeev equations with S- and P-wave non-local separable potentials. They find that with only S-waves $(a_0^2)^2-(a_0^0)^2\approx 2$, but by including the P-wave with $a_1^1\approx 0{\cdot}2$, the P-wave can

‡ This view is not necessarily correct. For example, current algebra (see Chapter 14) predicts a slowly-varying energy dependence for the weak interaction matrix element for $K\rightarrow 3\pi$ which already gives a good representation of the data, completely ignoring final-state interactions (see e.g. Neveu and Scherk (1970)). However, we will only consider the structureless possibility here.

§ Since the S-state for three pions is totally symmetric in isospin it can have total $I=1$ or 3. However, that with $I=1$ dominates that with $I=3$ (e.g. the rate for $\tau^+ >$ rate for τ'^+), and so it is conventional to assume the validity of the $\Delta I=\frac{1}{2}$ rule (see Kellett (1971) for a discussion of this rule). Under this assumption only the S-wave $I=0$ and 2 $\pi\pi$ amplitudes can be studied.

‖ The specific models of current algebra and the Lovelace–Veneziano form will be considered in Chapters 14 and 12, respectively.

become dominant, and fits can be achieved with $(a_0^2)^2 < (a_0^0)^2$. These P-wave solutions are sensitive to a cutoff used in the calculations, but are interesting in that they show the possible dangers inherent in neglecting the P-wave.

Finally, there is the dispersion theoretic approach of Khuri and Treiman (1960), and Sawyer and Wali (1960). This attempts to input information on the analytic properties of the decay amplitudes, and is probably the most attractive method of studying $\pi\pi$ scattering from final state interactions in K and $\eta \to 3\pi$ decays. We will describe briefly the method, and the results that have been obtained using it.

(ii) *Khuri–Treiman Method*

If we assume that the final state is pure $I = 1$, which is in agreement with experiment, then it is convenient to pretend that the kaon has isospin one and that isospin is conserved in the decay. This leads to a compact notation and involves no error. Then, if we denote the isospin indices of the K and the final state pions by ρ, and α, β, γ, respectively, the matrix element for the decay

$$K_\rho(K) \to \pi_\alpha(k_1) + \pi_\beta(k_2) + \pi_\gamma(k_3),$$

may be written (cf. equation (2.17))

$$M_{\rho,\alpha\beta\gamma} = A\delta_{\rho\alpha}\delta_{\beta\gamma} + B\delta_{\rho\beta}\delta_{\gamma\alpha} + C\delta_{\rho\gamma}\delta_{\alpha\beta},$$

where A, B, C are functions of the scalar invariants $s_i \equiv -(K - k_i)^2$ $(i = 1, 2, 3)$‡ with the constraint

$$\sum_i s_i = M^2 + 3\mu^2 \equiv 3s_0.$$

Matrix elements for the physical decays of (5.2.24) then follow from the crossing relations of equation (2.24), e.g.

$$\begin{aligned} M(++-) &= A + B \\ M(00+) &= C \\ M(+-0) &= C \\ M(000) &= A + B + C. \end{aligned} \tag{5.2.26}$$

Khuri and Treiman (1960) obtained dispersion relations for the decay amplitudes A, B and C by firstly writing dispersion relations for the scattering-like process $a + K \to b + c$, and then making an analytic continuation in the variable s_3. In the approximation where only S-wave $\pi\pi$

‡ The quantity s_3 is reserved for the "odd" pion, i.e., π^- for τ^+ and τ'^- decays; π^+ for τ^- and τ'^+; and π^0 for K_L^0 decays.

intermediate states are retained in the unitarity condition for $\pi K \to \pi\pi$, the dispersion relations for A, B, and C can be written in the form

$$\begin{aligned} A(s_1, s_2, s_3) &= D_0 + U(s_1) + V(s_2) + V(s_3), \\ B(s_1, s_2, s_3) &= D_0 + V(s_1) + U(s_2) + V(s_3), \\ C(s_1, s_2, s_3) &= D_0 + V(s_1) + V(s_2) + U(s_3), \end{aligned} \tag{5.2.27}$$

where D_0 is the value of A, B and C at the symmetric point $s_1 = s_2 = s_3 = s_0$, and $U(s)$ and $V(s)$ satisfy the following coupled integral equations

$$U(s) = \frac{(s-s_0)}{\pi} \int_{4\mu^2}^{\infty} ds' \frac{\bar{A}(s')T_0^{0*}(s') + \frac{1}{3}[\bar{B}(s') + \bar{C}(s')][T_0^{0*}(s') - T_0^{2*}(s')]}{(s'-s_0+i\varepsilon)(s'-s+i\varepsilon)}, \tag{5.2.28}$$

$$V(s) = \frac{(s-s_0)}{\pi} \int_{4\mu^2}^{\infty} ds' \frac{\frac{1}{2}[\bar{B}(s') + \bar{C}(s')]T_0^{2*}(s')}{(s'-s_0+i\varepsilon)(s'-s+i\varepsilon)}, \tag{5.2.29}$$

with

$$\bar{A}(s_1) \equiv \tfrac{1}{2} \int_{-1}^{+1} d\cos\theta_{13} A(s_1, s_2, s_3), \text{ etc.}$$

The angle θ_{13} is that between particles 1 and 3, in the centre-of-mass system of 2 and 3, and $T_l^I(s)$ is the usual $\pi\pi$ partial wave amplitude. This formalism includes all (elastic) pairwise $\pi\pi$ interactions in the sub-channels, but only includes multiple scattering effects successively projected into S-states for the scattering pairs. The full analytic structure associated with multiple scattering, for example the special singularities associated with the triangle diagram, are not included.

In their original paper Khuri and Treiman (1960) took the first iteration of an iterative solution to these equations, i.e., they used $A = B = C = D_0$ under the integrals. The $\pi\pi$ amplitudes were parameterized in scattering length forms. The solution for the decay matrix elements is then easily obtained from equations (5.2.25)–(5.2.28). (The subtraction constant can be factored out as an overall normalization constant.) By fitting the measured slopes of the Dalitz plots for τ^+ and τ'^+ as found from the empirical form

$$|M|^2 \propto 1 + g(s_3 - s_0), \tag{5.2.30}$$

Khuri and Treiman deduced that $a_0^0 - a_0^2 \approx -0{\cdot}7$, which is totally inconsistent with currently accepted estimates of this quantity (see Chapter 6).

Other methods of approximate solution of the Khuri–Treiman equations are possible, for example by using the Omnès function discussed in Chapter 4 (cf. equation (4.1.9, and Section 9(b)). As an example of these calculations, Graves-Morris (1971) has constructed "D-function" forms consistent with the $I=0, 2$, S-wave phase shifts of Morgan and Shaw (1970), and has shown that the resulting decay amplitudes give a reasonable fit to data for both K and $\eta \rightarrow 3\pi$ decays.

In the absence of a rigorous analytic solution, a full numerical solution is probably the next best thing. Some progress has been made in this direction by Mathews (1971), who has produced numerical solutions for both K and $\eta \rightarrow 3\pi$, including P-wave interactions.‡ Using the data of Mast *et al.* (1969) for K-decay, and those of Cnops *et al.* (1968) for η-decay, Mathews confirms the Khuri–Treiman result that with S-waves alone (parameterized by scattering lengths) it is difficult to get a steep enough slope for the Dalitz plot distributions unless $a_0^0 - a_0^2 \approx -1$. Using a form for T_0^0 corresponding to the Up–Down type of phase shift (see Chapter 3) does not improve the situation greatly, in apparent contradiction to the findings of Graves-Morris (1971). However, fits can be achieved with acceptable (Up–Down) forms for δ_0^0 if a P-wave contribution is included (in practice the parameterization of Olsson (1967) was used with $a_1^1 \approx 0{\cdot}15$), but this result is not unique, and a Down–Up form is also acceptable if the P-wave is suitably adjusted. The sensitivity of the solutions to the P-wave contribution confirms the work of Barbour and Schult (1967a,b).

In conclusion, these various studies show that although present data on $K \rightarrow 3\pi$ and $\eta \rightarrow 3\pi$ are certainly *consistent* with currently favoured forms for the low-energy $\pi\pi$ amplitudes this result is very model-dependent, and certainly one cannot *deduce* information about $\pi\pi$ scattering from these decays, as is often claimed.

5.3. Electron–Positron Colliding Beams

Electron–positron colliding beams are one of the potentially more important sources of information on the $\pi\pi$ system. Just as K_{l4} decay allows us to study the $I=0$ $\pi\pi$ S-wave at low energies in a relatively unambiguous manner, $\pi^+\pi^-$ production by one photon annihilation allows us to study the $I=1$ P-wave phase shift over the whole region where inelasticity is negligible. Like K_{l4} the advantage lies in the fact that the two pions are the only hadrons involved in the process. Such reactions are, of course, an extremely important supplement to, and check on, information inferred in more complicated ways from purely hadronic phenomena, e.g. $\pi N \rightarrow \pi N$, $\pi N \rightarrow \pi\pi N$. For this reason we shall discuss the analysis of the reaction $e^+e^- \rightarrow \pi^+\pi^-$ in some detail, although the data are at present not of sufficient quality for convincing results to be obtained. A second topic of importance is one-photon annihilation into hadrons at higher energies. Here the

‡ He has also considered the effects of a non-flat bare weak interaction matrix element, and the effects of mass differences between the pions.

advantage is that the J^P of the hadronic system is necessarily 1^-, making it an excellent way to search for higher mass vector mesons. In most other reactions these will be produced, if they exist, along with states of other quantum numbers, rendering analysis difficult. Again, this will be an area of importance in the future, although at present data are rather scarce. Finally, two-photon annihilation, which can be observed at higher energies, allows one to study $C=+1$ hadronic systems also. We shall discuss each of these topics in turn.

(a) $e^+e^-\to\pi^+\pi^-$: One-Photon Annihilation

This process is, as we have already noted, a potent source of information on the $\pi\pi$ P-wave. To lowest order in the electromagnetic interaction, the amplitude is given by the one-photon annihilation diagram of Fig. 5.3.1, the

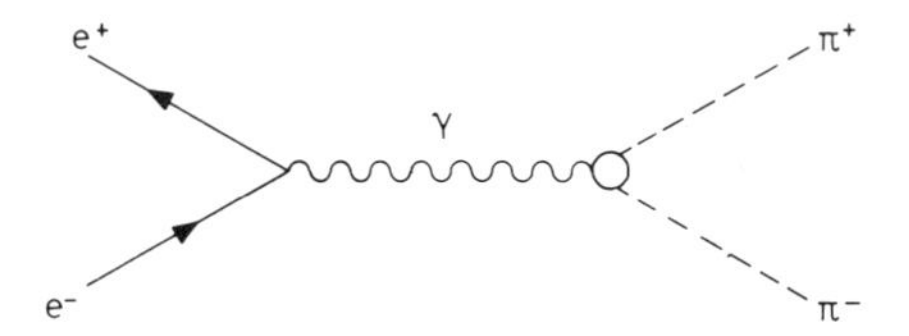

FIG. 5.3.1. $\pi^+\pi^-$ pair production via one-photon annihilation.

pions being produced in the $I=J=1$ state with no other strongly interacting particles involved. The cross-section thus gives a direct measure of the magnitude of the pion form factor, the phase of which is related by time reversal and unitarity to the $\pi\pi$ elastic phase shift δ_1^1. Inferences are made from the former to the latter by dispersion relations, as we shall discuss.‡

The differential cross-section for annihilation to $\pi^+\pi^-$ is given by

$$\frac{d\sigma}{dz}=\frac{\pi\alpha^2}{4}\frac{(t-4)^{\frac{3}{2}}}{t^{\frac{5}{2}}}(1-z^2)|F_\pi(t)|^2 \tag{5.3.1}$$

so that

$$\sigma_{\text{tot}}(e^+e^-\to\pi^+\pi^-)=\frac{\pi\alpha^2}{3}\left(\frac{t-4}{t}\right)^{\frac{3}{2}}\frac{1}{t}|F_\pi(t)|^2. \tag{5.3.2}$$

Here α is the fine structure constant ($\alpha\approx 1/137$), $z\equiv\cos\theta$, with θ the centre-of-mass production angle, and t is the square of the total CM energy

‡ Cf. also the review of Gourdin (1974).

$2E$. The pion form factor, defined by

$$(2\pi)^3(4p_1^0p_2^0)^{\frac{1}{2}}\langle\pi(p_2)|J^\mu(0)|\pi(p_1)\rangle$$
$$= e(p_1+p_2)^\mu F_\pi[t=-(p_1-p_2)^2], \qquad (5.3.3)$$

is normalized to $F_\pi(t=0)=1$, and can be studied in the process under discussion in the range $4<t<\infty$. As discussed in Section 5.4 below, information on this form factor can also be obtained in the spacelike region from pion electron scattering (for very small t-values), and, more indirectly, from analyses of single pion electroproduction. Such information can be used as a constraint upon forms used to analyse data in the time-like region, our present concern. The reason that this region is so important for a study of $\pi\pi$ scattering arises from the relation

$$\operatorname{Im} F_\pi(t) = \frac{q}{\omega} T_1^1(t)^* F_\pi(t) = e^{-i\delta_1^1} \sin \delta_1^1(t) F_\pi(t), \qquad (5.3.4)$$

(where $t=4+4q^2$, $\omega^2=1+q^2$) which for the region below the 4π production threshold $4\leqslant t<16$ follows from unitarity and T-invariance. If we denote the phase of $F_\pi(t)$ by $\delta(t)$, i.e.

$$F_\pi(t) = |F_\pi(t)| \exp[i\delta(t)],$$

then equation (5.3.4) implies

$$\delta(t) = \delta_1^1(t), \qquad 4\leqslant t<16 \quad (\text{modulo } \pi). \qquad (5.3.5)$$

This result, exact for $4\leqslant t<16$, should be a good approximation over the whole elastic region for $\pi\pi$ scattering, $t\leqslant 40\mu^2$ (see Chapter 3), and is customarily used over this wider region. Thus a determination of the form factor phase at low t-values is equivalent to a determination of the δ_1^1 phase shift. Unfortunately, the phase of $F_\pi(t)$ is not directly measured (cf. equations (5.3.1) and (5.3.2)), and the theoretical apparatus of dispersion relations has to be employed. This we now review, first giving some general results, and then discussing the problem in terms of certain models, before summarizing the present experimental results.

(i) *General Results on $F_\pi(t)$*

We assume that the form factor has the analytic structure indicated by perturbation theory (see e.g. Eden, *et al.* (1966)). That is, it is a real analytic function of t with a single branch cut along the real axis $4\leqslant t<\infty$. If we further assume that it has a finite number of zeros, n, in the complex t-plane, we can introduce a new function

$$G(t) \equiv F_\pi(t)/P_n(t), \qquad (5.3.6)$$

where $P_n(t)$ is an nth-order polynomial with its zeros situated so as to cancel those of $F_\pi(t)$, and

$$P_n(0) = F_\pi(0) = 1. \tag{5.3.7}$$

The function $G(t)$ is thus free of zeros, and by writing a dispersion relation for $\ln G(t)$, we immediately arrive at the phase representation

$$F_\pi(t) = P_n(t) \exp\left\{\frac{t}{\pi}\int_4^\infty \frac{\delta(t')}{t'(t'-t)}\, dt'\right\}. \tag{5.3.8}$$

This representation exists if $F_\pi(t)$ is polynomially bounded in t for large t, and in fact can be written provided it does not rise faster than $\exp(a\sqrt{t})$. The asymptotic behaviour is related both to the phase and the number of zeros. If $\delta(t) \to \beta\pi$ as $t \to \infty$ such that the integral

$$\int_4^\infty dt' \left[\frac{\delta(t') - \beta\pi}{t'}\right],$$

exists, it follows immediately from (5.3.8) that

$$F_\pi(t) \sim t^{n-\beta}, \qquad \text{as } t \to \infty. \tag{5.3.9}$$

This representation solves the problem of expressing $F_\pi(t)$ in terms of a given $\delta(t)$, provided the locations of all the zeros of $F_\pi(t)$ are given. In most models it is assumed that there are no zeros, so that $P_n(t) \equiv 1$, although this does not follow from any general principles. It is, however, possible to test this assumption by the use of sum rules involving only experimentally measurable quantities. This emerges from studying how the phase $\delta(t)$ can be expressed as an integral over the measurable quantity $|F_\pi(t)|$ (Bowcock and Kannelopoulos, 1968; Truong, Vinh Mau and Yem, 1968).

We consider firstly the case when $F_\pi(t)$ has no zeros. We can then write a dispersion relation for the function $\ln F_\pi(t)/(4-t)^{\frac{1}{2}}$ and arrive immediately at the desired result ($F_\pi(0) = 1$ gives one subtraction)

$$\ln F_\pi(t) = \frac{t(t-4)^{\frac{1}{2}}}{i\pi}\int_4^\infty \frac{\ln|F_\pi(t')|\, dt'}{t'(t'-t)(t'-4)^{\frac{1}{2}}}, \qquad \text{for } t<4 \tag{5.3.10}$$

and

$$\delta(t) = \frac{t(t-4)^{\frac{1}{2}}}{\pi} P\int_4^\infty \frac{\ln|F_\pi(t')|\, dt'}{t'(t'-t)(t'-4)^{\frac{1}{2}}}, \qquad \text{for } t>4. \tag{5.3.11}$$

This latter equation expresses $\delta(t)$ directly in terms of the measurable quantity $|F_\pi(t)|$. Furthermore, by letting $t \to \infty$, we deduce the sum rule

$$0 = \int_4^\infty dt' \frac{\ln|F_\pi(t')|}{t'(t'-4)^{\frac{1}{2}}}. \tag{5.3.12}$$

If there are zeros, then we should instead work with the function $G(t)$ defined in equation (5.3.6), i.e. write a dispersion relation for $\ln G(t)/(4-t)^{\frac{1}{2}}$. In this case, the relations (5.3.10)–(5.3.12) are modified by the addition of terms dependent upon the zeros. Thus, equation (5.3.12) is both a test for the presence of zeros and a measure of their importance. For a more thorough discussion of these points, we refer to Truong, Vinh Mau and Yem (1968), and from now on we neglect the question of possible zeros.

(ii) *Rho Dominance Models*

In the region $t \approx m_\rho^2$, one would expect the form factor to be dominated, at least to a first approximation, by the rho-meson pole. The simplest representation of the form factor is thus a Breit–Wigner form (normalized to unity at $t=0$), corresponding to the same Breit–Wigner for the P-wave $\pi\pi$ partial wave amplitude. The cross-section is then proportional to

$$|F_\pi(t)|^2 = \frac{m_\rho^4[1+\Gamma_\rho^2/m_\rho^2]}{[t-m_\rho^2]^2+m_\rho^2\Gamma_\rho^2} \approx \frac{m_\rho^4}{[t-m_\rho^2]^2+m_\rho^2\Gamma_\rho^2}, \tag{5.3.13}$$

i.e. it would have a peak at $t=m_\rho^2$ of width $2m_\rho\Gamma_\rho$, and height m_ρ^2/Γ_ρ^2. A peak with roughly these properties is observed experimentally as we shall see, so that the above equation does prove a reasonable qualitative guide. However it is clearly quantitatively unsatisfactory since it neglects the deviation of the $I=J=1$ scattering amplitude from a Breit–Wigner form. (The latter does not even have the correct threshold behaviour.)

To improve this we have to discuss the properties of the $I=1$ P-wave scattering amplitude $T_1^1(t)$. It is convenient to divide out the P-wave threshold behaviour and work with the reduced amplitude

$$f_1^1(t) = \frac{T_1^1(t)}{q^2} = \frac{\omega}{q^3} e^{i\delta_1^1} \sin \delta_1^1. \tag{5.3.14}$$

Then elastic unitarity takes the form

$$\operatorname{Im} f_1^1(t) = \left(\frac{q^3}{\omega}\right)|f_1^1(t)|^2,$$

or (5.3.15)

$$\operatorname{Im}[f_1^1(t)^{-1}] = -(q^3/\omega).$$

We now write $f_1^1(t)$ as a quotient in the usual way (Chew and Mandelstam, 1960) (see Chapter 9),

$$f_1^1(t) = N(t)/D(t), \tag{5.3.16}$$

where $N(t)$ has only the left-hand cut $-\infty < t \leq 0$, and $D(t)$ only the right-hand cut $4 \leq t < \infty$, $f_1^1(t)$ having no other singularities. Thus, if $D(t)$ is normalized to unity at $t=0$, it satisfies the dispersion relation

$$D(t)=1+\frac{t}{\pi}\int_4^\infty dt'\frac{\operatorname{Im} D(t')}{t'(t'-t)}, \tag{5.3.17}$$

and the function

$$F_\pi(t)=[D(t)]^{-1}, \tag{5.3.18}$$

has all the properties required of the form factors (and no zeros) aside from the possible deviation of arg F_π from $\delta_1^1(t)$ at large t, which for the moment we neglect. Using equations (5.3.14) and (5.3.15) the above equation can be written

$$D(t)=1-\frac{t}{\pi}\int_4^\infty dt'\left(\frac{q'^3}{\omega'}\right)\frac{N(t')}{t'(t'-t)}. \tag{5.3.19}$$

Thus given $N(t)$, which satisfies

$$N(t)=\frac{1}{\pi}\int_4^\infty dt'\frac{\operatorname{Im} N(t')}{t'-t}, \tag{5.3.20}$$

the form factor $F_\pi(t)$ is determined. Alternatively, the coupled equations (5.3.19) and (5.3.20) can be solved for a given left-hand cut discontinuity of $f_1^1(t)$ (see e.g. Lyth (1971a)).

Convenient parametric forms for $F_\pi(t)$ for fitting to data are achieved by assuming simple forms for the N-function governed by the rho-position and width parameters defined by

$$\cot\delta_1^1(t=m_\rho^2)=0, \tag{5.3.21}$$

and

$$\left[\frac{d}{dt}\cot\delta_1^1(t)\right]_{t=m_\rho^2}=(m_\rho\Gamma_\rho)^{-1}. \tag{5.3.22}$$

One possibility is to represent $N(t)$ by a single pole on the negative real axis (Frazer and Fulco, 1959). Another is to assume $N(t)$ constant and replace equation (5.3.19) by a twice-subtracted relation (Gounaris and Sakurai, 1968). In both cases, there are two parameters which can be related to m_ρ, Γ_ρ using equations (5.3.21) and (5.3.22), and the final expressions are almost identical. In the latter case one obtains

$$F_\pi(t)=F_\rho(t)\equiv\frac{m_\rho^2+dm_\rho\,\Gamma_\rho}{(m_\rho^2-t)-im_\rho\Gamma_\rho(q/q_\rho)^3(m_\rho/\sqrt{t})+l(t)}, \tag{5.3.23}$$

where

$$l(t) = \Gamma_\rho \frac{m_\rho^2}{q_\rho^3}\{q^2[h(t) - h(m_\rho^2)] + q_\rho^2 h'(m_\rho^2)(m_\rho^2 - t)\},$$

$$h(t) = \frac{2}{\pi}\frac{q}{\sqrt{t}} \ln\left(\frac{\sqrt{t} + 2q}{2}\right),$$

and d is a constant, dependent on the rho-mass,

$$d = \frac{3}{\pi}\frac{1}{q_\rho^2}\ln\left(\frac{m_\rho + 2q_\rho}{2}\right) + \frac{m_\rho}{2\pi q_\rho} - \frac{m_\rho}{\pi q_\rho^3} \approx 0{\cdot}48. \tag{5.3.24}$$

Near $t = m_\rho^2$, the term $l(t)$, being proportional to $(m_\rho^2 - t)^2$ can be neglected and the expression reduces to

$$F_\rho(t) \approx \frac{m_\rho^2[1 + d\Gamma_\rho/m_\rho]}{(m_\rho^2 - t) - im_\rho\Gamma_\rho(q/q_\rho)^3(m_\rho/\sqrt{t})}. \tag{5.3.25}$$

This "Gounaris–Sakurai" form has frequently been used to fit electron–positron annihilation data in order to determine m_ρ and Γ_ρ. In the region of the rho-meson, it is similar to the simpler form (5.3.13), but there are quantitative differences. For example, for reasonable values of the rho-parameters (e.g. $m_\rho = 775$ MeV, $\Gamma_\rho = 130$ MeV), the position of the peak in $|F_\pi(t)|^2$ is shifted downwards by about 14 MeV and its height increased by about 15% (Gounaris and Sakurai, 1968).

The above form is commonly used in data fitting, and it is important to keep in mind the limitations stemming from its derivation. Firstly, the $I = J = 1$ $\pi\pi$ phase shift $\delta_1^1(t)$, for which the only physical input is the rho-mass and width, has been constrained to a very specific form—for example one has implicitly imposed the condition $\delta(\infty) = \pi$. This part of the calculation has been improved by Lyth (1971a), who has estimated the discontinuity across the left-hand cut numerically using phenomenological estimates of the various $\pi\pi$ scattering amplitudes involved. This produces changes in $|F_\pi(t)|^2$, as compared to the Gounaris–Sakurai approximation, of about 10% in the height of the peak, but less than ~5% in the width at half maximum. The second ingredient is the assumption of elastic unitarity at all energies, so that $\delta(t) = \delta_1^1(t)$. This, together with the assumption of no zeros, fixes $F_\pi(t)$ uniquely via equation (5.3.8). However, the above phase relation will break down as inelastic contributions enter the unitarity relation. One can characterize the deviation by a phase $\tilde{\delta}$, so that we have

$$\delta(t) = \delta_1^1(t) + \tilde{\delta}(t),$$

where $\tilde{\delta}(t)$ is negligible for $t < \bar{t}$, and $\bar{t}$ marks the limit of applicability of elastic unitarity, generally assumed to be at least several half-widths above the rho-mass, i.e.

$$\bar{t} - m_\rho^2 \gg m_\rho \Gamma_\rho. \tag{5.3.26}$$

The resulting modification to $F_\rho(t)$ can be read off immediately from equation (5.3.8) to be

$$F_\pi(t) = F_\rho(t) \exp\left\{\frac{t}{\pi}\int_{\bar{t}}^{\infty} dt' \frac{\tilde{\delta}(t')}{t'(t'-t)}\right\}. \tag{5.3.27}$$

From equation (5.3.27), it is clear that, over the rho-region $m_\rho^2 + m_\rho\Gamma_\rho > t > m_\rho^2 - m_\rho\Gamma_\rho$, the correction factor is slowly varying, so that the main effect is again on the peak height. The same remark applies to the effect of an additional heavy vector meson (ρ'), or of a distant zero (cf. equation (5.3.8)). For more detailed examples of calculations including inelasticity, we again refer to Lyth (1971a).

The conclusion of the foregoing discussion is that given the mass and width of the rho-meson, the behaviour of $F_\pi(t)$ in the rho-region can be qualitatively understood, but that quantitative results are less reliable. In particular, the height of the peak is very model dependent. Thus any attempt to extract the mass and width of the rho from a two parameter (m_ρ, Γ_ρ) fit to limited data on $|F_\pi(t)|^2$ is bound to lead to different results depending on the form used. This problem can be avoided, to a large extent, by using the Gounaris–Sakurai form $F_\pi(t)$, equation (5.2.23) leaving the parameter d free, so that only the position and width of the peak are determined by m_ρ and Γ_ρ, and the value of the height is not used as a constraint. If corrections to the rho-dominance prediction for the height are small, i.e. if

$$(d - 0{\cdot}48)\Gamma_\rho / m_\rho \ll 1, \tag{5.3.28}$$

then it is presumably reasonable to neglect altogether the corrections to the position and width.

(iii) *Omega–Rho Mixing in $e^+e^- \to \pi^+\pi^-$*‡

In the previous section, we have been discussing the contribution, to lowest order in the fine structure constant α, of the rho-meson, represented symbolically in Fig. 5.3.2. If electromagnetic mixing corrections are neglected, the omega- and rho-mesons will be states of pure isospin 0 and 1, respectively, which we will denote by ω_0 and ρ_0. The decay $\omega_0 \to 2\pi$ is forbidden by G-parity so that no analogous omega contribution to $e^+e^- \to \pi^+\pi^-$ is possible. However, if electromagnetic interactions are included, $\omega_0 \to 2\pi$ is allowed to order α. Furthermore, the physical ω meson, of mass

‡ Omega–rho mixing, first emphasized by Glashow (1961), has manifestations in a number of processes. (For general reviews see Donnachie and Gabathuler (1970) and Goldhaber, G. (1970); cf. also Section 3.1(h).)

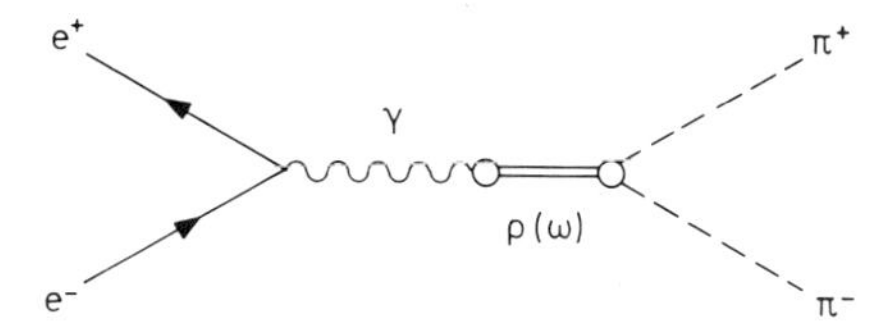

FIG. 5.3.2. Vector meson contributions to $\pi^+\pi^-$ pair production.

784 MeV and width 10·0 MeV, will not be a pure isospin state, but will contain an admixture of ρ_0 to order α, i.e.

$$|\omega\rangle = (1-\varepsilon^2)^{\frac{1}{2}}|\omega_0\rangle + \varepsilon|\rho_0\rangle \\ (\varepsilon = 0(\alpha)). \tag{5.3.29}$$

The physical ω can therefore decay to two pions to order α via $\rho_0 \to 2\pi$. This effect is provided for by adding an omega term to $F_\pi(t)$

$$F_\pi(t) \to F_\pi(t) + A\ e^{i\theta}\frac{m_\omega^2}{m_\omega^2 - t - im_\omega\Gamma_\omega}. \tag{5.3.30}$$

One would look to see a manifestation of this in the form of a narrow interference structure in the region of the ω-meson. At first sight, one would expect the effect to be very small, of order α, and this is certainly true of the $\omega_0 \to 2\pi$ contribution. However, the ρ, ω and hence the ρ_0, ω_0 states are almost degenerate in mass, so that the ρ_0, ω_0 mixing contribution will be enhanced. This argument leads one to expect quite a substantial interference term between the two parts of equation (5.3.30), as is observed. If all effects are attributed to mixing, and the non-enhanced $\omega_0 \to 2\pi$ part is omitted altogether, then the $\omega \to 2\pi$ decay proceeds solely through the $\rho_0 \to 2\pi$ contribution and will have the appropriate phase

$$\theta = \delta_1^1(m_\omega^2). \tag{5.3.31}$$

As we shall see, it is important to take omega–rho mixing into account in extracting information on the rho-parameters from the data.

(iv) *Experimental Results on $e^+e^- \to \pi^+\pi^-$ Below 1 GeV*

The data in the rho-region obtained at Orsay (Le Francois, 1971) and Novosibirsk (Ausländer *et al.*, 1969) are shown in Fig. 5.3.3. The two experiments are in agreement, and there is a clear ω–ρ interference signal in the Orsay results. Unfortunately, thc width of the peak is poorly determined, and an accurate value of the rho-width cannot be obtained. The Orsay data only have been fitted by the authors with the simplified Gounaris–Sakurai form of equation (5.3.25), and the parameter "*d*" left free to take account of the effects discussed in the last part of paragraph (ii).

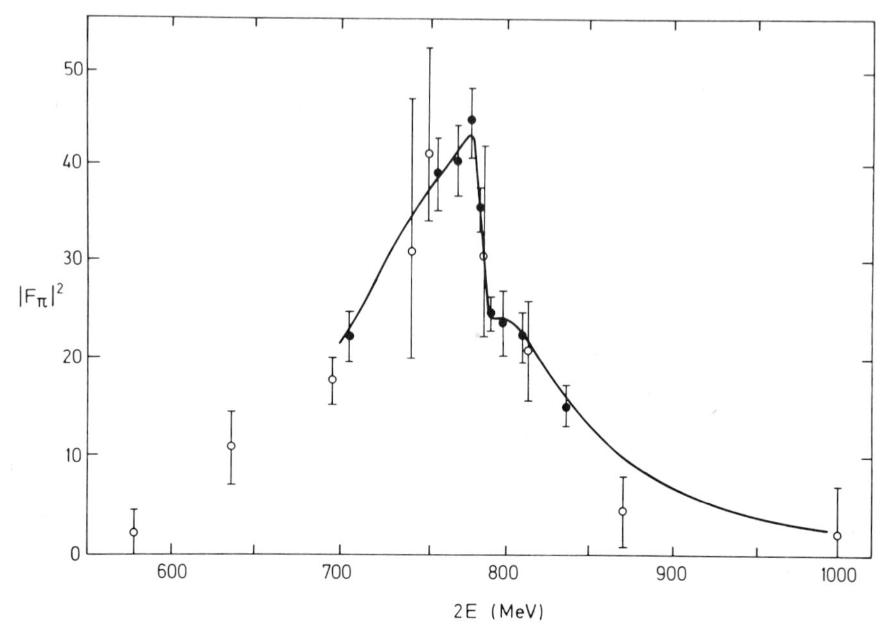

FIG. 5.3.3. The pion form factor deduced from $e^+e^- \rightarrow \pi^+\pi^-$ experiments for total centre-of-mass energies below 1 GeV. The closed circles are the Orsay data (Le Francois, 1971), the open circles the Novosibirsk data (Ausländer *et al.*, 1969). The solid line is the fit to the Orsay data described in the text.

The omega–rho interference term has been included according to equation (5.3.30). The parameters m_ω and Γ_ω are supplied from other data, and the remaining five parameters fitted, with the following results:

$$m_\rho = 780{\cdot}2 \pm 5{\cdot}9 \text{ McV}, \qquad \Gamma_\rho - 152{\cdot}8 \pm 15{\cdot}1 \text{ MeV}$$

$$d = 0{\cdot}66 \pm 0{\cdot}37, \qquad \theta = 87{\cdot}5 \pm 13{\cdot}8° \tag{5.3.32}$$

$$B^{\frac{1}{2}}_{\omega \rightarrow 2\pi} = 0{\cdot}20 \pm 0{\cdot}05.$$

The result for A of equation (5.3.30) has been converted into one for the branching ratio $B(\omega \rightarrow \pi\pi)$ using experimental information on $B(\omega \rightarrow e^+e^-)$ (see Le Francois (1971)). The fit is shown in Fig. 5.3.3. If the $\omega \rightarrow 2\pi$ effect is neglected, not only is the fit considerably worsened, but there is an appreciable shift in the rho-parameter obtained ($m_\rho = 768{\cdot}7 \pm 4{\cdot}4$ MeV, $\Gamma_\rho = 132{\cdot}2 \pm 11{\cdot}4$ MeV). In the fit including $\omega \rightarrow 2\pi$ the values of d and θ are both consistent with the theoretically expected values, equations (5.3.24) and (5.3.31), and that for d lies in a range such that equation (5.3.28) is indeed satisfied.

(b) Hadronic Production in the GeV Region

In this energy range there are two topics of direct relevance to our present interests. Firstly, higher energy data on $|F_\pi(t)|$, via the reaction

$$e^+e^- \to \pi^+\pi^-, \tag{5.3.33}$$

are of help in fixing the behaviour of $\delta_1^1(t)$ below the elastic threshold, since the integration in the dispersion integrals extend to all $t \geqslant 4$. Secondly, the study of this, and any other of the hadron channels, enables one to look for higher mass vector mesons, and, if they exist, to study their properties. The most interesting data are at present on the reaction

$$e^+e^- \to \pi^+\pi^+\pi^-\pi^-. \tag{5.3.34}$$

The experimental results for reaction (5.3.33) are shown in Fig. 5.3.4 as values of $|F_\pi(t)|^2$. As can be seen, there is no evidence for a resonance in this

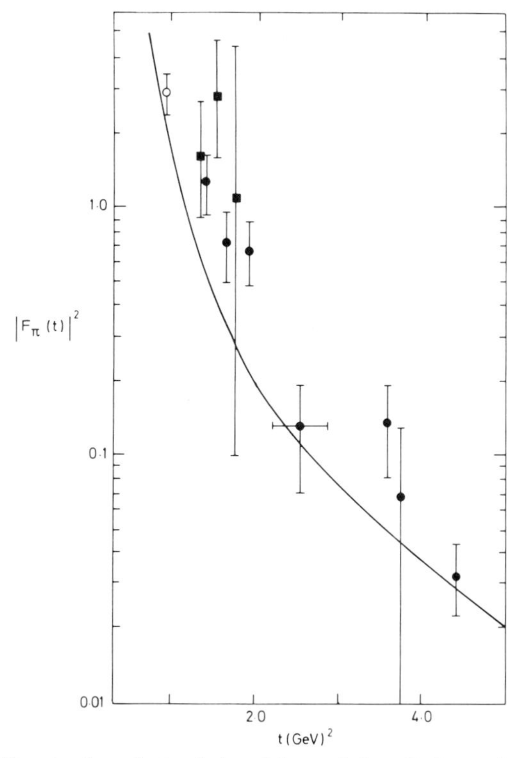

FIG. 5.3.4. The pion form factor deduced from $e^+e^- \to \pi^+\pi^-$ experiments for total centre-of-mass energies above 1 GeV. The data are taken from: ○ (Benaksas *et al.*, 1972); ■ (Balakin *et al.*, 1972); and ● (Barbiellini *et al.*, 1973; Bernardini *et al.*, 1973). The curve is the value of $|F_\pi|^2$ predicted from the Gounaris–Sakurai fit.

particular channel. The values are, however, larger than those expected from the rho-meson tail, and resonances weakly coupled to $\pi\pi$ cannot be excluded.

A much more striking result concerns the cross-section for reaction (5.3.34), the data on which are shown in Fig. 5.3.5. An obvious interpretation of these results is that one is exciting a heavy vector meson—a ρ'-resonance—with mass ≈ 1600 MeV, width ≈ 350 MeV. Although no effect

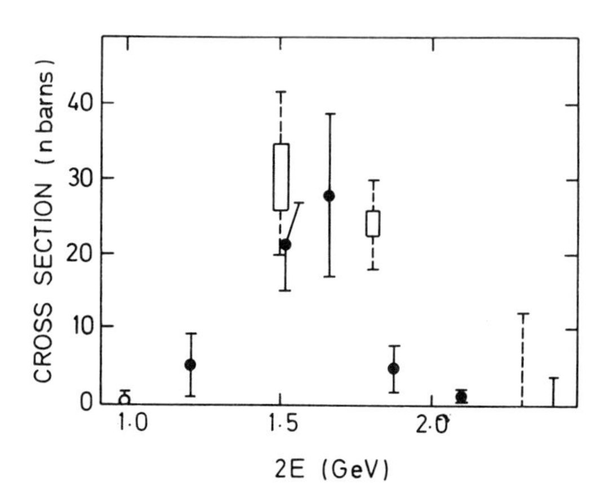

FIG. 5.3.5. The cross-section for $e^+e^- \to \pi^+\pi^-\pi^+\pi^-$. The solid circles are from Barbarino *et al.* (1972), the open circle from Le Francois (1971) and the boxes from Bartoli *et al.* (1972).

has been clearly established on other channels (e.g. $e^+e^- \to \pi^+\pi^- +$ neutrals), a similar effect has been seen in the related photoproduction process (cf. Section 5.4)

$$\gamma + p \to \pi^+\pi^+\pi^-\pi^- p.$$

There have also been suggestions of an inelastic $I = J = 1$ resonance of about this mass from dipion production data with pion beams, but the evidence on this point is somewhat unclear (cf. Section 3.3).

(c) Two-Photon Annihilation

The total cross-section for the one-photon annihilation process

$$e^+e^- \to \pi^+\pi^-, \qquad (5.3.35)$$

discussed above, decreases rapidly with increasing beam energy E, i.e.

from equation (5.3.2)

$$\sigma_t^\gamma(E) = \frac{\pi\alpha^2}{12}\left(\frac{E^2-1}{E^2}\right)^{\frac{3}{2}} \frac{1}{E^2}|F_\pi(t=4E^2)|^2. \tag{5.3.36}$$

However, pion pairs can also be produced via the two-photon exchange processes

$$e^+e^- \rightarrow e^+e^-\pi^+\pi^-, \tag{5.3.37}$$

illustrated in Fig. 5.3.6. The general topic of two-photon annihilation to $C=+1$ hadronic and leptonic systems, and the range of validity of the

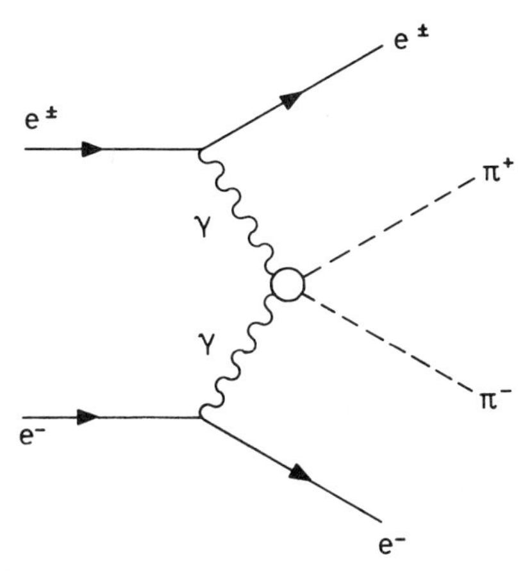

FIG. 5.3.6. $\pi^+\pi^-$ pair production via two photon annihilation.

Weizäcker–Williams approach, in which the virtual photons are replaced by real photons, has been reviewed by Brodsky (1971).‡ For the process of Fig. 5.3.6, if the scattered leptons are not observed, the total pion-pair production cross-section is approximately given by (Brodsky *et al.*, 1970):

$$\sigma_t^{\gamma\gamma}(E) = \frac{2\alpha^2}{\pi^2}\ln^2\left(\frac{E}{m_e}\right)\int_4^{4E^2}\frac{ds}{s}f\left[\left(\frac{s}{4E^2}\right)^{\frac{1}{2}}\right]\sigma_{\gamma\gamma}(s), \tag{5.3.38}$$

where

$$f(x) = (2+x^2)^2[\ln(x)]^{-1} - (1-x^2)(3+x^2), \tag{5.3.39}$$

and $\sigma_{\gamma\gamma}$ is the total cross-section for two (real) photon annihilation to give two pions. Although $\sigma_t^{\gamma\gamma}$ is of higher order in α than σ_t^γ (α^4 compared with α^2), it is slowly increasing with energy, rather than rapidly decreasing, and

‡ Cf. also: Proceedings of the International Colloquium on Photon–Photon Collisions in Electron–Positron Storage Rings, *Suppl. Jour. de Physique* 35 (1974) C-2.

so, although negligible at low energies, will ultimately dominate. In fact, if both equations (5.3.36) and (5.3.38) are calculated in Born approximation with point-like coupling ($F_\pi = 1$), Brodsky *et al.* find $\sigma_t^{\gamma\gamma} > \sigma_t^{\gamma}$ for $2E \gtrsim 1{\cdot}25$ GeV. There are two aspects to this—it could be an inconvenient background to the one-photon process, but it could also be an interesting process in its own right.

The first point is not so serious (until the two-photon process is overwhelmingly dominant) since the two pions produced in reaction (5.3.35) are colinear, with equal and opposite momenta, and together have invariant mass equal to $2E$ (whereas the pion pairs produced in the reaction (5.3.37) are not in general colinear) and, as can be seen from equations (5.3.38) and (5.3.39) their invariant mass is strongly peaked at the threshold value $m_{\pi\pi} = 2m_\pi$. Thus, the two processes should be easily distinguishable, even if the leptons are not observed.

On the second point, the process gives us an opportunity to study the production of neutral $\pi\pi$ systems of even charge conjugation (i.e. even I, J values) in a situation where no other hadrons are involved. As noted above, the dipion mass spectrum peaks strongly towards threshold, so that information will be principally on the $I = 0$ and 2 S-waves. Within the framework of the Weizäcker–Williams approximation (see the discussion in Brodsky (1971)), the observed cross-section is to be interpreted, via equation (5.3.38), in terms of that for the real two-photon process $\gamma\gamma \to \pi^+\pi^-$. Since the spectrum is peaked near threshold as noted, low-energy theorems (Abarbanel and Goldberger, 1968) would suggest that a reasonable first approximation will be given by the simple Born terms. However, since unitarity and time reversal invariance compel the production amplitudes $A_J^I(\gamma\gamma \to \pi\pi)$ to have the same phase as the corresponding $\pi\pi$ scattering amplitudes, there must be corrections to the (purely real) Born approximation. The magnitude of these has been estimated in terms of the $\pi\pi$ phase shifts using conventional dispersion theory methods by several authors (Lyth, 1971b, 1972; Goble and Rosner, 1972; Schierholz and Sundermayer, 1972). Another approach (Carlson and Tung, 1972) is to try and detect the interference terms between the S-wave amplitudes and the higher waves ($J \geq 2$) for which the Born approximation is retained. In both these papers, the process is studied as for real photon annihilation, and whether such approaches will lead to reliable information on the S-wave $\pi\pi$ phase shifts is not yet clear. There is at present little or no experimental information on reaction (5.3.37), or on the analogous reaction

$$e^+e^- \to e^+e^-\pi^0\pi^0,$$

which vanishes in Born approximation, and which would be needed if any model independent separation of the $I = 0$ and 2 states is to be made.

5.4. Other Electromagnetic Interactions

(a) Form Factors in the Spacelike Region

Historically, study of the isovector form factors of the nucleon has been an important source of information on the $\pi\pi$ system, since the existence of the rho-meson itself was first predicted from such an analysis (Frazer and Fulco, 1959, 1960a,b). Although the predicted mass was somewhat lower than that subsequently found by experiment, this is nonetheless an outstanding example of a prediction from the dispersion relation approach to hadron physics. Subsequent study of the form factors, which have been measured extensively in the space-like region $t<0$ for elastic electron–nucleon scattering, led to the prediction of one or more heavier mesons (rho-primes), which have also since been predicted both by the quark model and by dual models, as discussed in Chapter 13. The existence of such particles is one of the more important questions in the field of $\pi\pi$ scattering, so we will concentrate on this aspect in what follows.

Detailed information is at present available on four isovector electromagnetic form factors: the two nucleon form factors F_1^v and F_2^v, the pion form factor F_π; and the form factor G for the transition $\Delta^+(1232)\rightarrow\gamma p$. The best studied are the two nucleon form factors, defined by

$$(2\pi)^3\left(\frac{p_1^0p_2^0}{m^2}\right)^{\frac{1}{2}}\langle N(p_2)|J^\mu_{(0)}|N(p_1)\rangle = ie\bar{u}(p_2)\left[\gamma^\mu F_1(t)-\frac{\kappa}{2m}\sigma^{\mu\nu}q_\nu F_2(t)\right], \quad (5.4.1)$$

where κ is the anomalous magnetic moment ($\kappa_p=\mu_p-1=1{\cdot}79$, $\kappa_n=\mu_n=-1{\cdot}92$), and $t=-q^2=-(p_2-p_1)^2$. The isovector form factors which concern us here are defined by

$$F_i^v(t)=\tfrac{1}{2}[F_i^p(t)-F_i^n(t)], \qquad (i=1,2). \quad (5.4.2)$$

The form factors F_1 (Dirac) and F_2 (Pauli) have the role of invariant amplitudes, and like the pion form factor $F_\pi(t)$ discussed in Section 5.3 are analytic in the cut plane $4\leq t<\infty$. The Sachs' form factors, $G_M(t)$ and $G_E(t)$ are also often used. These correspond to the t-channel helicity amplitudes f_{++} and f_{+-} for the $N\bar{N}$ system in the crossed channel, and are related to the form factors F_1 and F_2 by the relations

$$\begin{aligned} G_E(t)&=F_1(t)+\frac{\kappa t}{4M^2}F_2(t),\\ G_M(t)&=F_1(t)+\kappa F_2(t). \end{aligned} \quad (5.4.3)$$

These form factors have the same analytic structure as F_1, F_2, but while $G_M(t)$ retains the invariant amplitude character of F_1 and F_2, $G_E(t)$ contains a kinematical factor leading to the constraint

$$G_E(t=4M^2)=G_M(t=4M^2). \tag{5.4.4}$$

Thus, if one wishes to follow the usual dispersion theory practice of using independent invariant amplitudes (for all t), any two of F_1, F_2 and G_M are appropriate, but not G_E. On the other hand, the cross-section formulae for elastic electron nucleon scattering take a simpler form in terms of G_E and G_M which are thus often used for discussing data. We shall therefore initially use G_E, G_M to summarize the data, switching to F_1, F_2 or F_1, G_M when considering their implications.

The available data have been reviewed many times (cf. e.g. Wilson (1971); Felst (1973); and Gourdin (1974)). The form factors $G_E^p(t)$, $G_M^p(t)$, $G_M^n(t)$ are reasonably well determined over the region $0\leqslant -t\leqslant 2\ (\text{GeV})^2$, where they approximately satisfy the scaling laws

$$G_E^p(t)\approx\frac{G_M^p(t)}{\mu_p}\approx\frac{G_M^n(t)}{\mu_n}\approx\frac{G_M^v(t)}{\mu_v},\qquad 0\leqslant -t\leqslant 2\ \text{GeV}^2. \tag{5.4.5}$$

The neutron form factor is known over a more limited region, where it is reasonably represented by (Galster *et al.*, 1971):

$$G_E^n(t)=\frac{-t\mu_n G_E^p(t)}{4M^2-5{\cdot}6t},\qquad 0\leqslant -t\leqslant 1\ (\text{GeV})^2. \tag{5.4.6}$$

We finally need to specify $G_M^v(t)$. A popular, but not very accurate, form for this is the so-called "dipole fit"

$$\frac{G_M^v(t)}{\mu_v}=\frac{G_M^p(t)}{\mu_p}=\left(\frac{0{\cdot}71}{0{\cdot}71-t}\right)^2, \tag{5.4.7}$$

which gives an approximate representation of the proton data over the much larger t-region $0\leqslant -t\leqslant 25\ (\text{GeV})^2$. However, over the smaller t-range with which we are primarily concerned, it is more accurately represented by the empirical form (Shaw, 1972)

$$\frac{G_M^v(t)}{\mu_v}=\frac{G_M^p(t)}{\mu_p}=\frac{m_0^2(m_0^2+1)}{(m_0^2-t)(m_0^2+1-t)},\qquad 0\leqslant -t\leqslant 2\ (\text{GeV})^2, \tag{5.4.8a}$$

where

$$m_0=656\ \text{MeV}, \tag{5.4.8b}$$

and the same parameter m_0 occurs in the empirical form

$$2F_1^v(t)=\frac{m_0^2}{m_0^2-t},\qquad 0\leqslant -t\leqslant 1\ \text{GeV}^2, \tag{5.4.9}$$

suggested to represent the data on $F_1^v(t)$. This is of course obtained by combining data on $G_E^{p,n}(t)$, $G_M^{p,n}(t)$.

The closely related form factor $G(t)$ for the electromagnetic transition $\Delta^+ \to p$, has been studied in the reaction

$$e^- + p \to e^- + \Delta^+,$$

by both missing-mass and coincidence measurements for $0 \leq -t \leq 2\ \text{GeV}^2$ (Bartel *et al.*, 1968; Albrecht *et al.*, 1971; Siddle *et al.*, 1971; Alder *et al.*, 1972). The results are in good agreement with the suggested relation (Shaw, 1972)

$$\frac{G(t)}{G(0)} = \frac{F_2^v(t)}{\kappa_v}, \qquad 0 \leq -t \leq 1\ (\text{GeV})^2. \tag{5.4.10}$$

The agreement extends out to 2 $(\text{GeV})^2$ if equation (5.4.9) is used out to this t-value in order to deduce $F_2^v(t)$.

The pion form factor F_π can be studied in the spacelike region by scattering pions from atomic electrons. Because of the small mass of the electron, a rather limited range of momentum transfers is accessible, even with very high energy pion beams, so that one is essentially limited to measuring the charge radius

$$r_\pi^2 = \left.\frac{-6\, dF_\pi(t)}{dt}\right|_{t=0}.$$

The most recent experiment (Adylov *et al.*, 1974) gives

$$r_\pi = 0{\cdot}78^{+0\cdot09}_{-0\cdot10}\ \text{fm}. \tag{5.4.11}$$

Information over a wider range of t can be obtained indirectly from the study of single-pion electroproduction, $e^-p \to e^-\pi^+n$, by separation of the pion pole shown in Fig. 5.4.1, but this is not a model-independent procedure. At intermediate energies, however, the process is dominated at forward pion production angles by a large excitation of this term by longitudinally polarized virtual photons. This is possible because the pion exchange contribution to the longitudinal terms, unlike that to the transverse terms, does not vanish kinematically in the forward direction. (Both vanish at zero momentum transfer to the nucleon, which coincides with the forward direction asymptotically.) Thus, in practice, a reasonable separation can be made (see, for example, the discussion of Devenish and Lyth (1972), and the reviews of Harari (1971c) and Gourdin (1974)), and the results obtained can be used as constraints upon forms used to analyse data in the timelike region obtained from e^+e^- colliding beams, as discussed in the preceding section. The results cover the region $0 < -t \leq 1\ (\text{GeV})^2$, and can

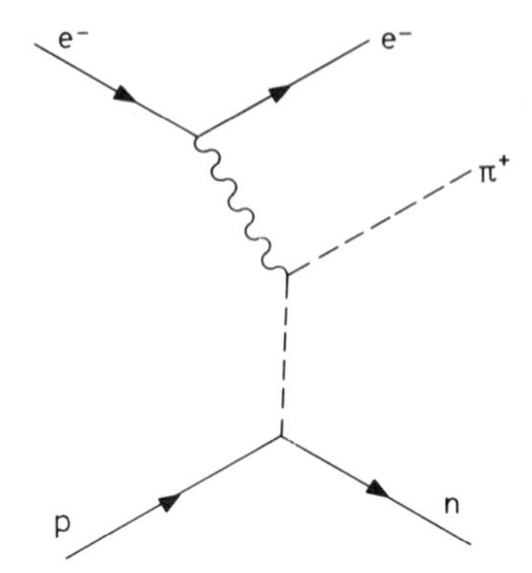

FIG. 5.4.1. The pion-pole contribution to single pion electroproduction.

be summarized by saying that $F_\pi(t)$ is qualitatively similar to the isovector Dirac form factor of the nucleon, i.e.

$$F_\pi(t) = 2F_1^v(t), \qquad 0 < -t \approx 1\ (\text{GeV})^2 \tag{5.4.12}$$

rather than to the more rapidly falling $G_M(t)$. (The previous result for r_π, equation (5.4.11), is compatible with either.) Of course, the results are to a certain extent model-dependent and should be regarded with some reservation. We note, however, that the approximate rho-dominance behaviour indicated by equation (5.4.12) (together with equations (5.4.8b) and (5.4.9)) agrees well with the behaviour in the timelike region discussed in Section 5.3.

So far, we have merely reviewed the empirical situation with regard to the form factors, in particular the isovector form factors, which can be summarized by equations (5.4.8)–(5.4.12). We now turn to the interpretation of these results in terms of narrow resonance models, where $I = 1$, $J^P = 1^-$ for the contributing resonances. On doing this we see that the form factors $F_1^v(t)$, $F_\pi(t)$ are in qualitative agreement with a simple rho-pole. The mass predicted ($m_\rho = m_0 \approx 656$ MeV) is somewhat light, but this is perhaps not unreasonable in view of the crudeness of the model. On the other hand, the form factors $G_M^v(t)$ and $G(t)$ fall off much more rapidly, and the suggestion of higher mass isovector, vector mesons—the rho-primes—arises from attempts to understand these strong deviations from rho-dominance. For example, one can write (Chan *et al.*, 1966)

$$G_M^v(t) = c_\rho \frac{m_\rho^2}{m_\rho^2 - t} + c_{\rho'} \frac{m_{\rho'}^2}{m_{\rho'}^2 - t}. \tag{5.4.13}$$

However, using a correct rho-mass ($m_\rho = 765$ MeV), a rather light rho-prime ($m_{\rho'} = 975$ MeV) was required to fit the data. Alternatively, one could interpret equations (5.4.8) and (5.4.9) directly in terms of a $\rho+\rho'$ model regarding the light rho-mass implied ($m_\rho = m_0 = 656$ MeV) as a measure of the error involved in the neglect of non-resonant background,‡ or higher mass resonances (Shaw, 1972). The rho-prime mass is then

$$m_{\rho'} = \sqrt{m_0^2+1} \approx 1200 \text{ MeV},$$

and equations (5.4.9) and (5.4.10) imply that it decouples from $F_1^v(t)$, and especially from the pion form factor $F_\pi(t)$. The most likely two-body decay mode would then be $\omega\pi$ (Shaw, 1972), and possible evidence for such a state has subsequently been discussed by other authors (Bramon and Greco, 1973; Bramon, 1973). A study of the reaction $e^+e^- \to \omega\pi^0$ is the obvious way to settle this issue.

More generally, the similarity of $F_\pi(t)$ to rho-dominance implies that if the behaviour of these various form factors is to be understood in terms of rho-primes their coupling to $\pi\pi$ must be much weaker than that to $N\bar{N}$. In this case there are better reactions to look for them than elastic $\pi\pi$ scattering (Gounaris, 1971).

A way of fitting the data with both the correct rho-mass, and a reasonable ρ'-mass, is to incorporate an infinite sequence of poles in the sum (5.4.13); this possibility also allows for a more flexible asymptotic behaviour. Such models normally incorporate the sequence of rho-primes predicted by the Veneziano model (see Chapter 12). For example, the formula

$$G_M^p(t) = \frac{\gamma\Gamma(1-\alpha(t))}{\Gamma(r+1-\alpha(t))},$$

can be used (di Vecchia and Drago, 1969; Frampton, 1970) to give a reasonable account of nucleon form factor data over a wide range of t. Here

$$\alpha(t) = \tfrac{1}{2} + t/2m_\rho^2,$$

so that one has the vector meson spectrum

$$m_n^2 = m_\rho^2(2n-1), \qquad n \geq 1$$

and the first ρ' occurs at $m_{\rho'} \approx 1300$ MeV. Whether all the higher mass poles should be interpreted as resonances, or as merely an approximation to the cut corresponding to the neglected continuum contributions, is ambiguous.

In summary, we see that although suggesting the existence of one or more higher mass rho-mesons, and probably the weakness of their couplings to the $\pi\pi$ channel, the above considerations do not lead to any very specific

‡ For the case of F_1^v, there are appreciable contributions from the 2π continuum near threshold which at least partly account for the downward shift in mass. Since these are not present in F_π, they imply deviations from the approximate relation (5.4.12) (Höhler and Pietarinen, 1975).

predictions of their masses. In fact, all of the models discussed above give the mass of the first rho-prime to be lower than the effect in colliding beams (see Section 5.3), photoproduction (see below), and possibly dipion production (see Section 3.3), a result which is reminiscent of the light rho-mass suggested by a literal interpretation of the data on $F_1^v(t)$, equation (5.4.9).

(b) Photoproduction of I = 1 Vector Mesons

According to the ideas of vector meson dominance, the photoproduction of rho-mesons on protons

$$\gamma + p \to \rho^0 + p,$$

proceeds predominantly via the diffractive mechanism of Fig. 5.4.2(a). The data indeed show that the process is predominantly diffractive. Hence, in the analogous photoproduction process on heavy nuclei, the amplitudes for photoproduction on individual nucleons will add coherently, so that very large cross-sections result. In this way one may hope to produce rho-mesons with very high statistics in a situation where S-wave backgrounds will be small, since these latter are presumably produced non-diffractively. Experimentally, however, the ρ mass distribution obtained always shows a rho-bump of very asymmetric shape, broad on the low-mass side and narrow on

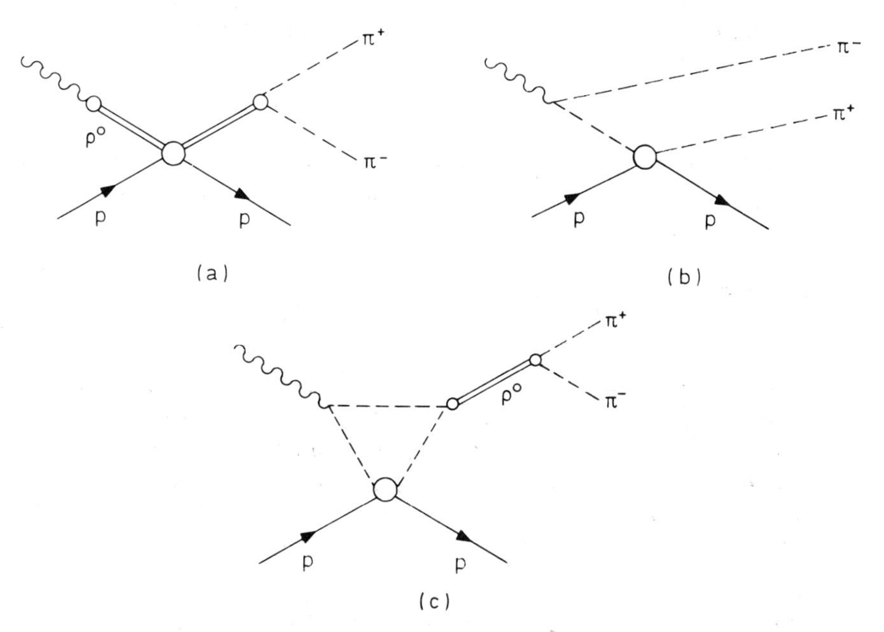

FIG. 5.4.2. Diagrams contributing to $\pi^+\pi^-$ pair photoproduction.

the high-mass side. This is normally interpreted as being due to an interference between the rho-dominance diagram, Fig. 5.4.2(a), and the Drell–Söding diagram, Fig. 5.4.2(b), which also produces almost purely P-wave pion pairs near the forward direction, but now with a smooth mass distribution. Unfortunately, there are complicated questions of double counting to consider. In addition to the usual term from Fig. 5.4.2(a) which treats the rho as a stable particle, an additional term from the diagram of Fig. 5.4.2(c) must also be present. The imaginary part of this term (after factoring out the phase of the πN scattering amplitude taken as i), to a large extent cancels the Drell–Söding term (Bauer, 1970). On combining these terms, excellent fits to the data from copper and lead targets are obtained, giving values of $\Gamma_\rho = 118 \pm 6$ MeV and 124 ± 5 MeV, respectively, for $m_\rho = 775$ MeV. These results, of course, depend on the correctness of the treatment of the interfering background. For a more detailed discussion of these points we refer to the review of Yennie (1971).

Although the above considerations fail to provide detailed model-independent information on the rho-width and shape, rho-mesons are copiously produced in this way, and one may hope that heavier vector mesons (other than ω, ϕ) may also be diffractively produced in photoproduction reactions. It is certainly a promising way to look for rho-primes. The best information to date is for the $\pi^+\pi^-$ mass spectrum, which has been measured for masses up to 2 GeV from nuclei at various energies (Alvensleben *et al.*, 1971; Bulos *et al.*, 1971a). The spectrum follows a simple extrapolation of the rho-tail except for a broad shoulder centred around 1500 MeV. However, because of the smallness of this effect, its significance is unclear.

More interesting results have been obtained for the reaction

$$\gamma + p \to \pi^+\pi^-\pi^+\pi^- p,$$

at energies between 4·5 and 18 GeV (Davier *et al.*, 1969, 1973). A strong ρ^0 signal is seen, and the $\rho^0\pi^\pm$ mass spectrum exhibits a peak at about 1100 MeV, i.e. in the region of the A_1 effect.‡ If events with the (3π) mass in this region are selected, the (4π) mass spectrum has a peak consistent with a πA_1-resonance with parameters

$$m = 1550 \pm 40 \text{ MeV}, \qquad \Gamma = 260 \pm 110 \text{ MeV}. \tag{5.4.14}$$

The peak is seen in the complete 4π mass spectrum, but on a larger background, much of which is associated with Δ^{++} production. The cross-section seems roughly constant over the energy range, consistent with a diffractive mechanism, so that an interpretation of this effect as a $1^-(\rho')$

‡Whether the A_1 effect is, or is not, a "resonance" (cf. Section 13(c)) is not particularly relevant.

resonance is natural. This interpretation has been greatly strengthened by an experiment at 9·3 GeV using linearly polarized photons (Bingham *et al.*, 1972a,b) when the decay distributions relative to the photon polarization can be used to check consistency with production of a 1^- meson with the s-channel helicity conservation expected for diffractive production. The analysis is somewhat complicated, and for details we refer to the original paper, contenting ourselves with showing the mass distributions in the 4π, 2π and K^+K^- channels for the appropriate angular distribution coefficient

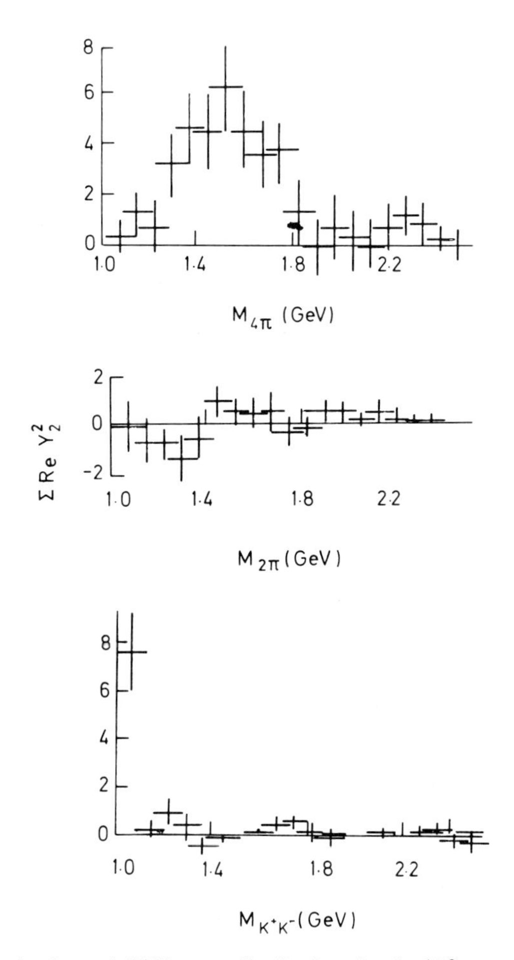

FIG. 5.4.3. 4π, 2π and K^+K^- mass distributions for the $\langle Y_2^2 \rangle$ moment for the reactions

$$\gamma + p \rightarrow (4\pi)p, (2\pi)p, K^+K^-p$$

taken from the data of Bingham *et al.* (1972a).

(Fig. 5.4.3). The peak in the 4π distribution occurs predominantly in the $\rho^0\pi^+\pi^-$ mode, and upper limits on the branching ratios into $\pi^+\pi^-$, K^+K^- of 20% and 40% respectively are quoted (Bingham *et al.*, 1972b). However, the mass and width parameters are found to be highly dependent on the parameterization used.

Taken together with the effect seen in the 4π mass spectrum from the e^+e^- colliding beam experiments (cf. Fig. 5.3.5) these experiments yield strong evidence for a broad peak in the $\pi^+\pi^-\pi^+\pi^-$ mass spectrum at around 1500–1600 MeV with the quantum numbers of a rho-prime meson, and with small branching ratio into $\pi^+\pi^-$, and K^+K^-. One would look to see an associated effect in the $\pi^+\pi^-$ channel from peripheral production experiments with pion beams, and some analyses have indeed found an inelastic state at this mass; however, the phase-shift solutions are not unique, so this is not incontrovertible support (cf. Section 3.3(e)). Notwithstanding this there§ is rather strong evidence in the electromagnetic processes for the existence of a ρ' state with small branching ratio into (2π), decaying mainly to (4π). While it is clear that the mass of this state is heavier than that predicted by the form factor considerations of the previous section, it fits in well with other theoretical expectations, as discussed in Chapter 13.

§ A clear $\pi^+\pi^-$ signal in ρ' photoproduction has recently been detected, with branching ratio, relative to $2\pi^+2\pi^-$ of order 2% (W. Y. Lee *et al.*, preliminary report to the International Symposium on Lepton and Photon Interactions at High Energies, Stanford, 1975).

Chapter 6

Summary

To round off Part II of this book, and for subsequent ease of access, the main results on $\pi\pi$ and $K\pi$ scattering discussed in the preceding chapters are here collected in summary form. Achieving this within the scope of a few pages entails a simplified presentation, and the omission of important qualifications. For these the reader is referred back to the source chapters. Of the methods and processes which have at various times been looked to for information on $\pi\pi$ scattering, we only include those which have at least pretensions to model independence. With the crucial exception of peripheral production, this leads us to exclude processes involving three-hadron final states—η, $K \to 3\pi$ decays (Section 5.2), $N\bar{N} \to 3\pi$ (Section 3.4), and $\pi N \to \pi\pi N$ at low energies (Section 3.2). In analysing these, the contemporary approach is to feed in $\pi\pi$ information, rather than trying to get it out. The complexity of the dynamics enforces the same approach with KN and NN scattering (Section 4.2). This leaves rather a short list of "reliable" sources of information—K_{e4} decay, $e^+e^- \to \pi^+\pi^-$ and K^+K^-, πN elastic scattering, and peripheral production processes—with the latter carrying the overwhelming work-load. With available statistics even these do not in practice admit completely model-independent analyses, and much of our confidence in the results obtained rests on the consistency found between different methods and processes.

We now proceed to our summary, which is in two parts, (a) methods, and (b) results.

(a) Status of the Principal Methods

(i) $\pi N \to \pi\pi X$ (*Sections 3.1, 3.2 and 3.3, Appendix B*)

By far the most important source of information on $\pi\pi$ scattering. The extraction of $\pi\pi$ phases by the separation of the one-pion exchange (OPE) term is subject to convincing cross-checks, both between different methods applied to the same reaction (e.g. $\pi^- p \to \pi^+\pi^- n$), and between different reactions. For example, $I=2$ phases from $\pi^+\pi^+$ and $\pi^+\pi^0$ final states must be the same, as must results for each isospin from $\pi N \to \pi\pi N$, $\pi N \to \pi\pi\Delta$, where the characteristics of both the OPE term and the background are different. The method also has the advantage, compared to others, of giving access to a wide range of energies, partial waves and final states. To date, it has provided us with extensive information on $\pi\pi$ phases for $500 \lesssim W \lesssim 1900$ MeV.

(ii) $KN \to K\pi X$ (*Sections 3.1, 3.2 and 3.3*)

All the advantages cited above for the pion-induced reaction apply here also. Despite somewhat lower statistics, a convincing pattern, with successful cross-checks, has emerged. With large-scale spark chamber experiments now beginning to produce results, a good deal of progress in deciding detailed questions on $K\pi$ scattering is to be expected.

(iii) K_{e4} *Decay* (*Section 5.1*)

A model-independent method of determining the low-energy behaviour of the $I=J=0$ phase shift, in particular the scattering length a_0^0. This supposes the normal selection rules for weak decays, some of which are subject to test in this decay itself. Present experiments are somewhat in conflict, and further data are required.

(iv) e^+e^- *Annihilation*: *One-Photon Exchange* (*Section 5.3*)

Study of the $\pi^+\pi^-$ channel constitutes an almost model-independent method of determining the $I=J=1$ phase shift over the region where elastic unitarity is a good approximation ($W \lesssim 1$ GeV). Unfortunately, precise data, other than in the region of $\rho-\omega$ interference, are not at present available. At higher energies, the study of both 2π and other channels furnishes an unambiguous method of investigating the $J^{PG}=1^{--}$ resonance spectrum.

(v) $\pi N \to \pi N$ (*Section 4.1*)

The extraction of information on $\pi\pi$ scattering from this process requires a complicated theoretical analysis, so that the assessment of errors demands great care. As a method, its prime use is in yielding stringent consistency tests on the phase shifts resulting from other methods, especially (i) above.

(vi) $N\bar{N} \to \pi^+\pi^-$ (*Section 3.4*)

This process offers a method of studying meson resonances in formation above the $N\bar{N}$ threshold. Simple arguments would suggest that the contribution of low-lying (daughter) resonances should be enhanced. The analysis of existing differential cross-sections is highly model-dependent. This problem should be considerably eased, although perhaps not eliminated, when polarization data become available.

(b) Information Obtained

(i) *The Threshold Region*

The threshold region is conveniently characterized by the parameters a_l^I. b_l^I defined in equation (2.40). Useful relations between these parameters follow from crossing symmetry and dispersion relations, together with the experimental knowledge of the phase shifts above 500 MeV summarized below. Indeed, the finding is that all the low-energy $\pi\pi$ partial waves are thereby determined in terms of a single parameter, conveniently taken to be the $I=0$ S-wave scattering length a_0^0 (cf. Chapter 10, especially Fig. 10.3.3 and Table 10.3.1). Alternative estimates of this parameter are cited in Table 6.1. In principle, the best method is the analysis of K_{e4} decay, but present data do not lead to a decisive result (see Table 6.1). Extending the observation of peripheral dipion production ($\pi N \to \pi\pi X$) down to low masses is another promising approach, although separation of the OPE signal from background may be harder (Sections 3.1 and 3.2) than for the range $W \geqslant 500$ MeV hitherto explored. Such investigations also yield the low-energy P-wave phase shift, and thus determine the associated scattering length a_1^1. Dispersion relation predictions for this quantity are rather definite ($0{\cdot}03 \leqslant a_1^1 \leqslant 0{\cdot}05$) (see Chapter 10); contrary findings (Männer, 1974) raise misgivings as to the associated S-wave determination (Table 6.1), as discussed in Section 3.3, and it will be important to cross-check with other measurements (preliminary results of which (Lemoigne *et al.*, 1974) are listed in the table). Less directly, the analysis of dipion exchange in πN elastic scattering, especially the low-energy D-waves, leads to inferences on a_0^0 as shown (cf. Section 4.1). Other methods are in our opinion too model-dependent to yield convincing information. We have made one exception to this in listing the value of a_0^0 inferred from the study of single-pion production, $\pi N \to \pi\pi N$, very close to threshold, where the special circumstances (a pronounced minimum at the low-mass

TABLE 6.1. Estimates of S- and P-Wave $\pi\pi$ Scattering Lengths

a_0^0	a_1^1	Reference	Method
$0{\cdot}60 \pm 0{\cdot}25$		Zylbersztejn *et al.* (1972)	
$0{\cdot}17 \pm 0{\cdot}13$		Beier *et al.* (1973)	K_{e4}
$0{\cdot}26 \pm 0{\cdot}08$		Tryon (1974a)[b]	
$0{\cdot}31 \pm 0{\cdot}10$[d]		Extermann *et al.* (1975)	
$0{\cdot}44 \pm 0{\cdot}10$	$\approx 0{\cdot}1$	Männer (1974)	Peripheral $\pi N \to \pi\pi N$
$0{\cdot}27 \pm 0{\cdot}05$[a]		Lemoigne *et al.* (1974)	for low $M_{\pi\pi}$
$0{\cdot}06 \pm 0{\cdot}20$		Nielsen and Oades (1974)	Analysis of πN elastic scattering
$-0{\cdot}06$ to $+0{\cdot}03$		Jones *et al.* (1974)	$\pi N \to \pi\pi N$ close
$0{\cdot}18 \pm 0{\cdot}02$		Bunyatov (1974)[c]	to threshold

[a] Preliminary result. The reported measurement is $|a_0^0 + \frac{1}{2}a_0^2| = 0{\cdot}26 \pm 0{\cdot}05$; we have converted it to a determination of a_0^0 *assuming* the relation between a_0^0 and a_0^2 predicted by dispersion relations (cf. Fig. 10.3.3).
[b] Reanalysis of data of Zylbersztejn *et al.* (1972) and Beier *et al.* (1973) using a different parameterization.
[c] Uses soft pion theory (cf. Section 14.4(b)).
[d] Preliminary result.

end of the spectrum) somewhat diminish the usual reliance on the isobar model (but see Morgan and Pennington (1975)).

The exotic S-wave scattering length, a_0^2 is of equal interest, and one would like to check the dispersion relation predictions (Chapter 10), on its correlation with a_0^0, especially in view of the above-mentioned discrepancy with a_1^1. Present estimates rest on rather low statistics. Indications on these and the analogous low-mass πK phase shifts, and $\pi\pi$ D-waves are reported in Section 3.3.

(ii) *$\pi\pi$ Phase Shifts*

$500 \leq W \leq 950$ MeV. The phases are rather well established in this region, the results of different analyses of peripheral dipion production being in good agreement with each other (cf. Sections 3.1, 3.2 and 3.3), and with subsidiary information from $e^+e^- \to \pi^+\pi^-$ (Section 5.3) (and also, to some extent, πN elastic scattering (Section 4.1)). The behaviour of the various phase shifts in this region are shown in Figs 3.3.3, 3.3.4 and 3.3.6.

$950 \leq W \leq 1400$ MeV. Despite some residual ambiguities, the main characteristics of the phases are well established over this region, and are shown in Figs 3.3.8 and 3.3.9.

1400 ≤ W ≤ 2000 MeV. Although the behaviour of the higher D- and F-waves is reasonably definite, there exist considerable uncertainties as to the form of the S- and P-waves (Figs. 3.3.8 and 3.3.9).

TABLE 6.2. Resonant Effects with $\pi\pi$ or $K\pi$ quantum numbers

(A) $I=0, 1; G=+; J\geq 1$

State	$J^{PC}(I)$	Mass (MeV)	Width (MeV)	Branching ratio to $\pi\pi$ (%)
ρ (770)[a]	$1^{--}(1)$	775 ± 10	153 ± 10	100
f (1270)[b]	$2^{++}(0)$	1275 ± 10	185 ± 20	≈ 83
f' (1516)[c]	$2^{++}(0)$	1516 ± 5	40 ± 15	≈ 0 ($K\bar{K}$ only mode seen)
ρ' (1600)[d]	$1^{--}(1)$	Broad effect seen in 4π channel in electromagnetic interactions—mass ≈ 1600 MeV —coupling to $\pi\pi$ small		
g (1686)[e]	$3^{--}(1)$	1686 ± 20	180 ± 30	26
h (2035)[f]	$4^{++}(0)$	2035 ± 25	200 ± 60	$\pi\pi$, $K\bar{K}$ seen

(B) $I=J=0$

State	$J^{PC}(I)$	Comments
S^* (1000)	$0^{++}(0)$	Consensus for second-sheet pole at about $E^{\text{II}}\approx 1000-35i$ MeV (cf. Table 3.3.4) with some evidence (Section 3.3(f)) for an additional sheet III pole.
ε (900) ε' (1250)	$0^{++}(0)$	The $I=J=0$ $\pi\pi$ phase shift transits 90° at about 900 MeV and 270° at about 1250 MeV with $\eta_0^0\approx 1$ in both cases. Possibly different aspects of a single broad elastic resonance, $M\sim 1000$ 1300 MeV, $\Gamma\approx 600$ MeV, as discussed in Section 3.3(f).

(C) $I=\frac{1}{2}, S=+1$

State	J^P	Mass (MeV)	Width (MeV)	Branching ratio to $K\pi$ (%)
K^* (892)[c]	1^-	$892{\cdot}2\pm 0{\cdot}2$	$49{\cdot}8+1{\cdot}1$	≈ 100
κ (1250)[g]	0^+	≈ 1250	≈ 500	Inelastic modes assumed unimportant
K^* (1421)[c]	2^+	1421 ± 5	100 ± 10	≈ 55
K_N (1800)	3^-	Details in dispute (cf. Section 3.3(g))		

(iii) *$K\pi$ Phase Shifts*

While information is less detailed than in the $\pi\pi$ case, the main characteristics of the phase shifts are established over the range $800 \leq W \leq 1400$ MeV. The $I=\frac{1}{2}$ P- and D-waves are dominated by the K^* (892) and K^* (1421) resonances, respectively (cf. (iv) below), and the corresponding $I=\frac{3}{2}$ waves are small. Results on the $I=\frac{1}{2}$ and $\frac{3}{2}$ S-waves are shown in Fig. 3.3.12.

(iv) *Information on Resonances*

Comments on the status and parameters of the resonances previously alluded to which either do, or could, couple to $\pi\pi$ or πK are collected in Table 6.2.

---

[a] Estimate based on the cited values of Table 3.3.2 and equation (5.3.32).

[b] Estimate based on the cited values of Table 3.3.3.

[c] Value from the Particle Data Group (1974).

[d] Seen in $e^+e^- \to 4\pi$ (cf. Fig. 5.3.5), and photo-production (Fig. 5.4.3). Possibly also in dipion production (cf. Section 3.3(e)) (see footnote, p. 192).

[e] Value from Particle Data Group (1974). Note, however, that a recent phase shift analysis (Hyams *et al.*, 1973) indicates a somewhat higher value for the mass and width (cf. Section 3.3(e)).

[f] Estimate based on the cited values of Section 3.3(e).

[g] These estimates assume that the 'Down' branch for $\delta_0^{\frac{1}{2}}$ is followed both at 900 and 1400 MeV (cf. Fig. 3.3.12).

Part III

Theory and Models of $\pi\pi$ Scattering

Chapter 7

Rigorous Results

Before proceeding to discuss specific theories and phenomenological models for $\pi\pi$ scattering, it is useful to summarize the more important results that have been rigorously established from general field theory. This is done in this chapter, in which we first indicate briefly the extent to which the analytic properties assumed in Chapter 2 have been established, and then go on to describe some of the results that have been derived rigorously from these analytic properties, using also unitarity and crossing symmetry. Our treatment will be rather brief. A more extended account of several of the topics discussed can be found in the book of Martin and Cheung (1970).

(a) Rigorous Analyticity

The analytic properties given in Section 2(e) are based on the Mandelstam representation, and exceed those which have been rigorously established from general field theory, in the form, for example, proposed by Wightman (1956). The singularities of the Mandelstam representation are, as noted in Chapter 2, those suggested by the simple box diagrams of perturbation theory. The absence of further singularities has not been proved even in perturbation theory, to all orders, although for a simple process like $\pi\pi$ scattering it is perhaps not implausible. However, the simultaneous polynomial boundedness in s and t is an additional assumption, which may well break down, thereby invalidating the representation (2.41), even if the singularity structure is correct. This happens in some popular models of high-energy scattering in which the Regge trajectories rise indefinitely for increasing values of s (see Chapter 12 and Appendix C).

What properties then are established rigorously? These fall into two classes: those which use "positivity", and the classical results of field theory, which do not. These latter are:

(i) At fixed-s, the amplitude is analytic in $\cos \theta_s$ within an ellipse, focii ± 1, whose extent was established by Lehman (1958). In particular, the partial wave expansion equation (2.12) will converge within this ellipse. For large s, the extremities are

$$\cos \theta_s = \pm \left[1 + \frac{64}{s}\right]^{\frac{1}{2}}, \tag{7.1}$$

which reduce to the physical region $0 > t > -s$ as $s \to \infty$. For finite s, the range is larger, and always encompasses $0 \geq t \geq -28$.

(ii) The amplitude satisfies fixed-t dispersion relations, equation (2.45), for all t in the latter range, $0 \geq t \geq -28$, (Symanzik *et al.*, 1954; Bogoliubov *et al.*, 1958; Bremermann *et al.*, 1958).

(iii) The possibility of an analytic continuation between the s- and u-channel amplitudes has been established *for arbitrary negative t* by Bros *et al.* (1965), thus establishing the crucial crossing property of the physical amplitudes discussed in Chapter 2.

The next important step was taken by Martin, who combined the above "classical" results with the "positivity" requirements

$$\text{Im } T_l^I(s) \geq 0, \tag{7.2}$$

which follow from the much stronger restrictions of unitarity

$$\left(\frac{\nu+1}{\nu}\right)^{\frac{1}{2}} \geq \text{Im } T_l^I(s) \geq \left(\frac{\nu}{\nu+1}\right)^{\frac{1}{2}} |T_l^I(s)|^2. \tag{7.3}$$

In this way Martin (1966a,b) was able to extend the ellipse of convergence,‡ and establish the validity of fixed-t relations for§

$$4 > t > -28, \tag{7.4}$$

the extension to positive values of t being crucial for what follows. It can also be shown that if the forward relations converge with N subtractions, the number required for $0 < t < 4$ is N, if N is even, and $N+1$, if it is odd (Jin and Martin, 1964). In fact $N = 2$, as we shall show below. For an account of the derivation of these results, we refer to the review of Martin and Cheung (1970).

‡ The new domain is sometimes called the "Martin–Lehman" ellipse, to distinguish it from the "Lehman ellipse" of classical field theory.

§ This should be compared with the range $4 > t > -32$ resulting from Mandelstam analyticity, equation (2.45).

(b) High-Energy Behaviour

(i) *Froissart Bound*

The usefulness of extending the validity of fixed-t dispersion relations to positive values of t is that it enables an important result to be established on the total cross-section. This is the Froissart bound, which was originally derived from the Mandelstam representation, (Froissart, 1961).

Consider the partial-wave expansion

$$T^I(s, t, u) = \sum_l (2l+1) P_l(\cos\theta_s) T_l^I(s), \tag{7.5}$$

for some point within the range $0 < t < 4$, i.e.,

$$1 < \cos\theta < 1 + \frac{8}{s-4}, \tag{7.6}$$

and therefore within the Martin–Lehman ellipse (all s). If the fixed-t relation converges with N subtractions, then, since $\mathrm{Im}\ T_l^I(s)$ is greater than zero, each term in the expansion must be positive, and thus bounded on average by s^N, for large s. Thus

$$(2l+1)\ \mathrm{Im}\ T_l^I(s) P_l(1 + t/2q^2) < s^N. \tag{7.7}$$

Now, for large l, and $x > 1$

$$P_l(x) > \frac{c}{(2l+1)^{\frac{1}{2}}} [1 + (2x-2)^{\frac{1}{2}}]^l, \tag{7.8}$$

so that

$$\begin{aligned} \mathrm{Im}\ T_l^I(s) &< \frac{s^N}{c(2l+1)^{\frac{1}{2}}} [1 + (t/q^2)^{\frac{1}{2}}]^{-l} \\ &< c' \exp\left\{ -l\left(\frac{4}{s}\right)^{\frac{1}{2}} (1 - \varepsilon(s)) + N \ln\left(\frac{s}{s_0}\right) \right\}, \end{aligned}$$

where $\varepsilon(s) \to 0$ as $s \to \infty$, and c, c' and s_0 are all constants. Hence, for large s, all terms in equation (7.5) may be neglected for

$$l > L = c'' \sqrt{s} \ln(s/s_0), \tag{7.9}$$

where c'' is a constant. The result on the total cross-section follows directly from the Optical Theorem

$$\sigma_{\mathrm{tot}}^I(s) = \frac{8\pi}{[\nu(\nu+1)]^{\frac{1}{2}}} \mathrm{Im}\ T^I(s, t=0),$$

and equations (7.3) and (7.5). Thus

$$\sigma^I_{\text{tot}}(s) \leq \frac{8\pi}{\nu} \sum_{l=0}^{L} (2l+1) = \frac{8\pi}{\nu}(L+1)^2,$$

and so, from equation (7.9), we have, for large s

$$\sigma^I_{\text{tot}}(s) \leq 8\pi c'' \ln^2 (s/s_0).$$

This is the Froissart bound, and establishes that $N = 2$, as mentioned in Section 7(a) above. Using $N = 2$ in equation (7.7) enables the constant c'' to be determined (Łukaszuk and Martin, 1967), giving the final result

$$\sigma^I_{\text{tot}}(s) \leq \left(\frac{\pi}{\mu^2}\right) \ln^2 (s/s_0). \tag{7.10}$$

A crucial part of the argument that gave this result is that there exists a minimum mass in the problem. If $\mu \to 0$, the domain of analyticity shrinks to $t = 0$, so that the condition $\cos\theta_s > 1$ used in equation (7.8) fails, and no bound exists.‡ However, $\mu \neq 0$, and what results is an effective cut-off in l, or, in terms of an impact parameter representation, that the effective range of interaction at high energies cannot increase faster than $\ln(s/s_0)$.

It is interesting that this bound is the same as that which results from a Yukawa distribution of *fixed* "range" b, at least according to the following argument, traditionally ascribed to Feynman. Suppose the target particle has a probability density function

$$P(r) = P_0 \exp(-br),$$

at a distance r from its mean position. If we assume that the probability of an interaction is bounded by a polynomial in s, then the probability of an interaction at a distance r between the incident particle and the target satisfies the inequality

$$P(s, r) < P_0 s^N \exp(-br).$$

Thus, the interaction will be negligible for

$$r > r_0 = \frac{N}{b} \ln(s/s_0).$$

Putting $\sigma_{\text{tot}} < \pi r_0^2$ gives a result of the same form as equation (7.10). Furthermore, equating the two results with $N = 2$ yields $b = 2\mu$, a result which is in accord with naive intuition.

‡ Unless further assumptions are made. See e.g. Auerbach *et al.* (1973).

(ii) *Phase-Energy Relations*

Crossing symmetry and analyticity result in a rigorous relation between the asymptotic phase and energy dependence of scattering amplitudes. If we restrict ourselves to the simple cases of power and logarithmic dependences on $z(=(s-u)/4\mu^2)$ at fixed t, then for *crossing-even* functions.

$$T(t, z) \sim C_+(-iz)^\alpha \ln(-iz)^\beta, \tag{7.11}$$

and, for *crossing-odd* functions

$$T(t, z) \sim iC_-(-iz)^\alpha \ln(-iz)^\beta, \tag{7.12}$$

where $C_\pm$ are real constants. The proof is simple, and can be found in, for example, the book of Eden (1967).

(iii) *Pomeranchuk Theorems*

The Pomeranchuk theorems (Pomeranchuk, 1958) relate the total cross-section for a given particle reaction $\sigma_{\text{tot}}(AB)$ to that for the crossed channel antiparticle reaction $\sigma_{\text{tot}}(\bar{A}B)$, under the *assumption* that

$$\frac{\text{Re } T(z)}{\text{Im } T(z) \ln z} \to 0, \tag{7.13}$$

as $z \to \pm\infty$. From this assumption, and the previous results of this section, it follows that:

(a) If $\sigma_{\text{tot}}(z)$ and $\bar{\sigma}_{\text{tot}}(z)$ have finite, non-zero limits σ and $\bar{\sigma}$, respectively, then these limits are equal.

(b) If the ratio $\sigma_{\text{tot}}(z)/\bar{\sigma}_{\text{tot}}(z)$ has a finite limit as $z \to +\infty$, this limit is unity. Again, for the proof, we refer to Eden (1967).

(c) Froissart–Gribov Representation

The existence of fixed-t (fixed-s) dispersion relations with no more than two subtractions in the range $4 > t > -28$ $(4 > s > -28)$ allows a representation to be established for the t- (s-) channel partial waves for $l \geq 2$ in this region. If the twice-subtracted dispersion relations

$$T_t^I(z, t) = a_I(\delta_{I0} + \delta_{I2}) + a_1 z \delta_{I1} + \frac{z^2}{\pi} \int_{1+t/4}^{\infty} dz' \frac{\text{Im } T_t^I(z', t)}{z'^2} \left[\frac{1}{z'-z} + \frac{(-1)^I}{z'+z} \right], \tag{7.14}$$

are substituted into the projection formula

$$T_l^I(t) = \tfrac{1}{2} \int_{-1}^{1} d(\cos\theta_t) P_l(\cos\theta_t) T_t^I(\cos\theta_t, t),$$

where

$$z = \frac{s-u}{4\mu^2} = \left(\frac{t-4}{4\mu^2}\right) \cos \theta_t, \tag{7.15}$$

then by using the orthogonality properties of $P_l(\cos \theta)$, and the Neuman integral

$$\frac{1}{2}\int_{-1}^{1} dx \frac{x^n P_l(x)}{y^n(y-x)} = Q_l(y) = (-1)^{l+1} Q_l(-y), \qquad (l \geqslant n)$$

one obtains, for $l \geqslant 2$

$$T_l^I(t) = \left[\frac{1+(-1)^{I+l}}{\pi}\right] \frac{4}{t-4} \int_{1+t/4}^{\infty} dz \text{ Im } T_t^I(z, t) Q_l\left(\frac{4z}{t-4}\right). \tag{7.16}$$

By changing variables in (7.16) we obtain the familiar form of the Froissart–Gribov representation

$$T_l^I(t) = \left[\frac{1+(-1)^{I+l}}{\pi}\right] \frac{2}{t-4} \int_{4}^{\infty} ds \text{ Im } T_t^I(s, t) Q_l\left(1+\frac{2s}{t-4}\right). \tag{7.17}$$

Similarly, in the s-channel

$$T_l^I(s) = \left[\frac{1+(-1)^{I+l}}{\pi}\right] \frac{2}{s-4} \int_{4}^{\infty} dt \text{ Im } T_s^I(s, t) Q_l\left(1+\frac{2t}{s-4}\right). \tag{7.18}$$

These representations are valid for $l \geqslant 2$ and $4 > t > -28$, $4 > s > -28$, respectively. They are of great importance in studying the properties of the partial-wave amplitudes in both the complex energy and complex angular momentum planes, as we shall see. If Mandelstam analyticity is assumed, the ranges can be extended to

$$4 > t > -32, \qquad 4 > s > -32.$$

(d) Low-Energy Bounds

We have seen that the Froissart bound can be interpreted as a limit on the effective cut-off range R with

$$\sigma_{\text{tot}} < \pi R^2.$$

One might expect there to be an analogous result to the potential theory bound

$$a > -R,$$

on the scattering length. Such results have indeed been rigorously derived by the use of dispersion relations for inverse amplitudes (Goebel and Shaw, 1968; Shaw, 1968). To do this, it is necessary to establish rigorous results on the zeros of the amplitudes analogous to the results on subtractions in the forward relations. This has been done, and will be discussed in Section 8.1 together with the resultant bounds. However, these results, like the Froissart bound, are not absolute, and to give them numerical content some phenomenological information must be used. This presumably arises from the fact that we are not dealing with distributions that vanish identically outside a range R, but instead fall off exponentially with a range determined by the lightest mass.

Nonetheless, one might still think that absolute bounds are possible, since if the strength of an attractive interaction was to increase indefinitely, a bound state would appear, violating the requirement that two pions are the lightest exchange possible. Clearly unitarity, as the only non-linear constraint available, will play a crucial role in arriving at such results.

Such absolute bounds have been rigorously derived. For example, the coupling constant λ has been bounded (Łukaszuk and Martin, 1967), and the result is

$$-2{\cdot}6 \leqslant \lambda \leqslant 17, \tag{7.19}$$

and for the $\pi^0\pi^0$ S-wave scattering length

$$\begin{aligned} a_0^{00} &\equiv T_s^{00}(s-4, t=0) \\ &= \tfrac{1}{3}[T^0(s=4\mu^2, t=0) + 2T^2(s=4\mu^2, t=0)], \end{aligned}$$

the result

$$a_0^{00} \geqslant -1{\cdot}75, \tag{7.20}$$

has been obtained (Bonnier and Vinh Mau, 1968; Common, 1969; Bonnier, 1975; Lopez and Mennessier, 1975).

These bounds can be strengthened by assuming that the D-wave scattering lengths are negligible (Common and Wit, 1971). However, with assumptions like this stronger results can be obtained more simply by the method of Goebel and Shaw, discussed in Section 8.1. It has also been shown (Common, 1970; Yndurain, 1970) that some of these results can be exploited to convert the Froissart bound into an absolute bound on the asymptotic cross-section.

Unfortunately, the derivation of all these results is somewhat complicated, and since the actual numbers obtained are too large to be of phenomenological interest, we will content ourselves with merely drawing attention to them. The fact that they exist at all is itself rather remarkable.

(e) Crossing and Positivity

We turn now to a topic of more practical importance where rigorous results are useful. This arises from the problem of trying simultaneously to impose the requirements of crossing symmetry and unitarity. Whereas it is easy to impose crossing symmetry on the full amplitude, by means of various expansions, unitarity is rather complicated. On the other hand unitarity takes a very simple form for the partial-wave amplitudes. However, in this case, it is crossing symmetry which is a complicated constraint, because the regions of convergence of the partial-wave expansions in the different physical channels do not overlap. An exception to this is the Mandelstam triangle $s>0$, $t>0$ and $u>0$, within which the inequality $-1<\cos\theta<+1$ holds in all three channels simultaneously, so that not only do the three expansions trivially converge, but the ranges over which projections of the full amplitude are made to yield the partial-wave amplitudes are wholly contained in this region.

In this section we shall describe two ways in which the above facts can be exploited to yield conditions on the partial-wave amplitudes in the sub-threshold region $(0\leqslant s\leqslant 4)$ which follow from crossing alone, or from crossing and positivity. These conditions find their principal use as a means of imposing at least some of the content of crossing on unitary models for partial-wave amplitudes which do not necessarily satisfy these requirements. They are, of course, redundant in models or calculations in which crossing and positivity are explicitly incorporated, in particular the approach via the Roy equations (Roy, 1971) and their developments (Mahoux *et al.*, 1974; Auberson and Epele, 1974; Roy and Singh, 1974; Lopez, 1973) which are described in Chapter 8.

(i) *Balachandran–Nuyts–Roskies* (*BNR*) *Relations*

These relations were first derived in general by Balachandran and Nuyts (1968), and applied to $\pi\pi$ scattering by Roskies (1969, 1970a). They take the form of integral equalities between different partial waves, and were later derived in the following rather simple way (Roskies, 1970a).

Consider the totally crossing symmetric amplitude for $\pi^0\pi^0$ scattering, $T^{00}(s, t, u)$. If this is multiplied by an anti-symmetric polynomial in s, t and u, and then integrated over the symmetric domain $s>0, t>0, u>0$, the integral must vanish. Thus

$$\iint ds\,dt\,t^P(s-u)^{2N+1}T^{00}(s,t,u)=0,$$

where P and N are integers. If we now insert the s-channel partial-wave series in this expression we have

$$0=\int_0^4 ds\int_{-1}^1 d\cos\theta\left(\frac{4-s}{2}\right)\left(\frac{4-s}{2}\right)^P(1-\cos\theta)^P$$
$$\times\left[s-\left(\frac{4-s}{2}\right)(1+\cos\theta)\right]^{2N+1}\sum_l(2l+1)P_l(\cos\theta)T_l^{00}(s),$$

which may be written

$$\int_0^4 ds \sum_{l=0}^{2N+2P+1} C_l^{2N+1,P}(s) T_l^{00}(s) = 0, \tag{7.21}$$

where

$$C_l^{2N+1,P}(s) = (2l+1)\left(\frac{4-s}{2}\right)^{P+1} \int_{-1}^{1} d\cos\theta P_l(\cos\theta)(1-\cos\theta)^P \times \left[s - \left(\frac{4-s}{2}\right)(1+\cos\theta)\right]^{2N+1}.$$

Equation (7.21) is now a set of conditions between different partial waves. The functions $t^P(s-u)^{2N+1}$ form a complete set anti-symmetric in s and u, and it can be shown that the conditions (7.21) are sufficient to guarantee exact $s \leftrightarrow u$ symmetry, provided the partial-wave amplitudes T_l fall off exponentially in l for $0 \leqslant s \leqslant 4$, and are well-behaved near $s=0$ and $s=4$ (Roskies, 1970a). If $u \leftrightarrow t$ symmetry is built in by keeping l even initially, then complete crossing symmetry is guaranteed.

The argument can be used for all amplitudes with definite $s \leftrightarrow u$ crossing properties, i.e.

$$T^0(s,t,u) + 2T^2(s,t,u),$$

$$2T^0(s,t,u) + 3T^1(s,t,u) - 5T^2(s,t,u),$$

$$2T^0(s,t,u) - 9T^1(s,t,u) - 5T^2(s,t,u),$$

to extend the results to all charge states. If we restrict ourselves to those conditions involving S- and P-waves only, then five conditions are obtained:

$$\int_0^4 ds(s-4)R_0^0[2T_0^0(s) - 5T_0^2(s)] = 0, \tag{7.22}$$

$$\int_0^4 ds(s-4)R_1^0[T_0^0(s) + 2T_0^2(s)] = 0, \tag{7.23}$$

$$\int_0^4 ds(s-4)R_1^0[2T_0^0(s) - 5T_0^2(s)] - 9\int_0^4 ds(s-4)^2 R_0^1 T_1^1(s) = 0, \tag{7.24}$$

$$\int_0^4 ds(s-4)R_2^0[2T_0^0(s) - 5T_0^2(s)] + 6\int_0^4 ds(s-4)^2 R_1^1 T_1^1(s) = 0, \tag{7.25}$$

$$\int_0^4 ds(s-4)R_3^0[2T_0^0(s) - 5T_0^2(s)] - 15\int_0^4 ds(s-4)^2 R_2^1 T_1^1(s) = 0, \tag{7.26}$$

where the functions R_i^j are defined by:

$$R_0^0 \equiv 1; \qquad R_0^1 \equiv 1$$

$$R_1^0 \equiv 3s-4; \qquad R_1^1 \equiv 5s-4$$

$$R_2^0 \equiv 10s^2-32s+16; \qquad R_2^1 \equiv 21s^2-48s+16;$$

$$R_3^0 \equiv 35s^3-180s^2+240s-64.$$

These will be used later, for example in Chapter 10.

(ii) *Martin Inequalities*

Although in the limit $l_{\max} \to \infty$ the above conditions are complete, in practice one always considers the subset of conditions obtained with a finite $l_{\max}$, so that it is useful to examine how more information can be thrown on to the lower partial waves. This can be done by adding the requirement of positivity to that of crossing (Martin 1967, 1968, 1969), to arrive at a set of inequalities for amplitudes and their derivatives. The crucial point is that the absorptive parts in fixed-t dispersion relations are positive for relations evaluated over the whole Mandelstam triangle, since for $4 \geqslant t \geqslant 0$, $\cos\theta_s \geqslant +1$ and $P_l(\cos\theta_s) \geqslant +1$, so that, from positivity,

$$\operatorname{Im} T_s^I(s,t) = \sum_l (2l+1) \operatorname{Im} T_l^I(s) P_l(\cos\theta_s) \geqslant 0. \tag{7.27}$$

Thus, if we consider for example the completely crossing symmetric amplitude $T^{00}(s,t,u)$, then it is symmetric about $s=u$, and in $0<s<4$ is completely concave, i.e. at $s=u$

$$\partial_s^{2n+1} T^{00}(s,t) = 0,$$

$$\partial_s^{2n} T^{00}(s,t) \geqslant 0.$$

Analogous results hold for fixed s and u relations by crossing symmetry.

Some of the results which follow from this can be seen from simple geometry. Consider Fig. 7.1. The S-wave amplitude $T_0^{00}(s)$ is obtained by averaging along lines of fixed s between $t=0$ and $u=0$, i.e. $\cos\theta_s = \pm 1$. By crossing symmetry, the values of T^{00} at A and B are equal to that at C, and at any point in AB, e.g. K, its value is less by the above concavity property. Hence immediately

$$T_0^{00}(s=4) \geqslant T_0^{00}(s=0). \tag{7.28}$$

Similarly, by almost as simple arguments, it can be seen that

$$T_0^{00}(4) > T_0^{00}(GF) > T_0^{00}(HI) > T_0^{00}(DE),$$

where, e.g. $T_0^{00}(GF)$ means T_0^{00} evaluated at the constant value of s along

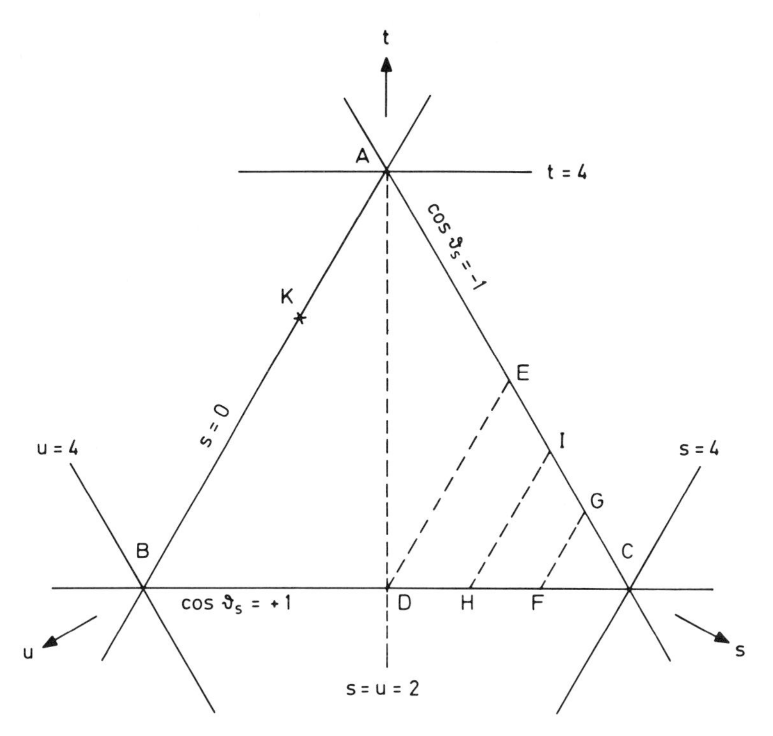

FIG. 7.1. Diagram to illustrate the derivation of some rigorous inequalities.

GF, so that $T_0^{00}(DE) \equiv T_0^{00}(2)$ and

$$\frac{dT_0^{00}(s)}{ds} \geqslant 0, \qquad 2 \leqslant s \leqslant 4. \tag{7.29}$$

Similar results can be obtained for higher partial waves, for example on the scattering lengths a_l defined in equation (2.40). For the case of $\pi^0\pi^0$ scattering

$$3a_l^{00} \equiv a_l^0 + 2a_l^2 > 0, \qquad \text{for } l \geqslant 2,$$

and by considering the amplitude $T^E(s, t, u)$ for $\pi^0\pi^0 \to \pi^+\pi^-$ the inequality

$$a_l^0 - a_l^2 > 0, \qquad \text{for } l \geqslant 2,$$

results. Thus

$$a_l^0 \geqslant 0 \quad \text{and} \quad a_l^2 \geqslant -\tfrac{1}{2}a_l^0, \quad \text{for } l \geqslant 2. \tag{7.30}$$

An alternative method of obtaining inequalities (Roskies, 1970b; Piguet and Wanders, 1969) is to note that a BNR relation for $l = 0, 2$ with the coefficient of the $l=2$ term positive becomes an inequality on the $l=0$ term if D-waves are neglected, and so on.

A very large number of inequalities have been derived, both by these and by more sophisticated arguments, all of which, however, rest on the same basic assumptions. For a review, we refer to the paper of Yndurain (1972). We will not make any attempt at completeness here, but just list a few of the more important relations, some of which will be used later, for example in Chapter 10.

The most interesting are for the $\pi^0\pi^0$ S-wave, for which we already have

$$T_0^{00}(4) \geqslant T_0^{00}(0). \tag{7.31}$$

Equation (7.29), for the derivative of this amplitude, can be improved somewhat to give (Martin, 1968):

$$T_0^{00\prime}(s) \equiv \frac{dT_0^{00}}{ds} > 0, \qquad 1{\cdot}7 < s < 4. \tag{7.32}$$

Other results for $\pi^0\pi^0$ scattering are (Martin, 1967, 1968, 1969; Common, 1968; Grassberger, 1972a):

$$T_0^{00\prime}(s) < 0, \qquad 0 < s < 1{\cdot}22 \tag{7.33}$$

$$T_0^{00\prime\prime}(s) \equiv \frac{d^2T_0^{00}(s)}{ds^2} > 0, \qquad 0 < s < 1{\cdot}7 \tag{7.34}$$

$$T_0^{00}(0) \geqslant T_0^{00}(3{\cdot}19), \tag{7.35}$$

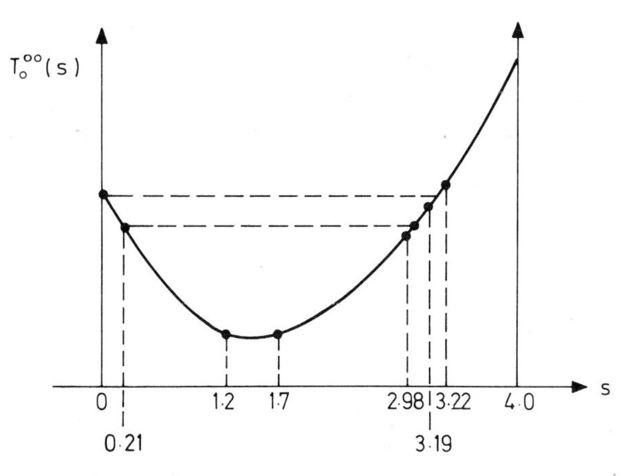

FIG. 7.2. Shape of the $\pi^0\pi^0$ amplitude in the region $0 \leqslant s \leqslant 4$ following from rigorous inequalities.

and

$$T_0^{00}(3\cdot205) \geq T_0^{00}(0\cdot2134) \geq T_0^{00}(2\cdot9863). \tag{7.36}$$

These results show that $T_0^{00}(s)$ has a minimum in the range

$$1\cdot219 \leq s \leq 1\cdot697, \tag{7.37}$$

which is unique by virtue of equations (7.32)–(7.34). Thus the amplitude must have a shape similar to that shown in Fig. 7.2 (Le Guillou *et al.*, 1971). Results can also be obtained on amplitudes and their derivatives for other charge states (Auberson *et al.*, 1970; Grassberger and Kuhnelt, 1973). Thus as well as the inequality

$$4T_0^{00\prime}(0) < -[2T_0^{00}(4) - T_0^{00}(2) - T_0^{00}(0)], \tag{7.38}$$

Auberson (1970) also gives the result

$$2T_0^{2\prime}(0) - 2T_1^{1\prime}(0) < T_0^2(0) - T_1^1(0) - \tfrac{2}{3}[T_0^0(4) - T_0^2(4)]. \tag{7.39}$$

Finally, we note that inequalities involving higher partial waves have also been derived (Martin, 1969; Grassberger, 1972a; Common and Pidcock, 1972; Common *et al.*, 1974; Pidcock, 1974) which, for example, restrict severely the possible form of the D-wave amplitudes.

(f) Atkinson's Method

In the previous paragraph, we have considered the large number of restrictions on the scattering amplitudes following from crossing and positivity. In this paragraph, we turn to the problem of the explicit construction of amplitudes which automatically satisfy these requirements (and, incidentally, many more). Since the methods are very technical, and we will not use them elsewhere, we will simply state the results that have been obtained, referring to the original papers (Atkinson, 1968a,b, 1969; Kupsch, 1969a,b, 1970a,b) and the review of Atkinson (1971) for further discussion.

What has been achieved is the demonstration of the existence of a convergent iteration procedure for the construction of amplitudes possessing a number of desirable properties. These are: crossing symmetry; Mandelstam analyticity; and the unitarity inequalities for elastic scattering, equation (7.3). Note that the latter constraints considerably exceed the positivity requirements (7.2). Finally, in the region between the elastic and inelastic thresholds, the elastic unitarity equalities are satisfied exactly. Thus, as noted above, all the equalities and inequalities of paragraph (e) are automatically satisfied. However, the most striking thing about the results is the great freedom allowed in constructing the amplitudes. The Mandelstam double spectral function ρ_{st} of equation (2.41) is used as a starting point for the procedure, but the iteration can be carried through for very large classes of such functions. In other words, amplitudes which satisfy the conditions above are far from unique, although, of course, the full content of coupled-channel unitarity has not been used. Nevertheless, it is clear that calculations which only use the above conditions will probably need a good deal of additional phenomenological information to attain uniqueness.

Finally, there is a limitation to the results obtained to date. In the initial paper (Atkinson, 1968a) the result was only established for $\pi^0\pi^0$ scattering and for the case without subtractions. Subsequent work (Atkinson, 1968b, 1969; Kupsch, 1969a,b, 1970a,b) has succeeded in removing these restrictions. However, it has not yet proved possible to construct an amplitude leading to a constant total cross-section at high energies, the results only being established if the total cross-section does not fall less rapidly than $(\log s)^{-2}$.

Chapter 8

Analyticity at Fixed Momentum Transfer

In this chapter, and in the two that follow, we will be concerned with exploiting the analytic properties (both proved and assumed) of scattering amplitudes to explore the phenomenology and dynamics of $\pi\pi$ scattering. In the first two of these chapters (8 and 9) we will develop the tools required for the phenomenological analyses to be discussed in Chapter 10, and also describe two applications currently of more theoretical than phenomenological interest. The three chapters have been arranged in this way to allow the phenomenological applications to be discussed in a more unified manner, without undue (and misleading) emphasis on calculational technique.

In this first chapter, we further develop ideas based on analyticity at fixed momentum transfer. In the first section we discuss the scattering length bounds referred to briefly in Chapter 7. In the second section we discuss the use of fixed-t dispersion relations to give conditions on partial wave amplitudes. The results of the latter discussion will be exploited extensively in Chapter 10.

8.1. Inverse Amplitudes and Scattering Length Bounds

We have seen in Chapter 7 that the idea of a "range of interaction" is closely related to the Froissart bound. Another important consequence of this concept is that it results, in the absence of bound states, in a negative lower bound on the S-wave scattering length. The existence of such bounds, and their interpretation, which was mentioned in Chapter 7, has been discussed by Goebel and Shaw (Goebel and Shaw, 1968; Shaw, 1968), and we shall follow their discussion here. However, before making any applications of inverse amplitude dispersion relations it is necessary to establish some results on the possible zeros of the amplitude, which play a role analogous to both poles and subtractions in the conventional relations.

It is convenient to work with the forward scattering amplitudes for physically realizable states, i.e. $\pi^+\pi^0$, $\pi^0\pi^0$ etc., rather than eigenstates of isospin. These are defined in terms of isospin amplitudes by equation (2.19), and we shall use a subscript P to denote a direct (s) channel charge state, and $\bar{P}$ to denote the corresponding crossed (u) channel charge state, so that if $P \equiv \pi^+\pi^+$, then $\bar{P} = \pi^+\pi^-$ etc. Since initially we will work at $t=0$, we will drop the t-variable, and use the crossing anti-symmetric variable $z \equiv (s-u)/4$. Thus, we will write

$$T_P(z) \equiv T_P\left(z = \frac{s-u}{4}, t=0\right). \tag{8.1.1}$$

The essential point about using physical (P) amplitudes is that the optical theorem guarantees positivity of the absorptive part, both on the direct (s) and crossed (u) channel cuts, thereby enabling the following crucial theorem to be established.

Theorem. The number of zeros, N_0 of a forward physical elastic scattering amplitude $T_P(z)$ is given by

$$2N_0 = 2N_P + 2 + \varepsilon + \bar{\varepsilon}, \tag{8.1.2}$$

where N_P is the number of poles, and ε and $\bar{\varepsilon}$ are the signs (± 1) of the amplitude at the direct and crossed channel thresholds, respectively, i.e.

$$T_P(z=1) = \varepsilon|T_P(z=1)|,$$

$$T_{\bar{P}}(z=1) = T_P(z=-1) = \bar{\varepsilon}|T_P(z=-1)|.$$

Proof. This follows directly from the result that the phase change around a closed contour is

$$2\pi(N_0 - N_P), \tag{8.1.3}$$

(Titchmarsh, 1939). Consider the behaviour of $T_P(z)$ as the closed contour of Fig. 8.1.1(a) is traced out in the complex z-plane. The general behaviour for the simple case of a crossing-symmetric amplitude with negative scattering length, so that $\varepsilon = \bar{\varepsilon} = -1$, is sketched in Fig. 8.1.1(b), the contour being traced out twice. The crucial point is that, since it is a physical amplitude, Im $T_P(z) > 0$ on AB and DE, and Im $T_P(z) < 0$ on CD and FA, so that the origin cannot be enclosed. Equation (8.1.2) follows directly from (8.1.3). The result follows similarly for the other cases.

Lemma (i). The result also holds for any fixed t in the range $0 \leqslant t < 4$, since $P_l(\cos\theta) > 1$ for $\cos\theta > 1$, so that the convergence of the partial wave

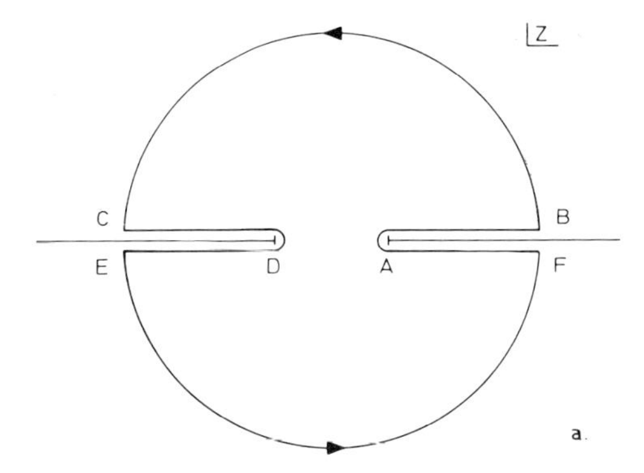

FIG. 8.1.1a. Contour of integration for proving theorem on zeros.

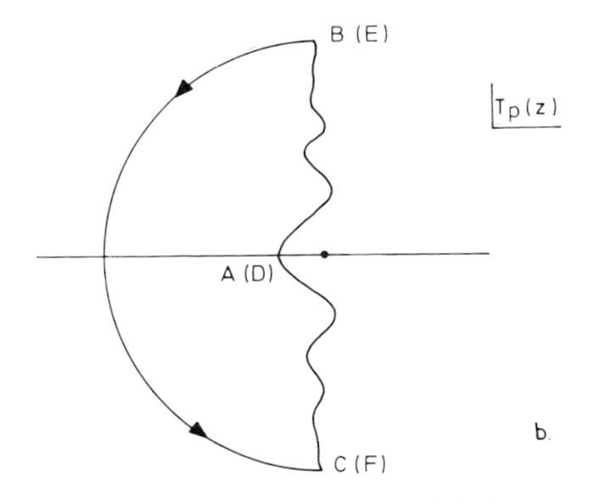

FIG. 8.1.1b. Behaviour of crossing-symmetric amplitude $T_P(z)$ with negative scattering length.

expansion (2.12) ensures the positivity of the absorptive parts. The relevant signs ε, $\bar{\varepsilon}$ are still defined by the threshold amplitudes, but the thresholds are now at

$$z = \pm z_0 = \pm(1 + t/4).$$

Lemma (ii). Since there are no poles in $\pi\pi$ scattering, $N_P = 0$ and N_0 is either 0, 1, or 2. The positions of these zeros are severely restricted by the requirement that $T_P(z)$ be a real analytic function. If there is only one zero, it must lie on the real axis in $-z_0 < z < z_0$. If there are two they must either both lie in this region, or be at complex conjugate values.

These results will be useful later.

(a) Sum Rule Bounds

We can now derive the bounds, but firstly we will indicate how they arise in potential theory, where they can be established by consideration of the wave function for S-wave scattering at threshold, $\psi(r, q=0)$. We denote the wave function obtained by solving the Schrödinger equation outside the range of interaction, $r>R$, by ψ_0. For $q=0$ this is linear in r, and may be written

$$\psi_0(r, q=0) = C_0 + C_1 r,$$

where C_0 and C_1 are constants. The scattering length a is defined by $\psi_0(-a)=0$. If $a<-R$.

$$0 = \psi_0(-a) = \psi(-a),$$

since $\psi=\psi_0$ for $r>R$. However, a zero of ψ implies the existence of a bound state, and hence *if there is no bound state*

$$a \geqslant -R, \tag{8.1.4}$$

which is the desired result. In the relativistic case there is no clearly defined range in the sense of the potential vanishing exactly outside some given region, so that the "range" used in the intuitive arguments below is not a well-defined concept. We do not pursue this since the results are given in terms of rigorous sum rule inequalities which do not require such considerations, although they are useful for physical illustration.

We shall derive a bound on the scattering length a_P, where

$$a_P = T_P(z=1), \tag{8.1.5}$$

and

$$a_{\bar{P}} = T_{\bar{P}}(z=1) = T_P(z=-1).$$

As noted above, the number of zeros of $T_P(z)$ is 0, 1, or 2, depending on whether neither, one, or both of the scattering lengths a_P and $a_{\bar{P}}$ are positive. Since if $a_P>0$ the lower bound is trivial, we need only consider the case $a_P<0$, when the inverse dispersion relation can be written

$$T_P^{-1}(z) = \frac{1}{\pi}\int_1^\infty dz' \frac{\operatorname{Im} T_P^{-1}(z')}{z'-z} + \frac{1}{\pi}\int_1^\infty dz' \frac{\operatorname{Im} T_{\bar{P}}^{-1}(z')}{z'+z} + \frac{(1+\bar{\varepsilon})x}{z_R - z}, \tag{8.1.6}$$

there being zero or one pole depending on whether $\bar{\varepsilon}=-1$ or $+1$, respectively, and, as noted above, $-1<z_R<+1$. From the relation

$$\operatorname{Im} T_P^{-1}(z) = -\frac{\operatorname{Im} T_P(z)}{|T_P(z)|^2}, \tag{8.1.7}$$

the absorptive part of the inverse amplitude is negative definite, and the integral contributions in equation (8.1.6) are negative for $-1\leqslant z\leqslant +1$. Thus, for $a_{\bar{P}}>0$ (when $\bar{\varepsilon}=+1$), evaluating at $z=-1$ gives $x>0$. Finally, evaluating at $z=+1$ we have

$$T_P^{-1}(z=1) = a_P^{-1} \leqslant -Q_P, \tag{8.1.8}$$

where

$$Q_P \equiv \frac{1}{\pi}\int_1^\infty \frac{dz}{z-1}\frac{\text{Im } T_P(z)}{|T_P(z)|^2} + \frac{1}{\pi}\int_1^\infty \frac{dz}{z+1}\frac{\text{Im } T_{\bar{P}}(z)}{|T_{\bar{P}}(z)|^2},$$

and the equality holds in the case $a_P < 0$ and $a_{\bar{P}} < 0$.

Since the integrands are positive definite, any knowledge of a non-vanishing lower limit on them for any finite range of z implies a negative lower bound on the scattering length

$$a_P \geq -I_P^{-1}, \tag{8.1.9}$$

where

$$Q_P \geq I_P > 0, \tag{8.1.10}$$

and again the equality in (8.1.10) implies that in (8.1.9), if both a_P and $a_{\bar{P}} < 0$.

These sum rule inequalities are now proved under the assumption that (8.1.6) converges, i.e. under the assumption that $T(z) > 0(1)$ as $z \to \infty$ (and that $\sigma_{\text{tot}} > c/z$ as $z \to \infty$). For $T(z) < 0(1)$ the forward relation converges without subtraction, so that

$$a_P = \frac{1}{\pi}\int_1^\infty dz \frac{\text{Im } T_P(z)}{z-1} + \frac{1}{\pi}\int_1^\infty dz \frac{\text{Im } T_{\bar{P}}(z)}{z+1}, \tag{8.1.11}$$

and the result follows trivially. This leaves only the possibility $T(z) \sim$ constant for large z to be considered. This is the case which occurs in potential scattering. In this situation, a constant c must be added to (8.1.11), and c^{-1} to (8.1.6). If $c \geq 0$ the result follows from (8.1.11), and if $c \leq 0$ from (8.1.6), so that the result is established independently of the high-energy behaviour, completing the proof.

Finally, it is useful to slightly re-arrange the bound for phenomenological use. Using the Optical Theorem equation (2.11) we have

$$a_P \geq -Q_P^{-1}, \tag{8.1.12}$$

where

$$Q_P \equiv \frac{2}{\pi}\int_1^\infty dz \frac{q}{W}\frac{R_P(z)}{z-1} + \frac{2}{\pi}\int_1^\infty dz \frac{q}{W}\frac{R_{\bar{P}}(z)}{z+1}, \tag{8.1.13}$$

$$R_P(z) \equiv \frac{\sum (2l+1)\alpha_{Pl}|T_l^l(z)|^2}{|\sum (2l+1)\alpha_{Pl}T_l^l(z)|^2}$$

$$= \frac{W^2}{32\pi}\frac{\sigma_{\text{tot}}^P(z)}{|T_P(z)|^2} \geq \frac{W^2}{32\pi}\frac{\sigma_{\text{el}}^P(z)}{|T_P(z)|^2}, \tag{8.1.14}$$

and the isospin coefficients α_{PI} are defined by

$$T_P(z) = \sum_I \alpha_{PI} T^I(z).$$

The numerical values of these coefficients may be read off from equation (2.19).

The connection with the idea of a range can clearly be seen. A finite range corresponds to a finite cut-off in l, say L. Then (ignoring isospin for simplicity)

$$R(z) \geq \frac{\sum^L (2l+1)|T_l|^2}{|\sum^L (2l+1)T_l|^2} \geq \frac{1}{\sum^L (2l+1)} = \frac{1}{(L+1)^2}, \tag{8.1.15}$$

so that a lower bound on R, and hence on the scattering length, results. More directly, one can see from equation (8.1.14) that R is bounded by the forward peak of the elastic amplitude. We will not pursue this here, but merely draw attention to the result demonstrated by Goebel and Shaw that the sum rule, in the case of a finite range and a heavy target $M \gg R^{-1}$, does lead to the intuitive result (8.1.4) if a factor of order unity is neglected. For a light target $MR \ll 1$, the result is weakened to

$$a \leq -M^{-1} \ln(1/MR).$$

(b) Bounds on λ

The coupling constant λ is defined by the values of the $\pi\pi$ amplitudes at the symmetry point $s = u = t = \frac{4}{3}$, i.e. $z = 0$, $t = \frac{4}{3}$ (see equation (2.25)). Since the theorem on zeros holds for $t = \frac{4}{3}$ also, it is possible to derive bounds on λ by use of the inverse dispersion relations at fixed $t = \frac{4}{3}$, evaluated at $z = 0$ (Shaw, 1968). However, since λ is not a threshold amplitude, we cannot exclude the case where both threshold amplitudes are positive, so that by (8.1.2) there are two zeros. Consequently, although in essence the same, the argument is a little more complicated, and *bounds are obtained for $s \leftrightarrow u$ crossing symmetric amplitudes only.* We simply state the result:

$$-n_P\lambda \equiv T_P(z=0, t=\tfrac{4}{3}) \geq -H_P^{-1}, \tag{8.1.16}$$

where

$$H_P \equiv \frac{4}{\pi} \int_1^\infty \frac{dz}{z} \frac{q_s}{W_s} S_P, \tag{8.1.17}$$

$$S_P \geq \frac{\sum (2l+1) P_l(\cos\theta_s) \alpha_{PI} |T_l^I|^2}{|\sum (2l+1) P_l(\cos\theta_s) \alpha_{PI} T_l^I|^2} > 0, \tag{8.1.18}$$

and $n_P = (1, 3)$ for $P \equiv (\pi^+\pi^0, \pi^0\pi^0)$. The reader is referred to the original paper for the details.

(c) Phenomenological Results

The above equations, although rigorous, do not yield numerical bounds as they stand. However, their phenomenological usefulness stems from the fact that, with only weak additional assumptions, stringent results can be obtained. We will illustrate this by considering the $\pi^0\pi^0$ amplitude

$$T^{00}(z)=\tfrac{1}{3}T^0(z)+\tfrac{2}{3}T^2(z), \tag{8.1.19}$$

which leads to the most powerful results. We consider three cases for the additional input information.

(i) *Absence of D-waves*

The weakest results follow from the sole assumption that D-waves are negligible below some centre-of-mass energy W_s, when from equations (8.1.14), (8.1.18) and (8.1.19) the results

$$R_{00}\geq 1 \quad \text{and} \quad S_{00}\geq 1, \tag{8.1.20}$$

follow. If we take $W_s=960$ MeV (see Chapter 3), the results

$$\lambda<0{\cdot}25, \qquad a_0^{00}>-0{\cdot}55, \tag{8.1.21}$$

are obtained. These can be compared with the best rigorous absolute bounds, equations (7.19) and (7.20), i.e.

$$-2{\cdot}6\leq\lambda\leq 17, \qquad a_0^{00}\geq-1{\cdot}75. \tag{8.1.22}$$

(ii) *Information in the Rho Region*

Below 560 MeV we retain the assumption that D-waves are absent, but in the region of the rho, 560–960 MeV, we incorporate information obtained from analyses of dipion production (see Chapter 3). Thus, we will assume that δ_0^2 is constant between $-10°$ and $-20°$, and for δ_0^0 we consider three alternatives: (a) that it rises linearly from 45° to 135°; (b) that it rises linearly from 30° to 150°; (c) that it resonates at 765 MeV with a width between 200 MeV and 400 MeV. For all these cases the above results yield bounds in the ranges

$$\lambda<0{\cdot}10, 0{\cdot}15; \qquad a_0^{00}>-0{\cdot}26, -0{\cdot}38. \tag{8.1.23}$$

The addition of a reasonable value for the D-wave, e.g. $\delta_2^0=\pm 2°$ at the rho mass, has a negligible effect on these results.

(d) Extensions of the Method

The method of Goebel and Shaw aims at obtaining stringent bounds from rather little information in a simple way. If, however, one wishes to obtain bounds from a specific set of phase shifts, known over the range $z_1 \leq z \leq z_2$, more powerful results can be obtained by modifying the method (Basdevant *et al.*, 1973; see also Bonnier and Procureur, 1973; Bonnier, 1974). If instead of working with $T^P(z)$ one considers

$$F(z) = T^P(z) + H(z), \tag{8.1.24}$$

where $H(z)$ is a known function, analytic in the z-plane, and such that

$$\operatorname{Im} F(z) \geq 0, \qquad z_1 \leq z \leq z_2, \tag{8.1.25}$$

is still satisfied, the argument goes through as before giving‡

$$F(1) = a_P + H(1) > -\left[\frac{1}{\pi}\int_{z_1}^{z_2} \frac{dz}{z-1} \frac{\operatorname{Im} F(z)}{|F(z)|^2}\right]^{-1}. \tag{8.1.26}$$

This negative lower bound on $F(1)$ implies a lower bound on a_P, which may be either positive or negative, depending on the value of $H(1)$. Furthermore, for a given $T^P(z)$ in the range $z_1 \leq z \leq z_2$, $H(z)$ may be varied, subject to equation (8.1.25), to obtain the best bound.

Basdevant *et al.* (1973) have applied this technique to the $\pi^0\pi^0$ scattering amplitude using for $H(1)$ cubics in the variable which maps the z-plane, cut along $z_1 \leq z \leq z_2$ into the unit circle. Using the data of Baton *et al.* (1967a,b, 1970a,b) over the range $0{\cdot}45 \leq W \leq 1$ GeV, and of Hyams *et al.* (1973) for $1{\cdot}0 \leq W \leq 2{\cdot}0$ GeV, the result

$$a_0^{00} > -0{\cdot}1, \tag{8.1.27}$$

is obtained. While this result applies only for a rather specific input, it illustrates the strength of the bounds on a_0^{00} which follow from assuming specific phase shifts at energies appreciably above threshold.

8.2. Partial Waves from Fixed-*t* Dispersion Relations

Fixed-*t* dispersion relations have the advantages, compared to the partial wave dispersion relations to be discussed in the next chapter, that s–u crossing is explicitly built in, and that the discontinuities are often more directly related to physically accessible quantities (the total cross-section in the case of forward dispersion relations). Apparent disadvantages are that procedures have to be devised for extracting partial wave amplitudes in order that unitarity may be simply implemented, and t–u crossing has to be imposed. However, in practice, neither requirement gives difficulty, at least when working in the low-energy region. We will first discuss an approximate treatment assuming a truncated partial wave series; in this case the forward

‡ For $H(z) = 0$, this result is a specific case of equation (8.1.9).

relation, and a finite number of its derivatives with respect to t, suffice to determine the partial waves considered. We then sketch the more systematic approaches involving partial wave projections on the fixed-t relations. The resulting equations (Roy, 1971) appear now to provide the best vehicle for discussing the phenomenology of low energy scattering.

(a) Fixed-t Relations and their Derivatives at $t=0$

In Chapter 3, we saw that partial waves with $l \leqslant 2$ were sufficient to provide a rather accurate description of $\pi\pi$ scattering for $W \leqslant 1$ GeV. If the weaker approximation is made, that the scattering in this region can be represented by just S-, P-, D- and F-waves, then a knowledge of the scattering amplitude and its derivative with respect to t (both evaluated at $t=0$) is sufficient to determine all the partial waves. This, together with fixed-t relations, can be used to give a set of equations for the partial waves themselves, as we now show.

In this context it is convenient to work in terms of t-channel isospin eigenstates, written as functions of the crossing variable z and t. These are related to eigenstates of s-channel isospin by the crossing matrix $C_{II'}^{(st)}$ defined in Chapter 2, i.e.

$$T_t^I(z, t) = \sum_{I'} C_{II'}^{(st)} T_s^{I'}(s, t, u). \tag{8.2.1}$$

The fixed-t dispersion relations (written without subtractions) take the form

$$T_t^I(z, t) = \frac{1}{\pi} \int_{1+t/4}^{\infty} dz' \operatorname{Im} T_t^I(z', t) \left[\frac{1}{z'-z} + \frac{(-1)^I}{z'+z} \right], \tag{8.2.2}$$

so that s–u crossing is explicitly imposed. The above relation, and its derivative with respect to t, at $t=0$, yield equations for the quantities

$$T_t^I(z, t=0), \qquad \left[\frac{\partial T_t^I}{\partial t}(z, t) \right]_{t=0},$$

which can be expressed in terms of s-channel partial wave amplitudes via the relations

$$T_t^I(z, 0) = \sum_{I'} C_{II'}^{(st)} \sum_l (2l+1) T_l^{I'}(\nu), \tag{8.2.3}$$

$$\left[\frac{\partial T_t^I}{\partial t}(z, t) \right]_{t=0} = \sum_{I'} C_{II'} \sum_l (2l+1) \left\{ \frac{T_l^{I'}(\nu)}{2\nu} \frac{l(l+1)}{2} - \frac{1}{8} \frac{\partial T_l^I(\nu)}{\partial \nu} \right\}, \tag{8.2.4}$$

where $\nu=(z-1)/2-t/8, (\mu\equiv 1)$. If the full amplitude is assumed to be approximated by S-, P-, D- and F-waves over the range of z (not z') investigated, these equations suffice to determine the partial waves. This approximation has been used (Morgan and Shaw, 1969, 1970) to study the energy range $W_s \lesssim 1$ GeV (cf. Chapter 10), and this range could easily be extended by including more partial waves and derivatives in an obvious manner. So far the restrictions due to t–u crossing have not been imposed. However, subtractions must be made in equation (8.2.2) in practice, and these can be exploited to induce an appreciable measure of tu crossing over the triangle region by the explicit imposition of the threshold factors $(t-u)^l$ as discussed by the above authors.

(b) Partial Wave Projections and Roy's Equations

A more systematic approach to this problem is to perform a partial wave projection on equation (8.2.2). This involves expanding the absorptive part Im $T_t^I(z, t)$ in a partial wave series, and then projecting out partial waves from both sides of the equation using the orthogonality properties of Legendre polynomials. This procedure results in a coupled set of integral equations for the partial wave amplitudes, and has been used extensively in πN scattering and photoproduction since its introduction by Chew *et al.* (1957a,b). However, it was only relatively recently that it was realized (Roy, 1971) that in the case of $\pi\pi$ scattering this approach could be considerably extended by the use of s–t crossing to yield a set of equations of great power and usefulness. Their advantage is that while being relations between partial wave amplitudes in the physical region (so that unitarity is easy to impose), they nonetheless exhaust the requirements of s–t as well as s–u crossing, provided that a set of well-defined subsidiary conditions is imposed.

The derivation of these relations goes in four stages which we will illustrate for the case of $\pi^0\pi^0$ scattering. For more details, and the general isospin case, we refer to the original literature (Roy, 1971; Basdevant, Le Guillou and Navelet, 1972). The first step is to write a twice subtracted fixed-t dispersion relation

$$T^{00}(s, t, u)=t^{00}(t)+\frac{1}{\pi}\int_4^\infty \frac{ds'}{s'^2}\left[\frac{s^2}{s'-s}+\frac{u^2}{s'-u}\right]\text{Im } T^{00}(s', t, 4-s'-t). \quad (8.2.5)$$

Because of the requirement of s–u crossing, there is just one, as yet undetermined, subtraction function for each eigenstate of t-channel isospin, and for the exhibited case of $\pi^0\pi^0$ scattering. The second stage is to eliminate this undetermined subtraction function in favour of a constant by using st crossing for a fixed value of s, taken to be $s=0$.

$$T^{00}(s,t)=[T^{00}(s,t)-T^{00}(0,t)]+T^{00}(t,0),$$
$$=[T^{00}(s,t)-T^{00}(0,t)]+[T^{00}(t,0)-T^{00}(4,0)]+a_0^{00}. \quad (8.2.6)$$

This trick, which was used earlier in a different context (Bonnier and Vinh Mau, 1968), leaves only one arbitrary constant, which has been taken to be the S-wave scattering length a_0^{00}. For the case of $\pi^0\pi^0$ scattering, this **yields**

$$T^{00}(s,t)=a_0^{00}+\frac{t(t-4)}{\pi}\int ds'\frac{(2s'-4)\,\mathrm{Im}\,T^{00}(s',0)}{s'(s'-t)(s'+t-4)(s'-4)}$$
$$-\frac{su}{\pi}\int ds'\frac{(2s'+t-4)\,\mathrm{Im}\,T^{00}(s',t)}{s'(s'-s)(s'-u)(s'+t-4)}, \quad (8.2.7)$$

with $u=4-s-t$. The third stage is to project out partial waves using only half the interval in the scattering angle, as if s–t crossing symmetry was already implemented, i.e. define

$$T_l^{00}(s)=\frac{2}{s-4}\int_{(4-s)/2}^{0} dt\,T^{00}(s,t)P_l\left(1+\frac{2t}{s-4}\right). \quad (8.2.8)$$

The fourth, and final, step is to express $\mathrm{Im}\,T^{00}(s',t)$ itself as a sum of partial wave contributions,

$$\mathrm{Im}\,T^{00}(s',t)=\sum_{l'\,\mathrm{even}}(2l'+1)\,\mathrm{Im}\,T_l^{00}(s',t)P_l\left(1+\frac{2t}{s'-4}\right). \quad (8.2.9)$$

The desired relations now follow from equations (8.2.7)–(8.2.9). Their range of validity may be found by considering equations (8.2.8) and (8.2.9). The expansion (8.2.9) converges for t in the range $-28\leqslant t\leqslant 4$ for all $s'\geqslant 4$, and consequently, since the range of t in equation (8.2.8) is $(4-s)/2\leqslant t\leqslant 0$, the resulting equations are valid for $-4\leqslant s\leqslant 60$.‡

For an eigenstate of isospin, they take the form

$$T_l^I(s)=\lambda_l^I(s)+\sum_{I'=0}^{2}\sum_{l'=0}^{\infty}(2l'+1)\int_4^{\infty} ds'\,G_{ll'}^{II'}(s,s')\,\mathrm{Im}\,T_{l'}^{I'}(s'), \quad (8.2.10)$$

where $-4\leqslant s\leqslant 60$. The subtraction terms $\lambda_l^I(s)$ contribute to S- and P-waves only, explicit expressions in terms of the two S-wave scattering lengths being

$$\lambda_0^0=a_0^0+(2a_0^0-5a_0^2)(s-4)/12,$$
$$\lambda_0^2=a_0^2-(2a_0^0-5a_0^2)(s-4)/24, \quad (8.2.11)$$
$$\lambda_1^1=(2a_0^0-5a_0^2)(s-4)/72.$$

‡ If Mandelstam analyticity is assumed (cf. Chapter 2), equation (8.2.9) converges for $-32\leqslant t\leqslant 4$, so that the equations are valid in the larger range $-4\leqslant s\leqslant 68$.

These terms alone constitute the most general linear form in s, t and u for the scattering amplitude (cf. the discussion in Chapter 14). The kernels $G_{ll'}^{II'}(s, s')$ are known functions which behave like $(s')^{-3}$ for large s', ensuring good convergence of the integrals in equation (8.2.10). For the specific example of $\pi^0\pi^0$ scattering considered above, the subtraction constant $a_0^{00} = (a_0^0 + 2a_0^2)/3$, and the kernel for the S-wave contribution to the integral takes the form

$$G_0^{00}(s, s') = \frac{1}{\pi}\left[\frac{s-4}{(s'-4)(s'-s)} - \frac{2}{s'} + \frac{2}{s-4}\ln\left(\frac{s-4+s'}{s'}\right)\right]. \quad (8.2.12)$$

The first term contributes a right-hand cut, and the third a left-hand cut, to $T_0^{00}(s)$. We return to this in the next chapter.

Equations (8.2.10) do not exhaust the constraints from crossing symmetry; s–u crossing is enforced but not t–u crossing, although this has been used in projecting out the physical partial waves wanted by Bose statistics. A further constraint has therefore to be applied to the input partial wave absorptive parts to ensure that the unwanted partial waves do not appear. For the example already considered of $\pi^0\pi^0$ scattering, one requires

$$T^{00}(s, t) = T^{00}(s, 4-s-t). \quad (8.2.13)$$

If we write (8.2.7) in the form

$$T^{00}(s, t) = a_0^{00} + \frac{1}{\pi}\int_0^\infty ds'[K(s', s, t)\,\text{Im}\,T^{00}(s', t)$$
$$- K(s', t, 0)\,\text{Im}\,T^{00}(s', 0)], \quad (8.2.14)$$

where

$$K(s', s, t) = \frac{s(4-s-t)(2s'+t-4)}{s'(s'-s)(s'-4+s+t)(s'+t-4)}, \quad (8.2.15)$$

then the required condition is

$$\int_4^\infty ds'[K(s', s, t)\,\text{Im}\,T^{00}(s', t) - K(s', s, 4-s-t)\,\text{Im}\,T^{00}(s', 4-s-t)$$
$$+\{K(s', 4-s-t, 0) - K(s', t, 0)\}\,\text{Im}\,T^{00}(s', 0)] = 0. \quad (8.2.16)$$

These equations, and the analogous ones for the other isospin states, together with equations (8.2.10), guarantee complete crossing symmetry (Roy, 1971). These constraints have also been studied in various forms by Wanders (1969) and Roskies (1970c).

An important property of the Roy equations is that these supplementary crossing conditions do not involve S- and P-wave absorptive parts. Thus, in the approximation where only these absorptive parts are retained, a fully crossing symmetric output is guaranteed. In particular, solutions will automatically satisfy the BNR relations and (provided positive absorptive parts are used) the Martin inequalities for S- and P-waves (cf. Chapter 7). This makes the Roy equations especially useful tools for investigating the low-energy region where these waves dominate. In any case, applications must be restricted to the range $-4 \leq s \leq 60$, i.e. $W_s \lesssim 1080$ MeV, in which the equations are valid.

The latter restriction has been partially overcome by Mahoux, Roy and Wanders (1974). Instead of using s and t, these authors consider the variables

$$x \equiv -(st + tu + us)/4; \qquad y \equiv stu/64,$$

and obtain sets of coupled integral equations for the partial wave amplitudes starting from dispersion relations written along straight lines in the (x, y) plane. Each set is valid in a finite interval, and the union of the sets spans the range $-28 \leq s \leq 125{\cdot}3$, i.e., $W_s \lesssim 1560$ MeV. Starting from dispersion relations along hyperbolae yields an even larger region, $-28 \leq s \leq 164{\cdot}7$ (Auberson and Epele, 1974), and more generally, Roy and Singh (1974) have established the existence of physical region equations for $-28 < s < \infty$ (cf. also Lopez (1973)). Unfortunately, these extensions of the range are achieved at the cost of a considerable increase in complexity, and for further details we refer to the original papers.

Chapter 9

Analytic Properties of Partial Wave Amplitudes

In this chapter we discuss the analytic properties of partial wave amplitudes, and in particular, partial wave dispersion relations. The operational advantage of these, compared to fixed-t dispersion relations, is that unitarity is easy to incorporate. Crossing, however, is less straightforward. Furthermore, in establishing partial wave dispersion relations, one has to go beyond the domain of rigorously proven results and assume, for example, the Mandelstam representation.‡

The material is laid out as follows. In the first section, we discuss the basic properties of partial wave amplitudes: their analytic structure, and the derivation of dispersion relations; the restrictions from crossing; and the classification of the dispersion relation solutions by the N/D equations. In the second section, we turn to more specific dynamical questions, and in particular, efforts to understand the dynamics of the rho-meson. Early attempts to construct a hadron dynamics were based on trying to generalize from non-relativistic potential theory, using partial wave dispersion relations and the N/D equations as tools. The hope of achieving a simple predictive scheme along these lines rested heavily on the belief that nearby singularities should dominate. Although this sort of approach has been more or less abandoned as a scheme for a complete hadron dynamics, it achieved some partial success, and retains interest as a link between the language of analyticity, and the language of potential theory. More phenomenological

‡ This furnishes statements which are needed on the location of singularities *and* on the behaviour at infinity.

applications, in which the existence of the rho-meson is regarded not as a fact to be explained, but as a constraint to be imposed, are treated in Chapter 10.

9.1. Partial Wave Dispersion Relations

(a) Derivation and Crossing

The partial wave amplitudes are obtained from the full amplitude by the standard projection formula equation (2.13), which, rewritten in terms of s, or u gives

$$T_l^I(s) = \frac{1}{s-4}\int_{4-s}^{0} dt\, P_l\left(1+\frac{2t}{s-4}\right) T^I(s, t, u), \tag{9.1.1a}$$

$$= \frac{(-1)^l}{s-4}\int_{4-s}^{0} du\, P_l\left(1+\frac{2u}{s-4}\right) T^I(s, t, u). \tag{9.1.1b}$$

According to the Mandelstam representation, the singularities of T^I arise solely from the physical thresholds in s, t and u, and the branch points induced in $T_l^I(s)$ in the complex s-plane can be read off from equation (9.1.1). Thus, the s-channel threshold leads to a "right-hand" cut $4 \leq s < \infty$. The t-channel threshold ($t = 4$) enters the integration region to the left of $s = 0$ giving rise to a "left-hand" cut $-\infty < s \leq 0$. Higher thresholds give rise to branch points lying further along the negative real axis, and the u-channel structure is identical. Thus, according to the Mandelstam representation, the partial wave amplitudes have only the above right and left-hand cuts,‡ so that if the partial wave amplitudes are polynomially bounded in the complex s-plane, a dispersion relation can be written with a finite number of subtractions. The unsubtracted form is

$$T_l^I(s) = \frac{1}{\pi}\int_{-\infty}^{0} ds' \frac{\text{Im } T_l^I(s')}{s'-s} + \frac{1}{\pi}\int_{4}^{\infty} ds' \frac{\text{Im } T_l^I(s')}{s'-s}, \tag{9.1.2}$$

where we have exploited the real analyticity of $T_l^I(s)$ (which follows from the reality of $T^I(s, t, u)$ in the triangle $0 < s, t, u < 4$) to express the discontinuities in terms of the absorptive parts. For $l \geq 1$, l parameter-free subtractions can be achieved by using the threshold factor ν^l to write a dispersion

‡ In the general case of unequal masses, the non-linear relationship between q^2 and s makes the problem of translating the branch points to the s-variable very complicated (Kennedy and Spearman, 1962). For the simpler case of elastic scattering, e.g. $\pi K \to \pi K$, the angular projection of the u-channel singularities induces an s-channel cut $s \leq (m_K - m_\pi)^2$, and the t-channel singularities induce an s-channel cut $s \leq 0$, and a circle cut $|s| = m_K^2 - m_\pi^2$. (These result from the appropriate expressions for t and u as functions of q^2 and $\cos\theta$.) The resulting cut-structure is analogous to that of the more familiar case of $\pi N \to \pi N$ (cf. Fig. 4.1.3), but without the singularities due to the nucleon Born term which occur in that case.

relation for $T_l^I(\nu)/\nu^l$. It is in fact often more convenient to write the dispersion relation in terms of ν rather than s as above, when it reads

$$T_l^I(\nu)=\frac{1}{\pi}\int_{-\infty}^{-1} d\nu' \frac{\text{Im } T_l^I(\nu')}{\nu'-\nu}+\frac{1}{\pi}\int_0^{\infty} d\nu' \frac{\text{Im } T_l^I(\nu')}{\nu'-\nu}. \tag{9.1.3}$$

Before discussing these equations further, it is worth emphasizing that the above derivation is based on the validity of the Mandelstam representation, which goes beyond what has been rigorously established from field theory. However, although the above analytic properties have not been proved for the whole complex plane, they have been proved over a large region (Martin, 1966b). (See also the review Martin and Cheung (1970).) This includes, but is much larger than, the domains which can be established straightforwardly from the Roy equations (see Section 8.2) or the Froissart–Gribov representation (see below), and is certainly large enough for phenomenological applications, when the amplitude is approximated by its nearest singularities plus a slowly-varying background (cf. Chapter 10). In contrast, the question of polynomial boundedness in s, which follows from the simultaneous polynomial boundedness in s and t of $T^I(s, t, u)$, is, from the viewpoint of rigorous field theory, a completely open one.

In order to make the $\pi\pi$ partial wave dispersion relation, equation (9.1.3), a useful result we must incorporate two further principles—unitarity and crossing symmetry. The former allows the discontinuity on the right-hand cut to be expressed as

$$\text{Im } T_l^I(\nu)=\left(\frac{\nu}{\nu+1}\right)^{\frac{1}{2}}|T_l^I(\nu)|^2\, R_l^I(\nu), \qquad 0\leqslant\nu<\infty, \tag{9.1.4}$$

where (recall equation (2.38))

$$R_l^I(\nu)\equiv\frac{\sigma^I_{\text{tot},l}(\nu)}{\sigma^I_{\text{el},l}(\nu)}=1+\frac{1-(\eta_l^I)^2}{1-2\eta_l^I\cos 2\delta_l^I+(\eta_l^I)^2}\geqslant 1, \tag{9.1.5}$$

and the equality strictly applies to the elastic region $0\leqslant\nu<3(M_{\pi\pi}<560 \text{ MeV})$, although, in practice, inelasticity appears to be very small below 900 MeV, (see Section 3.3).

It now remains to incorporate crossing, the first consequence of which is to give an expression for the discontinuity across the near left-hand cut in terms of the physical region absorptive parts. Consider the Froissart–Gribov formula equation (7.18) obtained by projecting out partial waves from a fixed-s dispersion relation ‡

$$T_l^I(s)=\frac{2[1+(-1)^{I+l}]}{\pi(s-4)}\int_4^{\infty} dt\, Q_l\left(1+\frac{2t}{s-4}\right)\text{Im } T^I(s, t). \tag{9.1.6}$$

‡ For the purpose of calculating the discontinuity, we can ignore subtractions.

This relation is valid for $-32<s<4$ according to the Mandelstam representation, and is rigorously established for $-28<s<4$, as noted in Chapter 7.‡ The left-hand cut discontinuities are now explicitly exhibited in the Q_l factor. Recalling (Erdélyi *et al.*, 1953) that

$$\frac{1}{ei}\operatorname{Disc}[Q_l(x)]=\frac{\pi}{2}P_l(x)\theta(1-x^2), \tag{9.1.7}$$

one obtains for the left-hand cut discontinuity

$$\frac{1}{2i}\operatorname{Disc}T_l^I(s)=\frac{[1+(-1)^{I+l}]}{s-4}\int_4^{4-s}dt\,P_l\left(1+\frac{2t}{s-4}\right)\operatorname{Im}T^I(s,t). \tag{9.1.8}$$

If we now use the crossing relation

$$\operatorname{Im}T^I(s,t)=\sum_{I'}C_{II'}^{(st)}\operatorname{Im}T_t^{I'}(s,t), \tag{9.1.9}$$

together with the partial wave expansion

$$\operatorname{Im}T_t^{I'}(s,t)=\sum_{l'}(2l'+1)P_{l'}\left(1+\frac{2s}{t-4}\right)\operatorname{Im}T_{l'}^{I'}(t), \tag{9.1.10}$$

we obtain

$$\operatorname{Im}T_l^I(s)=\frac{[1+(-1)^{I+l}]}{s-4}\sum_{I'}\sum_{l'}(2l'+1)C_{II'}^{(st)}$$
$$\times\int_4^{4-s}dt\,P_l\left(1+\frac{2t}{s-4}\right)P_{l'}\left(1+\frac{2s}{t-4}\right)\operatorname{Im}T_{l'}^{I'}(t), \tag{9.1.11}$$

or, written in terms of the ν-variable,

$$\operatorname{Im}T_l^I(\nu)=\frac{[1+(-1)^{I+l}]}{\nu}\sum_{I'}\sum_{l'}(2l'+1)C_{II'}^{(st)}$$
$$\times\int_0^{-\nu}d\nu'P_l\left(1+\frac{2(1+\nu')}{\nu}\right)P_{l'}\left(1+\frac{2(1+\nu)}{\nu'}\right)\operatorname{Im}T_{l'}^{I'}(\nu'). \tag{9.1.12}$$

Thus, crossing determines the discontinuity on the nearby left-hand cut in the range $-32<s<4(-28<s<4)$ in terms of integrals over the partial wave absorptive parts in the physical region. In deriving this result the series equation (9.1.10), in which the variable s features as a momentum transfer, has been used outside its physical region $4-t<s<0$. Nonetheless, the series converges over just this limited range $-32<s<4(-28<s<4)$ as discussed in Chapter 2.§

‡ We shall henceforth in this section quote the "Mandelstam" domain with the rigorously established domain in brackets, i.e. $4>s>-32(4>s>-28)$.

§ The range in s for which the discontinuity across the left-hand cut is calculable can be extended by the use of conformal mapping techniques (Frazer, 1961; Atkinson, 1962).

We turn now to the question of imposing crossing on the real part. The difficulty of constructing a compact and complete enactment of this constraint, even formally, is intimately connected with the non-convergence of the partial wave series over all except the nearby portion of the left-hand cut. Lacking a general formalism for building crossing into the partial wave equations, one is driven to various approximations.

The first approach is to use an approximation scheme in which the absorptive parts of the higher partial waves (say, for all $l > L$) are neglected. In this case the convergence problem is avoided, and exact crossing is realized at the expense of unitarity in the higher ($l > L$) waves, and of possibly distorting the more distant left-hand cut. This would not be very serious if the dynamics were determined by the nearby singularities; unfortunately, consistency problems arise if $L \geqslant 1$, as we shall see when we discuss applications of this approach in Section 9.2.

A second possibility is to parameterize the more distant parts (or even perhaps the whole) of the left-hand cut, for example by poles and subtractions, and to try and restrict the parameters by imposing crossing in the form of subsidiary conditions. In this approach it is a matter of judgement as to how many such conditions should be used in association with how many parameters. If too many conditions are used in proportion to the number of parameters, one is likely to be spuriously constrained; if too few parameters are used, then crossing may be seriously violated. Despite these difficulties, this type of approach has been extensively used in phenomenological models based on partial wave analyticity. Examples are discussed in Section 10.2 below.

What subsidiary conditions have been used? In early work a popular choice was to take the conditions imposed by crossing on the full amplitude $T^I(s, t, u)$ at the symmetry point $\nu = \nu_s = -\frac{2}{3}$, $\cos\theta = 0$, and to saturate them with a limited number of partial waves. For example, if one takes the exact relations (equations (2.25) and (2.26))

$$\begin{gathered} 2T^0(\nu_s, 0) = 5T^2(\nu_s, 0) \equiv -10\lambda, \\ \left.\frac{\partial T^0}{\partial \nu}\right|_{\nu=\nu_s} = -2\left.\frac{\partial T^2}{\partial \nu}\right|_{\nu=\nu_s} = 2\left.\frac{\partial (T^1/\nu)}{\partial \cos\theta}\right|_{\nu=\nu_s} \equiv 2\lambda_1, \end{gathered} \tag{9.1.13}$$

and saturates them with just the S- and P-waves, one obtains the *approximate* crossing conditions

$$\begin{aligned} 2T_0^0(\nu_s) = 5T_0^2(\nu_s) &= -10\lambda, \\ \left.\frac{\partial T_0^0(\nu)}{\partial \nu}\right|_{\nu=\nu_s} &= -2\left.\frac{\partial T_0^2(\nu)}{\partial \nu}\right|_{\nu=\nu_s} \\ &= \left.\frac{6T_1^1(\nu)}{\nu}\right|_{\nu=\nu_s} = 2\lambda_1. \end{aligned} \tag{9.1.14}$$

Further partial waves and higher derivatives can be included in these partial wave expansions about the symmetry point, leading to extra contributions to the above equations and additional relations. A more systematic approach is to use the rigorous conditions discussed in Chapter 7, i.e. the BNR relations, and the Martin inequalities. These have the advantage of being exact and, in the case of the BNR relations, become a sufficient set of conditions to guarantee exact crossing symmetry for the full amplitude in the limit that the number of partial waves retained tends to infinity. In practice only a finite set of partial waves is retained, and then the BNR relations are usefully supplemented by the Martin inequalities. It is an undoubted methodological advantage to operate with exact conditions; nonetheless, there are pitfalls in applying crossing conditions to parametric models by whatever method. It may be inappropriate and misleading to enforce high-order crossing conditions on a model of restricted scope (see Chapter 10). For further discussion, and the derivation, of the BNR and Martin conditions we refer back to Chapter 7.

(b) Solution of the Partial Wave Equations: The *N*/*D* Method

We have summarized the basic analytic properties of the partial wave amplitudes, and have indicated how one can pass from a given set of right-hand cut discontinuities to a set of left-hand cut discontinuities, at least approximately, by crossing. Clearly, if one can pass back from these left-hand cut discontinuities to the right-hand cut by an independent method, one has the basis of a self-consistent calculational scheme. With this in mind, we consider the following well-defined mathematical problem:

Given a real analytic function $F(\nu)(\equiv T_l^I(\nu))$, with a prescribed left-hand cut discontinuity $\operatorname{Im} F_L(\nu)$ for $-\infty<\nu\leq-1$, and a right-hand cut $0\leq\nu<\infty$ governed by unitarity (equation (9.1.4) with $R(\nu)$ given), and with no other singularities, what are the possible solutions for $F(\nu)$?

A convenient discussion of this problem is given by the N/D method (Chew and Mandelstam, 1960) which exploits the simple form taken by unitarity equation (9.1.4) for the inverse amplitude

$$\operatorname{Im}[F^{-1}(\nu)]=-R(\nu)\left(\frac{\nu}{\nu+1}\right)^{\frac{1}{2}}. \tag{9.1.15}$$

(One can also consider the case where instead of $R(\nu)$ the inelasticity parameter $\eta(\nu)$ is specified (Frye and Warnock, 1963)). If $F(\nu)$ is written as the ratio $N(\nu)/D(\nu)$, where $N(\nu)$ is defined to have only the left-hand cut, and $D(\nu)$ only the right-hand cut, then equation (9.1.15) gives

$$\operatorname{Im} D(\nu)=-R(\nu)N(\nu)\left(\frac{\nu}{\nu+1}\right)^{\frac{1}{2}},\qquad 0\leq\nu<\infty, \tag{9.1.16}$$

so that one can immediately write down the dispersion relations

$$N(\nu)=\frac{1}{\pi}\int_{-\infty}^{-1} d\nu' \frac{D(\nu')\,\mathrm{Im}\,F_L(\nu')}{\nu'-\nu}, \tag{9.1.17}$$

$$D(\nu)=1-\left(\frac{\nu-\bar{\nu}}{\pi}\right)\int_0^{\infty} d\nu' \frac{R(\nu')N(\nu')}{(\nu'-\nu)(\nu'-\bar{\nu})}\left(\frac{\nu'}{\nu'+1}\right)^{\frac{1}{2}}. \tag{9.1.18}$$

The possibility of such a decomposition follows from the existence of the Omnès function $\mathscr{D}(\nu)$ (Omnès, 1958) defined in equation (9.1.23) below. The obvious multiplicative ambiguity between N and D has been removed by normalizing $D(\nu)$ at a subtraction point $\nu=\bar{\nu}$. Provided the inelasticity $R(\nu)$ behaves at infinity such that $R(\nu)/\nu$ decreases, the above subtraction in the dispersion relation (9.1.18) for D allows $N(\nu)$ to behave as a constant at infinity. Thus a subtraction can be made in the dispersion relation (9.1.17) for N. This allows an extra parameter to be introduced, giving some scope for varying the effects attributed to distant singularities (Chew and Mandelstam, 1960) (see Section 9.2).

These "N/D" equations can be solved by eliminating either N or D between them. For example, eliminating $D(\nu)$ one obtains the following non-singular integral equation for $N(\nu)$,

$$N(\nu)=B(\nu)+\frac{(\nu-\bar{\nu})}{\pi}\int_0^{\infty} d\nu' \frac{R(\nu')N(\nu')}{(\nu'-\nu)(\nu'-\bar{\nu})}\left(\frac{\nu'}{\nu'+1}\right)^{\frac{1}{2}}[B(\nu')-B(\nu)], \tag{9.1.19}$$

where $B(\nu)$ is obtained from the specified left-hand cut discontinuity‡ by

$$B(\nu)=\frac{1}{\pi}\int_{-\infty}^{-1} d\nu' \frac{\mathrm{Im}\,F_L(\nu')}{\nu'-\nu}. \tag{9.1.20}$$

For a not unduly restricted class of inputs, $\mathrm{Im}\,F_L(\nu)$ and $R(\nu)$, equation (9.1.19) is of the Fredholm type, and has a unique solution (Atkinson and Morgan, 1966). Once N is determined, D may be calculated from equation (9.1.18), and $F(\nu)$ follows from $F(\nu)\equiv N(\nu)/D(\nu)$. Bound states and resonances appear as zeros of $D(\nu)$. In the latter case, the phase shift $\delta=\pi/2$ where $\mathrm{Re}\,D(\nu)=0$; the condition $D(\nu)=0$ gives the position of the second sheet resonance pole. Resonance and bound state parameters (masses, widths, coupling constants) emerge from the calculation, and are determined by the input information. Such states are termed "composite" or "dynamical", in contrast to other states whose existence is not explained within the dynamics, and which thus feature as "elementary" particles.

‡ The nomenclature $B(\nu)$ is intended to bring to mind the concept of the Born approximation—however see Section 11.1.

The latter are catered for through the existence of the CDD ambiguity (Castillejo, Dalitz and Dyson, 1956). In the present context this rests on the observation that the input conditions are still satisfied if a sum of poles is added to the formula for $D(\nu)$;

$$D(\nu) \to D(\nu) + \sum_{i=1}^{n_c} \frac{a_i}{\nu - b_i}. \tag{9.1.21}$$

This introduces n_c pairs of additional parameters, and enables n_c resonance positions and widths to be adjusted‡ (Chew and Frautschi, 1961a,b; Gell-Mann and Zachariasen, 1961). For a reasonably wide class of left-hand cuts and inelasticities, the alternatives conform to Levinson's theorem, according to which the change of phase from threshold to infinity is given by

$$\frac{\delta(\infty) - \delta(0)}{\pi} \leqslant n_c - n_b, \tag{9.1.22}$$

where n_c is the number of CDD poles, and n_b the number of bound states (*not* resonances). The relation is an inequality to allow for the possibility of "extinct" bound states (zeros of D cancelled in T by coincident zeros of N). Within the assumed framework, the result (9.1.22) can be seen by forming the Omnès function

$$\mathscr{D}(\nu) \equiv \exp\left[\frac{-\nu}{\pi} \int_0^\infty d\nu' \frac{\delta(\nu')}{\nu'(\nu' - \nu)}\right], \tag{9.1.23}$$

which, by construction, has the phase $-\delta(\nu)$, no poles or zeros, and the asymptotic behaviour§

$$\mathscr{D}(\nu) \sim \text{const. } \nu^{\delta(\infty)/\pi}, \tag{9.1.24}$$

(cf. the representations of $F_\pi(t)$ in Section 5.3(a)). The D-function of the N/D representation can differ from $D(\nu)$ by at most a multiplicative rational function whose numerator zeros are the bound state positions, and whose denominator zeros are the CDD poles. However, the D function of equation

‡ There are a number of equivalent ways of introducing the CDD parameters. Equation (9.1.21) is the classic formulation in which they are associated with zeros of the actual amplitude, consequently poles of D. Alternatively, simultaneous poles at arbitrary positions can be introduced into N and D with the pole residues as the genuine parameters. On factoring out the common poles in N and D, this latter formulation is equivalent to making pairs of subtractions in N and D (cf. Frautschi, 1963; Atkinson and Morgan, 1966). In all of these formulations, the correspondence with the intuitive idea of adding extra pairs of parameters belonging to additional elementary particles (analogous to the extra mass and coupling constants of a field theory description) is somewhat indirect, just because one is manipulating S-matrix elements for strong interactions. However, consider the procedure of (9.1.21). Poles in D will generally induce zeros of D (close by if the poles are weak), and consequently, non-dynamical bound states and resonances.

§ There can be additional logarithmic factors.

(9.1.18) has been constrained to tend asymptotically to a constant; hence (9.1.22) follows by balancing powers of ν.‡ Thus, we see that each CDD pole induces a "once and for all" rise in the phase shift of π, whereas for a dynamical resonance the phase shift eventually falls again. Using this, one might hope that in favourable cases smooth extrapolation of phase shift data at finite energies could supply information on whether a given resonance is "dynamical" or not.§ It is interesting at this point to compare the situation with that in non-relativistic potential scattering. At least for a superposition of Yukawa potentials, the partial waves have the same analytic structure as in the relativistic case, except that the left-hand cut now arises from the potential rather than crossing. If we consider the single-channel case ($R(\nu)=1$), the original form of Levinson's Theorem (Levinson, 1949) holds with an equality sign in equation (9.1.22) and $n_c = 0$. In other words, the Schrödinger equation selects the solution without CDD poles. This gives a powerful motivation for terming resonances and bound states occurring in this solution "dynamical", as opposed to "elementary". The analogy between the N/D equations with a given left-hand cut, and the non-relativistic equation with a given potential, was an influential factor in attempting dynamical calculations based on the N/D scheme. (The analogy is not, however, perfect; being given the potential is not precisely equivalent to being given the left-hand cut—see Section 11.1.)

Unfortunately, the classification of solutions based on CDD poles is only unambiguous in the case of elastic scattering. For inelastic scattering, there are a number of different ways of introducing inelasticity into the N/D equations. Besides the "R-method" used above, an adaption of the method for the case where the inelasticity is prescribed via the parameter $\eta(\nu)$ has been given (Frye and Warnock, 1963). Alternatively, coupling to other two-body channels can be explicitly treated via a multichannel $\mathbf{ND}^{-1}$ method (Bjorken, 1960). These different procedures are *not* equivalent, and in particular, a given resonance may be "dynamical" in one scheme, and "elementary" in another (Bander, Coulter and Shaw, 1965; Atkinson, Dietz and Morgan, 1966). For example, the rho-meson might be "dynamical" in say a two-channel $\pi\pi$, $\pi\omega$ calculation, but be "elementary" in a one-channel $\pi\pi$ calculation with inelasticity $R(\nu)$ calculated from the two-channel case. For further details on these questions we refer to the review of Collins and Squires (1968). In fact, most early calculations were carried out

‡ A number of the steps in this argument need careful justification. We remind the reader that they are valid for a wide class of, *but not all*, possible $R(\nu)$ and Im $F_L(\nu)$. Further details may be found in Atkinson and Morgan (1966).

§ Frautschi (1963) gives a most illuminating heuristic "proof" of Levinson's Theorem based on counting the numbers of degrees of freedom in a box as the interaction is switched on adiabatically. The question of what is the appropriate way to "count" hadron states, e.g. for SU(3) comparisons, has also been discussed by Dashen and Kane (1974).

in the approximation of elastic unitarity, so that this problem did not arise. We discuss these calculations in Section 9.2 below.

9.2. Attempts at a Partial Wave Dynamics

In the previous section we have summarized the basic features of partial wave dispersion relations. We will now discuss attempts to construct a complete partial wave dynamics based on them, including attempts to understand the dynamics of the rho-meson.

The first assumption underlying this approach is that the essential role in the dynamics is played by the nearby singularities, that is, loosely speaking, by the long-range forces. This means that the important part of the left-hand cut, at least for determining *changes* of the partial wave amplitudes with energy, is that which can be calculated by crossing. Effects of distant singularities can be in part allowed for by making a subtraction in the equation for N. As already mentioned, this only introduces extra parameters for the S-waves. The second assumption is that over an appreciable section of the near right-hand cut elastic unitarity is a good approximation: if this is the dynamically important part, the distant cuts being relatively unimportant, it should be a reasonable approximation to use elastic unitarity along the whole cut (the above-mentioned subtraction will again help). In this case the alternatives associated with the CDD classification are clear-cut, and the solution with no CDD poles is chosen as the "dynamical" solution by analogy with potential scattering. In the case of S-waves, a single adjustable parameter does however remain. The other basic idea of the calculations is the use of self-consistency criteria. A typical scheme would be to assume a set of right-hand cut discontinuities, calculate the left-hand cut discontinuities from them by crossing, and then to pass back from the left-hand to right-hand cuts via the N/D equations (with no CDD poles). If the output and input right-hand cuts are consistent, one has a solution of the equations, and ideally, one would hope to find only one such self-consistent solution, identical with the real world. This is the essence of the *bootstrap* idea, although in actual calculations the various stages are not always clearly distinguished from each other.

We turn now from these matters of motivation to discuss some specific calculations.

(a) Chew and Mandelstam's S-Wave Dominant Approximation

Chew and Mandelstam (1960) developed an approximation scheme in which all but the lowest partial wave absorptive parts were neglected, allowing the left-hand cut to be calculated explicitly by crossing using equation (9.1.11). In the original paper they retained S- and P-waves, but the inclusion of P-waves was strictly inconsistent, as will be discussed below. However, the inconsistency eluded their numerical calculation, which achieved very small P-waves. The end results were just as if the P-waves had been omitted, and we shall describe them as if that had been the case.

If only S-wave absorptive parts are retained on the right-hand cut ($4 \leqslant s < \infty$), the full amplitudes can be expressed in the following explicitly crossing symmetric form (Cini and Fubini, 1960):

$$T^I(s, t, u) = -\lambda^I + S^I(s) + \sum_{I'} [C_{II'}^{(st)} S^{I'}(t) + C_{II'}^{(su)} S^{I'}(u)] \qquad (9.2.1)$$

with $\lambda^I = 5\lambda$, 0, 2λ for $I = 0$, 1, 2, respectively, and

$$S^I(s) \equiv \frac{(s - s_c)}{\pi} \int_4^\infty ds' \frac{\operatorname{Im} T_0^I(s')}{(s' - s)(s' - s_c)}, \qquad I = 0, 2, \qquad (9.2.2)$$

where $s_c = \frac{4}{3}$ is the symmetry point and $S^1 \equiv 0$. Projecting out partial waves from (9.2.1) gives

$$\begin{aligned} T_0^I(s) = -\lambda^I &+ \frac{(s - s_c)}{\pi} \int_4^\infty ds' \frac{\operatorname{Im} T_0^I(s')}{(s' - s)(s' - s_c)} \\ &+ \frac{2}{\pi} \sum_{I'} C_{II'}^{(st)} \int_4^\infty dt \operatorname{Im} T_0^{I'}(t) \\ &\times \left[\frac{2}{s - 4} Q_0\left(1 + \frac{2t}{s - 4}\right) - \frac{1}{t - s_c} \right], \end{aligned} \qquad (9.2.3)$$

and for $l \neq 0$

$$T_l^I(s) = \frac{4}{\pi(s - 4)} \sum_{I'} C_{II'}^{(st)} \int_4^\infty dt \operatorname{Im} T_0^{I'}(t) Q_l\left(1 + \frac{2t}{s - 4}\right). \qquad (9.2.4)$$

The higher partial waves are thus given by the Froissart–Gribov integral equation (9.1.6) saturated by the S-wave absorptive parts. The S-wave equation (9.2.3) can also be rewritten in the form of a partial wave dispersion relation

$$T_0^I(s) = T_0^I(s_c) + \frac{(s - s_c)}{\pi} \left\{ \int_4^\infty ds' \frac{\operatorname{Im} T_0^I(s')}{(s' - s)(s' - s_c)} + \int_{-\infty}^0 ds' \frac{\operatorname{Im} T_0^I(s')}{(s' - s)(s' - s_c)} \right\}, \qquad (9.2.5)$$

with (cf. equation (9.1.11))

$$\operatorname{Im} T_0^I(s) = \frac{2}{s - 4} \sum_{I'} C_{II'}^{(st)} \int_4^{4-s} dt \operatorname{Im} T_0^{I'}(t), \qquad -\infty < s \leqslant 0, \quad (9.2.6)$$

and the subtraction constants $T_0^I(s_c)$ given in terms of λ^I by

$$T_0^I(s_c) = -\lambda^I + \frac{2}{\pi}\sum_{I'} C_{II'}^{(st)} \int_4^\infty dt \, \text{Im}\, T_0^{I'}(t)$$
$$\times\left[\frac{2}{s_c-4}Q_0\left(1+\frac{2t}{s_c-4}\right)-\frac{1}{t-s_c}\right]. \tag{9.2.7}$$

The subtraction constants are thus given in terms of the single constant λ ($\lambda^{0,2} = 5\lambda, 2\lambda$, respectively), plus a correction which can be calculated iteratively if λ is small. The final equations, (9.2.4)–(9.2.6), could have been derived directly from (9.1.2) and (9.1.11); the starting point (9.2.1) was adopted merely to ensure that the subtractions introduced into the S-waves do not spoil the exact crossing symmetry.

The (purely real) higher waves can be calculated directly from equation (9.2.4) once the S-waves are given, so it remains to solve the equations for the latter. This was done in the approximation of elastic unitarity ($R(\nu) = 1$ in equation (9.1.4)). Since the equations are subtracted, the N/D equations must be modified to introduce the subtraction constant; this was done by making a subtraction in the equation for N. The equations were solved numerically by an iterative method (Chew, Mandelstam and Noyes, 1960), both subtraction constants being determined in terms of the single free parameter λ by equation (9.2.7). For the range of λ considered ($-0{\cdot}46 < \lambda < 0{\cdot}30$), the solution approximately conforms to the Chew–Mandelstam formula

$$\left(\frac{\nu}{\nu+1}\right)^{\frac{1}{2}} \cot \delta_0^0 \approx \frac{1}{T_0^I(\nu_c)} + h(\nu) - h(\nu_c), \qquad \nu_c = -\tfrac{2}{3}, \tag{9.2.8}$$

which would be exact if $[T_0^I(\nu)]^{-1}$ had no left-hand cut. Here, $h(\nu)$ denotes the function whose discontinuity is $-\theta(\nu)[\nu/(\nu+1)]^{\frac{1}{2}}$, and which therefore naturally figures in effective range formulae:

$$h(\nu) \equiv -\frac{\nu}{\pi}\int_0^\infty d\nu' \frac{[\nu'/(\nu'+1)]^{\frac{1}{2}}}{\nu'(\nu'-\nu)},$$
$$= \frac{2}{\pi}\left(\frac{\nu}{\nu+1}\right)^{\frac{1}{2}} \ln\left[\sqrt{\nu}+\sqrt{\nu+1}\right], \qquad (\nu \geq 0),$$
$$= \frac{2}{\pi}\left(\frac{-\nu}{\nu+1}\right)^{\frac{1}{2}} \arctan\left[(\nu+1)/(-\nu)^{\frac{1}{2}}\right], \qquad (-1 \leq \nu \leq 0),$$
$$= \frac{1}{\pi}\left(\frac{\nu}{\nu+1}\right)^{\frac{1}{2}} \ln\left[\frac{1+(\nu+1)/(\nu)^{\frac{1}{2}}}{1-(\nu+1)/(\nu)^{\frac{1}{2}}}\right], \qquad (\nu < -1).$$

The higher partial waves were also solved for numerically using (9.2.4) and found to be small, as is of course required for consistency.

The above type of solution is aptly referred to as "S-wave dominant". The S-wave scattering lengths satisfy the approximate relation

$$2a_0^0 = 5a_0^2, \tag{9.2.9}$$

and for positive scattering lengths, the phase shifts rise quickly to rather small maxima so that the cross-section tends to be concentrated in the very low-energy region. (For the approximation (9.2.8) the maximum occurs for $\nu = \pi/2a_0^0$). However, while this type of solution is of interest as a theoretical possibility, and as an illustration of the ideas used in partial wave dynamics, it is not the solution chosen by nature. In particular, the existence of the rho-meson has been explicitly excluded from the start by the neglect of P-wave absorptive parts.

(b) *P*-wave Dominant Models: The Rho Bootstrap

The existence of a P-wave resonance is the dominant feature of low-energy $\pi\pi$ scattering, and it was a prime objective of early work to achieve the rho-resonance as a predicted dynamical entity in the sense of Section 9.1. Obviously its exchange, i.e. its contribution to the left-hand cut discontinuity, is a key thing to be taken into account for all partial waves. In the case of the P-wave itself, it was hoped that once a P-wave resonance was inserted on the left-hand cut, using crossing, it would generate itself on the right-hand cut via the N/D equations. In other words, rho-exchange would provide the force to bind the rho itself. This is the origin of the bootstrap idea which we defined earlier, and which we now discuss.

As already noted above, the original discussion (Chew and Mandelstam, 1960, 1961) sought to include S- and P-wave absorptive parts and no others. Just as for the case with S-waves alone, a fully crossing symmetric amplitude can be written down (Lovelace, 1961) (cf. equation (9.2.1)). It is

$$\begin{aligned} T^I(s,t,u) = {} & -\lambda^I + S^I(s) + 3(t-u)P^I(s) \\ & + \sum_{I'} C_{II'}^{(st)}[S^{I'}(t) + 3(s-u)P^{I'}(t)] \\ & + \sum_{I'} C_{II'}^{(su)}[S^{I'}(u) + 3(t-s)P^{I'}(u)], \end{aligned} \tag{9.2.10}$$

where $P^I(s)$ $(I=1)$ is a suitably subtracted dispersion integral over the P-wave absorptive part such that $P^I(s=4)=0$, and the other quantities are as previously defined. The Chew–Mandelstam equations result from projecting out partial waves. As in the S-wave dominant case, they are partial wave dispersion relations with the left-hand cut determined from equation (9.1.11) and the subtractions performed in such a way as to retain exact crossing, but now with absorptive parts retained for both $l'=0$ and 1. The S-waves contain subtraction constants as before, but the P-wave equation can be written in the form of an unsubtracted relation for $T_1^1(\nu)/\nu$, and so contains no arbitrary subtraction constant.

The above equations, together with unitarity, would seem at first sight to be an ideal basis for a bootstrap. Unfortunately, it has been shown that they are internally inconsistent in that they lead to unbounded partial wave real parts unless $\operatorname{Im} T_1^1(\nu)\equiv 0$ (Lovelace, 1961). This trouble is associated with the behaviour of the left-hand cut integral $B(\nu)$ defined in equation (9.1.20). If we substitute equation (9.1.11) we obtain

$$B_l^I(s)=\frac{4}{\pi(s-4)}\int_4^\infty dt\, Q_l\left(1+\frac{2t}{s-4}\right)$$
$$\times\sum_{I'} C_{II'}^{(st)}\sum_{l'}(2l'+1)P_{l'}\left(1+\frac{2s}{t-4}\right)\operatorname{Im} T_{l'}^{I'}(t). \qquad (9.2.11)$$

This is usually viewed as the contribution to the full amplitude of the assumed partial wave exchanges. It is of course a purely formal expression since the partial wave series has been substituted over the range $-\infty<s\leqslant 0$ in equation (9.1.20), whereas it only converges over a limited part of the range. However, in the Chew–Mandelstam approximation, $l'=0, 1$ only are retained, and this problem is ignored. If we now consider rho-exchange in the narrow width approximation, i.e.

$$\operatorname{Im} T_1^1(t)\approx\pi\frac{\Gamma_\rho m_\rho^2}{2q_\rho}\delta(t-m_\rho^2), \qquad (9.2.12)$$

and substitute into (9.2.11), we have

$$[B_l^I(s)]_\rho=\frac{12\Gamma_\rho m_\rho^2}{2q_\rho(s-4)}C_{I1}^{(st)}Q_l\left(1+\frac{2m_\rho^2}{s-4}\right)\left(1+\frac{s}{2q_\rho^2}\right), \qquad (9.2.13)$$

where $q_\rho^2=m_\rho^2/4-1$. We recall that the crossing matrix elements are $C_{I1}^{(st)}=\frac{1}{3}, \frac{1}{2}, -\frac{1}{6}$ for $I=0, 1, 2$, respectively, so that the pattern of forces coincides with that suggested by naïve interpretation of the empirical phase shifts (cf. Section 3.3), i.e. attractive in $I=0, 1$, repulsive in $I=2$. Unfortunately the optimism raised by this is promptly dashed on noting that one has the asymptotic behaviour $B_l^I(s)\sim c\ln s$ with c a constant (for the exchange of angular momentum l the behaviour is $s^{l-1}\ln s$). In general with this asymptotic ($\ln s$) behaviour, the N/D equations, in particular equation (9.1.19) for $N(\nu)$, are then of a specific non-Fredholm type. Solutions depend on two arbitrary constants, and have the distinctive property $\delta(\infty)\neq 0$, i.e. they do not satisfy Levinson's Theorem (9.1.22) (Dietz and Domokos, 1964; Morgan, 1965; Auberson and Wanders, 1965; Atkinson and Contogouris, 1965).

Furthermore, for the full Chew–Mandelstam equations, in which the $l'=0, 1$ contributions to equation (9.2.11) are exactly related to the right-hand cut discontinuities (with exact elastic unitarity) by crossing, it has been shown (Lovelace, 1961), as noted above, that there are no solutions, even of this type, unless $\text{Im}\ T_1^1(\nu)\equiv 0$.

The Chew–Mandelstam approximation to the far left-hand cut $-\infty < s \leqslant -28$ is based on using a truncated series outside its range of convergence, and so has no formal validity. The difficulties mentioned above show clearly that it is an unsatisfactory procedure unless $\text{Im}\ T_1^1(\nu)\equiv 0$. Another indication is given by comparing with the results obtained by saturating the Roy equations with just S- and P-wave absorptive parts (Basdevant, Le Guillou and Navelet, 1972). The Roy equations contain two subtraction constants, and the subtraction terms contain contributions linear in s. The Chew–Mandelstam equations are formally recovered if one demands that these linear subtraction terms be cancelled by corresponding contributions from the dispersion integrals. (This is necessary if one is to have an acceptable representation of the bounded quantity $T_l^I(s)$ for all $s \geqslant 4$. The Roy equations do not claim to do this, being valid only for $s \leqslant 60$). However, cancellation of the term linear in s implies the special sum rule (Basdevant, Le Guillou and Navelet, 1972)

$$2a_0^0 - 5a_0^2 = \frac{4}{\pi}\int_4^\infty \frac{ds'}{s'(s'-4)}[2\ \text{Im}\ T_0^0(s') + 27\ \text{Im}\ T_1^1(s') - 5\ \text{Im}\ T_0^2(s')], \tag{9.2.14}$$

which contradicts a sum rule following directly from saturating fixed-t dispersion relations with $l' \leqslant 1$:

$$2a_0^0 - 5a_0^2 = \frac{4}{\pi}\int_4^\infty \frac{ds'}{s'(s'-4)}[2\ \text{Im}\ T_0^0(s') + 3\ \text{Im}\ T_1^1(s') - 5\ \text{Im}\ T_0^2(s')]. \tag{9.2.15}$$

(This is the Olsson sum rule to be discussed in Chapter 10.) From these equations we again find that the Chew–Mandelstam approximation is only consistent if $\text{Im}\ T_1^1(s)\equiv 0$. Generalizations to any finite number of partial waves likewise fail.

We thus see that although the Chew–Mandelstam approximation hopefully gives an accurate representation of the near left-hand cut, the representation of the far left-hand cut is inadequate, and leads to serious problems. Practical bootstrap calculations usually try to avoid this difficulty by the use of a cut-off. There have been a number of schemes to implement this, either via approximations to the N/D method (e.g. Zachariasen, 1961), or more straightforwardly, by solving the exact equations with a cut-off. (e.g. Collins, 1966). The latter procedure means simply solving equations (9.1.17) and (9.1.18) with the integrals taken between $\nu = 0$ and $\nu = \nu_1$, a cut-off, instead of infinity. The resonance parameters implied by the output N and D functions are then compared to the input. To give some examples of specific calculations, Zachariasen (1961) achieved a self-consistent rho ($s_\rho^{\text{in}} = s_\rho^{\text{out}} = m_\rho^2$; $\Gamma_\rho^{\text{in}} = \Gamma_\rho^{\text{out}}$) with mass and width parameters $m_\rho = 350$ MeV, $\Gamma_\rho = 300$ MeV. Collins (1966) fixed the exchange rho-mass at its physical value and varied the input width to achieve the correct output rho-mass. For his choice of the cut-off parameter $\nu_1 = 49 (\sqrt{s_1} \approx 2$ GeV), the input width was required to be $[\Gamma_\rho]_{\text{in}} = 320$ MeV, and the output width came out much larger than this (already large) value. This trend of very large self-consistent widths compared to experiment was found to continue in certain enlargements of the scheme to include other channels (e.g. Fulco, Shaw and Wong, 1965).

It would seem that the exchange of the physical rho does not produce enough attraction to fulfil the bootstrap requirement, at any rate in simple schemes. In fact, Tryon (1972a, 1974b) has shown how to estimate moments of the left-hand cut discontinuity in terms of physical region data, and argued on the basis of these that the left cut discontinuity is too weak to generate a resonance using the N/D equations and elastic unitarity.‡ In this case, the rho-meson must appear in a one-channel calculation in association with a CDD pole.

This would account for the disappointing results of the above attempts to generate it dynamically. It would undermine attempts to carry on with this predictive dynamical programme by inserting more ingredients—specifically Regge poles and analyticity in l. These lead to the strip approximation to be discussed in Section (11.1). As we shall see, the resulting extension of the bootstrap idea does not appear to be successful.

In contrast to the dynamical considerations of partial waves discussed in this chapter, another line of development has been to use the partial wave equations merely as consistency conditions without commitment to the dynamical status of the various resonances appearing. This approach is considered in the following chapter.

‡ This conclusion reinforces an earlier argument (Lyth, 1971c) in relation to the Reggeized ρ bootstrap which is discussed in Section 11.1.

Chapter 10

Uses of Analyticity—Models and Phenomenology

10.1. Introduction

We now examine systematic applications to phenomenology and model building using the techniques developed in the two preceding chapters. The attempt to employ the constraints of S-matrix theory to achieve a complete dynamics is abandoned. Instead, experimental data are inserted into some more flexible framework which, nonetheless, retains good theoretical constraints. The same techniques are also used in building models with special theoretical boundary conditions, for example as suggested by current algebra or duality.

The constraints of analyticity, crossing and unitarity‡ (ACU) have an exceptionally dominant role in $\pi\pi$ phenomenology and model building below 1 GeV.§ This arises not only from the obvious theoretical circumstances of having a fully crossing symmetric system with small gaps between the physical regions (small on the scale of hadronic phenomena), but also from the circuitous way in which experimental parameters are extracted (see Chapter 3). Such basic experimental parameters as the scattering lengths are hard to obtain, unlike the situation with conventional hadron channels studied in formation; yet these parameters are of outstanding interest for comparisons with current algebra and dual models. The unique combination of theoretical simplicity and rather poor and incomplete data has dictated the strategy for doing $\pi\pi$ phenomenology. As compared say to the analysis of πN scattering, the methods may appear

‡ We here understand *elastic unitarity*.
§ We discuss applications to the analysis of πK scattering in Section 10.4.

rather roundabout. Furthermore, the task of implementing ACU, although well defined, admits a number of alternative procedures. As regards the dependability of the results, this is a source of strength; however it necessarily forces a survey such as this to be somewhat a catalogue of methods.

Since the position and width of the rho are the best determined experimental parameters, a major question is, given these, what are the possible behaviours of the S and other waves; what extra information is needed, in addition to ACU, to achieve unique predictions for all three I-spin states over the low-energy region.‡ By this latter phrase, we mean the energy region below the $K\bar{K}$ threshold (988 MeV), where inelasticity can be neglected. Uniqueness is of course only sought within the context of reasonable "smoothness" assumptions. For example, the possibility of resonances in the exotic $I=2$ channel is usually excluded, in accord with experiment. Likewise, additional unobserved narrow $I=0$ and 1 resonances are ruled out. More importantly, something has to be assumed about the contributions of distant singularities. These are associated (very directly in the case of fixed-t dispersion relations) with medium- and high-energy scattering amplitudes. That only very approximate estimates can be made for these is not so serious as might at first be supposed. This is because, for any reasonable estimate of these contributions, they only give rise to slowly varying effects over the low-energy region, and are very small in subtracted dispersion relations. Thus the associated uncertainty can be absorbed into a few constants, occurring as subtraction parameters or as coefficients in a slowly varying "background" contribution.

Hence one may hope that, within reasonable assumptions of this kind, the ACU constraints will lead to a well-determined set of amplitudes when only relatively few pieces of information are supplied in addition to the rho-mass and width. Indeed, on the basis of work with partial wave models using constraints in the Mandelstam triangle (cf. Section 10.2), it has even been suggested that just the rho-parameters and an estimate of charge exchange scattering at high energies was sufficient (Le Guillou *et al.*, 1971). This extreme predictability has not been borne out by other work (Morgan and Shaw, 1969, 1970; Tryon 1971a; Basdevant *et al.*, 1972a,b, 1974), and typically three or four parameters are needed in addition to m_ρ and Γ_ρ. Nonetheless, significant correlations remain. For example, given information on the $I=0$ S-wave, one can predict the $I=2$ S-wave. These correlations open the way to an out-and-out phenomenological approach (cf.

‡ Obviously, given exact information on a single I-spin state over a finite interval, all amplitudes can be generated over the whole stu-plane by analytic continuation using crossing and unitarity. Bowcock and John (1969) examined how far this could be translated into a practical programme by performing an analytic continuation from experimental information on the $I=1$ amplitude in the rho-region. Despite uncertainties which are instructively exposed in this approach, a broad ε effect in the $I=0$ S-wave was strongly indicated.

Section 10.3) in which all available experimental information is used, and the object is to infer quantities on which direct evidence is lacking—for example threshold parameters and D-wave phase shifts. This has proved rather successful. Given information on the $I=J=0$ and $I=J=1$ phases above 500 MeV, where data are plentiful, the D-waves can be predicted and the threshold region amplitudes determined, except for one parameter which remains to be fixed from experiment. This is important because, as already noted, the values of the threshold parameters are of great theoretical interest, yet direct information on them is difficult to obtain. In this respect, further K_{e4} experiments (cf. Section 5.1), which can fix just this one parameter, would be particularly valuable.

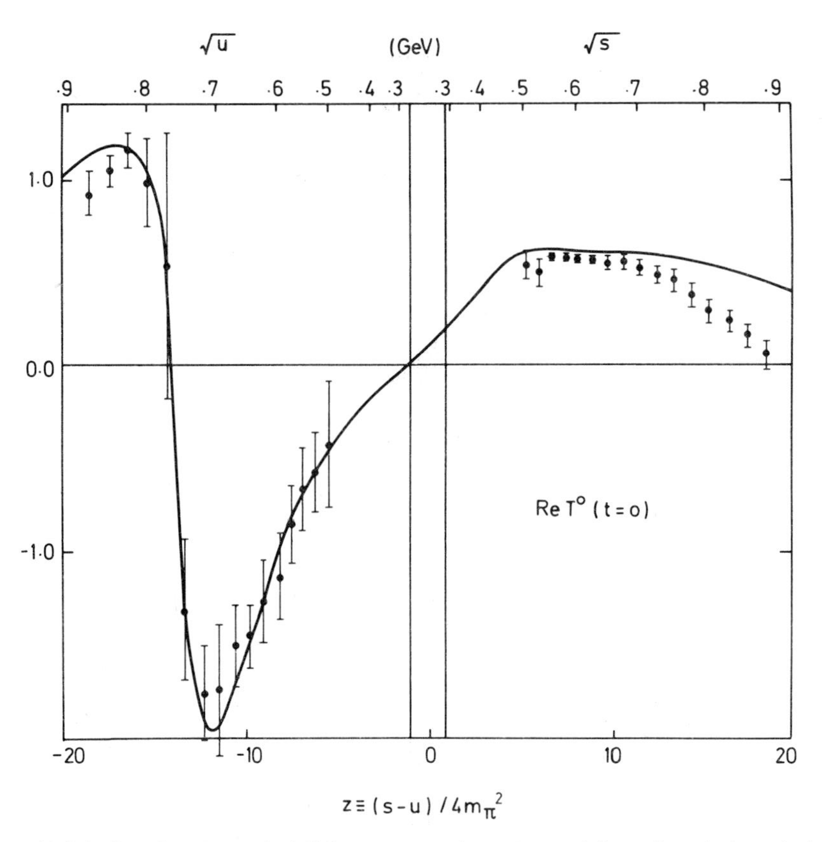

FIG. 10.1.1. Crossing plots—the full line corresponds to a forward dispersion relation solution ("BD1" for δ_0^0, cf. Section 10.3) from Morgan and Shaw (1970); the points with error bars are computed from Baton *et al.* (1970b). (a), (b), (c): Re $T^I(s, 0, 4-s)$ and its continuation $\sum_{I'} C_{II'}^{(su)}$ Re $T^{I'}(4-s, 0, s)$ for $s<0$ (for $I=1$ divided by q^2 (from Morgan (1972)).

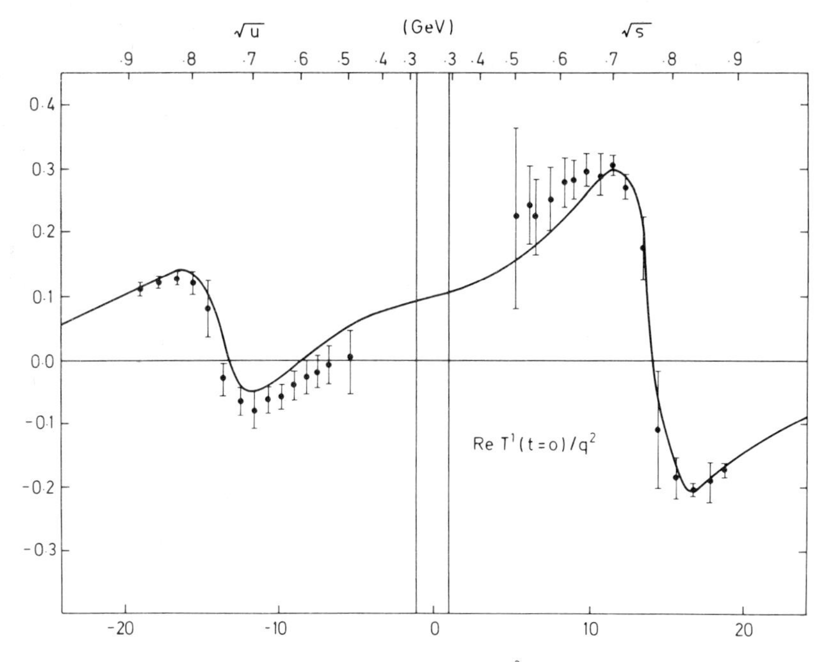

FIG. 10.1.1(b).

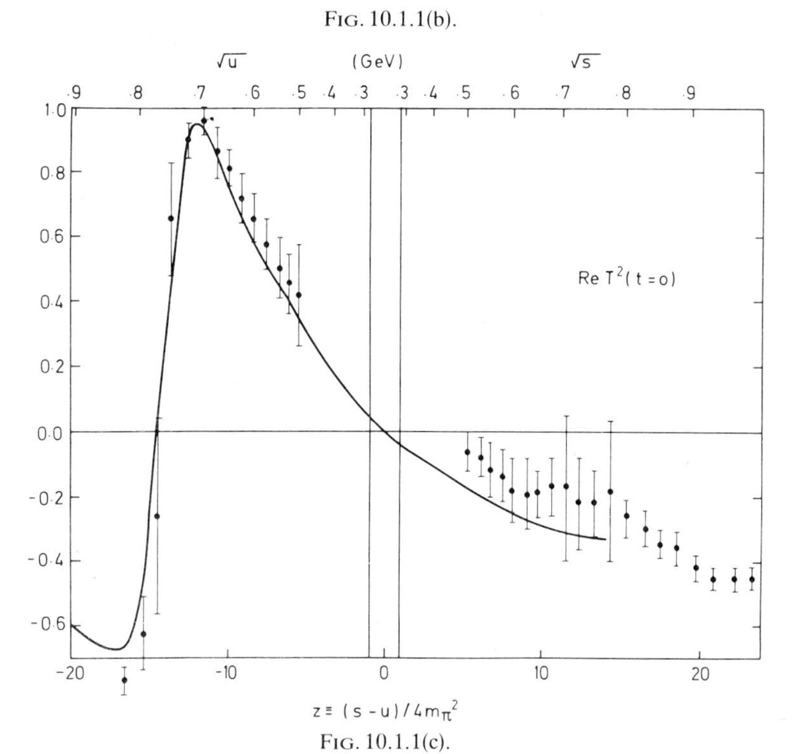

FIG. 10.1.1(c).

The residual freedom within an ACU framework, once m_ρ and Γ_ρ are fixed, can also be used to construct models which embody interesting theoretical predictions—for example from current algebra or the Veneziano model.

The three main aims of the work to be described in the present chapter may thus be summarized: Firstly, to explore the full range of possible solutions, given the parameters of the rho; secondly, to develop methods of embodying theoretically interesting predictions within an ACU framework (these will be further exploited in Chapters 12 and 14), and, finally, to effect phenomenology. This latter work is similar in spirit to the well-known use of dispersion relations to deduce values of the πNN coupling constant, scattering lengths, and small partial waves from πN scattering data.

We have already noted that pion–pion scattering is an especially suitable process for which to attempt this sort of programme. Detailed reasons are firstly, the empirical observation that only five partial waves ($T_0^{0,2}$, T_1^1, $T_2^{0,2}$) are required to describe the data for $W \leqslant 1$ GeV, and that, over this energy region (up to the $K\bar{K}$ threshold), inelasticity is negligible (cf. Chapter 3); elastic unitarity can therefore be used to good approximation over this whole region. Secondly, crossing is, as already noted, a key constraint for $\pi\pi$ scattering, not only because all three channels are identical, but also because the gap between the physical regions is small on the scale of hadronic phenomena. The power of crossing as a constraint on the low-energy $\pi\pi$ system is brought out visually in Figs (10.1.1 and 10.1.2). The quantity plotted in Fig. 10.1.1(a),(b),(c) (from Morgan (1972)) is the real part of the forward scattering amplitude Re $T^I(s, 0, 4\mu^2 - s)$, and its continuation $\sum_{I'} C_{II'}^{(su)} T^{I'}(4\mu^2 - s, 0, s)$ for $s < 0$, computed from the $I = 0$, 1 and 2 phase shifts of Baton *et al.* (1970b), and exhibited as a function of the crossing symmetric variable $z = (s - u)/4\mu^2$ (Re $T^1(z, t)$ is shown with the factor q^2 divided out). Crossing asserts that the results for $z < -1$ are an analytic continuation of those for $z > 1$. The curves are from an explicitly s–u crossing symmetric forward dispersion calculation (Morgan and Shaw, 1970) (see Section 10.3). To be specific, they correspond to the BD1 solution, with similar δ_0^0, δ_1^1, but somewhat different δ_0^2, δ_2^0 from those of Baton *et al.* The detailed behaviour of the interpolating curve reflects consequences of analyticity and unitarity, and assumptions on high-energy behaviour. The general qualitative structure is evidently necessitated by crossing—in particular the occurrence of zeros in the even I-spin amplitudes in the region of the direct and cross channel thresholds. The occurrence of these zeros causes effective range expansions to be unsuitable parameterizations for the S-wave phase shifts. A further point brought out by these diagrams is that, when continuing data from the rho-region down to threshold within some analytic model, one is, because of crossing, perform-

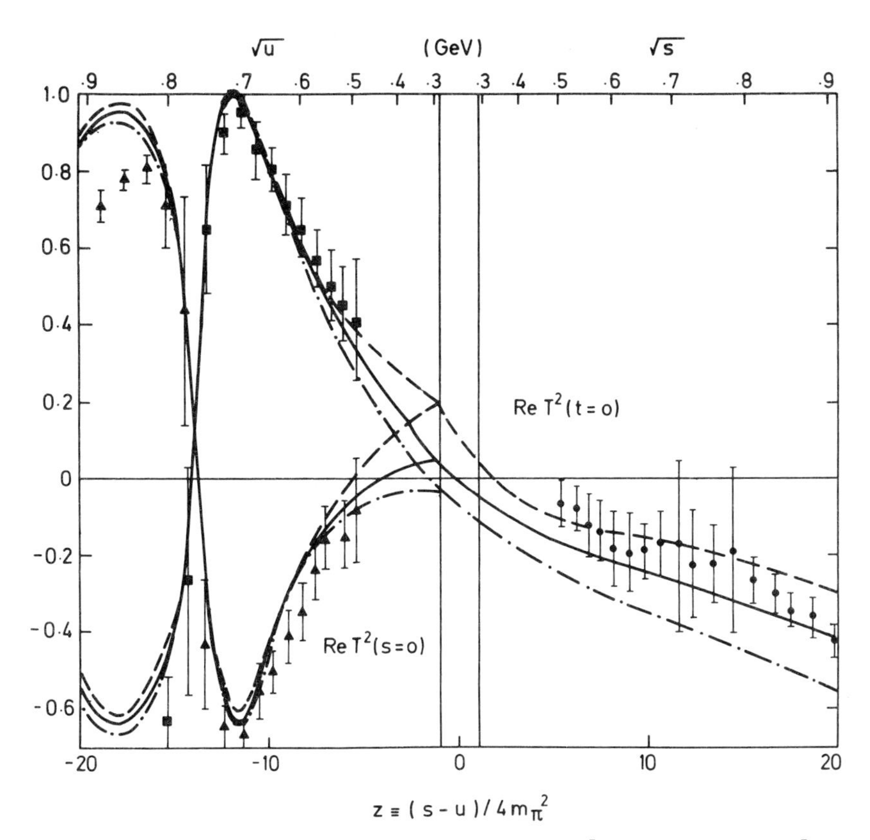

FIG. 10.1.2. Alternative crossing plot: Re $T^2(s, t=0, 4-s)$ (●), Re $T^2(4-u, 0, u)$ (■) and Re $T^2(0, 4-u, u)$ (▲), calculated from the same data sets of phase shifts given by Basdevant *et al.* (1972b), Solution 1 —·—·—, Solution 2 ———, Solution 3 — — — (see Section 10.3).

ing an interpolation rather than an extrapolation, with an attendant reduction in the uncertainty of the results. It should of course be stressed (and this will be a major theme of this chapter) that the data admit a number of alternative interpolations. As an illustration of this, Fig. 10.1.2 depicts Re $T^2(s, t=0, u=4-s)$, Re $T^2(s=4-u, t=0, u)$, and Re $T^2(s=0, t=4-u, u)$‡ calculated from the same data and for three specimen phase shifts given by Basdevant *et al.* (1972b) (to be discussed below).

From the above, it is clear that in the threshold region, crossing is the crucial constraint. As one increases the energy, it becomes less powerful

‡ This affords an alternative non-redundant way of displaying the crossing information entering Figs 10.1.1; i.e. given the above three quantities all the forward amplitudes $I=0$, 1 and 2 can be calculated in all channels.

owing to the longer (and hence more uncertain) analytic continuation required to connect the two reactions. On the other hand with phase shifts becoming large, it becomes more important to enforce unitarity exactly. Both requirements are thus necessary constraints in a significant way, so that one has to aim to enforce both simultaneously. It is here that the main technical problem of model-making lies. If one works directly with the partial wave amplitudes, unitarity is easy to enforce, but crossing presents difficulties. Conversely, if one works with the full amplitudes, crossing is easy to impose, but unitarity and the extraction of partial wave amplitudes, in terms of which data and results are usually presented, becomes complicated. These problems, and methods for solving them, have already been discussed in Sections 8.2 and 9.1. Another technical point to keep in mind is that the range of possible solutions compatible with ACU must necessarily be found by exploration rather than by some deductive method. Because of this, methods which base themselves on some form of parameterization or expansion are, as we shall see, in great danger of finding a spurious uniqueness. Open-ended computational procedures in which one attempts to construct solutions with given properties satisfying the constraints, and either succeeds or fails, are much more satisfactory in this respect. All this makes it valuable to explore and compare different methods of imposing ACU, even, and indeed especially, when the empirical properties aimed at are the same.

10.2. Some Analytic Models

In this section, we discuss several approaches for constructing analytic models of low-energy $\pi\pi$ scattering. As well as being of interest in their own right, they illustrate techniques used in "unitarizing" current algebra and Veneziano calculations, as discussed in Chapters 12 and 14. They are also useful for exploring the range of solutions possible in an ACU framework (given the parameters of the rho).

(a) Models using Crossing Symmetric Expansions

In the threshold region, one may hope to obtain a crossing symmetric representation of the amplitude by the simple device of expanding the amplitude A occurring in the A, B, C representation of (2.17) in terms of s, t, u as in equation (2.25). That is, displaying only the linear terms,

$$A = -\lambda + \tfrac{1}{4}\lambda_1(s - \tfrac{4}{3}). \tag{10.2.1}$$

If the expansion for A is constructed to satisfy $t \leftrightarrow u$ crossing (equation (2.24a)), and the B, C amplitudes are constructed using equation (2.24b,c), we have an exactly crossing symmetric amplitude. However, since such expansions are purely real, they should only be used in the triangle region $0 < s, t, u < 4$. They can be extended above threshold by expanding instead in terms of the variables

$$Q_s \equiv \tfrac{1}{2}(4-s)^{\frac{1}{2}} = -iq_s, \quad \text{etc.} \tag{10.2.2}$$

That is

$$A = x_1 + x_2 Q_s + x_3(Q_t + Q_u) + x_4 Q_s^2 + \cdots. \tag{10.2.3}$$

This introduces appropriate branch points at the elastic thresholds, and the coefficients can be adjusted so that unitarity is approximately satisfied.

Threshold region expansions of these types (and their off-mass shell extensions) find their most important application in current algebra calculations. Their use and properties are discussed in some detail in Chapter 14.

Such expansions, suitably augmented with resonance pole contributions, have also been applied far above threshold. A number of shortcomings are then exposed. Firstly, such representations fail to incorporate an essential singularity which necessarily occurs on the second sheet at $s = 0$ (Martin, 1967); however, this is a small effect occurring only in high partial waves ($l \geq 2$), thus probably not an important failing. More significantly, there are additional singularities associated with the second-sheet left-hand cut (for a simple discussion, see Morgan and Shaw (1972)). Consequently, when applied over the energy range $W \lesssim 1$ GeV, the curtailed series (plus resonance poles) is being used well outside its region of convergence, albeit with the most prominent singularities included. It is therefore a somewhat inflexible parameterization for use over such a large domain in s, t, u. (In contrast, the current algebra application only applies the series (without poles) over a modest region about threshold—say $0 \leq s, t, u < 5$.)

Of models which actually exploit this approach, the most interesting is perhaps that of Arbab and Donohue (1970). They used a crossing symmetric approximation for the $\pi\pi$ amplitudes up to 1 GeV consisting of two components—the first is a sum of ρ and f^0 pole contributions, the second an expansion in the channel momenta up to fifth power. Solutions were generated by requiring approximate unitarity over a set of mesh points, and by relating low-energy properties to assumptions on high-energy behaviour via finite energy sum rules.‡ Two sum rules each were used for $I_t = 2$ and 1 exchange with no high-energy contributions for the former and Reggeized rho-exchange for the latter. Thus, very stringent assumptions on the medium- and high-energy behaviour are made (especially in view of the low cut-off energy). The mass and width of the f^0 were fixed, and the rho-position constrained to lie near the experimental value. Under these

‡ Finite energy sum rules are discussed in detail in Chapter 12.

assumptions, a solution was found with $\Gamma_\rho = 80$ MeV and with S-waves qualitatively similar to those obtained from experiment, i.e. with the $I = 2$ wave small and negative and the $I = 0$ S-wave large and positive, rising slowly to 90° at about 1100 MeV. It is perhaps interesting to note that this behaviour is not associated with a second sheet pole. The scattering lengths obtained are small ($a_0^0 = 0{\cdot}11$, $a_0^2 = -0{\cdot}026$) with their ratio similar to that predicted by current algebra considerations. Provided the amplitudes are approximately linear in s, t, u in the triangle region, as turns out to be the case, this guarantees that the zero contours for $A(s, t, u)$ will be similar to those of Weinberg in the triangle region (cf. the discussion of Section 14.1), a fact noted and stressed by the authors.

Dilley (1971) has also considered a crossing symmetric expansion up to fifth powers in the channel momenta without additional terms. Unitarity is enforced over the more limited region $0 < \nu < 1$, and a variety of solutions constructed and classified according to the behaviours of the S-wave scattering lengths. For the given input form, unitarity is adequately satisfied for two families of solutions. The first is an S-dominant set with $2a_0^0 \approx 5a_0^2$. The second class has a_0^0 and a_0^2 of opposite sign, as appears to be the case experimentally, with, at least for small scattering lengths, a ratio in the neighbourhood of the Weinberg value (cf. Section 14.1). It is interesting that such a value is selected by this simple formalism; however it is unclear what would happen if more parameters were allowed. Dilley's continuum of solutions embracing the null solution affords a nice illustration of the general results of Atkinson (1968a,b, 1969, 1970) and Kupsch (1969a,b, 1970a,b) concerning the range of solutions admitted by ACU.

(b) Partial Wave Models using the BNR and Martin Conditions

We here consider models in which analyticity and unitarity are explicitly embodied in parametric forms for the partial wave amplitudes. Crossing is then introduced by adjusting the parameters so that the BNR relations and Martin inequalities involving the low partial waves ($l \leq 1$ or 2) are satisfied. (Relevant examples are discussed in Chapter 7.) This approach was initiated by Auberson *et al.* (1968), and since their work a succession of models has appeared using increasingly flexible parameterizations. For example, Le Guillou *et al.* (1971) use eleven parameters to approximate the S- and P-waves. In addition to m_ρ and Γ_ρ (which are essentially fixed because of the requirement that the experimental P-wave should be reproduced over the range $500 \leq W_s \leq 1100$ MeV), two further parameters are used for the P-wave. Of these, one controls the energy dependence of the rho-width and so fixes a_1^1, and the other fixes the strength of the rho-exchange contribution to $T_1^1(s = 0)$. For each reasonable choice of these P-wave parameters, five of the S-wave parameters are fixed by using the five crossing sum rules equations (7.22)–(7.26) and the remaining two parameters are found to be tightly constrained by the rigorous inequalities, equations (7.31)–(7.39), following from positivity. Le Guillou *et al.* give a "central" solution with the low-energy parameters

$$a_0^0 = 0{\cdot}206, \qquad a_0^2 = -0{\cdot}073, \qquad a_1^1 = 0{\cdot}045,$$

but a range of values is allowed. However, all solutions have a δ_0^0 behaviour of the "Between–Down" or "Up–Down" type, the phase rising towards, but not passing through 90°. In contrast to the model of Arbab and Donohue (1970) discussed above, this behaviour *is* in this case associated with a second sheet pole, its position corresponding to the parameters $M_\varepsilon = 420$ MeV, $\Gamma_\varepsilon = 380$ MeV for the central solution. The authors claimed that this S-wave behaviour, including the selection of the "Down" branch above the rho, was a consequence of the rigorous conditions applied, and was essentially independent of the parameterization. This claim has not been sustained by subsequent work, as we shall see.

An alternative analysis of this type has been performed by Bonnier and Gauron (1970, 1972). In the first of these papers, S- and P-waves only were considered, with three parameters for each S-wave, and one (a_1^1) for the P-wave in addition to m_ρ, Γ_ρ. A special feature is that the partial wave absorptive parts are required to have the correct behaviour at the crossed threshold, $s = 0$.‡ This work led to three distinct classes of solutions. In all three the $I = 2$ S-wave is unsatisfactory, becoming positive in the rho-region. In the second paper, the S-wave parameterization was improved, and the $I = 0$ and 2 D-waves included with four parameters each. With D-waves explicitly allowed for there is a total of fifteen BNR relations, so that the problem is determined without recourse to the inequalities. The authors then obtained a narrow band of solutions of the Between–Up type—i.e. δ_0^0 rises towards 180° above the rho, corresponding to the parameters $m_\varepsilon = 700$ MeV, $\Gamma_\varepsilon = 260$ MeV. The following values are quoted for the threshold parameters (Bonnier and Gauron, 1972):

$$a_0^0 = 0{\cdot}22 \pm 0{\cdot}04, \qquad a_0^2 = -0{\cdot}10 \pm 0{\cdot}02, \qquad a_1^1 = 0{\cdot}052 \pm 0{\cdot}012.$$

These values are similar to those of Le Guillou *et al.* (1971) and of Kang *et al.* (1971) who use similar techniques to "unitarize" the Veneziano model, as discussed in Section 12.3(e). The results are also similar to those of Arbab and Donohue (1970) quoted above.

‡ For example, if Im $T_l^I(s)$ is expanded as a power series in s about $s = 0$, the first term vanishes, and there are conditions on the higher coefficients. We refer to Bonnier and Gauron (1970, 1972) for further details.

Bonnier and Gauron (1972) were less optimistic than Le Guillou *et al.* concerning the independence of choice of parameterization of their results above the rho, but believed that they had at least achieved a model independent description of the threshold region. Even this hope has not been confirmed by subsequent study. For example, Piguet and Wanders (1972) have carefully discussed the limitation set on the physical amplitudes by the usual specific constraints in the Mandelstam triangle in the context of quasi-linear models. (That is, ones in which the S- and P-waves are almost linear in the Mandelstam triangle.) They find that if the rho is imposed *and* the behaviour of the $I=2$ S-wave fixed, then the $I=0$ S-wave is strongly restricted from threshold up to about 450 MeV, but thereafter a considerable freedom is allowed, so that by the rho-region there is no significant restriction. This finding is in accord with intuition—crossing is powerful near threshold, but weakens in predictive power as one moves away from the Mandelstam triangle. We shall also anticipate results to be discussed later by noting that Tryon (1971a) (cf. Section 12.3(e), and also 10.2(c) below) and Basdevant *et al.* (1972a,b, 1974) (cf. Section 10.3) have constructed explicit counter-examples to the above uniqueness claim (i.e. amplitudes which satisfy the triangle region crossing constraints with scattering lengths different from the above values), anticipating and confirming the view of Piguet and Wanders summarized above.

We have been stressing how a subset of the BNR and Martin relations can be over-selective when used in conjunction with too restrictive a parameterization. One should however recall that a truncated series of relations is really *underconstraining* (a set of necessary but not sufficient conditions for securing crossing and positivity). As we shall see below, the "unique" solutions actually fail to satisfy the Roy equations (see Section 8.2(b)).

(c) Models using Partial Wave Dispersion Relations

Partial wave dispersion relations figured prominently in the bootstrap program (cf. Section 9.2). In that context, subtractions were, in keeping with the bootstrap philosophy, kept to a minimum. For model building, one wants more flexibility. As a first example we consider the work of Tryon (1968, 1969b) who used twice-subtracted dispersion relations for S-, P- and D-waves. This introduces two parameters for each S-wave and one for the P-waves; (subtraction constants in higher waves are eliminated using the ν^l threshold factors). Three of these subtraction constants are eliminated by saturating the crossing conditions at the symmetry point, equation (9.1.13), by just S-, P- and D-waves, leaving two free parameters which are taken to be the S-wave scattering lengths. Crossing is also used to calculate the left-hand cut discontinuities, using the expansion equation (9.1.11). This expansion converges over the range $s > -32$ ($\nu > -9$); in the calculation, however, it is terminated at $l = 2$ and used to estimate the whole left-hand cut. As the relations are heavily subtracted, the approximation involved in estimating the far left-hand cut in this way may not be too serious, and the estimation of the near left-hand cut should be accurate.

Having set up these equations, which should guarantee crossing to a good approximation, Tryon goes on to construct simultaneous solutions to them and the unitarity equations. Solutions were investigated with $M_\rho = 768$ MeV, and Γ_ρ values in the range 120 ± 30 MeV. The search revealed large numbers of solutions for each choice of the free parameters a_0^0, a_0^2; for example, for the values $a_0^0 = 0{\cdot}23$, $a_0^2 = -0{\cdot}03$, the $I = 0$ S-wave could be either resonant or non-resonant. However, the different solutions are very similar over the low-energy region $W \lesssim 450$ MeV; with the above scattering lengths, and the information that δ_0^0 does not reach 90° below 740 MeV, the following range of values at $W = m_K = 494$ MeV is obtained

$$\delta_0^0(m_K) = 38 \pm 7°, \qquad \delta_0^2(m_K) = -8{\cdot}5 \pm 2{\cdot}0°.$$

The scattering length values imposed above are of the type suggested by current algebra considerations. Another partial wave model which enforces these particular scattering lengths is that of Kang and Lee (1971). These authors use once-subtracted dispersion relations for S- and P-waves with the subtraction constants adjusted to fit current algebra predictions. Solutions are generated by the N/D method with the left-hand cut approximated by poles, the parameters of which are determined by using the crossing sum rules, equations (7.22)–(7.26). The S-wave phase-shifts predicted are in qualitative agreement with experiment, and a P-wave resonance is predicted at about the rho-mass. However, the latter has a lower half width of 150 MeV, and the phase shift never reaches 135°, falling slowly to zero immediately after passing through resonance. Furthermore, Tryon (1972b) has shown that the solution grossly violates both crossing for the absorptive parts and a simple forward sum rule, so that it cannot be regarded as a satisfactory solution to the ACU constraints.

Tryon (1971a,d) has also constructed a model by applying the techniques of partial wave dispersion relations to the difference of the partial wave amplitudes, and their values as calculated from the simple un-unitarized Veneziano model. This calculation is discussed in detail in Section 12.3(e). Its interest for us here lies in the fact that to the 2% accuracy to which the equations used are numerically satisfied, unitarity and all the crossing conditions in the Mandelstam triangle are satisfied. Nonetheless, for a given m_ρ, Γ_ρ a large number of solutions are found, in contradiction to the "uniqueness" claim of LeGuillou *et al.* (1971) discussed above. In fact,

Tryon found that, given m_ρ and Γ_ρ, it was necessary to fix three further parameters, for example the values at which δ_0^0 passes through 90° and 135°, and the parameter a_0^0 (or λ). All remaining aspects of the amplitudes, in particular the $I=2$ S-wave, are then determined. Also, once a_0^0 is fixed, the dependence of a_0^2 on the other parameters (excepting m_ρ and Γ_ρ) is rather slight, so that, as a_0^0 (or λ) is varied, a band of solutions is found in the (a_0^0, a_0^2) plane. This reproduces the "universal curve" found earlier in calculations using fixed-t dispersion relations ((Morgan and Shaw, 1970); cf. Section 10.3).

(d) Inverse Partial Wave Amplitudes

An alternative approach to that outlined in the preceding sections is to write dispersion relations for the inverse of the partial wave amplitudes (Moffat, 1961). These have the same cut structure as the partial wave amplitudes themselves, and the right-hand cut integral is calculated very easily, because of the simple form taken by unitarity for the inverse amplitudes within the elastic region (cf. equation (9.1.15)). In addition to the cuts, there is the possibility of poles arising from zeros of the original partial wave amplitudes. It is in this way that the CDD ambiguity found in the N/D method (Section 9.1) enters the problem—adding a pole to the D-function is equivalent to adding a pole to the inverse amplitude. The most extensive analysis along these lines is that of Gore (1972)‡ which discusses partial wave dispersion relations for the inverse S- and P-wave amplitudes. For each S-wave, Gore allows one subtraction constant and one pole term; the inverse P-wave amplitude $\nu T_1^1(\nu)^{-1}$ is twice subtracted, giving eight parameters in all. The number of poles (i.e. zeros of T_l^I) is chosen so that the zero pattern suggested by current algebra for the region of the Mandelstam triangle is a possible one. This choice restricts the allowed solutions somewhat. In particular, δ_0^0 is not allowed to rise through 180° above the rho (which would mean introducing another zero); the allowed behaviour for δ_0^0 above the rho is thereby somewhat restricted in that the solution cannot head towards 180°. However, this should have little effect in the lower energy region. Of the eight parameters, two are fixed from the rho-characteristics m_ρ and Γ_ρ, and a further four from approximate crossing conditions at the symmetry point. Three of these are the conditions which we discussed in the previous chapter (9.1.14), and the fourth is a second-derivative condition from which D-waves, assumed to retain their threshold behaviour out to the symmetry point, are eliminated. Sometimes, an analogous third derivative condition, with two-parameter D-waves and one-parameter F-waves eliminated, is

‡ Earlier references can be found in this paper. See also the work of Johnson (1972) and references therein.

used to fix a further parameter; in other cases, this is fixed by requiring an epsilon resonance of mass 745 MeV, or by imposing the value of δ_0^2 (to be $-10°$, $-15°$ or $-20°$) at the rho-mass. In each case, a range of solutions is generated as a function of the single remaining parameter λ (cf. equation (2.25)). For each alternative procedure, the "universal curve" for a_0^0, a_0^2 (discussed in Section 10.3 below, (cf. Fig. 10.3.3)), is reproduced. The range of solutions is not so great as that given by Tryon (1971a), but, as a somewhat more restricted parameterization was used, this is not surprising. Finally, we note that Gore has also used his method to examine S-wave dominant solutions of the type discussed in Chapter 9.

10.3. Dispersion Relation Phenomenology

In this section, we consider some calculations in which ACU methods are used as tools in the analysis of experimental data.‡ A principal objective is to deduce information on the threshold parameters (scattering lengths, etc.) from data at somewhat higher energies, say $W \geq 500$ MeV. Methods using partial wave dispersion relations, such as we have just been discussing, could very well be used for this task. However, it so happens that almost all phenomenology has been based on analytic properties at fixed t (or s, or u), and it is on these that we shall concentrate. These methods, as well as furnishing extrapolations to threshold, also provide an alternative framework to that discussed in the preceding section for exploring the range of possible solutions compatible with minimal experimental assumptions (m_ρ, Γ_ρ, plus high-energy behaviour). For this latter purpose, the fixed-t techniques have one clear advantage over partial wave methods. This concerns the uncertainties in low-energy amplitudes due to contributions of distant singularities. For partial wave dispersion relations, such contributions are difficult to estimate. For fixed-t relations they are simply given in terms of the medium- and high-energy scattering amplitudes (at fixed t); thus estimates of their importance can easily be tested against reasonable models for high-energy scattering. A second, operational, advantage is that crossing is more easily incorporated than in partial wave models. A potential disadvantage is that partial waves must still be extracted in the elastic region in order conveniently to impose unitarity. Two methods for achieving this have already been discussed in Section 8.2: the use of fixed-t relations and their derivatives at $t = 0$; and the later and more systematic approach via the Roy equations. We shall discuss applications of both these methods in turn below after first considering the simpler question of $t = 0$ sum rules.

‡ By "experimental data" we mean on-shell $\pi\pi$ phase shifts, inferred from dipion production data (cf. Chapter 3).

(a) Forward Sum Rules

An obvious approach for estimating threshold parameters is to seek out rapidly convergent sum rules for them. For example, if we write a forward dispersion relation for $T^1(\nu, t=0)/\nu$ (exploiting the P-wave threshold factor), we have

$$\frac{T^1(\nu, t=0)}{\nu}=\frac{1}{\pi}\int_0^\infty d\nu' \frac{\text{Im } T^1(\nu', t=0)}{\nu'(\nu'-\nu)}-\sum_I \frac{C_{1I}^{(su)}}{\pi}\int_0^\infty d\nu' \frac{\text{Im } T^I(\nu', t=0)}{(\nu'+\nu+1)(\nu'+1)}. \tag{10.3.1}$$

Evaluating this at $\nu=0$, we obtain the Olsson sum rule (Olsson, 1967):

$$3a_1^1=\frac{1}{\pi}\int_0^\infty d\nu\left\{\frac{\text{Im } T^1(\nu, t=0)}{\nu^2}-\sum_I C_{1I}^{(su)}\frac{\text{Im } T^I(\nu, t=0)}{(\nu+1)^2}\right\}. \tag{10.3.2}$$

Alternatively, evaluating it at $\nu=-1$ we obtain

$$\begin{aligned}\frac{2a_0^0-5a_0^2}{6}&=\frac{1}{\pi}\int_0^\infty \frac{d\nu}{\nu(\nu+1)}\left\{\text{Im } T^1(\nu, t=0)-\sum_I C_{1I}^{(su)} \text{Im } T^I(\nu, t=0)\right\}\\&=\frac{1}{\pi}\int_0^\infty \frac{d\nu}{\nu(\nu+1)}\sum_I C_{1I}^{(st)} \text{Im } T^I(\nu, t=0).\end{aligned} \tag{10.3.3}$$

This is just the sum rule obtained by evaluating the forward dispersion relation for $I_t=1$ exchange at threshold. According to conventional Regge phenomenology (cf. Appendix C), the integrands of both sum rules converge as $\nu^{-\frac{3}{2}}$, and, if the difference between them is taken, a very rapidly convergent sum rule for the quantity $(2a_0^0-5a_0^2-18a_1^1)$ is obtained (Wanders, 1966). Early attempts to evaluate these sum rules were made by Olsson (1967) and Meiere and Sugawara (1967). The best results come from equation (10.3.2), from which Olsson estimated

$$a_1^1=0{\cdot}040\pm0{\cdot}005. \tag{10.3.4}$$

Of this result, about 50% came from the rho-contribution, 25% from the low-energy S-waves and 25% from medium- and high-energy contributions. In view of the very large uncertainty concerning the latter, the error estimate is perhaps somewhat overoptimistic, within the terms of the calculation. However, the central value is upheld by subsequent calculations (cf. Table 10.3.1). One may also attempt to apply a threshold sum rule for $I_t=2$ exchange, i.e.

$$\frac{2a_0^0+a_0^2}{6}=\sum_I C_{2I}^{(st)}\frac{1}{\pi}\int_0^\infty d\nu\frac{(2\nu+1)\text{ Im } T^I(\nu, t=0)}{\nu(\nu+1)}, \tag{10.3.5}$$

but even the convergence of this integral is unclear, since in conventional

TABLE 10.3.1. Scattering Lengths from the Analyses of: Morgan and Shaw (1970) (MS); Basdevant *et al.* (1972b) (BFP)[a]; Pennington and Protopopescu (1973) (PP).

Solution	a_0^0	a_0^2	a_1^1	a_2^0	a_2^2	Comment
MS-"Favoured"	0·16 ±0·04	−0·05 ±0·01	0·035 ±0·002	0·0016 ±0·0002	0·0003 ±0·0002	$a_0^0/a_0^2 = -3{\cdot}2 \pm 1{\cdot}0$ imposed
MS-"DU 1"	−0·21	−0·17	0·031	0·0014	0·0001	
BFP I	−0·05	−0·115 ±0·005	0·034 ±0·002	0·0014 ±0·0001	$(-2{\cdot}5 \pm 1{\cdot}0)10^{-4}$	—
BFP II	0·16	−0·048 ±0·003	0·035 ±0·002	0·0014 ±0·0002	$(-1{\cdot}5 \pm 1{\cdot}0)10^{-4}$	—
BFP III	0·60	0·043 ±0·004	0·041 ±0·002	0·0018 ±0·0002	$(5{\cdot}3 \pm 2{\cdot}0)10^{-4}$	—
PP	0·15 ±0·07	−0·053 ±0·028	0·036 ±0·002	0·0017 ±0·0001	$(0{\cdot}7 \pm 3{\cdot}0)10^{-5}$	—

[a] Fit to δ_0^0 phases of Baton *et al.* (1970b); other inputs are discussed in Basdevant *et al.* (1974).

Regge theory, the leading singularity gives, ignoring log terms, a $\nu^{2\alpha_\rho(0)-1}$ behaviour at high energies, with $\alpha_\rho(0) \approx 0{\cdot}5$. In fact, it turns out (Tryon, 1969a) that the sum rule converges, if at all, much too slowly to be useful, unless some very specific model for the behaviour above the region of the g (1690) meson is adopted.

One conclusion to be drawn from these sum rules is that in more comprehensive schemes which embody them, such as those to be described below, the parameters a_1^1, and to a somewhat lesser extent $2a_0^0 - 5a_0^2$, are likely to be much better determined that the other S-wave combination $2a_0^0 + a_0^2$. The greater uncertainty in the last-mentioned quantity feeds into the determination of the individual S-wave scattering lengths.

(b) Forward and Derivative Dispersion Relations

The most extensive application of forward relations to $\pi\pi$ phenomenology is that of Morgan and Shaw (1969, 1970) on which we shall concentrate.‡ These authors adopted the approach to extracting partial waves and imposing elastic unitarity discussed in Section 8.2(a). Thus they wrote forward dispersion relations for the quantities

$$T_t^I(z) \equiv T_t^I(z, t), \qquad D_t^I(z) \equiv \left[\frac{\partial}{\partial t} T_t^I(z, t)\right]_{t=0}. \tag{10.3.6}$$

Assuming that only S-, P-, D- and F-waves contribute over the elastic region, knowledge of the above quantities is sufficient to extract values of all the partial waves, on which unitarity can then be imposed (cf. Section 8.2). The dispersion relations were evaluated over the range $0 \leqslant z \leqslant z_\rho$, where z_ρ corresponds to the rho-mass, and over this range were written in the approximate form

$$T_t^{0,2}(z) = \frac{2}{\pi}\int_1^{z_2} dz' \frac{z' \operatorname{Im} T_t^{0,2}(z')}{z'^2 - z^2} + e_0^{0,2} + e_2^{0,2}\left(\frac{z}{z_\rho}\right)^2, \tag{10.3.7a}$$

$$T_t^1(z) = \frac{2z}{\pi}\int_1^{z_2} dz' \frac{\operatorname{Im} T_t^1(z')}{z'^2 - z^2} + e_1^1\left(\frac{z}{z_\rho}\right) + e_3^1\left(\frac{z}{z_\rho}\right)^3, \tag{10.3.7b}$$

where z_2 is a cut-off taken above the f^0 meson. For the D_t^I of (10.3.6) which must be considered if waves with $l = 2$ and 3 are to be included, the similar relations

$$D_t^{0,2}(z) = \frac{2}{\pi}\int_1^{z_2} dz' \frac{z' \operatorname{Im} D_t^{0,2}(z')}{z'^2 - z^2} + f_0^{0,2}, \tag{10.3.8a}$$

‡ For earlier references, see Morgan and Pišút (1970).

$$D_t^1(z) = \frac{2z}{\pi}\int_1^{z_2} dz' \frac{\operatorname{Im} D_t^1(z')}{z'^2 - z^2} + f_1^1\left(\frac{z}{z_\rho}\right), \tag{10.3.8b}$$

were used. The structure allowed to the subtraction terms in the parameters e_i^j embodies the "smoothness" assumptions concerning high-energy contributions. On comparing, for example, equation (10.3.7a) with a subtracted dispersion relation, we find

$$e_0^{0,2} = T_t^{0,2}(z=0) - \frac{2}{\pi}\int_1^{z_2} dz' \frac{\operatorname{Im} T_t^{0,2}(z')}{z'}, \tag{10.3.9}$$

and

$$\frac{2z^2}{\pi}\int_{z_2}^{\infty} dz' \frac{\operatorname{Im} T_t^{0,2}(z')}{z'(z'^2 - z^2)} = e_2^{0,2}\left(\frac{z}{z_\rho}\right)^2 + e_4^{0,2}\left(\frac{z}{z_\rho}\right)^4 + \cdots. \tag{10.3.10}$$

Note that the approximation $e_4^{0,2} = 0$ etc. has been made in equation (10.3.7) over the range $0 \leq z \leq z_\rho$. Thus, while the magnitude of the high-energy contributions is left unrestrained, their energy dependence over this limited region is restricted. To check on this, one can estimate values of e_4^0, e_4^2 etc. in reasonable models, noting that the amplitudes to which they contribute are roughly of order unity. For example, on considering the $I_t = 0$ case, and using the explicit Regge pole model of Chiu *et al.* (1968) to evaluate the contributions from above 1·5 GeV, values of 0·04 and 4×10^{-4} respectively were obtained for e_2^0 and e_4^0. The g (1960) meson (assuming $\Gamma_{g\to 2\pi} = 120$ MeV) gave 0·04 and 2×10^{-3}; if the f^0 (1270) had not been included explicitly but parameterized in the above manner, extra contributions to e_2^0 and e_4^0 of magnitude respectively 0·04 and 5×10^{-4} would have been required. Thus the neglect of e_4^0 and higher terms is indeed a reasonable method of imposing the "smoothness" requirement on the high-energy contributions. Note however that this has been done at the price of introducing the nine extra parameters e_i^j, f_i^j. Three of them can be eliminated by imposing the threshold conditions which follow from Bose statistics (equivalently from t–u crossing)

$$T_s^1(z=1, t=0) = 0 = \frac{\partial T_s^{0,2}}{\partial t}(z=0, t)\big|_{t=0}. \tag{10.3.11}$$

To fix the other six parameters, it was demanded that a number of predetermined characteristics of the amplitudes at $z = z_\rho$ should be reproduced—the mass and width of the rho-meson, the values of $\delta_0^{0,2}(z = z_\rho)$, and the rate of change of $\delta_0^0(z)$ at $z = z_\rho$. Provided that F-wave contributions to T_s^1 can be neglected, these are sufficient to determine all the parameters. The authors point out that the above adjustments secure an important measure of s–t crossing, at least over the Mandelstam triangle. Since s–u crossing is explicitly imposed from the beginning, a fair measure of full s–t–u crossing is achieved.

The last step is to impose unitarity on the solutions of equations (10.3.7) and (10.3.8) selected by requiring the above attributes in the rho-region. This was done iteratively, starting with an explicit form for the imaginary parts with the appropriate rho-region properties. From this, real parts were calculated using equations (10.3.7) and (10.3.8) with the subtraction constants adjusted to secure the required behaviour in the rho-region. New imaginary parts were then constructed using unitarity, and the procedure cycled until adequate consistency was achieved. The equations were solved in this way up to the rho-region and the amplitudes kept fixed at their input values above this. The join between the two regions is smooth since the values and (for the large waves) the derivatives of the amplitudes at $z = z_\rho$ are, by construction, unchanged from the input values. Output phase shifts were constructed by identifying

$$\mathrm{Re}\,\{[(\mathrm{Re}\,T_l^I)^{\mathrm{out}} + i(\mathrm{Im}\,T_l^I)^{\mathrm{out}}]^{-1}\} \quad \text{with} \quad \left(\frac{\nu}{\nu+1}\right)^{\frac{1}{2}} \cot \delta_l^I(\nu).$$

The resulting solutions were only retained if they satisfied the dispersion relations to within one or two per cent up to about 900 MeV.

In this way, it proved possible to construct a large number of solutions (corresponding to different input behaviours in the rho-region) which satisfied unitarity and the dispersion relations up to 900 MeV, and had exact s–u crossing, and approximate s–t crossing over the triangle region. The authors did not attempt to investigate the space of all possible solutions, but restricted themselves to candidates of phenomenological interest at that time. For the better determined input parameters, the values $m_\rho = 765$ MeV, $\Gamma_\rho = 120 \pm 20$ MeV and $\delta_0^2(m_\rho^2) = -15 \pm 5°$ were employed. The main variation from case to case concerned the behaviour of δ_0^0 which was, in each instance, specified as an input over the whole rho-region ($600 \leqslant W \leqslant 900$ MeV). In the first paper (Morgan and Shaw, 1969), a range of fairly symmetrical epsilon resonances was considered with $600 \leqslant m_\varepsilon \leqslant 900$ MeV, $200 \leqq \Gamma_\varepsilon \leqslant 800$ MeV. In the second paper, a variety of solutions compatible with alternative determinations of Scharenguivel *et al.* (1969) was examined. Some examples are shown in Figs 10.3.1(a),(b). The nomenclature used (Down–Up, Between–Down etc.) is based on the alternatives of Scharenguivel *et al.* (also shown) in an obvious manner. The predicted form of δ_0^2 corresponding to the Between–Down type of solution is shown in Fig. 10.3.2, which brings out the fact that, below 600 MeV, the behaviour is fairly independent of the value of $\delta_0^2(m_\rho^2)$ ($-20° \leqslant \delta_0^2(m_\rho) \leqslant -10°$) chosen.

The output of the forward dispersion relations consists essentially of the low-energy behaviour of S- and P-waves, predictions for the D-waves and

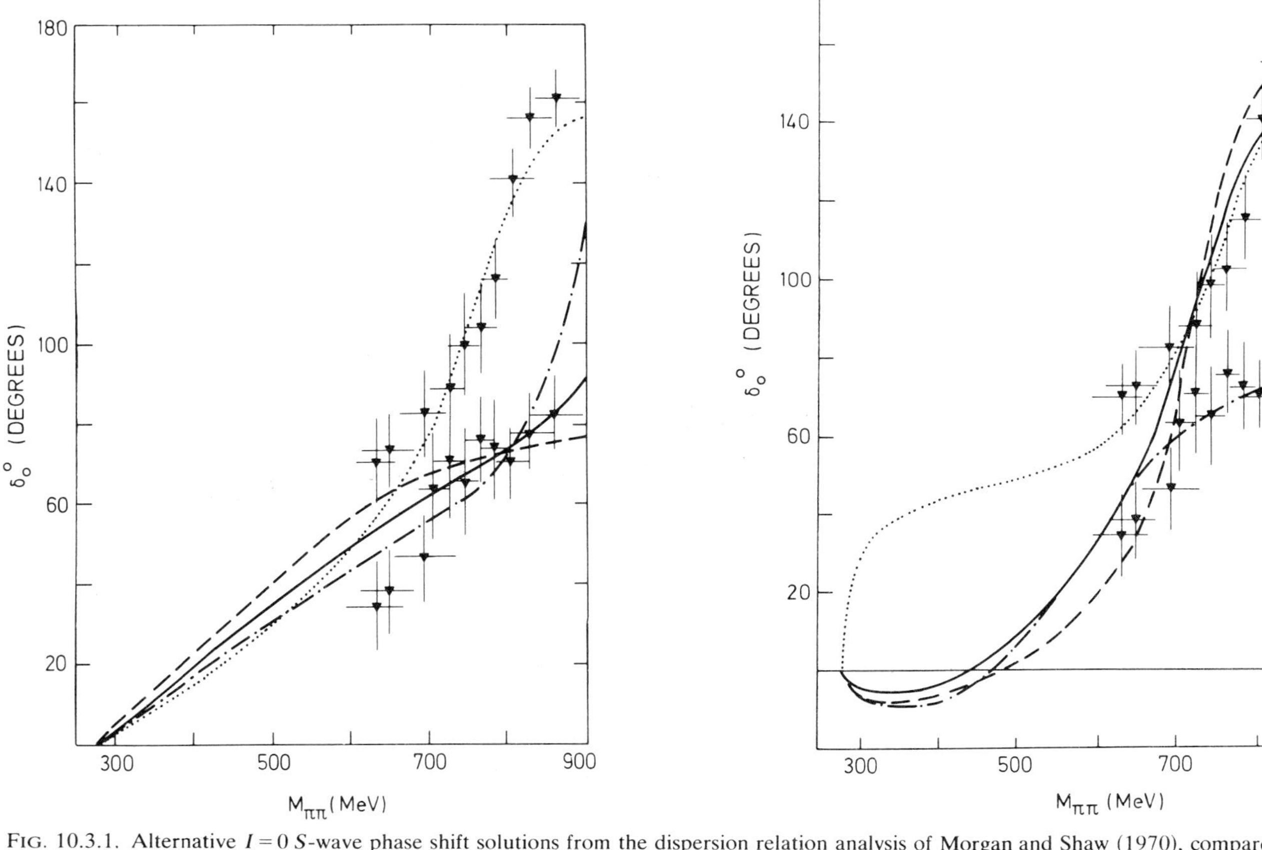

FIG. 10.3.1. Alternative $I = 0$ S-wave phase shift solutions from the dispersion relation analysis of Morgan and Shaw (1970), compared to the data of Scharenguivel *et al.* (1969) (▼). The solutions shown are referred to in the text as: (a) Between–Down 1(BD1 ———), Between–Down 2 (BD2 — · — · —), Up–Down (UD — — —), Between–Up (BU· · · · ·); (b) Down–Up 1 (DU1 ———), Down–Up 2 (DU2 — — —), Up–Up (UU· · · · ·), Down–Down (DD — · — · —).

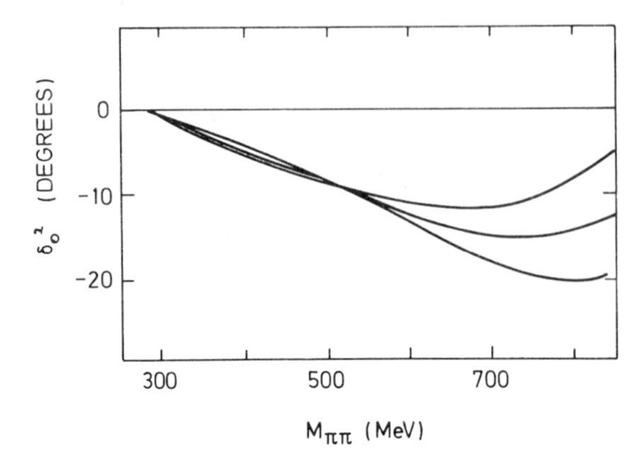

FIG. 10.3.2. Predicted forms for δ_0^2 for $\delta_0^2(m_\rho)=-10°, -15°$ and $-20°$ respectively from Morgan and Shaw (1970) corresponding to the Between–Down type of solution for δ_0^0 (cf. Fig. 10.3.1).

values of the subtraction parameters. We have already noted how these latter are related to high-energy contributions to dispersion integrals; one can therefore ask whether the values obtained are reasonable. This did enable the authors to restrict the range of solutions somewhat: e.g., for the BU solution, they found $e_2^0<0$, in contradiction to equation (10.3.10) and the optical theorem. Nevertheless, a wide variety of solutions which adequately resembled the empirical δ_0^0's still remained. Application of BNR relations and Martin inequalities to the S-waves did not exclude any of the remaining solutions. Despite the non-selectivity of these theoretical criteria, the authors did arrive at a "favoured" band of solutions, centred about the BD solutions of Fig. 10.3.1, by appealing to data on the reaction $\pi^- p \to \pi^0\pi^0 n$ and estimates then current of the scattering length ratio a_0^0/a_0^2.‡

All this reinforces the importance of finding relations applying to the whole range of acceptable solutions. The most interesting results concern the threshold parameters, which are listed for the "favoured" band of solutions, and for the DU 1 solution in Table 10.3.1. This illustrates the relative stability of the P- and D-wave scattering lengths against changes in the rho-region. It was furthermore noted by Morgan and Shaw that when the S-wave scattering lengths corresponding to various solutions are plotted as points in the (a_0^0, a_0^2) plane, they are all found to lie on a narrow band—the so-called "universal curve" (see Fig. 10.3.3). It was also found that, to a

‡ The status of these two additional sources of information has changed with the passage of time. As regards $\pi^0\pi^0$ production, there have been a number of new experiments, often disagreeing as to details, but all tending to support the Down type solution above the rho-region. Concerning the other input assumption, the two techniques which had been used for estimating the scattering length ratio now seem much less compelling. All these questions are discussed in more detail in Chapter 3.

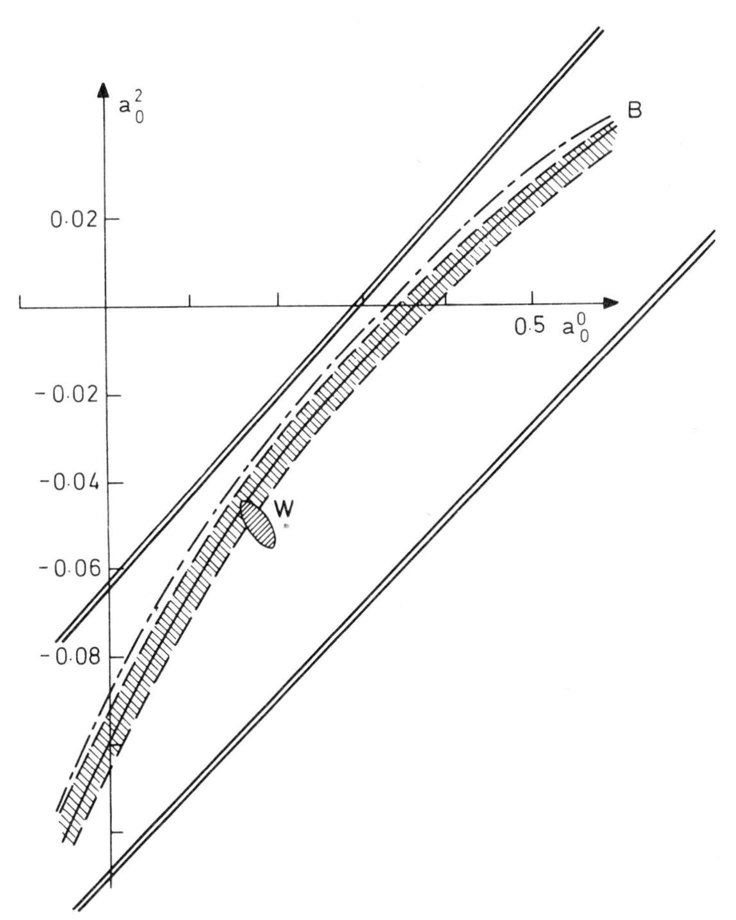

FIG. 10.3.3. Universal curve correlating the two S-wave scattering lengths a_0^0 and a_0^2 (from Basdevant *et al.* (1973); cf. also Morgan and Shaw (1970)). The region between the double lines is that explored by Basdevant *et al.* (1972a; 1974) before imposing detailed resemblance to data on δ_0^0. The solid curve (B) shows the result of fitting to the δ_0^0 information from Baton *et al.* (1970b); the shaded area indicates the estimated error, due mainly to uncertainty in the rho-width, and also allows for the effect of using the δ_0^0 phase shifts of Protopopescu *et al.* (1973). The dash–dot curve corresponds to using an alternative model for the S^*-region (for details see the original reference). The shaded ellipse (W) indicates the current algebra prediction of Weinberg (1966) (see Chapter 14).

good approximation, the values of the other threshold parameters (a_1^1, $a_2^{0,2}$, $b_0^{0,2}$) depend only on the locations of (a_0^0, a_0^2) on the curve. These results have subsequently been obtained by other techniques and using more recent data‡ (Tryon, 1971a,d; Gore, 1972; Basdevant *et al.*, 1972a,b, 1974) (see

‡ The position of the "universal curve" depends to some extent on the input data; we return to this in Section (d) below.

Sections 10.2 and 10.3(c) below). An important consequence of these systematics is that, given information from say above 500 MeV, it is sufficient to measure just one parameter in the threshold region (for example a_0^0 or a_0^0/a_0^2) to determine the threshold amplitudes completely. In their favoured solution, Morgan and Shaw (1970) adopted the value $a_0^0/a_0^2 = -3{\cdot}2 \pm 1{\cdot}0$ from early model-dependent estimates (Cline *et al.*, 1970; Gutay *et al.*, 1969). This led to the results given in Table 10.3.1. We return to this point below.

(c) Application of the Roy Relations

A more modern approach to implementing the ACU constraints is to apply the Roy equations discussed in Section 8.2. A very thorough investigation has been performed by Basdevant, Froggatt and Petersen (Basdevant *et al.*, 1972a,b, 1974). They write the Roy equations (cf. equation (8.2.10)) in the form

$$T_l^I(s) = \lambda_l^I(s)[\delta_{l0} + \delta_{l1}] + \phi_l^I(s) + \sum_{I'=0}^{2} \sum_{l'=0}^{1} (2l'+1) \int_4^N ds'\, G_{ll'}^{II'}(s, s')\, \mathrm{Im}\, T_{l'}^{I'}(s'), \qquad (10.3.12)$$

where

$$\phi_l^I(s) = \sum_{I'=0}^{2} \left\{ \sum_{l'=2}^{\infty} \int_4^{\infty} \cdot + \sum_{l'=0}^{1} \int_N^{\infty} \cdot \right\} ds'(2l'+1) G_{ll'}^{II'}(s, s')\, \mathrm{Im}\, T_{l'}^{I'}(s') \qquad (10.3.13)$$

and the $\lambda_l^I(s)$ are determined in terms of the scattering lengths a_0^0, a_0^2 by equation (8.2.11). To guarantee complete crossing symmetry, these equations must be supplemented by a set of subsidiary conditions on the absorptive parts (cf. equation (8.2.16)). However, as noted in Section 8.2, these conditions impose no restrictions on the S- or P-wave absorptive parts, so that if the ϕ_l^I are known, or negligible, over some region, equation (10.3.12) expresses completely the restrictions of crossing for S- and P-waves over this region.

Basdevant *et al.* estimated the "driving terms" $\phi_l^I(s)$ (in which they included also the S- and P-wave contributions above a cut-off $s = N = 110\, m_\pi^2$) from appropriate Regge-pole and resonance contributions. On doing this, using reasonable estimates for the parameters involved, they found that the driving terms were small for $W \lesssim 1$ GeV.‡ Furthermore, at low energies, and especially for $0 \lesssim s \lesssim 4$, the subsidiary crossing conditions (equation (8.2.16) etc.) on D- and higher waves are satisfied to good

‡ As previously noted in the discussion of forward dispersion relations, conventional high-energy absorptive parts lead to smooth effects at low energies.

accuracy. With the driving terms and subtraction constants supplied, (10.3.12) is an equation in which the unknown functions are the S- and P-wave partial wave amplitudes (up to $s = N$). The above conditions guarantee that the solutions will, to good accuracy, automatically satisfy the BNR relations and, provided that positivity has been secured, also fulfil the Martin inequalities. That the converse is not necessarily the case has been shown by Bonnier and Gauron (1973). They have tested a number of models incorporating the BNR and Martin conditions against the Roy equations and found varying degrees of success. However, while some models fail to reproduce themselves, e.g. that of Le Guillou *et al.* (1971) above 600 MeV, the results are very sensitive to details of the input, especially to the values assumed for the scattering lengths; for all the cases considered, good solutions to the Roy equations can be found by making quite small adjustments to the original solutions.

To return to Basdevant *et al.* (1972a,b), these authors went on to construct simultaneous solutions to unitarity and equation (10.3.12). Suitable parametric forms were supplied for all the S- and P-waves, in particular two very flexible types of representation for the problematic $I = 0$ S-wave were explored—one based on a two-channel K-matrix, the other on a conformal mapping—to check that the results were not dependent on details of the parameterization. In each case, the parameters were adjusted so that the S- and P-wave Roy equations, and also elastic unitarity below the $K\bar{K}$ threshold, were simultaneously satisfied to within a given error (of the order of 1%). Once the S- and P-waves are known, the higher waves can be calculated directly from equation (10.3.12). One thus obtains solutions which will to a good approximation satisfy fixed-t dispersion relations, unitarity, and crossing over the range $s \leq 60\ m_\pi^2$. This has the advantage over the earlier work based on fixed-t dispersion relations (Morgan and Shaw, 1969, 1970) that $s-t$ crossing is more systematically and accurately treated. However, this improvement is not found to bring out a significant reduction in the variety of solution admitted (given m_ρ, Γ_ρ), as had been hoped by some authors (Le Guillou *et al.*, 1971; Bonnier and Gauron, 1972). Basdevant *et al.* (1972a, 1974) investigated the wide band of possibilities in the (a_0^0, a_0^2) plane shown in Fig. 10.3.3, and found that for a_0^0, a_0^2, m_ρ and Γ_ρ fixed it was still necessary to assign δ_0^0 at some point, for example to specify the energy at which it went through 90°. However, this single further piece of information was sufficient to determine the amplitude.‡ As regards the number of pieces of information needed to fix the amplitude, the results are essentially similar to those of Piguet and Wanders (1972), Tryon (1971a, d) and Morgan and Shaw (1970). There is however somewhat more predictive power than that

‡ The results found are largely independent of variations in the driving terms, which were allowed to vary within estimated errors of 30% on ϕ_0^0 and 50% on ϕ_1^1 and ϕ_0^2.

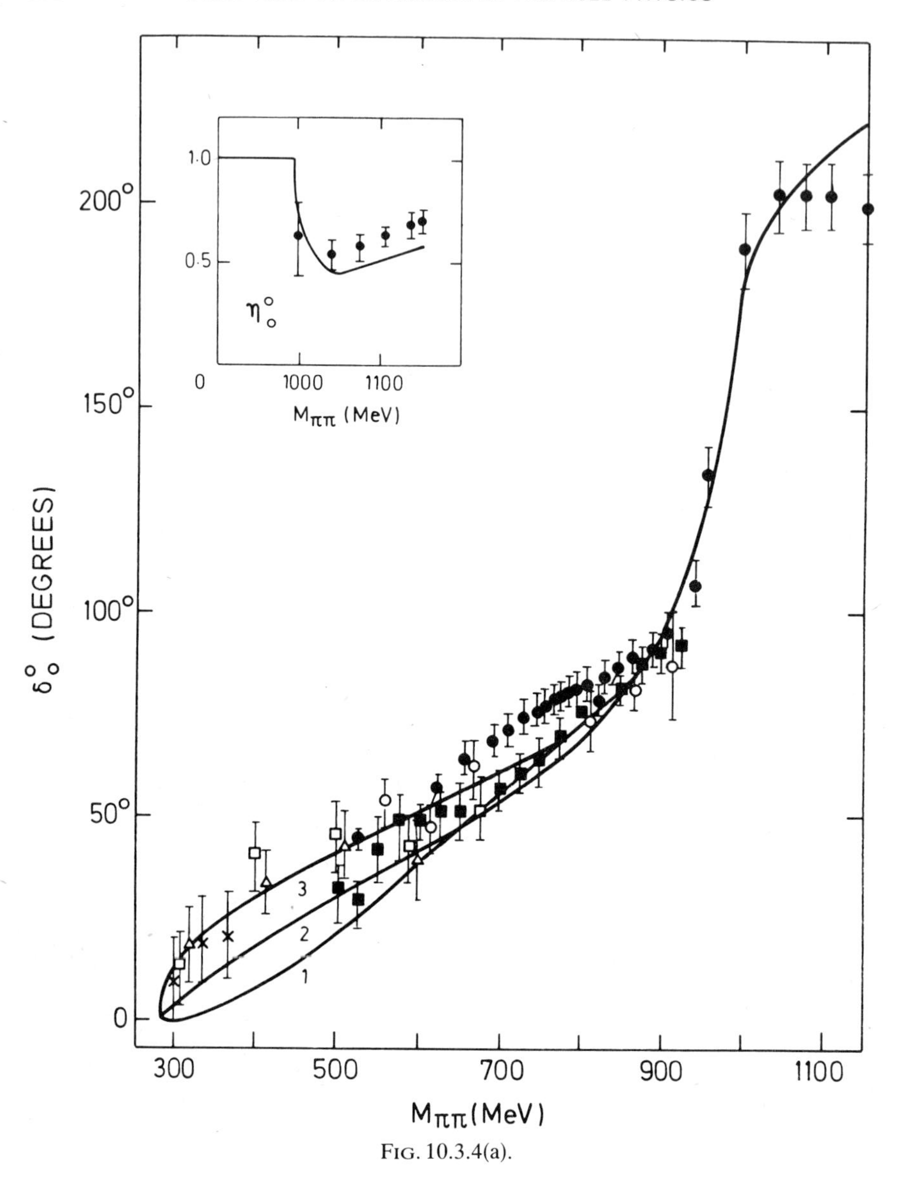

FIG. 10.3.4(a).

FIG. 10.3.4. Examples of S-wave phase shifts from the analysis of Basdevant *et al.* (1972b) (fits to δ^0_0 of Baton *et al.* (1970b); for the analogous examples fitted to the phases of Protopopescu *et al.* (1973) see Basdevant *et al.* (1974)). (a) δ^0_0: Data points, ● Protopopescu *et al.* (1973), ○ Baillon *et al.* (1972), ■ Baton *et al.* (1970b), □ Sonderegger and Bonamy (1969), △ Bensinger *et al.* (1971), × Zylbersztejn *et al.* (1972); η^0_0 from Protopopescu *et al.* (1973). (b) δ^2_0: Points from ■ Baton *et al.* (1970b), ○● Colton *et al.* (1971), □ Katz *et al.* (1969), △ Walker *et al.* (1967).

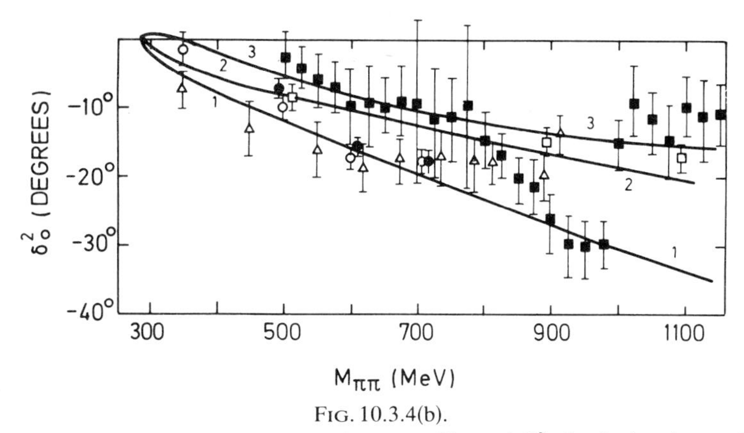

FIG. 10.3.4(b).

found by the last authors in that, given m_ρ, Γ_ρ and δ_0^0, the behaviour of δ_0^2 is predicted over the whole low-energy region, rather than just up to 600 MeV. This accords with the result found by Tryon (1971a,b) from his model using partial wave dispersion relations.

After discussing the range of possible solutions, Basdevant *et al.* (1972b, 1974) move on to the detailed fitting of information below, through and above the rho-region in order to make more precise inferences on the behaviour around threshold. With $m_\rho = 765$ MeV, $\Gamma_\rho = 135$ MeV fixed, they demanded a fit to: (a) the phase shifts of Baton *et al.* (1970a,b) over the energy region 500–900 MeV; this has both Between–Down and Between–Up type solutions, but the latter was ruled out on demanding: (b) a fit to the behaviour of the inelasticity, η_0^0 near the $K\bar{K}$ threshold found from fitting two-channel $\pi\pi$ and $K\bar{K}$ data in this region (cf. Section 3.3). With the above requirements, the solutions are found to be restricted to a narrow band in the (a_0^0, a_0^2) plane (Fig. 10.3.3), with one phase shift behaviour corresponding to each choice of (a_0^0, a_0^2). This band in (a_0^0, a_0^2) is closely similar to the universal curve of Morgan and Shaw (1970). The shaded area in Fig. 10.3.3 indicates estimated errors, due mainly to the ρ-width and the difference of phase between Baton *et al.* (1970b) and Protopopescu *et al.* (1973). The phase shifts corresponding to these particular solutions are shown in Figs 10.3.4(a),(b) and the corresponding values for the threshold parameters are listed in Table 10.3.1. As can be seen, these are, for any given a_0^0 value in the range considered, in very good agreement with those obtained from the analysis via forward dispersion relations discussed previously. In particular, the prediction for a_1^1 is rather stable, and is always in the range $a_1^1 \approx 0{\cdot}039 \pm 0{\cdot}006$ (cf. equation (10.3.4)). Once again, the result that the threshold region can be characterized by a single parameter, e.g. a_0^0, is obtained. (Provided of course, that the δ_0^0 values above say 500 MeV are established.)

A very similar analysis, again using the Roy equations has been carried out by Pennington and Protopopescu (1973), fitting the δ_0^0 phase shifts of Protopopescu *et al.* (1973). These authors find the threshold parameters given in Table 10.3.1, similar to the central of the three example solutions of Basdevant *et al.*, and the "favoured" solution of Morgan and Shaw. However, the narrow delineation of a_0^0 values rests on demanding a very close fit to the data. In view of the differences of several degrees or more between the data used here (Protopopescu *et al.*, 1973) and other δ_0^0 determinations (cf. Section 3.3(d), especially Figs. 3.3.6(a), (b)), this should be viewed as just one of the range of possible solutions found by Basdevant *et al.* (1972b).

(d) Remarks on Relations among the Threshold Parameters

It should be stressed that the threshold relations we have been discussing—the "universal curve" relating a_0^0 and a_0^2 etc.—are dependent on the empirical input information. It so happens that Morgan and Shaw's analysis using the data of Scharenguivel *et al.* (1969), and subsequent analyses, for example those of Basdevant *et al.* (1972b) and Protopopescu *et al.* (1973), all gave essentially the same "universal curve". In fact, Morgan and Shaw's Between–Down solutions, which they favoured, are very similar to Baton *et al.*'s phases below 900 MeV. If the phases were appreciably different, the position of the curve would, in general, be shifted. Thus, for example, in the analysis of Estabrooks and Martin (1973b), the δ_0^0 phases lay some 10° above previous estimates (an effect now attributed to a misleading D-wave signal (Estabrooks and Martin, 1974a)), and would entail a displacement of the universal curve by $\Delta a_0^2 \approx -0{\cdot}015$ (Basdevant *et al.*, 1973). The empirical basis of the threshold relations should therefore be kept in mind.

Fitting ACU candidate solutions in one region, then drawing conclusions in another (here the threshold) is a somewhat complicated procedure, because of the intricate interplay of the adjustable parameters (whether using forward dispersion relations or the Roy equations). One can gain a semi-quantitative idea of what is going on by considering suitably subtracted threshold sum rules. An especially relevant example is a once-subtracted (say at $z = z_\rho$) dispersion relation for $T_t^1(z)/z$ evaluated at threshold to give

$$a_0^0/3 - 5a_0^2/6 = T_t^1(z_\rho)/z_\rho + \frac{2}{\pi}(1 - z_\rho^2)\int_1^\infty dz' \frac{[\frac{1}{3}\,\mathrm{Im}\,T_s^0 + \frac{1}{2}\,\mathrm{Im}\,T_s^1 - \frac{5}{6}\,\mathrm{Im}\,T_s^2]}{(z'^2 - z_\rho^2)(z'^2 - 1)}.$$

Consider now a family of phase-shift solutions which fit data in the rho-region and have assigned high-energy behaviour. The major variation from case to case of contributions to the sum rule will come from the low-energy region, especially from the $I = 0$ S-wave. Near threshold, $\mathrm{Im}\,T_s^{0,2}$ will have the form (for convenience rewritten as a function of $\nu = \frac{1}{2}(z-1)$)

$$\operatorname{Im} T_s^{0,2}(\nu) = \left(\frac{\nu}{\nu+1}\right)^{\frac{1}{2}} \left[(a_0^{0,2} + b_0^{0,2}\nu + \cdots)^2 + \left(\frac{\nu}{\nu+1}\right)(a_0^{0,2} + b_0^{0,2}\nu + \cdots)^4 + \cdots \right]. \tag{10.3.14}$$

This is obtained by substituting the standard threshold expansions for $\operatorname{Re} T_s^{0,2}(\nu)$ (equation (2.40)) into the equation for elastic unitarity (2.36). The associated low-energy portion of the sum-rule integral will thus contain contributions proportional to $(a_0^0)^2(a_0^0 b_0^0)$ etc. with positive coefficients. The slope parameter b_0^0 turns out to be only rather weakly dependent on a_0^0, so that in this way one can understand the qualitative shape of the (a_0^0, a_0^2) plot (Fig. 10.3.3) with its slope somewhat larger than $\frac{2}{5}$ and upward curvature. One can likewise understand the dependence on the fitted empirical phase shifts—increasing $\operatorname{Im} T_s^0$ (through larger δ_0^0 phase shifts below resonance) or $\operatorname{Im} T_s^1$ (increasing Γ_ρ) will lower the (a_0^0, a_0^2) curve.

The foregoing is a sketch but not a proof of how the universal curve for (a_0^0, a_0^2) arises, and of its dependence on the data upon which it is based. One can give similar qualitative discussions of how the other threshold parameters depend on a_0^0. However, to achieve anything approaching a demonstration, a number of loose ends would have to be tied down—for example, attaining an adequate number of subtractions and avoiding the need to know the $I_s = 2$ amplitudes in the rho-region. There is also the question of how the assumed threshold behaviour joins on to the region where one has empirical information; finally, the question of what is to count is a fit to the assumed phase shift data. Above all, it has not been shown that the threshold coefficients are functions of just one parameter. All these requirements are neatly secured in the discussion of Pennington (1974a) by making suitable assumptions on the forms of the $I_s = 1$ and 2, and $I_t = 2$ absorptive parts. However, the quantitative basis for all these relations still rests on the detailed calculations (using various forms of dispersion relations or the Roy equations) described earlier in this chapter.

10.4. Analysis of πK Scattering

The experimental and theoretical situation on πK phase shifts rather closely parallels that obtaining for the $\pi\pi$ phases. (Experimental results are discussed in Chapter 3, theoretical predictions respectively from the Veneziano model and current algebra in Chapters 12 and 14.) It is therefore natural to attempt an analogous ACU phenomenology. The channel structure is somewhat more complicated than for $\pi\pi$ scattering—πK scattering has $\pi\pi \to K\bar{K}$ as the third (t) channel from crossing the external lines. This latter is of interest in its own right, as a major source of inelasticity in $\pi\pi$ scattering. One programme is therefore to calculate the $\pi\pi \to K\bar{K}$ amplitudes, assuming the form of the $\pi\pi$ and πK elastic scattering. The most clear-cut results are for the $I = 1$ P-wave, and lead to a determination of the rho coupling, $G_{\rho K\bar{K}}/G_{\rho\pi\pi} = 1{\cdot}3 \pm 0{\cdot}3$ (Nielsen and Oades, 1973), in agreement with SU(3). The $I = 0$ S-wave amplitude, whose detailed structure near the $K\bar{K}$ threshold is a sensitive indicator of the pole structure of the S^* resonance (cf. Section 3.3(f)), is somewhat harder to tie down. Johannesson and Petersen (1974) impose a one-pole form for the S^*; however, their results, above the $K\bar{K}$ threshold, differ from those of the more predictive calculation of Hedegaard-Jensen (1974) which is more suggestive of a two-pole form (again cf. Section 3.3(f)). An alternative programme is to survey possible solutions for low-energy πK scattering, compatible with ACU and established experimental facts, for example the parameters of the K^* (890). Such a programme has been embarked on by Ader *et al.* (1973) (AMB), using twice subtracted fixed-s and fixed-t dispersion relations. Absorptive parts are approximated by their S- and P-wave contributions over the low-energy domain ($M_{\pi K} = \sqrt{s} \leqslant \sqrt{s_M} = 1{\cdot}4$ GeV, $\sqrt{t} \leqslant 2m_K \approx 990$ MeV), and by Regge forms elsewhere.‡ The t-channel amplitudes are calculated by the Omnès method from assumed $\pi\pi$ phase shifts (the authors use the "preferred" solution of Bonnier and Gauron (1973)) and πK absorptive parts. This enables the $\pi\pi \to K\bar{K}$ partial wave absorptive parts to be eliminated to yield equations for the πK partial waves of the form

$$f_l^I(s) = \hat{f}_l^I(s) + \frac{1}{\pi}\int_{(M_K+m_\pi)^2}^{s_M} ds' \sum_{l'}\sum_{I'} (2l'+1)\,\mathrm{Im}\, f_{l'}^{I'}(s')\, W_{ll'}^{II'}(s', s). \quad (10.4.1)$$

The $\hat{f}_l^I$ are defined in terms of the S- and P-wave scattering lengths

$$a_0^{\pm},\ a_1^{\pm},\ \left(3a_l^{\pm} = a_l^{\frac{1}{2}} + \begin{pmatrix}+2\\-1\end{pmatrix} a_l^{\frac{3}{2}}\right) \qquad \text{(cf. equation (2.28)).}$$

The assumed $\pi\pi$ phase shifts, and Regge parameters; the kernels $W_{ll'}^{II'}$ are functions of the assumed $\pi\pi$ phase shifts.

Suitable parametric representations for the πK partial waves are fitted to equation (10.4.1), just as in the analogous $\pi\pi$ calculations. In this way, at least a sample of the possible range of solutions is scanned. General conclusions are, firstly, that the spectrum of allowed solutions is very large (Fig. 10.4.1) and that, in particular, the up–down ambiguity for $\delta_0^{\frac{1}{2}}$ is not resolved. Secondly, notwithstanding this large scope, all solutions are found

‡ For the fixed-s relation, the partial wave series for the absorptive part does not strictly converge up to the above s-value (cf. Lyth, 1965; Oades, 1966); one may nonetheless hope that the S- and P-wave contributions give an adequate approximation.

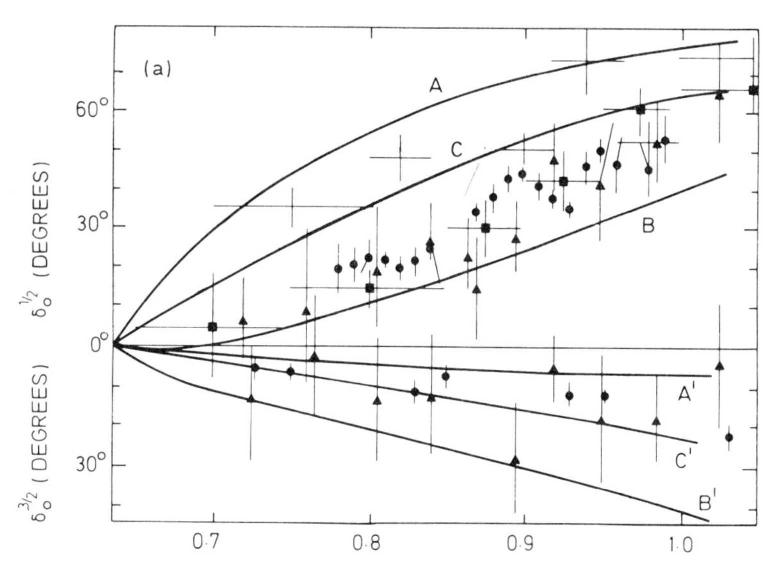

FIG. 10.4.1. AMB's dispersion relation analysis of πK scattering (Ader *et al.*, 1973). Data are from Bingham *et al.* (1972a) (●), Mercer *et al.* (1971) (▲), Yuta *et al.* (1971) (■), and Firestone *et al.* (1972) (+). Curves (A, B) and (A′, B′) depict the extreme cases which are admitted. Curves C and C′, correspond to a particular intermediate solution with scattering lengths: $a_0^{\frac{1}{2}} = 0{\cdot}16$, $a_0^{\frac{3}{2}} = -0{\cdot}07$, $a_1^{\frac{1}{2}} = 0{\cdot}016$, $a_1^{\frac{3}{2}} = 0{\cdot}003$.

to have S-wave scattering lengths which lie in a band of finite extent on a $(a_0^{\frac{1}{2}}, a_0^{\frac{1}{2}})$ plot (Fig. 10.4.2(a), cf. Fig. 10.3.3). Furthermore, the S- and P-wave scattering lengths for charge exchange to good approximation obey the proportionality relation

$$a_0^- \approx 6 m_K a_1^-, \tag{10.4.2}$$

as shown in Fig. 10.4.2(b). This relation can be deduced using a once-subtracted dispersion relation for $A^-(s, t, u)$ provided all dispersion integral contributions are small. As regards the $\pi\pi \to K\bar{K}$ amplitudes, especially for the S-wave, a wide range of possibilities is admitted. Nielsen and Oades' findings for the P-wave are upheld.

All in all, the situation is closely similar to that in the $\pi\pi$ case. This similarity would be even more marked if, with the incorporation of the improved πK phase shifts now in prospect (cf. Section 3.3), the allowed band in the $(a_0^{\frac{1}{2}}, a_0^{\frac{3}{2}})$ plane became narrower (cf. Figs. 10.3.3 and 10.4.2(a)). As in the $\pi\pi$ case, this would then leave a one-parameter ambiguity in the threshold region, which only further experiments could remove.

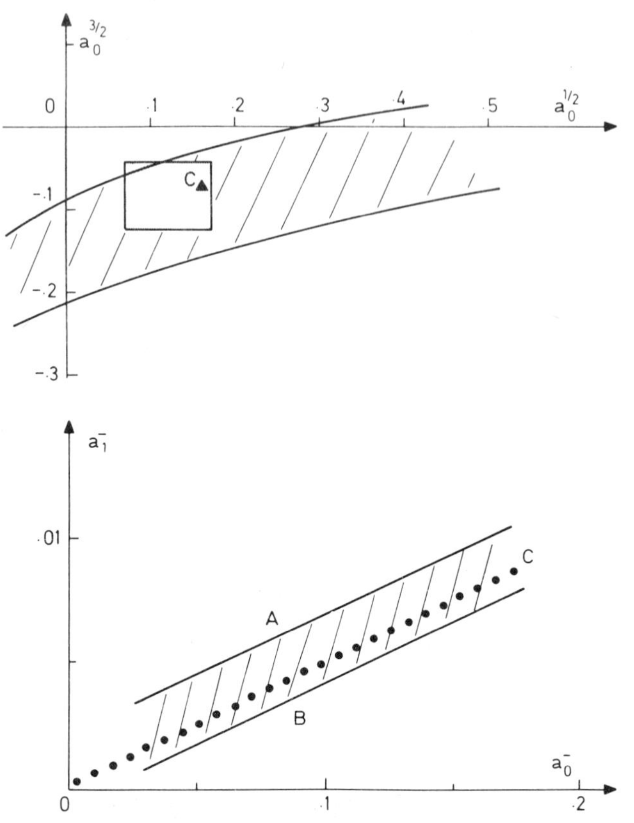

FIG. 10.4.2. AMB's dispersion relation analysis of $K\pi$ scattering (Ader *et al.*, 1973). (a) Band of allowed values for the S-wave scattering lengths. Point C corresponds to the solution (C, C′) of Fig. 10.4.1(a). The shaded area around C contains the values predicted by current algebra models (Ader *et al.*, 1972; Komen, 1973). (b) Plot of the correlation $a_0^- \approx 6m_K m_\pi a_1^-$. Curves (A, B) are the limiting loci for a_0^- and a_1^- found in AMB's solutions. (The dotted line C corresponds to the exact relation, $a_0^- = 6m_K m_\pi a_1^-$.) In all cases, AMB report for the P-waves: $a_1^{1/2} = (0{\cdot}02 \pm 0{\cdot}02)$, $a_1^{3/2} = (0{\cdot}0 \pm 0{\cdot}007)$, and below 1 GeV $|\delta_1^{3/2}| \leqslant 4°$.

10.5. Concluding Remarks

In this chapter, we have explored the results of applying the basic ACU constraints on $\pi\pi$ scattering in the low-energy region ($W \leqslant 1$ GeV), and the analogous domain for πK scattering. In order to do this, we have examined a number of different calculational techniques, and, at the end, there is a reassuring amount of agreement between the various approaches. While it is clear that, even with m_ρ and Γ_ρ fixed, the ACU constraints do not completely determine the amplitudes, they do impose considerable restrictions on them. Within a reasonable framework of assumptions, if m_ρ, Γ_ρ, and the form of δ_0^0 below 1 GeV are given, it seems possible to calculate all other aspects of the partial waves—the form of the P-wave phase shift, δ_1^1, down to threshold, and the behaviour of the $I=2$ S-waves and the higher waves ($l \geq 2$) over the whole low-energy region.

Phenomenologically, the most important result concerns the threshold parameters. As has been emphasized, the crucial ingredient upon which there has been uncertainty, seemingly now resolved, is the detailed form of the experimental $I=0$ S-wave phase shifts below and through the region—say for $500 \leq W \leq 1100$ MeV. Given reliable information on this, one can make the following statement: if δ_0^0 for candidate ACU solutions is constrained to agree with the experimental phases over the above region, then all the threshold parameters are predicted in terms of one parameter, which is conveniently taken to be a_0^0. The latter could in principle be inferred from improved phase shift determinations over the mass range $W > 500$ MeV; however, the accuracy required is not at present realistic. The best hope lies either in further refinement of K_{e4} experiments (cf. Section 5.1), or in dipion production experiments specially designed to explore the threshold region. Once given a reliable determination of a_0^0, and of δ_0^0 over the enlarged rho-region referred to above, predictions for all the other threshold parameters can be read off directly from the appropriate universal curves—Table 10.3.1 and Fig. 10.3.3, or their analogues if appreciably different input data are used. As discussed in Section 3.3(d), there has recently been a considerable acquisition of low-mass dipion production data (Männer, 1974) leading to quite precise determinations of a_0^0 and a_1^1, given certain assumptions on the production mechanism. Unfortunately, the P-wave result, $a_1^1 \approx 0{\cdot}1$, lies completely outside the range, $a_1^1 \approx 0{\cdot}040 \pm 0{\cdot}005$, indicated by all the theoretical estimates discussed in this chapter (all based on fixed-t dispersion relations).‡ This casts doubt on the reliability not only on the experimental P-wave, but also on the associated S-wave determinations, since the two are correlated in the observed experimental quantities. This discrepancy will have to be resolved before the new results can be used with confidence.

As was discussed in Section 10.4, the study of low-energy πK scattering is moving into a similar position to that obtaining for $\pi\pi$ scattering, and again the next word lies with the exploitation of improved experimental data.

‡ cf. also Basdevant *et al.* (1975).

Chapter 11

Dynamical and Field Theoretic Models

11.1. The Strip Approximation

(a) Introduction

The strip approximation (Chew and Frautschi, 1961a,b; Chew and Jones, 1964) is an interesting attempt to achieve a predictive dynamics of hadrons based on Regge poles. In Chapter 9, we have already encountered attempts to achieve approximate bootstraps with particular partial waves. The introduction of the concept of Regge poles seemed to offer the prospect of a much more systematic approach. Particle resonances were to be arranged in families of ascending spin with increasing mass, interpolated by Regge trajectories $J = \alpha(t)$; exchange of resonances was to be governed by the same trajectories continued to negative t. This enables the exchange of high spin resonances to be accommodated without violating the Froissart bound; the behaviour from such an exchange is $s^{\alpha(t)}$ and the associated trajectories are required to have $\alpha(t) < 1$ for $t < 0$. All this is now of course a well-established part of hadron theory. (A brief account of Regge-pole formalism and the phenomenology based on it is given in Appendix C.)

In the strip approximation, it is supposed that Regge trajectories have just a few resonances of relatively low mass lying on them, so that s-channel Regge poles are confined to a strip in energy (Fig. 11.1.1) and that above that energy amplitudes are given by t- and u-channel exchanges. For the $\pi\pi$ system with its complete crossing symmetry, this immediately opens the way to a bootstrap scheme with the t- and u-channel exchanges giving the forces to produce the Regge poles in the s-channel. In the resulting approximation,

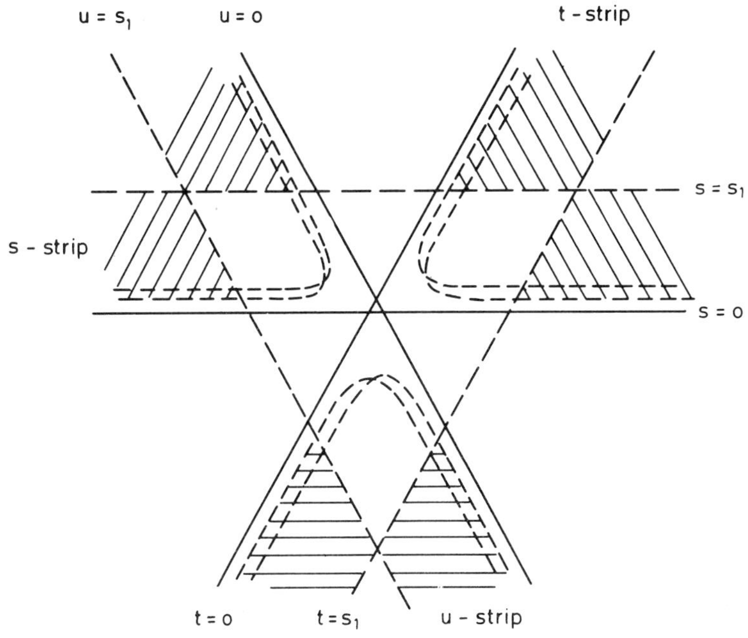

FIG. 11.1.1. Schematic Mandelstam diagram indicating s-, t- and u-channel strips (double spectral function boundaries shown as dashed lines) (cf. Fig. 2.2).

only a very simple subset of thresholds is considered. In the s-strip, elastic unitarity induces, via the usual convolution integral, a sequence of t-singularities. By crossing, these give a sequence of singularities in s, associated with multi-pion inelastic processes. These are assumed to exhaust inelastic contributions to the amplitude. Other coupled channels are usually ignored, although there have been calculations including $\pi\pi \to K\bar{K}$ and the related channels (Bali and Chiu, 1967).

Present thinking would disfavour the special assumptions of the strip approximation. The first assumption, the strip concept, is at variance with subsequent duality ideas and with indications from experiment. Of crucial importance is the indication of indefinitely rising straight Regge trajectories. This also bears against the second assumption that only a few nearby thresholds need be considered. If trajectories rise indefinitely, it would seem, on the contrary, that an infinite number of s-channel thresholds is necessary (Mandelstam, 1968a). Other pointers to the relative unimportance of nearby thresholds are the observed regularities under change of internal quantum numbers—SU(3) multiplets and the like. All these are trends which were perceived after the strip programme was proposed, and which diverted speculation into new channels such as duality and the quark model which we discuss in the following chapters. For the present, we leave these difficulties on one side.

There are two versions of the strip approximation, the one working directly with double spectral functions (cf. equation (2.41)), the other employing the N/D method for continuous J. In both methods, the heart of the calculation consists in solving for the full amplitude in the s-channel strip, with the t- and u-channel exchanges supplying the given "potential".‡ In the first approach, the so-called "old strip approximation", the problem is solved by the procedure of "Mandelstam iteration" (Mandelstam, 1958a), whereby the discontinuities in t and u of the input potential determine those of the full amplitude by an analytic continuation of elastic unitarity. The alternative "new strip approximation", whilst assuming the same underlying equations, manipulates the partial wave amplitudes for continuous angular momentum J, solving by an N/D method. The usual procedure is to use as input the Born term implied by the assumed input potential, ignoring modifications to the left-hand cuts induced by Mandelstam iteration. This circumvents the problem associated with the so-called AFS cuts (Amati, Fubini and Stanghellini, 1962) which arises on iterating t-channel Regge exchanges through s-channel elastic unitarity. This leads to a divergent behaviour towards the strip boundary (s large, not t large). From analysis of perturbation theory amplitudes, such cuts should be cancelled by additional, non-elastic, contributions to unitarity (Mandelstam, 1963); however no practical scheme of calculation has been developed to allow for this. Instead, a cut-off is applied at the strip boundary, introducing a further approximation. Thus, in the old strip approximation, the procedure of generating higher thresholds by Mandelstam iteration is truncated (towards the strip boundary), and in the new strip approximation it is ignored. In both cases, the position of the strip boundary is a parameter.

In the practical calculations to be discussed, the emphasis is on the leading Regge trajectories and the attainment of self-consistent values for their parameters. There have been calculations with a ρ input and with a ρ and P (Pomeron) input. It is natural to extend the discussion to the latter case, since rho-exchange produces, besides an output $I=1$ trajectory, a higher-lying $I=0$ trajectory. (In the strip schemes, one is of course committed to a pole rather than a cut description of the Pomeron, since the mechanisms which would generate cuts are excluded.) Calculations have been performed on both strip methods and reasonably self-consistent solutions found for the parameters of the ρ and P trajectories. In almost all cases, the self-consistent rho-width comes out much too large compared to experiment. An exception is the calculation of Collins and Johnson (1969a,b,c) which reported very

‡ The concept of the "potential" in this context will be explained further below.

satisfactory low-energy parameters. However this result has been criticized and requires further investigation (see below).

Overall, the results are not good, and this has played a part in the development of alternative approaches. The work remains interesting as the only attempt at a full-scale bootstrap, and also for the methods which are used. The techniques of analytic continuation via the double spectral functions are, of course, not tied to the strip approximation, and could well be of service for obtaining different objectives—for example for extending the range of validity of partial wave approaches, or for discussing the physically observed Regge trajectories.

We now go on to discuss the general formalism, and then examples of applications (Collins and Johnson, 1969a,b,c; Webber, 1971). References to earlier work can be traced back from these. For general reviews on applications of Regge poles and S-matrix dynamics see Chew (1966) and Collins and Squires (1968).

(b) Basic Formalism

The double dispersion representation of the $\pi\pi$ scattering amplitude can be cast in the form of a potential problem. Consider first the expression for the amplitude $T^I(s, t, u)$ in terms of its t and u discontinuities via a fixed s-dispersion relation

$$T^I(s, t, u) = \frac{1}{\pi}\int_4^\infty dt'\, D_t^I(t', s)\left[\frac{1}{t'-t} + \frac{(-1)^I}{t'-u}\right]. \tag{11.1.1}$$

The discontinuity function $D_t^I(t, s)$ is itself an analytic function of s, with discontinuity given by the double spectral function $\rho^I(s, t)$, determined from unitarity along with analogous u-discontinuities (cf. equation (2.42b)). There is an elastic contribution $\rho_{\text{el}}^I(s, t)$ and a remainder $\rho_{\text{in}}^I(s, t)$ containing all other possible contributions. The resulting fixed-t dispersion relation for D_t^I can be split into the following contributions

$$D_t^I(t, s) = V_t^I(t, s) + \frac{1}{\pi}\int_4^\infty ds'\, \rho_{\text{el}}^I(s', t)\left[\frac{1}{s'-s} + \frac{C_{II}^{(su)}}{s'-u}\right], \tag{11.1.2}$$

with the "potential" $V_t^I(t, s)$ given by the remaining terms

$$V_t^I(t, s) = \frac{1}{\pi}\sum_{I'}\int ds'\, \rho_{\text{inel}}^{I'}(s', t)\left[\frac{\delta_{II'}}{s'-s} + \frac{C_{II'}^{(su)}}{s'-u}\right]$$
$$+ \frac{1}{\pi}\sum_{I'\neq I} C_{II'}^{(su)}\int ds' \frac{\rho_{\text{el}}^{I'}(s', t)}{s'-u}. \tag{11.1.3}$$

For convenience, the contribution from the "ut" elastic double spectral function for $I' = I$ is added to the potential. The elastic double spectral function is itself determined from the discontinuity functions $D_t^I(t, s)$ by an analytic continuation of the convolution integral for elastic unitarity (Mandelstam, 1958a)

$$\rho_{el}^I(s, t) = \frac{2C(s)}{\pi q_s s^{\frac{1}{2}}} \int_4^{K=0} \int_4 dt_1\, dt_2 \frac{D_t^{*I}(t_1, s) D_t^I(t_2, s)}{K^{\frac{1}{2}}(s, t, t_1, t_2)}. \tag{11.1.4}$$

Here $C(s)$ is in fact unity, but will stand for a cut-off when the above equation is used in the strip approximation; q_s is the CM momentum, and K is the Mandelstam kernel of equation (2.44):

$$K(s, t, t_1, t_2) = t^2 + t_1^2 + t_2^2 - 2(tt_1 + tt_2 + t_1 t_2) - 4tt_1 t_2/(s-4). \tag{11.1.5}$$

The integrations in t are over the finite range in the t_1, t_2 plane governed by the boundaries $t_1 = t_2 = 4$, $K = 0$. For large s, the latter approaches $\sqrt{t_1} + \sqrt{t_2} = \sqrt{t}$, and generally the contributing t_1 and t_2 values are smaller than t. The effect is that one can move outwards in t, calculating $\rho_{el}(s, t)$ by iteration, passing back and forth between (11.1.2) and (11.1.4). This procedure is called Mandelstam iteration. It generates a sequence of Landau boundaries asymptotic as $s \to \infty$ to $t = 4m^2, 9m^2$ etc. where m^2 is the nearest t-singularity of V—here $m^2 = 4$ corresponding to 2π exchange. The inclusion of an infinite sequence of thresholds allows s-channel poles to be developed. It is interesting in a number of connections that asymptotic behaviour nonetheless usually ensues at relatively modest values of t.

If the potential $V_t^I(s)$ is given, the above procedure is the exact translation into the S-matrix framework of ordinary potential scattering.‡ Resonances and bound states are calculated without additional arbitrary constants (CDD poles—see Chapter 9). For the case of non-relativistic potential scattering via a Yukawa potential, it has been shown (Blankenbecler *et al.*, 1960) to be equivalent to the standard calculation using the Schrödinger equation.

In order to achieve a bootstrap, the potential $V_t^I(t, s)$ must itself be determined by t- and u-channel exchanges related to the output s-channel resonances by crossing. In the strip approximation, this is done by approximating $\rho_{inel}^I(s, t)$ by its peripheral component which, by crossing, is related to $\rho_{el}^{I'}(t, s)$

$$\rho_{inel}^I(s, t) = \sum_{I'} C_{II'}^{(st)} \rho_{el}^{I'}(t, s). \tag{11.1.6}$$

‡ For a general review see Chew (1966).

On substituting this into (11.1.3), one determines $V_t^I(t, s)$ as

$$V_t^I(t,s) = \frac{1}{\pi}\sum_{I'} C_{II'}^{(st)}\left\{\int ds'\, \rho_{\rm el}^{I'}(t,s')\left[\frac{1}{s'-s} + \frac{(-1)^{I'}}{s'-u}\right]\right\}$$

$$+\frac{1}{\pi}\sum_{I'\neq I} C_{II'}^{(su)}\int ds'\,\frac{\rho_{\rm el}^{I'}(s',t)}{s'-u}. \qquad (11.1.7)$$

This, together with the previous potential equations (11.1.1)–(11.1.4), defines a complete bootstrap scheme. In practice, self-consistency is only sought for the leading Regge pole contributions which are assumed to dominate the strip contributions to the double spectral functions. A Regge form is therefore assumed for the input double spectral function entering $V_t^I(t, s)$ and the parameters varied until the output parameters match the input. Output Regge poles are identified from the asymptotic behaviour at large t of the t-discontinuity $D_t^I(t, s)$ (and *its* s-discontinuity $\rho_{\rm el}^I(s, t)$)

$$D_t(t, s)(\equiv \text{Re}\, D_t + i\rho_{\rm el}),$$

$$\underset{t\to\infty}{\sim} \sqrt{\pi}\gamma(s)\frac{\Gamma[\alpha(s)+\frac{1}{2}]}{\Gamma[\alpha(s)+1]}t^{\alpha(s)},$$

$$\equiv \bar{G}(s)t^{\alpha(s)} \quad (\text{definition of } \bar{G}(s)). \qquad (11.1.8)$$

Here we have anticipated the definition of the Regge residue $\gamma(s)$ to be given below (see equation (11.1.22)). The above parameters are conveniently extracted by fitting to the logarithm of the above equations

$$\ln(D_t^I + i\rho_{\rm el}^I) \underset{t\to\infty}{\sim} \ln \bar{G}(s) + \alpha(s)\ln t. \qquad (11.1.9)$$

In the new strip approximation, output Regge trajectories are calculated using an N/D method for continuous J. Partial wave amplitudes for continuous J are defined by the continuation of the Froissart–Gribov representation, obtained by projecting out partial waves from (11.1.1). For this purpose, (11.1.1) is expressed as the sum of the "signatured" amplitudes (see Appendix C) arising respectively from the t and u discontinuities (for $\pi\pi$ scattering these are simply related by generalized Bose statistics) and the projections made on them. Thus, we consider

$$t_J^I(s) = \frac{1}{\pi q^{2J}}\int_4^\infty \frac{dt}{2q^2} D_t^I(t, s) Q_J(1 + t/2q^2). \qquad (11.1.10)$$

The factor q^{-2J} is inserted to eliminate the power branch point at $q^2 = 0$. The physical partial waves for integral J are then given by

$$T_J^I(s) = q^{2J}[1 + (-1)^{I+J}]t_J^I(s). \qquad (11.1.11)$$

The (elastic) unitarity relation continued in J now takes the form

$$\text{Disc}\,[t_J^I(s)] = 2i\rho^J |t_J^I(s)|^2, \tag{11.1.12}$$

with

$$\rho^J = 2q^{2J+1}/\sqrt{s}. \tag{11.1.13}$$

The Born term is obtained by substituting the potential contribution to D_t^I into (11.1.10)

$$B_J^I(s) = \frac{1}{\pi q^{2J}} \int_4^\infty \frac{dt}{2q^2} V_t^I(t, s) Q_J(1 + t/2q^2). \tag{11.1.14}$$

The discontinuity on the left-hand cut of the full amplitude $t_J^I(s)$ is inferred from (11.1.10) where it arises from the Q_J factor. This gives

$$\text{Im}\,(t_J^I)^{LHC} = \frac{1}{4(-q^2)^{J+1}} \int_4^{-4q^2} dt\, P_J(-1 - t/2q^2) D_t^I(t, s). \tag{11.1.15}$$

$B_J^I(s)$ likewise has a left-hand cut given by the analogous formula; in addition it has a right-hand cut arising from the s-discontinuity of $V_t^I(t, s)$. This latter arises from the t- and u-channel exchanges, and, in principle, starts at $s = 16$. However, in the new strip approximation it is associated with the t- and u-channel Regge poles which are arranged to start contributing at the strip boundary $s = s_1$. This is also the point up to which elastic unitarity is assumed. The assignment of this double role to the parameter s_1 is an *ad hoc* approximation, but perhaps not a serious one in relation to the other approximations. It has the convenient effect of avoiding overlap of the double spectral regions.

The resulting contribution to t_J^I from the left-hand and far right-hand cuts can now be calculated and takes the form

$$b_J^I(s) = B_J^I(s) + \Delta B_J^I(s). \tag{11.1.16}$$

The second term arises from the additional contributions to D_t^I in (11.1.15) from the Mandelstam iteration of the potential. It "balances" the integral over the near right-hand cut to achieve the large-s behaviour implied by the $(q^2)^{-J}$ factor in t_J^I. In practice, this extra contribution is omitted, or only calculated in part, since to evaluate it fully one must already know $D_t^I(t, s)$ which is essentially the quantity to be solved for. The earlier discussion in Chapter 9 of partial wave dynamics at fixed-J was presented with the analogue of $b_J^I(s)$ assumed to be a fixed given input, thus effectively identified with $B_J^I(s)$, as is conventional.

Given $b_J^I(s)$, the full amplitude t_J^I is to be constructed to satisfy

$$t_J^I(s)=b_J^I(s)+\frac{1}{\pi}\int_4^{s_1}\frac{ds'}{s'-s}\operatorname{Im} t_J^I(s'), \tag{11.1.17}$$

with $t_J^I(s)$ constrained to equal $b_J^I(s)$ at $s=s_1$. A novel feature is that Im $t_J^I(s_1)$ has to be matched to Im $b_J^I(s_1)$ to avoid a logarithmic singularity at $s=s_1$. The resulting integral equations can be accommodated to this requirement (Chew and Jones, 1964). Equation (11.1.17) is solved by the N/D method. Thus, one expresses t_J^I as the ratio

$$t_J^I=N_J^I/D_J^I, \tag{11.1.18}$$

with

$$N_J^I(s)=b_J^I(s)+\frac{1}{\pi}\int_4^{s_1}\frac{ds'}{s'-s}[b_J^I(s')-b_J^I(s)]\rho_J(s')N_J^I(s'), \tag{11.1.19}$$

and

$$D_J^I(s)=1-\frac{1}{\pi}\int_4^{s_1}\frac{ds'}{s'-s}\rho_J(s')N_J^I(s'), \tag{11.1.20}$$

where J is a continuous index.

The only J singularities in $t_J^I(s)$ are s-dependent poles arising from the zeros of $D_J^I(s)$, so the Regge trajectories are given by

$$D_{\alpha_i(s)}^I(s)=0. \tag{11.1.21}$$

By the implicit function theorem, the trajectory function $\alpha_i(s)$ (i denotes the ith pole) has essentially the same reality and singularity structure as $D_J^I(s)$ (except possibly where trajectories intersect, a contingency which is usually ignored). Likewise for the Regge pole residue function

$$\gamma_i(s)=\frac{N_{\alpha_i(s)}^I(s)}{[\partial D_J(s)/\partial J]_{J=\alpha_i(s)}}. \tag{11.1.22}$$

The bootstrap is completed by specifying how the potential, $V_t^I(t,s)$, is to be determined from the output trajectories. V_t^I is expressed via (11.1.7) in terms of the double spectral functions, which are assumed to be dominated by the leading Regge pole. It is therefore sufficient to define the Regge contributions to these, and the following form (Chew and Jones, 1964) is used

$$\rho_{\rm el}^I(t,s)=\sum_i \operatorname{Disc}_s[G_i(t)P_{\alpha_i(t)}(-1-s/2q_t^2)]\theta(s-s_1), \tag{11.1.23}$$

with

$$G_i(t) \doteq (\pi/2)[2\alpha_i(t)+1]\gamma_i(t)(-q_t^2)^{\alpha_i(t)}. \tag{11.1.24}$$

Note that the "contribution" from a Regge pole is a somewhat ambiguous concept, since the pole is really defined from its asymptotic behaviour. In the above form, a compromise is struck between the detailed observance of Mandelstam analyticity, and convenience.

The quantity $\gamma_i(s)$, being the residue of the reduced amplitude $t_J^I(s)$, has variable dimensions. A related quantity of fixed dimension is obtained by replacing $(q^2)^{-J}$ (in 11.1.10) by $(q^2/\bar{t})^{-J}$ thus supplying a scale for q^2. This leads to the modified residue

$$\bar{\gamma}_i(s) = \gamma_i(s)(\bar{t}^{\alpha_i(s)}). \tag{11.1.25}$$

The choice of $\bar{t}$ is far from trivial if a simple (i.e. slowly varying) behaviour is sought for $\bar{\gamma}_i$.

The functions $\alpha(s)$ and $\bar{\gamma}(s)$ satisfy the dispersion relations (we drop the suffix i)

$$\alpha(s) = \alpha(\infty) + \frac{1}{\pi}\int_4^\infty ds' \frac{\operatorname{Im}\alpha(s')}{s'-s}, \tag{11.1.26}$$

$$\bar{\gamma}(s) = \frac{1}{\pi}\int_4^\infty ds' \frac{\operatorname{Im}\bar{\gamma}(s')}{s'-s}. \tag{11.1.27}$$

In potential scattering $\alpha(\infty)$ is always a negative integer, some trajectories beginning at -1, some at -2 and so on. The relativistic elementary particle situation is more complicated; however, in their present form, the bootstrap equations yield trajectories with finite $\alpha(\infty)$, in practice $\alpha(\infty) = -1$, with consequent difficulties for obtaining realistic looking trajectories for negative α (for further details see (Collins and Squires, 1968).

This completes the discussion of the general formalism.

(c) Practical Calculations

Before we describe practical calculations, it is worth emphasizing the ambitious task upon which they embark, namely to determine self-consistent Regge trajectory and residue *functions* $\alpha(t)$, $\gamma(t)$. In practice, only approximate self-consistency is sought, and there is divergence of opinion as to how the input trajectories are best parameterized—choice of scale factor $\bar{t}$ etc. With the continuous-J N/D method, it has only proved practicable to identify the output trajectories for $t<0$. This limitation is not present in the approach via Mandelstam iteration.

We now look at some detailed examples.

(i) *The "New-Strip" Approximation*

There have been calculations with just a rho-trajectory as input (Collins and Teplitz, 1965). The self-consistent rho-width comes out several times larger than experiment; furthermore, a higher lying $I=0$ trajectory is achieved. It is natural to identify this with the P trajectory, and hence to consider calculations with a ρ and P trajectory as inputs. The latter gives problems because of the long range repulsion which it produces, as explained by Collins (1966).

Collins and Johnson (1969a,b,c) suggest that Mandelstam iteration of the potential be included to set the above problem right, but with the Born input maintained for $s > s_1$. The modification thus corresponds to including the "corners" of the double spectral function. This entails inelasticity in the s-domain over which the N/D calculations are performed, and this is incorporated via the method of Frye and Warnock (1963).

On doing this, Collins and Johnson achieve results with considerable resemblance to experiment, at any rate at low energies. (The Pomeron residue at $t=0$ corresponds to an asymptotic cross-section of 26 mb, to be compared to phenomenological estimates of 10–20 mb, using Regge phenomenology and factorization (Chiu *et al.*, 1968). Other high-energy predictions are not explicitly evaluated). The self-consistent rho-width comes out as $\Gamma_\rho = 143$ MeV and the associated S-wave scattering lengths are $a_0^0 = (0{\cdot}15 \pm 0{\cdot}05)$, $a_0^2 = (-0{\cdot}04 \pm 0{\cdot}015)$, closely similar to Weinberg's prediction (cf. Chapter 14).

Doubt has been cast on this (necessarily very intricate) calculation from two viewpoints:

(i) The results are not reproduced in an apparently analogous calculation using Mandelstam iteration (Webber, 1971), to be discussed below.

(ii) Lyth (1971c) has proposed a sum-rule test for partial wave bootstraps and finds, on evaluation, that the Collins–Johnson output fails the test.

The derivation of the test proceeds as follows. Consider the P-wave partial wave amplitude (modified above the inelastic threshold for convenience of applying the Frye–Warnock N/D technique)

$$F(s) = \eta \frac{e^{i\delta} \sin \delta}{\rho}, \tag{11.1.28}$$

where

$$\rho = \frac{(s-4)^{\frac{3}{2}}}{4s^{\frac{1}{2}}}. \tag{11.1.29}$$

Lyth defines a left-hand cut contribution

$$L(s) \equiv \operatorname{Re} F(s) - \frac{P}{\pi} \int_4^{s_2} ds' \frac{\operatorname{Im} F(s')}{s'-s}. \tag{11.1.30}$$

In the application, s_2 corresponds to 20 GeV^2, to conform to Collins and Johnson. Lyth then proves that suitable averages of $L(s)$ must be bounded, from the requirement that $F(s)$ satisfies the unitarity bounds

$$0 \leq |\mathrm{Re}\, F(s)|, \mathrm{Im}\, F(s) \leq 1/\rho(s). \tag{11.1.31}$$

Essentially, the extent to which $L(s)$ exceeds Re $F(s)$, which is bounded by unitarity, is itself bounded by unitarity. The $L(s)$ of Collins and Johnson appears to violate the bounds. (Im $F(s)$ is assumed to be small at small s (below the tail of the rho) as is the case experimentally; however this might not be the case in Collins and Johnson's calculation.) Lyth goes on to ask why, not withstanding this result, the correct rho-resonance ensued on processing $L(s)$ through the N/D calculations. He suggests that, either the calculation is insensitive to large changes in $L(s)$; or an additional unphysical feature, such as one or more pairs of spurious complex poles, occurs in the calculation. Clearly, further investigation is needed.

(ii) *Bootstrap Calculations using Mandelstam Iteration*

The most recent calculation is by Webber (1971), and earlier references can be found there. The method is the one already described. An input form for $\rho^I_{\mathrm{el}}(t, s)$, assumed to be dominated by the leading Regge poles, is inserted into $V^I(t, s)$, and the output trajectories calculated by Mandelstam iteration. The input form for ρ^I_{el} is very similar to that used in the new strip approximation (11.1.23) but with a smooth truncation at small s instead of an abrupt cut-off. Suitable parametric forms for the trajectory $\alpha(t)$ and the residue function are assumed, the latter with a cut-off at large t. The same cut-off function is employed as that used in the continued unitarity relation (the $C(s)$ of (11.1.4)) to quench the AFS cuts. This use of the same cut-off ensures that the strip width is the same for the output as the input.

The Mandelstam iteration is performed and output Regge parameters identified from the large-t behaviour of the solution for a variety of input parameters. A solution is found with approximately self-consistent input and output trajectories, firstly with just a rho-input, then with a rho- and P-input (in this latter case, the P residue is not satisfactorily consistent except at $t=0$).

The self-consistent rho-width in Webber's $\rho + P$ calculation comes out at 600 MeV. The Pomeron residue at $t=0$ corresponds to an asymptotic cross-section of 46 mb. (Compare Collins and Johnson's 26 mb and the rough estimate from phenomenology of 10–20 mb.) Other high-energy predictions are not explicitly evaluated.

(d) Conclusions

It seems that the truncated $\pi\pi$ bootstrap probably is not capable of reproducing the true physical situation, although the claim of Collins and Johnson clearly needs further investigation. The strip approximation in either form is explicitly an interference model in opposition to subsequent duality notions. Indefinitely rising Regge trajectories are not catered for. Furthermore, it is probably difficult to achieve sufficiently collimated residue functions. Comparison with the successes of the Veneziano model suggests that the whole notion of neglecting the interior double spectral region is mistaken. It would be interesting to see whether a workable and realistic scheme could be devised with additional free parameters to represent the neglected interior region.

11.2. Lagrangian Field Theory Models—Padé Approximants

(a) General Remarks on Lagrangian Models

In this section, we discuss applications of the Padé technique to effect approximate calculations of $\pi\pi$ scattering amplitudes implied by alternative Lagrangian field theories. The field theoretic approach was, of course, the original theoretical basis for particle studies, with the fields conceived as fundamental entities. With the ever-growing list of "elementary" particles, this latter viewpoint came to be disfavoured, and, for a time, field theories were almost wholly abandoned in favour of the S-matrix approach. Later, field theories came back into favour as heuristic devices, without commitment to any fundamental status for the fields, or their quanta. Progress in the understanding of weak interactions, especially the chain of developments leading to the postulation of PCAC (see Chapter 14) was an important spur to the revived interest in field theories.

Model field theories have the desirable property of embodying general axioms believed to hold in a wide class of models, and they provide a convenient language for discussing "off-shell" processes. A feature which has been much emphasized in the new wave of field theoretic investigations, following the S-matrix philosophy of stressing the importance of observable quantities, is the mathematical freedom in the definition of the fields and choice of Lagrangians to achieve the same S-matrix elements. This freedom is exploited, for example, to embody the PCAC condition in certain Lagrangian models. Within the V–A framework, matrix elements of the axial-vector current have the status of physically observable quantities. This allows an operational definition of the pion interpolating field through the relation

$$\partial_\lambda A^\alpha_\lambda(x) = \frac{f_\pi m_\pi^2}{\sqrt{2}} \phi^\alpha(x),$$

where f_π is the pion decay constant (see Appendix D).

There has of recent years been considerable use of *effective* Lagrangians, intended only to be evaluated in the tree approximation,‡ to serve as models for the off-shell extrapolation of physical matrix elements. These are not our present concern. We shall be discussing calculations based on the ϕ^4 theory, on the σ-model, and on the Yang–Mills ρ-exchange model. Each Lagrangian contains as one fundamental ingredient the pion field, with additional components in the latter cases.

A fuller evaluation of the content of field theories including loop diagrams, such as we are to discuss, leads to the standard problems of renormalization of wave functions, masses, and coupling constants. This is exploited in the treatment of the σ-model to enforce PCAC at each order in the coupling constant. The Yang–Mills theory used to appear intractable as regards renormalization but there have been subsequent developments (Veltman, 1968; t'Hooft, 1971a,b) with important bearing on possible theories of weak interactions (Weinberg, 1971; Salam and Ward, 1964).

Basdevant (1972) has stressed that the Padé approach shares the philosophy of the S-matrix programme in that it realizes the nearest singularity concept in a way compatible with unitarity. The Born term of the Lagrangian defines a property of the S-matrix in some domain, a pole singularity or the values at some point. Successive iterations, through the analyticity–crossing–unitarity mechanism, introduce the various unitarity cuts. Thus successive approximations will be valid (as solutions of the assumed Lagrangian) in increasingly larger regions around the points where the Born inputs are imposed. In this philosophy the Lagrangian is somewhat a matter of convenience. As we shall see, the σ-model and the ρ-model give very similar results. In general, different Lagrangians *may* be equivalent if one calculates high enough orders, the main difference in practice being the efficiency in reproducing particular features of the dynamics at low orders of approximation.

Finally, it should be emphasized that the physical content of reasonably realistic Lagrangian models is very imperfectly understood. The importance of the Padé approach in the field theory context lies in the possibility of making apparently realistic calculations. The method circumvents problems

‡ Higher order corrections to the "tree amplitudes" of non-linear field theories have recently been calculated using the "super-propagator" formalism (Lehman, 1972). For applications to $\pi\pi$ scattering see Ecker and Honnerkamp (1973) and Pervushin and Volkov (1974).

involved in perturbation expansions of field theory, certainly those to do with bound states, and very possibly the conjectured essential singularity at $g^2 = 0$.

(b) Perturbation Expansions and the Padé Approximant Method

The only method yet devised for practical calculations using Lagrangian field theory is to make a power series expansion in the coupling constant. Thus, if F is a scattering amplitude, and g the coupling constant,

$$F(g) = F^{(0)} + gF^{(1)} + g^2F^{(2)} + \cdots. \tag{11.2.1}$$

For strong interactions, however, this perturbation expansion converges slowly, if at all, and so constitutes only a formal solution for F. Nevertheless, the full information about F is contained in the coefficients $F^{(i)}$, and so what we seek is a technique for summing the series. The Padé approximation provides a method for doing this, and has been extensively used for making dynamical calculations starting from specific renormalizable Lagrangians. (For reviews see: Basdevant (1970, 1972); Zinn-Justin (1971a,b).)

The $[N, M]$ Padé approximant $F^{[N,M]}(g)$ to $F(g)$ is defined as the ratio of two polynomials in g, $P_N(g)$ and $Q_M(g)$, which has the same power series expansion in g as $F(g)$ up to order $N+M$, i.e.

$$F^{[N,M]}(g) \equiv \frac{P_N(g)}{Q_M(g)} = F(g) + O(g^{N+M+1}). \tag{11.2.2}$$

It follows that the coefficients of the polynomials $P_N(g)$ and $Q_M(g)$ are determined, up to a common factor, by the first $N+M+1$ coefficients $F^{(i)}$, and the $[N, M]$ Padé approximant is thus unique. For example,

$$F^{[1,1]}(g) = F^{(0)} + gF^{(1)}[1 - gF^{(2)}/F^{(1)}]^{-1}. \tag{11.2.3}$$

Padé approximants have many interesting and useful mathematical properties (for reviews see: Baker (1965); Zinn-Justin (1971b)), the most important of which concerns the so-called homographic transformation

$$f(z) \to \frac{af(z)+b}{cf(z)+d},$$

where a, b, c and d are arbitrary constants. For example, diagonal (i.e. $N = M$) approximants preserve homographic transformation properties. A most important result that can be shown to follow from this property (see e.g. Masson (1971)) is that certain Padé approximants preserve the unitarity condition. Thus, if we have a perturbation series for a partial-wave amplitude i.e.

$$T_l(s) = gT_l^{(1)} + g^2 T_l^{(2)} + \cdots,$$

and construct the Padé approximants

$$T_l^{[N,M]}(s) \equiv \frac{P_N(g)}{Q_M(g)} = T_l(s) + O(g^{N+M+1}), \qquad (11.2.4)$$

then these are unitary if $M = N$ (Gammel and McDonald, 1966; Bessis and Pusterla 1967, 1968). In fact, for the partial-wave amplitudes $T_l(s)$, this result is true for $M \geqslant N$ (Masson, 1967; Caser *et al.*, 1969), and in practice, the non-unitary $[M+1, M]$ approximants do not usually violate unitarity greatly (Basdevant *et al.*, 1968, 1969a; Caser *et al.*, 1969; Basdevant and Lee, 1969a). However, the total amplitude

$$T(s, t, u) = \sum_l (2l+1) P_l(\cos\theta_s) T_l(s),$$

has certain crossing properties which are lost in the amplitude defined by

$$T^P(s, t, u) \equiv \sum_l (2l+1) P_l(\cos\theta_s) T_l^{[N,M]}(s).$$

Moreover, $T^P(s, t, u)$ will not necessarily have the correct analytic structure in s at fixed t. In fact, for the equal mass case it will always have a left-hand cut starting at $s = 0$, and not at $s = -t$. It is therefore sometimes useful to consider Padé approximants for the full amplitude $T(s, t, u)$. These approximants satisfy crossing symmetry exactly, but unitarity will not be satisfied. For multichannel scattering, matrix Padé approximants can be constructed, and similar results hold (Basdevant *et al.*, 1968, 1969a).

From the approximate amplitudes of equation (11.2.4), various physical quantities may be extracted, such as phase shifts and resonance parameters. For example, the positions of poles in the complex plane may be obtained by finding the zeroes of $Q_M(s)$ for a fixed value of g. Those which are near the physical region are identified as bound states or resonances provided their parameters are stable as the order of the approximation is changed. The Padé approximants $T_l^{[N,M]}(s)$ can also be continued to complex values of l, and lead to predictions for Regge trajectories (Basdevant *et al.*, 1968, 1969a; Bessis and Pusterla, 1967, 1968).

There remains the problem of convergence. In potential theory, the method has been proved to converge (Chisholm, 1963; Garibotti and Villani, 1969), but no rigorous proof exists in field theory. The only direct way of studying convergence in this case is to see how stable are the results obtained in an actual calculation. The only example which has been studied in this way is the ϕ^4 theory for mesons, where the Feynman graphs up to fourth order in g have been calculated (Bessis and Pusterla, 1967, 1968). The results showed that the basic low-energy features (e.g. resonance positions) were extremely stable in going from third to fourth order, and lend support to the conjectured convergence of the Padé approximation.

There is also an indirect way to test convergence, and this is to study the crossing properties of the (unitary) Padé approximants to the partial wave amplitudes, i.e.

$$T_l^{[N,M]}(s) = T_l(s) + Rg^{N+M+1}.$$

This is conveniently done by using the BNR crossing conditions and the inequalities of Martin which were discussed in Chapter 7. This leads to estimates for the remainder term R, and hence of practical convergence (Basdevant *et al.*, 1969b; Basdevant and Lee 1969b, 1970; Basdevant and Zinn-Justin, 1971).

A different approach to the convergence question has been proposed by Johnson (1971). A functional relation is introduced which connects successive partial sums of a power series expansion, so that the sum to infinity of the series constitutes the solution to the fixed point equation under the functional equation. Johnson suggests that the Padé method estimates a solution by contraction, i.e. that the operation of constructing the Padé approximant is a contraction mapping. Basdevant (1972) conjectured that since the coefficients of the Feynman series are determined from one another by a functional equation embodying unitarity and crossing, it may therefore be possible to use techniques like Atkinson's (Atkinson, 1970) to prove that the contraction mapping suggested by Johnson does take place.

(c) Calculations Based on Specific $\pi\pi$ Lagrangians

The precise treatment of unitarity means that the Padé method is essentially a low-energy one, where the forces are taken into account exactly up to a given order in the coupling constant. The physical input lies in the choice of Lagrangian, and several have been studied.

(i) ϕ^4 *Theory*

The simplest Lagrangian that has been extensively studied is that of the ϕ^4 theory:‡

$$\mathscr{L} = \tfrac{1}{2}[(\partial_\mu \boldsymbol{\pi})^2 - \mu^2(\boldsymbol{\pi})^2] - \frac{g}{4}(\boldsymbol{\pi} \cdot \boldsymbol{\pi})^2, \tag{11.2.5}$$

‡ This theory is of historical interest in that it was this form of $\pi\pi$ interaction which was first discussed (Matthews, 1950, 1951a,b; Salam 1950; Ward 1951); see also the review of Matthews and Salam (1954).

Where, in the notation of Appendix A, $\boldsymbol{\pi} \equiv \boldsymbol{\phi}_\pi$, and g, the renormalized coupling constant is given by $g = -4\pi\lambda$. The first calculation with this Lagrangian (Alexanian and Wellner, 1965) used Padé approximants for the phase shifts, but this procedure is unsatisfactory as it generates essential singularities in the amplitudes. Later calculations computed the perturbation series up to third order and applied the Padé method to the K-matrix (Copley and Masson, 1967), and up to fourth order using the Padé method for the S-matrix (Bessis and Pusterla, 1967, 1968).

Some encouraging features were found in these calculations: (1) the Padé approximants were found to converge rapidly (Bessis and Pusterla, 1967, 1968). (2) The coupling could be adjusted to give a resonance in the $J = I = 1$ state with the physical rho-mass. With this value of g, an $I = 0$, $J^P = 2^+$ resonance is generated with mass approximately 1600 MeV, not too far above the mass of the f° resonance. (3) Rising Regge trajectories were found with reasonable intercepts, i.e. $\alpha(0) \approx 0{\cdot}7$. These results were, however, accompanied by a number of less satisfactory ones: (1) Both S-wave phase shifts were found to be negative, with scattering lengths $a_0^0 \approx -0{\cdot}5$, $a_0^2 \approx -0{\cdot}2$, and although this is not too bad for the $I = 2$ amplitude, it is unacceptable for the $I = 0$ amplitude. (2) All three isospin states had nearly degenerate Regge trajectories, which while reasonable for $I = 0$ and 1, leads to an exotic $I = 2$ resonance with mass approximately that of the f°. (3) The resonance widths were far too small. Typically, $\Gamma_\rho \approx 30$ MeV and $\Gamma_{f^\circ} \approx 50$ MeV.

We conclude that, although the ϕ^4 theory gives encouragement to applying Padé techniques, in its details the actual Lagrangian does not correspond to the physical world.

An attempt has been made to improve the above results by considering the coupled $\pi\pi$–$K\bar{K}$ problem (Basdevant *et al.*, 1968, 1969a). The exotic $I = 2$ resonance is found to be split from the f°, but the general features of the model remain similar to those of the simple ϕ^4 theory with pions alone. Thus, the scattering lengths are $a_0^0 \approx -0{\cdot}6$; $a_0^2 \approx -0{\cdot}3$, and the Regge intercepts are $\alpha_\rho(0) = 0{\cdot}5$; $\alpha_f(0) \approx 0{\cdot}6$.

(ii) *σ-Model*

In an attempt to remedy the defects of the ϕ^4 Lagrangian, it has been suggested that extra information should be included from current algebra (Lee, 1969). This is conveniently done by studying the Lagrangian of the σ-model (Gell-Mann and Levy, 1960).

$$\mathcal{L} = \tfrac{1}{2}[(\partial_\mu \boldsymbol{\pi})^2 + (\partial_\mu \sigma)^2] - \frac{\mu^2}{2}(\boldsymbol{\pi}^2 + \sigma^2) - \frac{g}{4}(\boldsymbol{\pi}^2 + \sigma^2)^2 + c\sigma, \tag{11.2.6}$$

where σ is the isoscalar, scalar field. The σ-model was developed in the aftermath of the discovery of the Goldberger–Treiman relation (see Appendix D) in order to provide a realization of the PCAC condition. This condition is satisfied, and the tree diagrams of this Lagrangian give precisely the results of current algebra in the soft-pion limit (Bardeen and Lee, 1969). Furthermore, it is possible to renormalize the theory such that current algebra constraints and the PCAC condition are preserved at each order of the perturbation expansion (Symanzik, 1969, 1970; Lee, 1969). The expansion is performed according to the number of loops (Symanzik, 1969, 1970). The parameter c may be related to the pion decay constant f_π by the PCAC condition; $c = f_\pi m_\pi^2$.

Present calculations using this model (Basdevant and Lee, 1969b, 1970) have been restricted to the one-loop approximation, i.e., only the [1, 1] Padé approximant has been calculated. The coupling constant g is fixed by requiring the $I = 0$ S-wave phase shift to pass through 90° at a specific energy. The resulting $I = 0$ and 2 S-wave phase shifts are shown in Fig. 11.2.1 for two such conditions. The improvement of these results over the ϕ^4

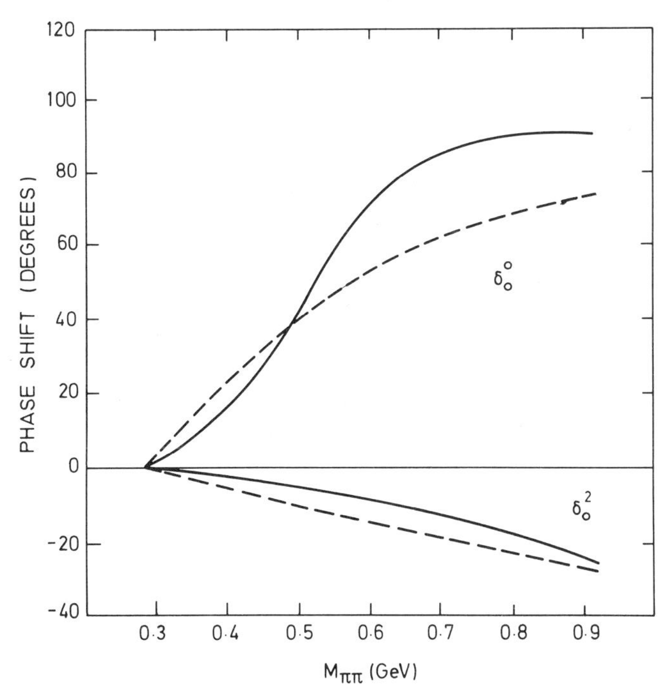

FIG. 11.2.1. S-wave $\pi\pi$ phase shifts from single-channel Padé calculations: ——— σ-model (Basdevant and Lee, 1969b, 1970) for $\delta_0^0(m_\rho^2) \approx 90°$; – – – – ρ-model (Basdevant and Zinn-Justin, 1971).

theory comes about because the dominantly S-wave forces of that theory are now supplemented by a σ-pole force. The latter is also dominantly S-wave, but, due to the Adler zero (see Chapter 14), the relative weights of the two terms are such as to cancel at low energies, leaving a strong P-wave force, which is known to be attractive in the $I=0$ state and repulsive for $I=2$. By examining the position of the complex pole on sheet II, the parameters

$$M_\sigma \approx 500\ \text{MeV}; \quad \Gamma_\sigma \approx 300\ \text{MeV},$$

are deduced. It is interesting that the position of the resonance is far from the 90° point of δ_0^0.

Having fixed the value of g in this way, the ρ and f° resonances are then generated dynamically with typical parameters (Basdevant and Lee, 1969b, 1970)

$$M_\rho = 780\ \text{MeV}; \qquad \Gamma_\rho = 35\ \text{MeV},$$

and

$$M_f = 1115\ \text{MeV}; \qquad \Gamma_f = 180\ \text{MeV}.$$

Thus, the ρ-resonance is still too narrow, as in the ϕ^4 theory. Also the exotic $I=2$ resonance remains, although it is now split from the f° by about 250 MeV.

Examination of the BNR relations and the Martin inequalities (see Chapter 7) shows that crossing symmetry is well approximated. What remains to be done is the calculation of higher Padé terms to see if the residual unsatisfactory features can be removed.

(iii) ρ-*Exchange Model*

A third model which has been studied considers ρ-exchange as the dominant force between pions, treating the ρ as a massive Yang–Mills field (Yang and Mills, 1954), universally coupled to the conserved isospin current, following the ideas of Sakurai (1961). The Lagrangian is

$$\begin{aligned}\mathscr{L} = {}&\tfrac{1}{2}[\partial_\mu \boldsymbol{\pi} + g\boldsymbol{\rho}_\mu \wedge \boldsymbol{\pi}]^2 - \frac{\mu^2}{2}(\boldsymbol{\pi}\,.\,\boldsymbol{\pi})\\ &-\tfrac{1}{4}[\partial_\mu \boldsymbol{\rho}_\nu - \partial_\nu \boldsymbol{\rho}_\mu + g\boldsymbol{\rho}_\mu \wedge \boldsymbol{\rho}_\nu]^2\\ &+\frac{m_\rho^2}{2}\boldsymbol{\rho}_\mu \cdot \boldsymbol{\rho}_\mu,\end{aligned} \tag{11.2.7}$$

where $\boldsymbol{\rho}_\mu$ is the isovector, vector field. The renormalized parameters m_ρ and g are directly related to the physical ρ mass and width. The first two orders of the perturbation series were calculated (Basdevant and Zinn-Justin, 1971).

Although the amplitudes following from such a Lagrangian are not strictly renormalizable, the behaviour of the theory in second order is the same as for a renormalizable one, in the sense that it requires no more subtractions than would be expected if it were renormalizable (see e.g. Veltman (1968)). Only one subtraction is necessary in the second order term. Since the Born terms explicitly satisfy the Adler zero condition (see Chapter 14)

$$A(s=t=u=m_\pi^2)=0,$$

this subtraction constant is determined by requiring that the second-order term vanishes at this point also. Thus the important requirement of PCAC (see Appendix D) is explicitly imposed.

As in the σ-model, only the [1,1] Padé approximant has been computed (Basdevant and Zinn-Justin, 1971). The phase shifts for both S-waves are shown in Fig. 11.2.1 the corresponding scattering lengths being

$$a_0^0 \approx 0{\cdot}17; \qquad a_0^2 \approx -0{\cdot}06.$$

The $I=0$ S-wave agrees well with experiment, and a pole appears on the second sheet with parameters

$$M_\sigma \approx 450 \text{ MeV}; \qquad \Gamma_\sigma \approx 500 \text{ MeV},$$

which could be interpreted as a σ-resonance provided these values remain stable on going to higher order approximants. The f° resonance is generated dynamically with parameters

$$M_f \approx 1280 \text{ MeV}; \qquad \Gamma_f \approx 250 \text{ MeV},$$

in fair agreement with experiment, and no exotic $I=2$ state now appears. Since the ρ-resonance is part of the input in this theory, there is no difficulty in fitting its parameters.

(iv) *Concluding Remarks on Single-Channel Calculations*

The experience gained from the above three Lagrangians is very encouraging. The extensive study of the $\lambda\phi^4$ theory, although not leading to results close to experiment, gives some grounds for confidence in the convergence of the Padé method, which even in lowest order leads to interesting results for the σ- and ρ-model Lagrangians. The phase shifts of the ρ-model bear a close resemblance to the experimental results, a particularly interesting feature being that the f° resonance is dynamically generated. The σ-model, although qualitatively successful, is less satisfactory; in particular the predicted rho-width is much too narrow, and there is an unwanted $I=2$ D-wave resonance at about 1400 MeV. It will be most interesting to see whether the calculation of higher order approximants improve matters. One possibility that has been raised (Basdevant, 1970) is that the results of the σ-

and ρ-models might converge if higher order approximants are included. In this case we could have a type of bootstrap situation, since in the σ-model a ρ-resonance is dynamically generated, and in the ρ-model a σ-resonance is produced. Alternatively, the differences may correspond to a genuine difference in the assumed dynamics, and may remain on calculating higher orders. The calculation of higher order terms is in any case important in both models to test the basic philosophy that the Padé series is rapidly convergent.

(d) Multichannel Calculations

Aside from the important problem of the convergence of the Padé series, there is the question of the effect of other coupled channels. This problem has been studied for the coupled $\pi\pi$–$K\bar{K}$ system in the ϕ^4 theory (Basdevant *et al.*, 1968, 1969a), but the $\pi\pi$ amplitudes differed little from those obtained in earlier single-channel calculations (Copley and Masson 1967; Bessis and Pusterla, 1967, 1968). More recently, a systematic study of this problem has been made by Iagolnitzer *et al.* (1973), and we shall consider their work here.

(i) *Extension to SU*(3)

A systematic study of this case in the vector meson exchange model has been made by Iagolnitzer *et al.* (1973). The Lagrangian is

$$\mathscr{L} = -\tfrac{1}{4}\,\mathrm{Tr}\,\{\partial_\mu V^\nu + ig[V^\mu, V^\nu]\}^2 + \frac{m^2}{2}\,\mathrm{Tr}\,V_\mu^2$$

$$+\tfrac{1}{2}\,\mathrm{Tr}\,\{\partial_\mu S + ig[V^\mu, S]\}^2 - \frac{\mu^2}{2}\,\mathrm{Tr}\,S^2, \qquad (11.2.8)$$

where V^μ and S are 3×3 Hermitian traceless matrices representing the fields associated with the octets of vector and pseudoscalar mesons, with SU(3) symmetric masses m and μ, respectively. The coupling constant g is fixed to within 20% by the widths of the vector mesons. The theory based on this Lagrangian is strictly not renormalizable,‡ but in the one-loop approximation, the results are the same as for a renormalizable theory owing to local cancellations of divergences (compare the discussion of the ρ-exchange Lagrangian given above). Nevertheless, the non-renormalizability introduces other technical difficulties (for the details see Iagolnitzer *et al.* (1973)), and to minimize these, physical masses for the vector mesons are only

‡ Iagolnitzer *et al.* (1973) also consider, without numerical details, a more complicated, but related, Lagrangian which is renormalizable, and show that the latter is expected to give similar results to the Lagrangian of equation (11.2.8).

introduced in the renormalization constants of the one-loop graphs. Like the ρ-exchange model, the P-waves, in this theory are essentially input, and the S- and D-waves dynamically predicted.

The values of m and μ were chosen to produce a reasonable P- and D-wave mass spectrum, although the results are sensitive to the precise choice. The actual values used were $m = 770$ MeV and $\mu = 250$ MeV. In addition, an independent subtraction constant is needed for each S-wave channel. In the $\pi\pi$ elastic channel, this constant was fixed in the same way as in the ρ-exchange model—the second-order amplitude is subtracted at the off-mass-shell point

$$s = t = u = m_\pi^2,$$

and the Adler zero imposed. For $\pi K \to \pi K$ and $\pi\eta \to \pi\eta$ the Born terms do not exactly satisfy the Adler condition, but in the spirit of the hypothesis that vector meson exchanges dominate the S-wave amplitudes at very low energies, the second-order amplitudes are subtracted at threshold and made to vanish there by adjusting the subtraction constants. In the $K\bar{K}$ elastic channel, the subtraction constant is exploited to impose the S^* at the $K\bar{K}$ threshold. We return to this point below.

(ii) *Padé Solution*

The [1, 1] Padé approximant was calculated both for the entire matrix of scattering amplitudes, and for each element separately. For the former, if

$$\mathbf{T}_l = G\mathbf{T}_l^{(1)} + G^2\mathbf{T}_l^{(2)},$$

is the second-order perturbation series for the matrix of coupled channels $\mathbf{T}_l$ ($G = g^2/16\pi^2$), then the [1, 1] Padé solution is

$$\mathbf{T}_l^{[1,1]} = G\mathbf{T}_l^{(1)}[\mathbf{T}_l^{(1)} - G\mathbf{T}_l^{(2)}]^{-1}\mathbf{T}_l^{(1)}.$$

In general, this approach is preferable to forming the Padé solution for each element separately (for example, only the Padé approximant to the matrix is unitary), although sometimes the latter has better analytic properties (this is true for $\pi\pi$ scattering, where the matrix form has a spurious left-hand cut produced by coupling to the $K\bar{K}$ channel).

The spectrum of resonances produced (or, more precisely of second-sheet poles of the matrix Padé approximant) is shown in Table 11.2.1 for the choice of parameters g, m, and μ given earlier. The masses of the 1^- states are input values, but the predicted widths are in good agreement with experiment, except for that of the rho, which is somewhat low. For the 2^+ states, a well-defined octet of resonances of reasonable masses is predicted, except for the K^{**} which is rather heavy (the experimental value is about 1410 MeV). However, except for the f°, the predicted widths are not very

TABLE 11.2.1. Resonance Spectrum Predicted from the SU(3) Coupled Channel Calculations of Iagolnitzer *et al.* (1973).

State	J^P	I^G	Mass (MeV)	Width (MeV)
ε	0^+	0^+	460	675
δ_N	0^+	1^-	775	610[a]
S^*	0^+	0^+	990	40
κ	0^+	$\frac{1}{2}$	665	840
ρ	1^-	1^+	764	83
K^*	1^-	$\frac{1}{2}$	845	52
ϕ	1^-	0^-	1022	≈ 0
f°	2^+	0^+	1365	165
f'	2^+	0^+	1536	8
A_2	2^+	1^-	1332	143
K^{**}	2^+	$\frac{1}{2}$	1539	5

[a] The position of this pole is very sensitive to the value of the $K\bar{K}$ subtraction constant.

satisfactory, those of the K^{**} (5 MeV) and f' (8 MeV) being especially bad. (The experimental values are about 110 MeV and 75 MeV, respectively.) As in the single-channel ρ-exchange model, there are no exotic states predicted.

The branching ratios of the various resonances are, of course, also given by the model. For the $I = S = 0$ states the $\eta\eta$ channel essentially decouples, whereas there is strong coupling between the $\pi\pi$ and $K\bar{K}$ channels. The main result is that the f and f' states couple relatively weakly to the $K\bar{K}$ and $\pi\pi$ channels, respectively, in accord with experiment. In the $I = \frac{1}{2}$, $S = 1$ state, the ηK channel decouples from the πK channel so that the K^{**} state is almost elastic. Of course, in nature, it may well decay into channels not included in this calculation.

Next we turn to the S-waves, and the question of 0^+ resonances. Of these the S^* has essentially been inserted via the subtraction procedure in the $K\bar{K}$ channel. The other experimentally established reasonably narrow 0^+ state in the region considered is the δ_N (975) with width about 60 MeV or less. The model fails to predict such a state, although there is a pole on the second sheet far away in the complex plane (cf. Table 11.2.1). There are also distant poles in the $I = 0$ $\pi\pi$ and $I = \frac{1}{2}$ πK S-waves. The interpretation of poles so far from the physical region is a controversial matter, and we shall therefore concentrate on the predictions for the physically observable phase shifts for these waves.

As mentioned previously, the $\pi\pi \to \pi\pi$ subtraction constant was fixed by imposing the Adler condition on the second-order terms, and for $\pi K \to \pi K$ and $\pi\eta \to \pi\eta$ by requiring the second-order contributions to vanish at threshold. For the $K\bar{K} \to K\bar{K}$ channel the subtraction constant was adjusted to reproduce the S^* effect in the $I=0$ channel, i.e. to insert a pole in the coupled-channel T-matrix close to the $K\bar{K}$ threshold. This is achieved at the cost of introducing a spurious left-hand $K\bar{K}$ cut in the $\pi\pi$ elastic amplitude, which prevents the matrix Padé method being used below 950 MeV. Above this energy the resulting $\pi\pi$ phase shift δ_0^0 is shown in Fig. 11.2.2. Including the $\eta\eta$ channel gives a slightly better fit to the inelasticity parameter η_0^0 without altering the phase shift appreciably. Below 950 MeV, the phase shift δ_0^0 obtained from a Padé approximant to the single $\pi\pi$ element is also shown in Fig. 11.2.2, and this solution should presumably interpolate smoothly to the former solution above the $K\bar{K}$ threshold. The $\pi\pi$ phase shift δ_0^2 obtained from the single-channel approximant is also shown. The corresponding scattering lengths are

$$a_0^0 = 0\cdot 148; \qquad a_0^2 = -0\cdot 043. \tag{11.2.9}$$

The πK S-waves (which are insensitive to the subtraction in $K\bar{K} \to K\bar{K}$) are shown in Fig. 11.2.3. For the $I=\frac{1}{2}$ amplitude, which is of the "Down" variety, there is only very weak coupling to the ηK channel, and the single- and two-channel results are almost identical. The corresponding scattering lengths are

$$a_0^{\frac{1}{2}} = 0\cdot 127; \qquad a_0^{\frac{3}{2}} = -0\cdot 059. \tag{11.2.10}$$

Finally, the Padé solution for $\pi\pi$, πK and $K\bar{K}$ elastic scattering, and $\pi\pi \to K\bar{K}$ was tested against the BNR relations (see Chapter 7), and all the relations involving only S-, P- and D-waves were found to be satisfied to within a few per cent.

(iii) *Concluding Remarks*

The coupled-channel analysis of Iagolnitzer *et al.* (1973) is the most extensive calculation of meson dynamics to date within the framework of a Lagrangian theory, and in view of the technical complexity involved in extending such calculations to higher order Padé approximants, it may well retain this status for sometime to come.‡ It is therefore interesting to consider to what extent it has succeeded in extending the successes of the single-channel $\pi\pi$ calculation (ρ-exchange) into an SU(3) framework. These successes are (cf. Section 11.2(c)(iii)):

(a) No exotic states are predicted.

(b) Given the rho-parameters, the existence, mass and width of the $f°$ (1280) are predicted.

‡ Aspects of the calculation which might perhaps be varied are procedures for surmounting the non-renormalizability of the input field theory, and the incorporation of subtractions and effects of mass splittings.

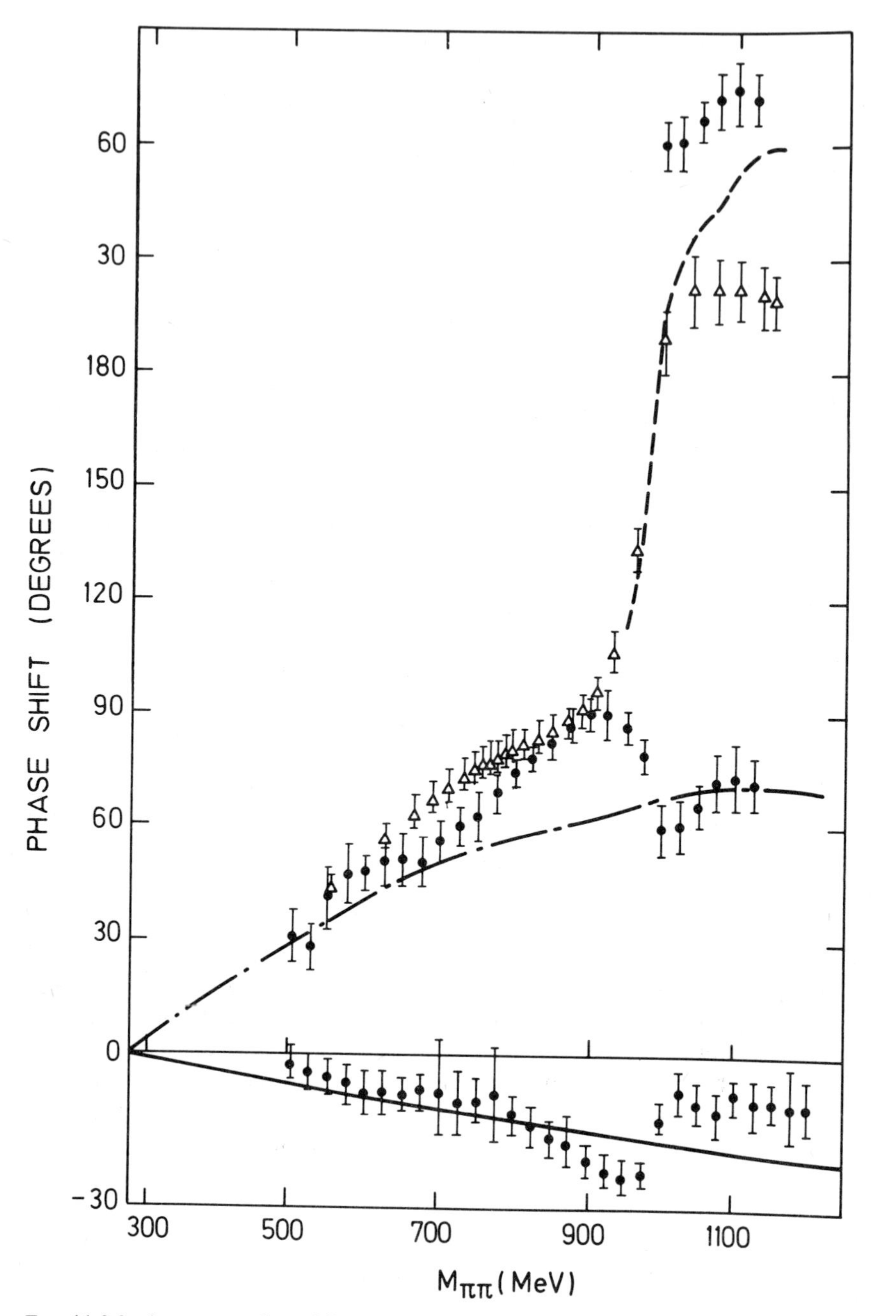

FIG. 11.2.2. *S*-wave $\pi\pi$ phase shifts from Padé calculations of Iagolnitzer *et al.* (1973). ——— δ_0^2; — — — δ_0^0 from Padé approximant to matrix of $\pi\pi$ and $K\bar{K}$ coupled channels; — · — · — δ_0^0 from Padé approximant to single $\pi\pi$ element. The data are from Baton *et al.* (1970b) (circles) and Protopopescu *et al.* (1973) (triangles).

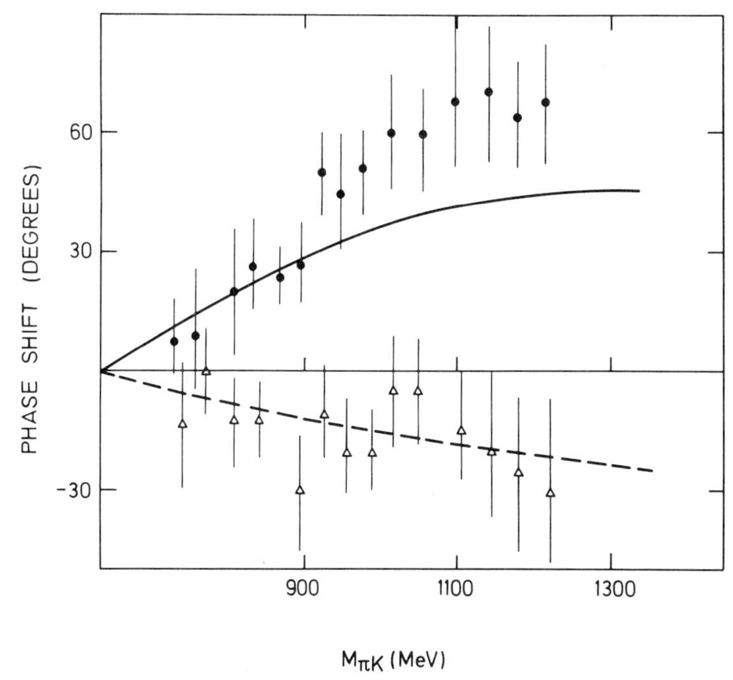

FIG. 11.2.3. S-wave πK phase shifts from Padé calculations of Iagolnitzer *et al.* (1973) ——— $\delta_0^{\frac{1}{2}}$; — — — $\delta_0^{\frac{3}{2}}$. The data are from Mercer *et al.* (1971).

(c) The S-wave phase shifts are predicted.

Of these, the result (b) is especially impressive.

Let us see how these points continue to the multichannel case: Feature (a) extended into the coupled-channel framework with complete success. Regarding (b), a well-defined multiplet of 2^+ states of roughly the correct masses is predicted although the prediction of the widths is much less successful. The main disappointment comes in extending (c). While the elastic, smoothly varying, phases are well predicted the treatment of the two narrow, well-established resonant effects in this region—the δ_N (975) and the S^* (1000) is much less satisfactory. No narrow state corresponding to the former appears. The latter is achieved by adjusting a free parameter, and in association with a quite spurious singularity which occurs in the same wave at a slightly lower energy. Whether these failings reflect a need for extra channels, extra terms in the Lagrangian, or higher order approximants is at present unclear. In any case, this failing should not be allowed to obscure the other successes of the model, which indicate that in its present form it already contains important aspects of low-energy meson dynamics.

Chapter 12

FESR's, Duality and the Veneziano Model

In this chapter, we trace the evolution from finite energy sum rules (FESR's) via general ideas of duality to specific dual models, in particular that of Veneziano; this latter we discuss in some detail, including applications. Our discussion will concentrate on the phenomenological aspects throughout. It is in this area that these ideas have their origin, and have met with most success. Parallel to the phenomenological work, and inspired by its successes, there has been an ambitious attempt to construct a complete hadron dynamics based on local duality. The object is to construct dual models which exactly incorporate general principles like crossing and unitarity, and correspond to a realistic spectrum of states. Although much progress has been made, the difficulties are formidable, and this theory has not yet reached a stage where it can have any appreciable impact on the interpretation of data. We shall therefore not pursue this aspect, referring instead to the many excellent reviews which are available (e.g. Schwarz, 1973; Veneziano, 1974; Mandelstam, 1974; Frampton, 1974).

12.1. Finite-Energy Sum Rules

(a) High-Energy Scattering and FESR

In the preceding chapters we have concentrated mainly on the behaviour of the $\pi\pi$ scattering amplitudes at low and intermediate energies, i.e. $W \lesssim 2$ GeV. However, as reliable amplitudes become available up to this energy, we may also hope to learn something about high-energy scattering via the use of finite-energy sum rules (FESR) (Igi, 1962; Igi and Matsuda, 1967; Logunov *et al.*, 1967; Dolen, Horn and Schmid, 1968). These arise as follows.

Let $T(z, t)$ be a crossing antisymmetric (symmetric) amplitude where z is the usual variable defined in equation (2.8). Applying Cauchy's Theorem to $z^n T(z, t)$ for n even (odd), using the semicircle contour of radius z_L shown in Fig. 12.1.1, we have, after using crossing symmetry ($z \to -z$ at fixed t)

$$2\int_{z_0}^{z_L} dz\, z^n \operatorname{Im} T(z, t) \approx i \int_{C_1} dz\, z^n T^{\infty}(z, t), \tag{12.1.1}$$

where z_0 is the threshold value of z, and we have assumed that $|z_L|$ is sufficiently large for the asymptotic form of the amplitude, $T^{\infty}(z, t)$, to be

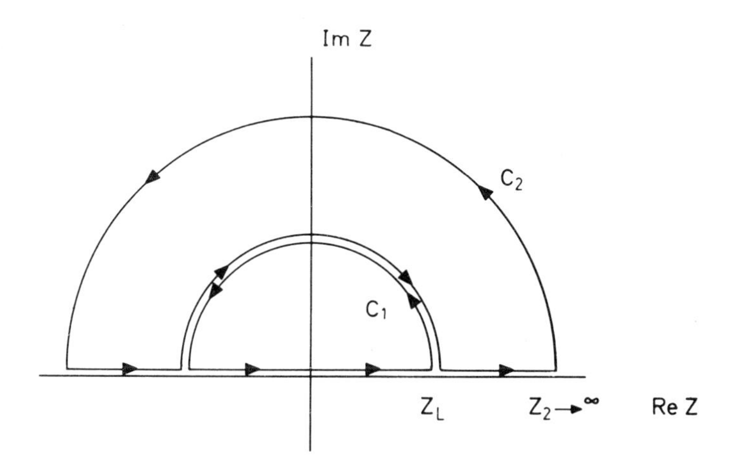

FIG. 12.1.1. The complex contours C_1 and C_2 (see text).

used to estimate the integral round the circle. Alternatively, if the low-energy amplitude is known, then the low-energy integral may be evaluated to obtain information on the high-energy amplitude.

In fact, the above relation leads to information only on the behaviour of the imaginary part of the amplitude. For example, if we consider the case for which the integrals converge as $z_L \to \infty$, the integral around C_1 can be reexpressed, using the Cauchy Theorem for the contour C_2, to give

$$\int_{z_0}^{z_L} dz\, z^n \operatorname{Im} T(z, t) \approx -\int_{z_L}^{\infty} dz\, z^n \operatorname{Im} T^{\infty}(z, t). \tag{12.1.2}$$

If we make the further specific assumption of Regge-pole dominance for $z \geqslant z_L$,

$$T^{\infty}(z, t) = \sum_i \beta_i(t) \frac{[1-\exp(-i\pi\alpha_i(t))]}{\sin \pi\alpha_i(t)} z^{\alpha_i(t)}, \tag{12.1.3}$$

then we obtain from (12.1.2)

$$\int_{z_0}^{z_L} dz\, z^n \operatorname{Im} T(z, t) \approx \sum_i \frac{\beta_i(t) z_L^{\alpha_i(t)+n+1}}{\alpha_i(t)+n+1}. \tag{12.1.4}$$

This relation can also be obtained independent of convergence properties directly from equation (12.1.1) by using the phase-energy relations discussed in Chapter 7. Equation (12.1.4) is an example of an FESR for Regge-pole parameters, but it is worth emphasizing that equation (12.1.1) gives important information on the high-energy amplitude *irrespective* of whether Regge pole behaviour is a good approximation, although, of course, unless a specific form for the high-energy behaviour is assumed the information is only of an integrated (or average) type.

Analogous information about high-energy real parts can be obtained simply by replacing $T(z, t)$ by the function (Gilbert, 1957)

$$G(z, t) \equiv iT(z, t)(z^2 - z_0^2)^{\frac{1}{2}}, \tag{12.1.5}$$

where, for example, z_0 is the threshold value of z. The square root is defined to be positive (negative) for $\operatorname{Re} z > z_0$ ($\operatorname{Re} z < -z_0$) and positive imaginary for $-z_0 < z < z_0$. Applying the same arguments as above leads to the result

$$2\int_{z_0}^{z_L} dz\, z^n (z^2 - z_0^2)^{\frac{1}{2}} \operatorname{Re} T(z, t) \approx -\int_{C_1} dz\, z^n (z^2 - z_0^2)^{\frac{1}{2}} T^{\infty}(z, t), \tag{12.1.6}$$

by analogy with equation (12.1.1). This relation now imposes restrictions on the real parts at high energies.

Since these sum rules are based on simple analytic properties, a wide range of analogous sum rules can be obtained by multiplying the scattering amplitude by other known functions before applying Cauchy's Theorem. In particular one can derive the continuous moment sum rules (CMSR) (Liu and Okubo, 1967) which contain both real and imaginary parts in the integrals. However, the basic idea is contained in the FESR's of equations (12.1.1) and (12.1.6).

Integrals over data in the low-energy region thus impose severe restrictions on high-energy models, and can be regarded as part of the "data" for which models of high-energy scattering must account, in addition to the high-energy observables themselves. The relations are particularly useful (and have been widely used in πN analyses) because they involve the amplitudes themselves, in contrast to differential cross-section and polarization measurements which are bilinear combinations of amplitudes. Even when there is only one spin amplitude (as in $\pi\pi$ scattering), they have the advantage of leading to information on its phase, as well as its modulus; this is very useful, since it is only at low energies that it is practicable to fix this from unitarity.

FESR integrals also have interesting applications independent of specific assumptions regarding high-energy models. The most important example is the study of zeros at fixed-t and fixed-u which are an important feature of high-energy scattering. Clearly, if one of the high-energy amplitudes vanishes at some fixed-t, the corresponding FESR integral must also vanish there.

In using FESR's, certain practical problems arise. For example, the sum rules are at some fixed-t, but the amplitudes are required for *all* $z \leqslant z_L$. Therefore, reconstruction from partial-wave amplitudes by the usual Legendre series expansion almost always involves an extrapolation outside the physical region, thereby introducing unknown errors. A more immediate problem, however, is the fact that the cut-off energy z_L is, in practice, the upper limit of the region in which we have phase shift analyses, or other partial wave data, and in general is well below the energy region to which models of high-energy scattering (e.g. Regge pole models) are thought to apply.‡ In practice this problem may not be as serious as might at first be supposed. The argument is as follows. If the asymptotic form $T(z, t)$ was *exact* for $|z| \geqslant |z_L|$, then it would also be exact for $|z| < |z_L|$. The FESR (equations (12.1.1) etc.) assert that if $T^\infty(z, t)$ is a good *approximation* for $|z| \geqslant |z_L|$ then it will hold on the *average* over the range $|z| < |z_L|$. (We are assuming that the asymptotic form chosen has the same cut structure at fixed-t as the true amplitude.) Since this is an average relation, it is not implausible (although it does not necessarily follow) that this feature holds more locally, e.g.

$$\int_{z_1}^{z_2} dz\, z^n \operatorname{Im} T(z, t) \approx \int_{z_1}^{z_2} dz\, z^n \operatorname{Im} T^\infty(z, t), \qquad (12.1.7)$$

where the range (z_1, z_2) spans several resonance structures. In this case the approach to Regge asymptotics would be as is schematically shown in Fig. 12.1.2, and it is clear that one can use a much lower cut-off in the sum rule when one is concerned with the area under the curve, than when a direct Regge fit to data is attempted. How low the cut-off can be taken will depend on the size of the structures relative to the average; this will differ from case

‡ Techniques have been devised for constructing FESR's which minimize the contribution from the unknown intermediate region (Elvekjaer and Pietarinen, 1972) but to date they have only been applied to πN scattering.

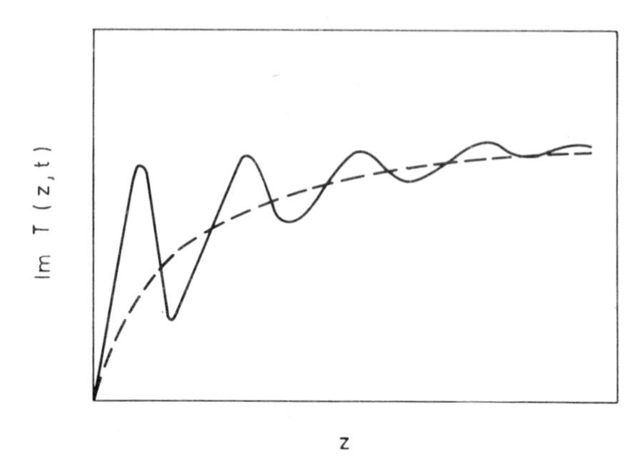

FIG. 12.1.2. Qualitative behaviour of the scattering amplitude (full line) compared to its Regge asymptotic form (dashed line).

to case, so that a degree of caution is always necessary when using FESR's. However, the type of behaviour indicated in Fig. 12.1.2 does occur. It was first noted in a study of πN charge-exchange scattering (Dolen, Horn and Schmid, 1968) and led directly to the ideas of duality, which we will discuss in Section 12.2 below. Before doing this, we will comment briefly on the application of FESR techniques to presently available $\pi\pi$ phase shift data.

(b) Application to $\pi\pi$ Scattering

At present, the use of FESR techniques to deduce features of the high-energy scattering amplitudes is limited by two factors. Firstly, the limited accuracy of our present information on the smaller partial waves gives rise to appreciable uncertainties in some of the FESR integrals, especially those involving real parts. Secondly, the method assumes that the cut-off used is high enough for the phase shift data to approximate the asymptotic behaviour in an average sense, as discussed above. There are strong reasons for supposing that this is not the case for the Pomeron contribution to the $I_t = 0$ exchange amplitude. Thus conventional, and in other cases successful, ideas on high-energy behaviour lead us to expect that this term should dominate the total cross-section for $\pi^+\pi^+$ scattering, which should be slowly varying and of order 17 mb over the energy range 5–30 GeV (see Section 12.2(a) and Appendix C). Over the region where phase shifts are available, the value is considerably less than this.

On the basis of these remarks, we conclude that the FESR over the imaginary part of the $I_t = 1$ exchange amplitude $T_t^1(z, t)$ is the most suitable for present study. The high-energy behaviour of this amplitude is expected to be dominated by the rho-meson Regge pole and its associated cuts; and the main interest focusses on the zero expected to occur at some fixed value of t ($t = t_0$) in the absorptive part of the amplitude at high energies, and hence on the behaviour of the FESR integrals

$$I_n^1(t) \equiv \int_{z_0}^{z_L} dz\, z^{2n} \,\mathrm{Im}\, T_t^1(z, t), \tag{12.1.8}$$

as functions of t. These integrals have been studied for the case $n = 0$ by Martin and Shaw (1973) using the parametric forms for the $\pi\pi$ phase shifts given by Hyams *et al.* (1973). The calculation was performed for cut-offs corresponding to $W_s = 1500$ and 1900 MeV, and the calculation repeated in the former case using the central values of the $I = 2$ phases given by Carroll *et al.* (1974). The position of the zero was found to be $t_0 = -0{\cdot}42$, $-0{\cdot}42$, $-0{\cdot}40$ GeV^2 in each of the three cases. The integrals have also been studied for the cases $n = 0, 1$ by Pennington (1974b) using the bands of phase shift values allowed by Hyams *et al.* (1973) and Durusoy *et al.* (1973). With a cut-off at 2000 MeV the zero was always found to lie in the range $t_0 = -0{\cdot}44 \pm 0{\cdot}05$ GeV^2.

Finally, because of the complete crossing symmetry of $\pi\pi$ scattering, it is possible to derive sum rules which exploit s–t as well as the s–u crossing employed in the simple FESR discussed above. Such sum rules have been studied by several authors, for example Tryon (1973, 1975), Basdevant and Schomblond (1973) and Pennington (1974b), and can in principle be used to supplement the normal FESR in studying high-energy behaviour. At present, however, these sum rules are primarily of use as consistency checks on the input and results of FESR calculations, rather than sources of additional information. We will therefore not discuss them further. An account of their evaluation and comparison with FESR, together with a critical summary of earlier work, can be found in the paper of Pennington (1974b).

12.2. Duality and FESR Bootstraps

In this section we introduce the ideas of duality, and discuss the bootstrap schemes that were the starting point for the specific dual models to be discussed in Section 12.3. Consideration of $\pi\pi$ scattering, and especially of the closely related process $\pi\pi \to \pi\omega$ played an important role in these latter developments.

The discussion is divided into two parts. In the first, we discuss the weaker and more phenomenological forms of duality (semi-local and global duality), and their implications. These are especially interesting for systems where one channel has "exotic"‡ quantum numbers, e.g. $\pi^+\pi^+$ scattering. In

‡ That is, they cannot be formed from systems of two quarks (cf. Chapter 13).

the second we discuss attempts to construct a new bootstrap dynamics based on a stronger form of duality (local duality), and the assumption of pure Regge pole exchange at higher energies. It is these latter attempts which lead directly to the even more specific Veneziano model discussed in Section 12.3 below.

(a) Two-Component Duality

In the discussion of FESR's in the previous section, we noted that a semi-local averaging might occur, and that the full amplitude might oscillate about the asymptotic form in the manner shown in Fig. 12.1.2. Thus, even in the sub-asymptotic region, the asymptotic amplitudes can be roughly estimated by taking a semi-local average, i.e.

$$\langle \text{Im } T(z, t)\rangle \approx \langle \text{Im } T^{\infty}(z, t)\rangle, \tag{12.2.1}$$

where

$$\langle \text{Im } T(z, t)\rangle \equiv \frac{1}{z_2 - z_1}\int_{z_1}^{z_2} dz \text{ Im } T(z, t), \tag{12.2.2}$$

and (z_2, z_1) contains several resonance structures. This idea seems to be borne out by experiment, and, when combined with the idea of resonance dominance of certain amplitudes, constitutes the main idea of *duality* (Dolen, Horn and Schmid, 1968). Equation (12.2.1) then becomes

$$\langle \text{Im } T_{\text{Res}}(z, t)\rangle \approx \langle \text{Im } T^{\infty}(z, t)\rangle, \tag{12.2.3}$$

where $T_{\text{Res}}(z, t)$ is the sum of resonance terms. This equation is an expression of *semi-local duality*, and enables the known properties of resonances to be used to make inferences about the properties of high-energy scattering amplitudes. In what follows, we shall usually argue from semi-local duality, as this is sufficient to embrace all the results, but in fact some are derivable from weaker forms of duality. (It is usually easy to see when a weaker form would be sufficient). Thus, when the averaging interval in equation (12.2.2) extends over the entire range $-z_L \leqslant z \leqslant z_L$ we refer to *global duality*, but if the range is restricted to the left and right cuts, $-z_L \leqslant z \leqslant 0$, $0 \leqslant z \leqslant z_L$ separately we speak of *semi-global duality*. These two forms are, of course, the same for amplitudes with definite crossing properties, but not for amplitudes of mixed crossing symmetry.

Two reservations must be made to the arguments above. Firstly, since the real part of a resonance amplitude vanishes at the resonance position but is non-negligible, and poorly known, some distance from this position, semi-local duality is not a useful idea for the real parts of amplitudes. Secondly, when considering the high-energy behaviour, we must exclude the contribution of the Pomeron (Schmid, 1968a), (for example, K^+p scattering exhibits a flat (Pomeron-dominated) total cross-section at high energies, but no resonances at low-energies). These restrictions lead to the hypothesis of *two-component duality* (Freund, 1968; Harari, 1968) in which the Pomeron contribution $T^P(z, t)$ is removed from $T^\infty(z, t)$ to leave (hopefully) just Regge pole terms and their associated cuts, i.e.

$$T_R^\infty(z, t) \equiv T^\infty(z, t) - T^P(z, t), \tag{12.2.4}$$

which is then asserted to be "dual" to resonances, i.e.

$$\langle \text{Im } T_{\text{Res}}(z, t) \rangle \approx \langle \text{Im } T_R^\infty(z, t) \rangle. \tag{12.2.5}$$

The Pomeron term is then required to be dual to the non-resonant background. Two-component duality implies that the non-resonant background at low energies should be smaller in amplitudes with internal quantum number exchange than in amplitudes to which the Pomeron contributes. This is clearly borne out in πN scattering, where the resonant circles in the partial-wave amplitudes sit on much smaller backgrounds in the $I_t = 1$ amplitudes than in the $I_t = 0$ amplitudes to which the Pomeron contributes (Harari and Zarmi, 1969).

These ideas have their most striking consequences when applied to reactions having exotic channels, and hence no (or at most very weak) resonances, e.g. $\pi^+\pi^+$, $K^+\pi^+$, K^+p, pp etc. (Schmid, 1968a; Harari, 1968). Firstly, the imaginary part of the forward amplitude, and hence the total cross-section for these processes, will be given entirely by the Pomeron, so that the total cross-section should be slowly varying even at moderate energies (this is in accord with experiments on K^+p and pp). Secondly, the reactions $\pi^-\pi^+$, $K^-\pi^+$, K^-p, $\bar{p}p$ will have the same Pomeron contributions as their exotic partners above, but have, in addition, resonances (which, from unitarity, must give positive contributions to the total cross-sections). Since these are dual to the non-Pomeron terms in the asymptotic behaviour, the total cross-sections for these reactions must approach those of their exotic partners from above. This is experimentally confirmed for K^-p and $\bar{p}p$ and is a prediction for $\pi^-\pi^+$ and $K^-\pi^+$.

Away from $t = 0$, since the exotic channels have no resonances duality predicts that the non-Pomeron contribution to their amplitudes are real. Thus for a reaction like $K^+ + n \to K^0 + p$, which is exotic and has no Pomeron contribution, the polarization should be zero. Even more striking results follow in such cases if simple Regge-pole dominance of the non-Pomeron part of the high-energy amplitudes is assumed. For example, in $\pi^+\pi^+$ scattering the leading non-Pomeron trajectories which should contribute are the ρ and P' (or f), so that for this amplitude

$$T_R^{\infty}(z,t)=\frac{1}{3}\beta_{P'}(t)\left\{\frac{1+\exp\left[-i\pi\alpha_{P'}(t)\right]}{\sin\left[\pi\alpha_{P'}(t)\right]}\right\}z^{\alpha_{P'}(t)}$$

$$\times-\frac{1}{2}\beta_\rho(t)\left\{\frac{-1+\exp\left[-i\pi\alpha_\rho(t)\right]}{\sin\left[\pi\alpha_\rho(t)\right]}\right\}z^{\alpha_\rho(t)}. \qquad (12.2.6)$$

Since there are no resonances in the $\pi^+\pi^+$ channel, duality implies that the imaginary part must vanish, leading to $P'-\rho$ *exchange degeneracy*

$$\alpha_{P'}(t)=\alpha_\rho(t)\equiv\alpha(t), \qquad (12.2.7a)$$

$$\tfrac{1}{3}\beta_{P'}(t)=\tfrac{1}{2}\beta_\rho(t)\equiv\beta(t). \qquad (12.2.7b)$$

Since the trajectory functions are dependent only on the quantum numbers of the exchanged particle, and not on the external particles, the result (12.2.7a) (*weak exchange degeneracy*) can be tested in other better known processes (e.g. πN scattering), and is found to hold at least approximately. The other result, equation (12.2.7b) (*strong exchange degeneracy*) is specific to $\pi\pi$ scattering, and is not tested. However, it does lead to interesting predictions regarding amplitude zeros, and associated dips in differential cross-sections, at the point for which $\alpha_\rho(t)=0$, which occurs at $t\approx-0{\cdot}6\ \mathrm{GeV}^2$ (Finkelstein, 1969). To avoid a pole in the P' real part at the point where $\alpha_{P'}(t)=\alpha_\rho(t)=0$, a zero must be introduced into $\beta_{P'}(t)$ at this point. By equation (12.2.7b) this induces a zero in $\beta_\rho(t)$ at the same point, so that the contribution of rho-exchange, which presumably dominates e.g. the reaction $\pi^+\pi^-\to\pi^0\pi^0$, will have a single zero in the imaginary part and a double zero in the real part at this point. Analogous arguments lead to zeros for other processes associated with exotic channels, and, by factorization of Regge pole residues, these structures (often referred to as "nonsense zeros") will propagate to non-exotic processes also, for example pion–nucleon charge-exchange. Such a structure is seen experimentally in the spin-flip amplitude in this latter case, but not in the non-flip amplitude, which instead exhibits a zero in the imaginary part (but not the real part) at the smaller t-value, $t\approx-0{\cdot}2\ \mathrm{GeV}^2$.

By simply repeating the above arguments for other exotic processes, the successful weak exchange degeneracy prediction, equation (12.2.7a) and the partially successful strong exchange degeneracy result, equation (12.2.7b) can be generalized to all the leading meson trajectories. Since the arguments are an exact repeat of those used in the $\pi\pi$ case, we shall summarize the results somewhat briefly. Thus, if we consider, for example, $K^+\pi^+$ scattering at fixed-u, then the relevant Regge exchanges are the K^* (890) and K^{**} (1420) trajectories, which are, of course, just the $S=1$ SU(3) partners of the ρ, P' trajectories. An exactly analogous argument to that given in the $\pi\pi$ case leads to the weak exchange degeneracy result

$$\alpha_{K^*}(t) = \alpha_{K^{**}}(t). \tag{12.2.8a}$$

If we turn to KK scattering ($I = 0, 1$), more Regge poles are allowed to contribute—the ρ, P', ω, A_2, ϕ and f' trajectories. In this case, demanding that the imaginary parts of both exotic amplitudes vanish leads directly to an extension of equation (12.2.7a)

$$\alpha_\rho(t) = \alpha_{P'}(t) = \alpha_\omega(t) = \alpha_{A_2}(t) = \alpha(t), \tag{12.2.8b}$$

and in addition

$$\alpha_\phi(t) = \alpha_{f'}(t). \tag{12.2.8c}$$

These various degenerate trajectories are shown in Fig. 12.2.1. We note that equation (12.2.8b) especially involves non-trivial predictions (e.g. $m_\omega = m_\rho$, $m_f = m_{A_2}$) which are in rather good agreement with experiment. A further

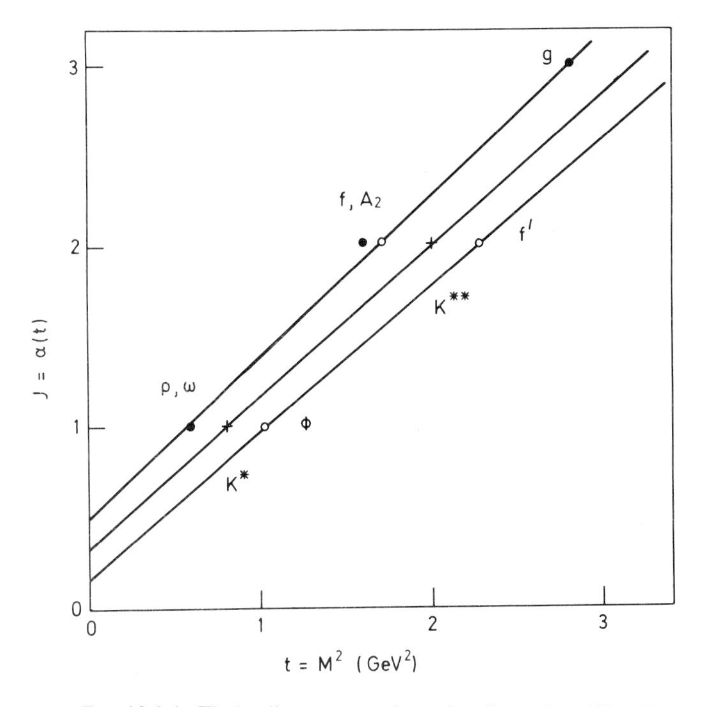

FIG. 12.2.1. The leading meson trajectories of equations (12.2.8).

point is that the f' trajectory is allowed, like the P', to couple to $\pi\pi$ giving an additional term

$$\frac{1}{3}\beta_{f'}(t)\left\{\frac{1+\exp\left[-i\pi\alpha_{f'}(t)\right]}{\sin\left[\pi\alpha_{f'}(t)\right]}\right\}z^{\alpha_{f'}(t)},$$

on the right-hand side of equation (12.2.6). However, in this case there is no known $I=1$ trajectory degenerate with $\alpha_{f'}$ to cancel the contribution to the imaginary part. Thus in order that the imaginary part should vanish

$$\beta_{f'}(t)=0$$

so that the f' trajectory must decouple from $\pi\pi$, and the decay $f'\to\pi\pi$ be forbidden. This result appears to be in good agreement with experiment.

All these results (except for the prediction of nonsense zeros, which rests on factorization, and hence pure Regge pole dominance) can alternatively, but equivalently, be obtained by the use of duality diagrams (Harari, 1969; Rosner, 1969), whereby reactions are represented as quark rearrangements. If the quark lines of such a diagram do not intersect, as in Fig. 12.2.2(a) the

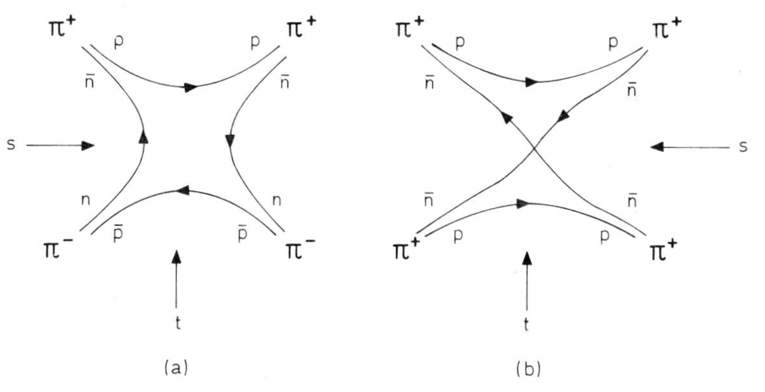

FIG. 12.2.2. Legal and illegal duality diagrams for (a) $\pi^+\pi^-\to\pi^+\pi^-$ and (b) $\pi^+\pi^+\to\pi^+\pi^+$, respectively.

diagram is referred to as "legal" or "planar", and gives a non-vanishing contribution to the absorptive part. If the quark lines cross, as in Fig. 12.2.2(b) it is called "illegal", and gives no contribution to the absorptive part, even though the t-channel quantum numbers are not exotic. The absence of legal duality diagrams for processes like $\pi^+\pi^+\to\pi^+\pi^+$ indicates a vanishing imaginary part, which implies the exchange degeneracy result, equation (12.2.7), for the Regge pole contributions. Similarly, the absence of legal duality diagrams for $K^+\pi^+$ and KK scattering leads to the results (12.2.8).

We see that the combination of duality with Regge pole exchange leads to many predictions. Of these, the weak exchange degeneracy results, equations (12.2.7a) are strikingly successful. The strong exchange degeneracy results, equation (12.2.7b), and the analogous results in πK, KK scattering, are not yet susceptible to direct test. However, together with factorization, they lead to the prediction of nonsense zeros at $\alpha(t)=0$ in processes dominated by the odd signature trajectories ρ, ω and ϕ. As discussed above, in the best known case of πN charge exchange scattering (ρ-exchange), this expectation is fulfilled for the spin-flip amplitude, but not for the non-flip amplitude; in fact the experimental behaviour of this latter amplitude is not well described by simple Regge pole behaviour. Thus our arguments about zeros, which depend crucially on this behaviour, break down in this case and must be reconsidered.

If Regge pole dominance of the high-energy amplitudes is not assumed, then inferences can only be made about zeros of the imaginary parts; this can be done irrespective of whether exotic channels are present, given sufficient knowledge of the resonance spectrum. For example, if we return to $\pi\pi$ scattering, and assume that the ρ-, f- and g-mesons dominate the imaginary parts in their respective energy regions, then the first zeros of the imaginary part in these regions will be at $t \approx -0{\cdot}25$, $-0{\cdot}32$ and $-0{\cdot}34$ GeV^2, respectively, corresponding to the first zeros of the respective Legendre polynomials. Via duality, this suggests a zero of the imaginary parts of the non-Pomeron high-energy amplitudes (especially for $I_t=1$ exchange) at $t \approx -0{\cdot}3$ GeV^2(Schmid, 1968b). However, as noted by Schmid, this zero can easily be pushed out to a larger t-value by the presence of "daughters".‡ As an illustration of this, if we assume the existence of S-wave daughters under the ρ- and f-mesons, and a P-wave daughter under the g-meson, with widths and elasticities equal to those of their parents, then the first zeros occur at $t \approx -0{\cdot}34$, $-0{\cdot}42$, $-0{\cdot}48$ GeV^2, respectively, closer to the value $t \approx -0{\cdot}4$ GeV^2 suggested by the (low cut-off) FESR calculations described in Section 12.1(b).

This approach has been taken further in the "dual absorption model" (Harari, 1971a) where it is combined with the idea of peripheral dominance of these amplitudes. The motivation for this is as follows. The imaginary parts of many helicity-flip amplitudes, to which the familiar vector exchanges (ρ, ω) contribute, vanish at $t \approx -0{\cdot}6$ GeV^2, in accordance with simple Regge pole ideas (leading to dips in differential cross-sections, e.g. in $\pi^- p \to \pi^0 n$ and $\gamma p \to \pi^0 p$). The analogous prediction of zeros at the same

‡ That is, states on Regge trajectories parallel to the leading, or "parent" trajectory, but lying one or more units below it, see Appendix C.

t-values for the corresponding non-flip amplitudes is, however, markedly less successful; in fact they often occur at much smaller t-values, (e.g. the well-known "cross-over" zeros in πN, $t \approx -0{\cdot}15\ \text{GeV}^2$ and NN, $t \approx -0{\cdot}20\ \text{GeV}^2$, scattering). To account for this, the following interpretation is suggested: that the predominant resonances, and hence by duality the predominant part of the high-energy amplitude, occur in partial waves corresponding to a fixed value of the impact parameter R. From this it follows that the t-distribution is given approximately by the Bessel function $J_n(R\sqrt{-t})$, where n is the total s-channel helicity flip. For a given R, the first zero appropriate for the non-flip amplitude (i.e. of $J_0(R\sqrt{-t})$) occurs at a much smaller t-value than those for amplitudes with $n = 1$. Further, since the dips in $n = 1$ amplitudes occur at about the same value of t ($\approx -0{\cdot}6\ \text{GeV}^2$), the radius R must also be roughly the same ($R \approx 1$ fermi), independent of the initial and final states concerned, and this value of R leads to a zero for the $n = 0$ amplitudes at $t \approx -0{\cdot}2\ \text{GeV}^2$. For a fuller discussion of these points we refer to the work of Harari (1971b). While this simple picture does not work for all exchanges,‡ it is in accord with observation so far for ρ and ω exchange; this suggests a zero in the imaginary part of $I_t = 1$ $\pi\pi$ scattering ($n = 0$) at $t \approx -0{\cdot}2\ \text{GeV}^2$, in contrast to the simple Regge pole plus exchange-degeneracy prediction of a zero at $t \approx -0{\cdot}6\ \text{GeV}^2$, As we have seen in Section 12.1(b), the FESR integral suggests an intermediate value, $t \approx -0{\cdot}4\ \text{GeV}^2$, although of course, the cut-off used is rather low. It will be of great interest to see if this result remains unchanged when higher energy amplitudes become available.

(b) Schmid Circles and FESR Bootstrap

We now turn from the phenomenological ideas of semi-local duality to the more ambitious attempts to construct dynamical schemes based on stronger (i.e. more local) forms of duality, which are brought out most clearly by a discussion of the reaction $\pi\pi \to \pi\omega$. It is first convenient to introduce separately the two main ingredients of this discussion: the ideas of the FESR bootstrap, and the local form of duality suggested by Schmid circles.

In addition to assuming that the low-energy amplitudes are dominated by resonances, we shall assume that the high-energy amplitudes are described by Regge pole exchanges. (The Pomeron part is excluded as before.) The FESR now enables a new type of bootstrap to be constructed (Dolen, Horn, and Schmid, 1968; Mandelstam, 1968a). By evaluating the FESR integrals, the low-energy s-channel resonances are related to the Regge exchanges in

‡ For example, it may not work for tensor exchanges (e.g. P', A_2), cf. Fox and Quigg (1973).

the crossed t-channel; i.e. the s-channel resonances "build" the t-channel Regge pole exchanges, which, if continued to $t>0$, yield the parameters of the t-channel resonances. If we choose a process in which the s- and t-channels have the same quantum numbers, e.g. $\pi\pi \to \pi\pi$, $\pi\pi \to \pi\omega$, then we have a "bootstrap" situation. This is the first of the ideas we shall need. Note that it requires only the use of global duality, and only for the imaginary parts.

The second idea suggests a more local form of duality, and extends it to include the real parts also. It stems from the work of Schmid (1968a), who considered the spin-flip amplitude $B^{(-)}(\bar{\nu}, t)(\bar{\nu} = (s-u)/4m)$ for πN charge-exchange scattering, which is known to be dominated at high energies by ρ Regge pole exchange. Taking the Regge pole parameters from high-energy fits over the range 6–18 GeV/c, Schmid continued the corresponding amplitudes down to the region $1{\cdot}0 < p_L < 2{\cdot}4$ GeV/c, where the physical amplitudes are known to have resonance structure, and projected out partial wave amplitudes directly. The Argand diagrams of the resulting partial-wave amplitudes $B_l^{(-)}(s)$ exhibit the now famous Schmid circles, reminiscent of the circles associated with resonances. Furthermore, by use of the relation

$$B_l^{(-)}(\bar{\nu}) \equiv \tfrac{1}{2}\int_{-1}^{1} d\cos\theta P_l(\cos\theta) B^{(-)}(\bar{\nu}, \cos\theta),$$

$$\approx \frac{4\pi}{E-M}\{f^{(-)}_{(l+1)-}(\bar{\nu}) - f^{(-)}_{(l+1)+}(\bar{\nu})\}, \tag{12.2.9}$$

the positions of the Schmid circles can be shown to correspond roughly to some of the prominent πN-resonances, suggesting that the relation between s-channel resonances and t-channel Regge exchanges is indeed rather close. The Schmid circles cannot, of course, precisely reproduce the resonance amplitudes since they are not generated by second-sheet poles; however, one can speculate that they do approximate the true resonance loops. This controversial conjecture (Schmid, 1968a, 1969) was the first form of local duality, and, as noted previously, it involves the use of duality for the real part as well as for the imaginary part, the occurrence of loops being closely connected with the Regge phase factor $e^{i\pi\alpha(t)}$ (Schmid, 1968b; Chiu and Kotanski, 1968). As we shall see it can be used to suggest rather directly a possible resonance structure to be associated with a specific high-energy model.

In pursuing the implications of these ideas, it is convenient to consider the process

$$\pi(p_1) + \pi(p_2) \to \pi(p_3) + \omega(e, q), \tag{12.2.10}$$

where e^μ is the polarization vector of the ω-meson. Only one invariant amplitude $A(s, t)$ exists for this process, given by

$$T(s, t) = \varepsilon_{\mu\nu\tau\sigma} e^\mu p_1^\nu p_2^\tau p_3^\sigma A(s, t). \tag{12.2.11}$$

This amplitude is completely crossing symmetric, and selection rules limit the quantum numbers of the intermediate states in all three channels to be $I = 1$, $J^{PC} = 1^{--}, 3^{--}, 5^{--}$ etc. Of the well-known trajectories, only the rho can contribute. Further, the amplitude has unit spin flip, and since the successes of simple Regge pole exchange models have occurred for such amplitudes, there is reason to hope that this latter approximation will be reasonably successful. Thus the process is particularly suitable to apply the ideas of an FESR bootstrap. This was realized by several authors (Gross, 1967; Ademollo *et al.*, 1967; Matsuda, 1968) who considered in particular the first moment sum rule

$$\int_0^{z_L} dz\, z \operatorname{Im} A(s, t) = \frac{\beta_\rho(t) z_L^{\alpha_\rho(t)+1}}{\alpha_\rho(t)+1}. \tag{12.2.12}$$

Both sides of this sum rule are assumed to be saturated by just the ρ trajectory, and the cut-off z_L placed somewhere between the ρ (765) and g (1690) mesons. Then, in the narrow-width approximation, independent of cut-off, the rho contribution is proportional to

$$2m_\rho^2 - m_\omega^2 - 3m_\pi^2 + t, \tag{12.2.13}$$

which vanishes at

$$t = m_\omega^2 + 3m_\pi^2 - 2m_\rho^2 = -0{\cdot}53 \text{ GeV}^2. \tag{12.2.14}$$

Hence, $\beta_\rho(t)$ must vanish at this point also, and since we have assumed simple Regge pole exchange, it is natural to assume that this corresponds to the zero associated with the vanishing of the trajectory function in exchange degenerate models. Thus we find

$$\alpha_\rho(t = -0{\cdot}53 \text{ GeV}^2) = 0, \tag{12.2.15}$$

in reasonably good agreement with experiment.

If a linear trajectory function is now assumed, i.e.

$$\alpha_\rho(t) = a + bt, \tag{12.2.16}$$

then the trajectory is fixed if the rho-mass is determined. One can attempt this by evaluating the sum rule at the point $\alpha_\rho(t) = 1$. Here the residue function is given in terms of the rho-width, which then cancels from both sides of the equation. The resultant equation, together with the expression for the zero can be solved to give both a and b, and hence the rho-mass. Unfortunately, the result depends on the cut-off, so that the rho-mass cannot in fact be calculated. However, if the cut-off is chosen to give the correct rho-position, the slope of the trajectory is determined correctly, as we have seen; it is then a trivial matter to evaluate the sum rule as a function of t to determine the residue function (up to an overall scale, since the equations are linear in the couplings).

So far the discussion has been confined to cut-offs between the $1^{-}(\rho)$ and $3^{-}(g)$ resonance positions. On considering what happened when the cut-off was increased to include the 3^{--}, 5^{--} etc. states in the FESR, it was found (Rubinstein *et al.*, 1968) that the resonance side of the relation became smaller than the Regge side. Since the rho-trajectory should become a better approximation to the Regge side as the cut-off energy is increased, it is reasonable to assume that additional structure is needed in the resonance spectrum. This was investigated (Rubinstein *et al.*, 1968) by analysing the Regge term into partial waves in the manner suggested by Schmid (Schmid, 1968a) as described above. The resulting Argand plots suggested the resonance spectrum shown in Fig. 12.2.3, the leading rho-trajectory being accompanied by a series of parallel daughters spaced two units below.

These results are very suggestive, and several points are worth emphasizing. The first of these is that we have gone beyond the empirically confirmed range of dual ideas—those which concern FESR's and approximate duality over an intermediate energy range where there is known resonance structure, but where the amplitudes are sufficiently smooth for the Regge exchanges also to give a rough description of the amplitudes. It is now being assumed that even at high energies the t-channel Regge pole exchange can be regarded as built from s-channel resonances, i.e. the descriptions of the amplitude in terms of *just* s-channel poles or *just* t-channel poles are alternatives at *all* energies. This is the extremely strong form of local duality we shall require in the next section. Given this, the necessity of lower lying trajectories is easy to understand. As we have seen in Section 12.2(a), for a simple Regge pole exchange with nonsense zeros, the dominant contributions to the imaginary parts of a spin-flip amplitude come from partial waves corresponding approximately to a fixed impact parameter R, so that the important J values increase like $\sqrt{s}$ rather than s, which is the behaviour of the leading trajectory (cf. Fig. 12.2.3).

If contributions from this impact parameter are to come from resonances, they must lie on lower trajectories. That these lower lying states correspond to a series of parallel daughters is indicated by the Schmid circles analysis, which suggests some other properties also. For example, as s increases for fixed J the elasticity of the resonances decreases extremely rapidly (Chiu and Kotanski, 1968) making it very unlikely that such resonances can be detected and identified even if they exist as predicted. Another point which has been emphasized (Eguchi and Igi, 1971) is that a Schmid analysis of the

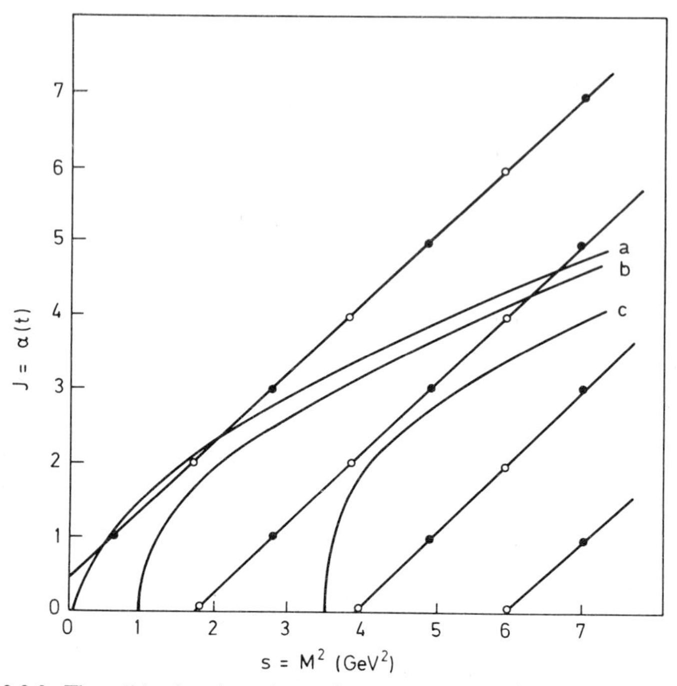

FIG. 12.2.3. The solid points show the predicted spectrum of $I = 1$ states coupling to $\pi\pi \to \pi\omega$. The open circles show the $I = 0$ states suggested by exchange degeneracy: these would couple to $\pi\pi$ but not $\pi\omega$. The curves show the J values corresponding to an impact parameter of about 0·8 fm, for the reactions $\pi\pi \to \pi\pi$ (a), $\pi\pi \to K\bar{K}$ (b), and $\pi\pi \to N\bar{N}$ (c).

leading t-channel Regge pole exchange (i.e. leading power in z) gives only doubly spaced (even) daughters. In order to cater for the existence of odd daughters in the s-channel, i.e. the existence of a unit-spaced spectrum, it is necessary to postulate the exchange of a t-channel daughter spaced one unit below the leading t-channel exchange. This question of the existence of the odd daughters assumes considerable importance in the discussion of the Veneziano model, as we shall see in the following section, as well as for resonance classification in general (see Chapter 13).

12.3. Dual Models

(a) Veneziano Model

The above general considerations quickly led to attempts at explicit realizations through analytic formulae. Before discussing such an explicit formula we summarize the properties required of resonance dominance models in general. First of all, we shall require that, (i) *the particles lie on linear Regge trajectories*

$$\alpha_i(t) = a_i + b_i t. \tag{12.3.1}$$

It was pointed out (Mandelstam, 1968a) that this implies the second property, (ii) *the resonances are of zero width.* This follows if we write a twice-subtracted dispersion relation for $\alpha(t)$

$$\alpha(t) = a + bt + \frac{t^2}{\pi}\int_{t_0}^{\infty} dt' \frac{\operatorname{Im}\alpha(t')}{t'^2(t'-t)}, \tag{12.3.2}$$

and impose equation (12.3.1), which implies that $\operatorname{Im}\alpha(t) \equiv 0$. With this property the amplitude can be written as a sum of poles, and we can then formulate precisely the third condition, (iii) *local duality.* Consider an amplitude for which, for example, the u-channel is exotic. Then we can adopt the following as a definition of local duality: that the amplitude can be written as a sum of either s-channel poles only, or as a sum of t-channel poles only. This is often summarized diagrammatically by Fig. 12.3.1. In the

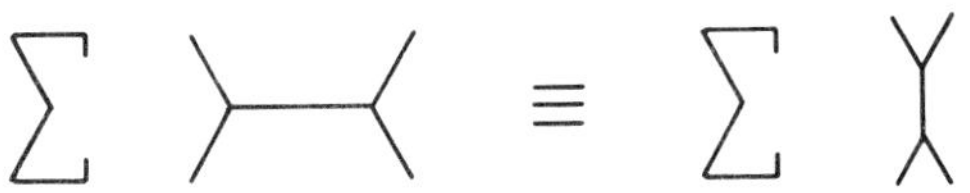

FIG. 12.3.1. Diagrammatic representation of the local duality hypothesis.

absence of exotic channels, we require that the amplitude can be written as a sum of s-channel and u-channel, or s- and t-channel, or t- and u-channel poles. Finally, we require: (iv) *asymptotic Regge pole behaviour,* and (v) *exact crossing symmetry.*

The obvious shortcomings of such a model will be the violation of unitarity associated with the narrow width approximation, and the complete neglect of the Pomeron (non-resonant) contribution.

The first model which satisfied all these requirements was that of Veneziano (1968). Although initially proposed for the reaction $\pi\pi \to \pi\omega$ it is convenient to consider firstly its application to $\pi\pi$ scattering (Shapiro and Yellin, 1968; Shapiro, 1969; Lovelace, 1968) in particular, to the u-channel exotic amplitude

$$V(s, t) \equiv T_u^2(s, t, u) = T_s^{+-}(s, t, u), \tag{12.3.3}$$

in the notation of Chapter 2. Crossing symmetry demands that

$$V(s, t) = V(t, s), \tag{12.3.4}$$

and using the crossing relations of Chapter 2, all the other $\pi\pi$ amplitudes can be expressed in the form

$$T^0(s, t, u) = \tfrac{3}{2}[V(s, t) + V(s, u)] - \tfrac{1}{2}V(t, u),$$
$$T^1(s, t, u) = V(s, t) - V(s, u),$$
$$T^2(s, t, u) = V(t, u). \tag{12.3.5}$$

Thus, if we construct an expression for $V(s, t)$ which satisfies equation (12.3.4) and form the other independent amplitudes from equation (12.3.5), the requirements of crossing symmetry are exactly fulfilled. Therefore, we need consider only this one amplitude, and the simplest Veneziano model for it is defined by the expression

$$V(s, t) = -\beta\frac{\Gamma[1-\alpha(s)]\Gamma[1-\alpha(t)]}{\Gamma[1-\alpha(s)-\alpha(t)]}, \tag{12.3.6}$$

where $\alpha(s)$ is an exchange degenerate $\rho - P'$ trajectory function, assumed to be linear

$$\alpha(s) = a + bs \approx 0{\cdot}5 + 0{\cdot}5(s/m_\rho^2). \tag{12.3.7}$$

This expression satisfies properties (i) and (v) above by explicit construction, and satisfies the other requirements by virtue of the properties of the gamma function (see e.g. Erdélyi *et al.* (1953)). We will briefly illustrate this.

The amplitude has no poles in the exotic u-channel, but does have poles in the s-channel corresponding to the poles of $\Gamma[1-\alpha(s)]$ at $\alpha(s) = N+1 \geq 1$. The residues of these poles are regular functions of t, the poles of $\Gamma[1-\alpha(t)]$ being cancelled by those of $\Gamma[1-N-\alpha(t)]$ and so the amplitude can be written explicitly as a sum of poles in s, i.e.

$$V(s, t) = \beta[\alpha(s)+\alpha(t)-1]\sum_{n=0}^{\infty}\binom{\alpha(t)+n-1}{n}\frac{1}{n+1-\alpha(s)}. \tag{12.3.8}$$

(The form of the residues given in equation (12.3.8) can be simply obtained by expanding $\Gamma[1-\alpha(s)]$ about the pole position and using the relation $\Gamma[1+x] = x\Gamma[x]$). Since $V(s, t)$ is explicitly symmetric in s and t, this is sufficient to establish that it can be expressed as a sum of t-channel poles also, so that the duality property (iii) is satisfied. The residue of the pole at $\alpha(s) = N+1$ is a polynomial in t, and hence in $\cos\theta_s$, of order $N+1$, so that in addition to the resonances along the leading trajectory at $\alpha(s) = J = N+1$, a series of daughter states with $N+1 > J \geq 0$ also occurs. The resulting spectrum is shown in Fig. 12.3.2. We will return to this point below.

The only one of our requirements remaining is that of asymptotic Regge pole behaviour ((iv), above). This is obtained from equation (12.3.6) in two

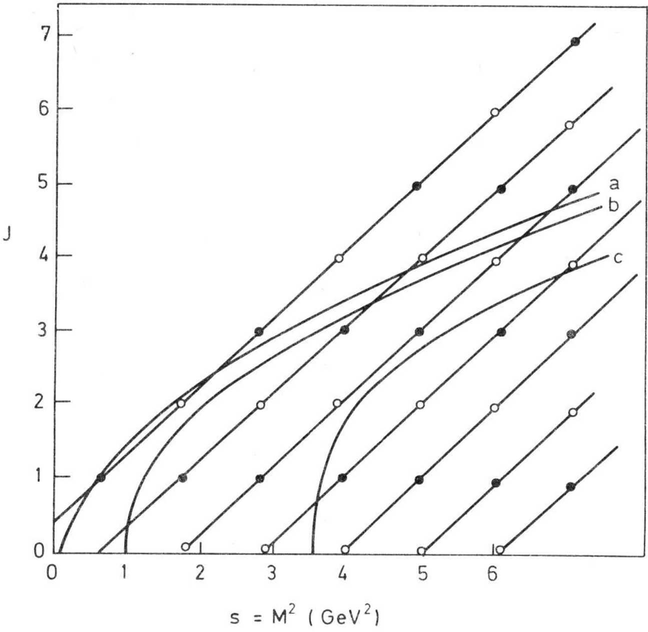

FIG. 12.3.2. The spectrum of $I=0$ (open circles) and $I=1$ (solid points) states predicted for $\pi\pi$ scattering by the single-term Veneziano model. The curve shows the J values corresponding to an impact parameter of 0·8 fm, for the reactions $\pi\pi\to\pi\pi$ (a), $\pi\pi\to K\bar{K}$ (b) and $\pi\pi\to N\bar{N}$ (c).

steps. Firstly, using the relation

$$\Gamma[x]\Gamma[1-x]=\pi \operatorname{cosec}(\pi x),$$

gives

$$V(s,t)=-\beta\pi[\cot \pi\alpha(s)+\cot \pi\alpha(t)]\frac{\Gamma[\alpha(s)+\alpha(t)]}{\Gamma[\alpha(s)]\Gamma[\alpha(t)]}.$$

Secondly, we use the asymptotic result

$$\Gamma[x+a]/\Gamma[x+b]\sim x^{a-b}, \qquad x\to\infty.$$

Applying this to $V(s,t)$ by letting $|s|\to\infty$ with t-fixed gives

$$V(s,t)\sim -\pi\beta\frac{[\cot \pi\alpha(s)+\cot \pi\alpha(t)]}{\Gamma[\alpha(t)]}[\alpha(s)]^{\alpha(t)}.$$

If we allow $|s|\to\infty$ along any ray except the real axis, then $\cot \pi\alpha(s)\to -i$, and using equation (12.3.7), we obtain

$$V(s,t)\sim -\pi\beta\frac{\exp[-i\pi\alpha(t)]}{\Gamma[\alpha(t)]\sin \pi\alpha(t)}(bs)^{\alpha(t)}, \tag{12.3.9}$$

i.e. Regge behaviour. Along the real axis, the real part is Regge behaved, but the discontinuity across the cut is given by a series of poles, so that Regge behaviour holds only on the average, in the FESR sense. We note that these results are only obtained if β is a constant; and that the scale parameter s_0 (cf. equation (C.2)) is specified by the inverse trajectory slope b^{-1}. A third point is that since the model satisfies the requirements of duality with an exotic channel, and consequently exchange degenerate trajectories, we would expect a zero structure at $\alpha(t)=0$ ($t\approx -0{\cdot}5\,\mathrm{GeV}^2$) of the type discussed in Section 12.2 for $I=1$ exchange in the t-channel. If we use equation (12.3.9) in equation (12.3.4) and (12.3.5) we have

$$T_{I_t=1}\sim\frac{-\pi\beta}{\Gamma[\alpha(t)]}\cdot\frac{-1+\exp[-i\pi\alpha(t)]}{\sin\pi\alpha(t)}\cdot(bs)^{\alpha(t)},$$

so that this structure does indeed occur.

So far, we have discussed some general features of the Veneziano model, in particular showing that it satisfies all the features required of a dual model. It is important to realize that none of these properties is affected if we add to the leading term, equation (12.3.6), a series of additional *satellite* terms of the form

$$\beta_{mnp}\left\{\frac{\Gamma[n-\alpha(s)]\Gamma[m-\alpha(t)]}{\Gamma[p-\alpha(s)-\alpha(t)]}+(s\leftrightarrow t)\right\}, \tag{12.3.10}$$

where

$$m+n\geqslant p\geqslant m\geqslant n\geqslant 1.$$

While the addition of such terms does not affect the couplings of the resonances along the leading trajectory, it does alter the elastic widths of the resonances (defined in terms of the pole residue) at daughter level. Thus there is a large degree of non-uniqueness in such models, at least in the narrow resonance approximation.

(b) $\pi\pi\rightarrow\pi\omega$: Singly or Doubly Spaced Trajectories?

We now turn to the other amplitude to which we have paid particular attention in our discussion of duality, namely that for the reaction $\pi\pi\rightarrow\pi\omega$ defined in equation (12.2.11). Because of its complete crossing symmetry, the leading Veneziano term is (Veneziano, 1968)

$$\begin{aligned}A(s,t,u)=\beta\{&\tilde{B}[1-\alpha(s),1-\alpha(t)]\\&+\tilde{B}[1-\alpha(t),1-\alpha(u)]+\tilde{B}[1-\alpha(u),1-\alpha(s)]\},\end{aligned} \tag{12.3.11}$$

where

$$\tilde{B}(x, y) = \Gamma(x)\Gamma(y)/\Gamma(x+y), \tag{12.3.12}$$

which satisfies all the required properties, but has the asymptotic behaviour $s^{\alpha(t)-1}$ appropriate to a spin flip amplitude. In this case symmetry considerations restrict the resonance quantum numbers to odd $J(P = C = -1)$. While this is automatically satisfied for the states along the leading trajectory, it is not true for daughter states. To eliminate the even-J states with the single-term formula above, it can be shown (Veneziano, 1968) that the condition

$$\alpha(s) + \alpha(t) + \alpha(u) = 2, \tag{12.3.13}$$

must be satisfied, which, on rearrangement, gives $\alpha(t) = 0$ at

$$t = m_\omega^2 + 3m_\pi^2 - 2m_\rho^2 = -0{\cdot}53 \text{ GeV}^2,$$

which is precisely the condition obtained from the FESR bootstrap (cf. equation (12.2.14). Furthermore the condition actually eliminates all of the states on the odd daughter trajectories so that the resonance spectrum suggested by the Schmid circle analysis is also regained (cf. Fig. 12.2.3).

Thus we have in this case a resonance spectrum with only even daughter states. The question that now arises is: do the resonances on the odd daughter trajectories not exist at all, or merely decouple from the reaction $\pi\pi \to \pi\omega$? Since the condition (12.3.13) does not lead to their decoupling in the $\pi\pi$ Veneziano formula (they do in fact couple rather strongly), the natural interpretation, within the framework of the one-term Veneziano formula, is that these resonances (e.g. the $\rho'(I = J = 1)$ daughter of the $f°$) do exist, couple strongly to $\pi\pi$, but decouple from the $\pi\omega$ state. There is however, the question of satellites to consider. While it is obvious that the addition of satellites can be used to remove particular resonances, it is not immediately obvious whether whole trajectories can be eliminated in this way. In fact this is possible, and alternate trajectories can be removed by satellite terms, independent of the trajectory function (Mandelstam, 1968b). For $\pi\pi \to \pi\omega$, this is achieved if we replace $\tilde{B}[1-\alpha(s), 1-\alpha(t)]$ in equation (12.3.11) by the formula

$$\tilde{B}[1-\alpha(s), 1-\alpha(t)] + \sum_{r=1}^{\infty} a_r \tilde{B}[r+1-\alpha(s), r+1-\alpha(t)], \tag{12.3.14}$$

where

$$a_r = (-1)^r \delta(\delta-1)\ldots(\delta-r+1)/r!,$$

and

$$2\delta = \alpha(s) + \alpha(t) + \alpha(u) - 2,$$

for arbitrary $\alpha(t)$. If equation (12.3.13) is satisfied, then $a_r = 0$. In the context of Veneziano models with satellites, in contrast to single-term models, we therefore have a choice as to whether odd trajectories are present or not. Most authors have in fact chosen to work with the unit-spaced spectrum, as will be apparent from the rest of this chapter. However, the doubly spaced spectrum has been advocated (Eguchi and Igi, 1971, 1972; Eguchi *et al.*, 1974). These authors have noted that the doubly spaced spectrum is the one suggested by Schmid circle analysis of the leading Regge terms if the nonsense zeros at $\alpha(t) = 0$ are present. The odd daughters can only arise from the Schmid circle analysis of non-leading exchange contributions, so that the argument for them is much less compelling. Furthermore, the doubly spaced pattern is much easier to reconcile with the quark model spectrum (cf. Chapter 13). As to which, if either, pattern is the one chosen by nature is for experiment to decide. On the present evidence (cf. Section 3.3) the singly-spaced alternative is probably excluded, whereas the doubly spaced spectrum is a possibility.‡ Should the doubly-spaced spectrum indeed be correct, there is a problem as to how the large $I = J = 0$ imaginary part under the rho-resonance is to be accommodated. Since this is not to be identified with a resonance pole it has to be some form of non-resonant background.§ Conversely, if a good description of low-energy $\pi\pi$ scattering is to be obtained by a model which neglects the non-resonant background, then one with singly spaced trajectories must be used. The most interesting of such models is that of Lovelace and Veneziano.

(c) Lovelace–Veneziano Model

A very interesting step was taken by Lovelace (1968) who considered the possible dependence of the single-term Veneziano formula

$$V(s, t) = -\beta \frac{\Gamma[1-\alpha(s)]\Gamma[1-\alpha(t)]}{\Gamma[1-\alpha(s)-\alpha(t)]}, \qquad (12.3.15)$$

on the external pion masses as these are extrapolated off-shell. This extrapolation must not alter the positions of the poles in s, t, u, since the internal states of the unitarity relation remain "on-shell". Thus the trajectory functions $\alpha(s)$ etc. can have no explicit dependence on the external masses, and any explicit dependence on external masses in equation (12.3.15) would

‡ See Chapter 13.
§ In Section 3.3(f) a phenomenological scheme is discussed in which the observed $I = 0$ S-wave phase shift below 900 MeV is attributed to the sum of three effects—the tail of an ε' (1100–1300), the tail of the S^* (1000), and a small background phase. Both the latter two contributions lie outside the present idealized scheme ($S^* \not\to \pi\pi$).

be contained in the multiplicative constant β. It follows that the amplitude, both on and off-shell, has a line of zeros along the locus

$$\alpha(s)+\alpha(t)=1. \tag{12.3.16}$$

However, PCAC (see Appendix D) demands a zero of the off-shell amplitude at $s=t=u=m_\pi^2$. (This is the Adler zero, a derivation of which is given in Chapter 14.) Assuming that this zero lies on the locus (12.3.16) leads immediately to the result $\alpha(m_\pi^2)=0{\cdot}5$, so that

$$\alpha(s)=0{\cdot}5+0{\cdot}5(s-m_\pi^2)/(m_\rho^2-m_\pi^2). \tag{12.3.17}$$

If we use $m_\rho^2\approx 30m_\pi^2$, this gives $\alpha(0)\approx 0{\cdot}48$, which is rather close to phenomenological estimates (e.g. $\alpha_\rho(0)=0{\cdot}56\pm 0{\cdot}02$ (Bolotov *et al.*, 1974)).

It finally remains to normalize the model, which is conveniently done by relating β to the rho-width Γ_ρ. If we consider the behaviour of $V(s, t)$ (which we recall is the $\pi^+\pi^-$ elastic scattering amplitude) near $\alpha(s)=1$, we have

$$V(s, t)\approx\beta\Gamma[1-\alpha(s)]\alpha(t). \tag{12.3.18}$$

From equation (12.3.17)

$$\alpha(s)=\frac{1}{2}+\frac{1}{2}\frac{(s-m_\pi^2)}{(m_\rho^2-m_\pi^2)}\approx\frac{1}{2}+\frac{s}{2m_\rho^2},$$

and

$$\alpha(t)=\frac{1}{2}+\frac{1}{2}\frac{(t-m_\pi^2)}{(m_\rho^2-m_\pi^2)}\approx\frac{1}{2}-\frac{s}{4m_\rho^2}(1-x),$$

so that neglecting terms of order $(m_\pi/m_\rho)^2$, we have, near the pole

$$V(s, t)\approx\frac{\beta(1+x)}{2(1-s/m_\rho^2)}=\frac{\frac{1}{2}\beta m_\rho^2(1+x)}{m_\rho^2-s}. \tag{12.3.19}$$

Identifying the term proportional to x with the rho-pole, we immediately obtain

$$\beta=\frac{3\Gamma_\rho m_\rho^2}{8k_\rho^3}=\frac{3\Gamma_\rho m_\rho^2}{(m_\rho^2-4)^{\frac{3}{2}}}. \tag{12.3.20}$$

This expression is exact to all orders in (m_π/m_ρ) (Osborn, 1969).

We now have a well-defined model—the Lovelace–Veneziano model—in terms of the constants m_ρ, Γ_ρ with $\alpha(m_\pi^2)$ determined by the Adler zero. We now discuss some of the properties of this model in more detail.

(i) *Scattering Lengths*

These can be obtained straightforwardly from equation (12.3.5) using equations (12.3.15) and (12.3.17). The results are

$$a_0^0 = 0{\cdot}395\beta; \qquad a_0^2 = -0{\cdot}103\beta$$
$$a_1^1 = 0{\cdot}080\beta;$$
$$a_2^0 = 2{\cdot}12\times10^{-3}\beta; \qquad a_2^2 = -8\times10^{-5}\beta. \tag{12.3.21}$$

Up to the question of overall normalization (i.e. considering only ratios), these values are very similar to the predictions of current algebra to be discussed in Chapter 14. Of course, one of the current algebra null conditions (the Adler zero) has been explicitly imposed; the way in which the other (the sigma condition) is automatically satisfied is discussed in a comparison between the two approaches in Chapter 14. In fact, if we normalize to the current algebra prediction, then

$$\beta = 0{\cdot}436, \tag{12.3.22}$$

and the rho-width predicted from equation (12.3.20) is 90 MeV. Conversely, if one wishes to normalize to a larger rho-width, say 140 MeV, then the scattering length values are larger than those predicted from current algebra. This is discussed further in Chapter 14.

(ii) *Resonance Widths*

The partial widths for the decays of the various resonances into the $\pi\pi$ channel can be obtained by calculating the pole residues and interpreting them in the narrow width approximation, as we have already done for the rho-pole, in the zero pion mass approximation. The values for the first nine towers are given in Table 12.3.1 (Shapiro, 1969). Along the leading trajectory the partial widths fall off rapidly as J increases, as expected. A second noteworthy feature is that the first daughter trajectory couples strongly to the $\pi\pi$ channel. In fact in the zero pion mass limit, equation (12.3.19) implies that the strength of the S-wave pole under the rho (i.e. the hypothesized ε-meson) has the same strength as the P-wave pole (ρ) itself, which leads immediately to

$$2\Gamma_\varepsilon = 9\Gamma_\rho. \tag{12.3.23}$$

Similarly, in the zero pion mass limit, the states in the f^0 tower (f^0, ρ', ε') have their widths in the ratio

$$\Gamma_{f^0} : \Gamma_{\rho'} : \Gamma_{\varepsilon'} = 9 : 10 : 0. \tag{12.3.24}$$

While the large S-wave phase shift under the rho can be interpreted as corresponding to such a broad ε, the situation in the f^0 region is less

TABLE 12.3.1. Partial Widths for $\pi\pi$ Decay Predicted by the Single-Term Veneziano Model. The widths are normalized to $\Gamma_{\rho\pi\pi} = 112$ MeV, and the ρ trajectory is $\alpha_\rho(t) = 0{\cdot}48 + 0{\cdot}90t$ (Shapiro, 1969)

	Mass (MeV)							
J^P	764	1300	1670	1980	2240	2480	2690	2890
8^+								5
7^-							6	4
6^+						13	10	12
5^-					14	11	10	10
4^+				34	27	20	21	14
3^-			38	30	16	19	10	13
2^+		96	82	27	39	16	25	12
1^-	112	112	14	36	10	20	8	14
0^+	565	−13	77	12	39	12	25	11

satisfactory. Here there is no trace of a ρ' coupling *strongly* to $\pi\pi$, whereas there is evidence for an ε' resonance which couples strongly to $\pi\pi$, in contradiction to equation (12.3.24).‡ It is clear that the detailed predictions of resonance couplings in this model cannot be taken too seriously.

(d) Zero Trajectories

The single-term Veneziano amplitude for $\pi^+\pi^-$ scattering, equations (12.3.6) and (12.3.15), has s-channel poles at $\alpha(s) = N \geq 1$, and t-channel poles at $\alpha(t) = M \geq 1$, arising from the gamma functions in the numerator. Where both conditions are satisfied, an *intersection zero* must necessarily occur to prevent a double (i.e. a product) pole. In the Veneziano model, these are provided by the lines of zeros resulting from the denominator gamma function at

$$\alpha(s) + \alpha(t) = K \geq 1. \qquad (12.3.25)$$

Owing to the assumed linear form of the trajectory functions, these correspond to lines of fixed u. For $K = N + M \geq 2$ they give the intersection zeros required, and in addition there is a further line of zeros for $K = 1$ which is not required to prevent the occurrence of double poles. This latter is the line of zeros (or zero trajectory) discussed earlier (cf. equation (12.3.16)) which passes through the Adler point $s = t = u = m_\pi^2$, provided $\alpha(m_\pi^2) = 0{\cdot}5$. If this condition is satisfied, the corresponding on-shell trajectories lie at the particular fixed-u values given by

$$u = 2m_\pi^2 - 2(K-1)(m_\rho^2 - m_\pi^2). \qquad (12.3.26)$$

‡ For alternative interpretations of the resonance content of the $I = J = 0$ channel see Section 3.3(f) and Chapter 13.

This prediction of fixed-u zeros for channels for which there is no known u-channel exchange is a striking feature of the Veneziano model. Such zero patterns have been extensively studied by Odorico (1970, 1971, 1972) who has speculated that these simple patterns of zeros linking intersection points might not depend on the special simplifying assumptions of the model, but persist into the real world. This is strikingly confirmed in the reaction $K^-p \to \bar{K}^0 n$ (Odorico, 1971) which has both an exotic channel, and no t-channel Pomeron exchange contribution. For $\pi^+\pi^-$ scattering, there is a Pomeron contribution which may complicate matters. However, if its contribution is small in the energy region considered ($\sigma(\pi^+\pi^-) \gg \sigma(\pi^+\pi^+)$) the predicted pattern may well survive. As can be seen from equation (12.3.25), the first intersection zero ($K=2$) would enter the physical region at $s \approx 1\ \mathrm{GeV}^2$, and Odorico (1972) has suggested that this zero may be associated with the sharp variation in the $\pi^+\pi^-$ cross-section near the $K\bar{K}$ threshold.

Turning to the zero trajectory $K=1$ (which is, as noted, not an intersection zero, but may be associated with the Adler zero at $s=t=u=m_\pi^2$), we see that on-shell this becomes the line $u=2m_\pi^2$, close to the backward direction $u=0$. Thus, in this particular model, not only does the absorptive part vanish on average for $u \approx 0$, as demanded by duality, but it vanishes at every s-value, as also does the real part. This additional property is not, as is sometimes stated, a consequence of local duality with exotic channels; it can for example be destroyed by the addition of one or more satellites. It is a specific property of the one-term Veneziano model for this process, and is closely related to the prediction of an S-wave daughter of the rho and a P-wave daughter of the f^0 (1270), since it demands that the sum of the resonance poles in each tower should cancel exactly at $u=2m_\pi^2$. In the zero mass pion limit, this leads exactly to equation (12.3.23) for example. In fact, there is good evidence against the existence of a P-wave daughter of the f^0 (1270) coupling strongly to $\pi^+\pi^-$, and the differential cross-section in this region shows a pronounced peak in the backward direction ($u=0$) rather than the predicted zero.

Zero trajectories can be computed from empirical $\pi\pi$ phase shifts. As an example, Fig. 12.3.3 (from Eguchi *et al.* (1974)) shows the trajectories for the process $\pi^+\pi^- \to \pi^+\pi^-$ computed from an energy dependent phase shift analysis of the 17 GeV/c CERN-Munich data (Hyams *et al.*, 1973). The quantities plotted are Re $t_0^{(i)}(s)$ where $t_0^{(i)}(s)$ are the complex zeros, $t=t_0^{(i)}$, of the full amplitude $T(s, t)$ for given s. Provided the corresponding Im $t_0^{(i)}$ are not too large, they will be associated with dips in the cross-section at $t=\mathrm{Re}\ t_0^{(i)}$ (cf. Section 3.3). As can be seen, the results bear a qualitative resemblance to the predicted pattern but with appreciable distortions. Thus, the zero which starts off at small u in the neighbourhood of the Mandelstam

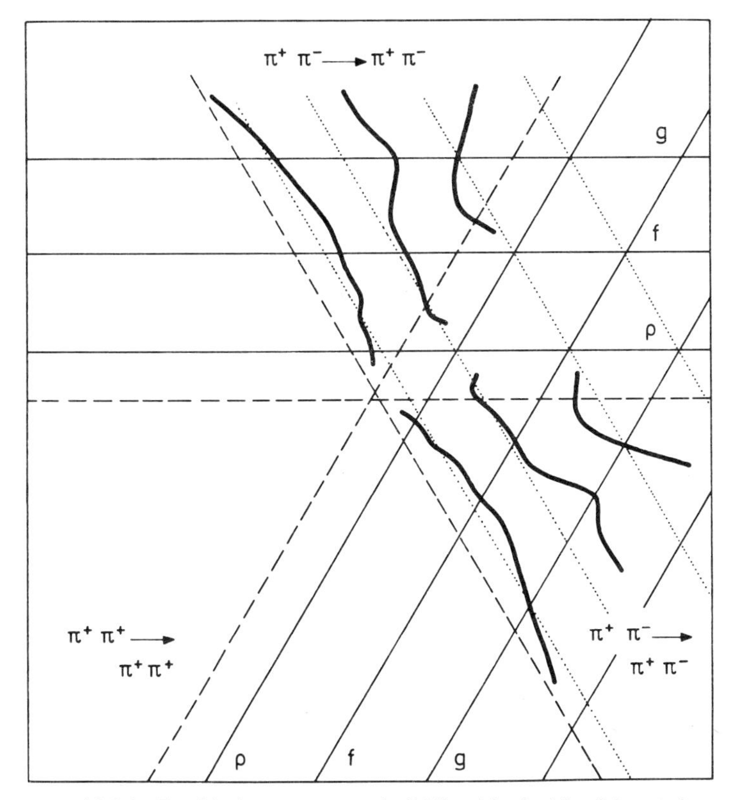

FIG. 12.3.3. Empirical zero contours (solid lines) in the Mandelstam plot.

triangle, and should presumably be associated with the Adler zero, soon swings away from its predicted locus $u = 0$ into the physical region. Likewise, the second trajectory which first enters the physical region at $s \approx 1\ \text{GeV}^2$ as Odorico would require, shows a considerable distortion from a simple fixed-u behaviour. Indeed at the energy values corresponding to the dominant peripheral resonances the zero positions are only slightly displaced from the corresponding Legendre zeros which would obtain if there were no daughter resonances or background at all. Similar presentations of the zero trajectories for $\pi^+\pi^-$ scattering have been given by the authors of the experiment, and by Estabrooks and Martin (1973b) following their own analysis of the same data. The results up to $\sqrt{s} = 1{\cdot}1$ GeV corroborate the earlier findings of Pennington and Protopopescu (1972), who also examined the zeros of the $\pi^-\pi^0$ scattering amplitude. In this, as in all the other charge modes except $\pi^+\pi^- \to \pi^+\pi^-$ with its exotic u-channel, quite different patterns are expected and obtained (Pennington, 1973; Eguchi *et al.*, 1974).

It will clearly be of interest to extend all these considerations to higher energies as more data become available,‡ both to test for the fixed-u patterns suggested for $\pi^+\pi^-$ scattering by Odorico; also to see how the fixed-t zero expected in the absorptive part for $I_t = 1$ exchange at high energies links to the low-energy structure.

Besides their theoretical interest, zero trajectories are, as already noted in Section 3.3(a), useful tools for discussing discrete ambiguities in phase shift analyses. Since the ambiguities concern the signs of the Im $t_0^{(i)}$ (cf. equation (3.3.2)), the emphasis is now on the Im $t_0^{(i)}$ (s) trajectories as a means of smoothly associating alternatives at different energies. This was originally shown in πN scattering by Barrelet (1972) and has been exploited in $\pi\pi$ scattering by Estabrooks and Martin (1973b, 1974b) and Männer (1974). To prefer smooth trajectories is of course to impose a "duality" pattern on the amplitudes; experience (cf. Fig. 12.3.3) would suggest that this is usually justified, but that there can be exceptions (for example, there are kinks in the first two trajectories of Fig. 12.3.3 at $s \approx 1$ GeV^2, usually associated with the S^* (1000) effect).

(e) Unitarization and Phenomenological Applications

The most obvious theoretical shortcoming of the model described above is the severe violation of unitarity, the absorptive part of the amplitude being represented solely by poles. In this section, we will discuss various attempts to construct models which overcome this difficulty, while still retaining the successful features of the Veneziano model. Frequently the motivation for these attempts is the desire to arrive at a form suitable for comparison with experimental data, and so it is on this aspect that we shall concentrate. However, it is worth noting that considerable theoretical effort has been made to use the Veneziano model as a starting point for a complete theory of hadrons. This programme, although of considerable interest, has not yet reached a stage where it has any appreciable impact on the interpretation of data. We therefore refer to the reviews cited earlier (Schwarz, 1973; Veneziano, 1974; Mandelstam, 1974; Frampton, 1974) for a discussion of this topic, and turn now to more phenomenological efforts. These tend to introduce unitarity (or some approximation to it) into the model at the expense of some other feature (usually crossing symmetry) which is considered to be less important for the specific application considered.

(i) *Complex Trajectory Functions: Application to 3π Decays*

The worst manifestation of the violation of unitarity is that the poles occurring in the amplitude lie on the real axis. A minimal requirement is to

‡ A speculative scheme to associate zeros of $p\bar{p} \to \pi^+\pi^-$ with those of $\pi^+\pi^-$ elastic scattering has been proposed by Bugg (1974).

move them on to the second sheet of the complex plane. This is easily achieved by introducing an imaginary part into the trajectory function, for example using

$$\alpha(s)=\alpha_0+\alpha_1 s+\lambda(s_0-s)^{\frac{1}{2}}, \tag{12.3.27}$$

where s_0 is the s-channel threshold. If crossing symmetry is to be retained, a similar modification must be made to $\alpha(t)$ and $\alpha(u)$. However, serious problems arise: in particular, the residues of the poles, being polynomials in $\alpha(t)$ (cf. equation (12.3.8)), are no longer polynomials in t (which is proportional to $\cos\theta$), so that poles in *all* partial waves occur, i.e. we have introduced high spin *ancestors* as well as daughters. For the first resonance tower (ρ, ε) only, this can be avoided by first rewriting equation (12.3.15) as

$$\begin{aligned} V(s,t) &= -\beta(\alpha(s)+\alpha(t)-1)\frac{\Gamma[1-\alpha(s)]\Gamma[1-\alpha(t)]}{\Gamma[2-\alpha(s)-\alpha(t)]}, \\ &= \frac{-0{\cdot}5\beta(s+t-2m_\pi^2)}{m_\rho^2-m_\pi^2}\cdot\frac{\Gamma[1-\alpha(s)]\Gamma[1-\alpha(t)]}{\Gamma[2-\alpha(s)-\alpha(t)]}, \end{aligned} \tag{12.3.28}$$

and then introducing the imaginary parts into $\alpha(s)$ and $\alpha(t)$ in equation (12.3.28), i.e. not in the first zero. Alternatively, the ancestor problem can be solved for s-channel scattering by introducing an imaginary part in $\alpha(s)$, as above, but not in $\alpha(t)$ and $\alpha(u)$. This of course, violates crossing symmetry, but at least leads to a more sensible s-channel form. However, it still has the serious fault of giving parent and daughter poles the same width, although they have different couplings.

The main application of this method of unitarization has been to various 3π decays, in particular $K\to 3\pi$ and $\eta\to 3\pi$ in the pion-pole model (Lovelace, 1968). In this model (Fig. 12.3.4) the decay amplitude is given, apart from an overall constant, by the $\pi\pi$ scattering amplitude in the appropriate kinematic region, but with one pion off-shell. Since in the Veneziano model the relative off-shell dependence is explicitly prescribed, the amplitude for the decay is completely specified, apart from an overall constant, which can be normalized to the decay rate. By using the single-term Veneziano model with the Adler zero imposed (equation (12.3.28)), and unitarizing by giving imaginary parts to $\alpha(s)$ and $\alpha(t)$ in equation

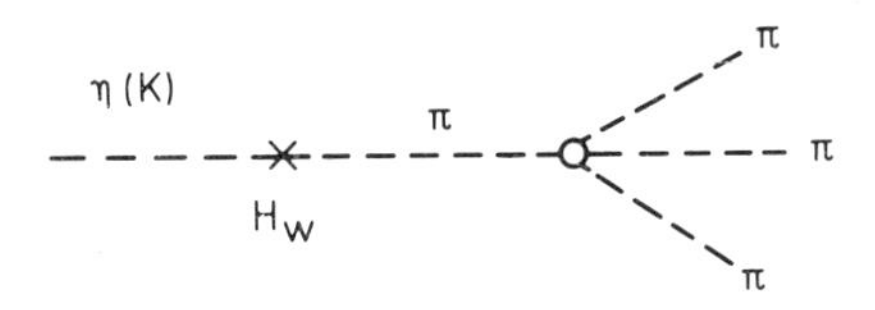

FIG. 12.3.4. Pion-pole diagram for η, $K\to 3\pi$ decays.

(12.3.28) according to the prescription of equation (12.3.27), Lovelace obtained predictions for the various K and η decays. The constant λ was fixed to correspond to an ε (and ρ) width of 280 MeV, as defined by the pole positions, and the predicted slopes of the decay distributions were in reasonably good agreement with experiment. For example, in Fig. 12.3.5 we show the decay spectrum for $\eta \to 3\pi$ decay.

Encouraged by this success Lovelace also considered the process

$$\bar{p}n \to \pi^+\pi^-\pi^-,$$

for stopped anti-protons (Anninos *et al.*, 1968). Assuming that the annihilation takes place predominantly from an ${}^1S_0$ state,‡ the initial state has the same quantum numbers as the pion. The reaction can thus be treated by the same pion-pole model in analogy to the K and $\eta \to 3\pi$ decays above, except that the extrapolation in the pion mass is now much greater. For empirical reasons Lovelace considered the satellite term

$$V(s,t) = -\beta \frac{\Gamma[1-\alpha(s)]\Gamma[1-\alpha(t)]}{\Gamma[2-\alpha(s)-\alpha(t)]}, \tag{12.3.29}$$

rather than the leading term of equation (12.3.15), so that the particles on the leading trajectory (ρ, f^0) do not couple. (Other authors have considered the sum of several terms, including the leading one.) Although the Dalitz

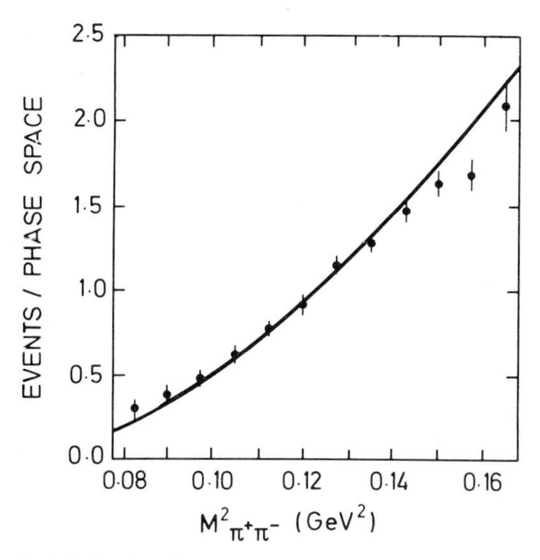

FIG. 12.3.5. Predicted decay spectrum for $\eta \to 3\pi$ decay (Lovelace, 1968).

‡ This assumption is now somewhat controversial. See the discussion in Section 3.4.

plot is quite complicated, one striking feature is a dearth of events in the centre of the plot, corresponding to the expected zero in the Veneziano formula at

$$\alpha(s)+\alpha(t)=3. \tag{12.3.30}$$

This is a very interesting feature in that it could easily be removed by the addition of sufficiently low-lying satellite terms. Taken in isolation, the success in explaining the salient features of $\bar{p}n \to 3\pi$ annihilation is impressive. Unfortunately, the unobserved daughter states (especially the ρ' daughter of the f^0) play a crucial role. For further discussion of this, and other aspects, of these processes we refer to Section 3.4(b).

(ii) *K-Matrix (Inverse Amplitude) Unitarization*

One possible method of producing unitary amplitudes from the Veneziano model is to interpret its partial wave projections not as those of the T-matrix, but as those of the K-matrix, defined by

$$T_l^I(s) \equiv \frac{K_l^I(s)}{1-i(\nu/(\nu+1))^{\frac{1}{2}}K_l^I(s)}. \tag{12.3.31}$$

Provided $K_l^I(s)$ is real, which it will be if we assume it is given by the Veneziano projections, the $T_l^I(s)$ will satisfy elastic unitarity exactly. However, it will violate crossing very badly, since all partial waves would be forced to vanish at $\nu=-1$, i.e. $s=0$. Lovelace (1969) therefore suggested generalizing equation (12.3.31) to

$$T_l^I(s)=\frac{V_l^I(s)}{1+\rho(s)V_l^I(s)}, \tag{12.3.32}$$

where $V_l^I(s)$ is the partial wave projection of the Veneziano formula. Again, $T_l^I(s)$ will obey elastic unitarity for real $V_l^I(s)$ provided

$$\operatorname{Im}\rho(s)=-\left(\frac{\nu}{\nu+1}\right)^{\frac{1}{2}}, \tag{12.3.33}$$

along the right-hand cut, $4 \leqslant s < \infty$. One can now make models of $\rho(s)$, subject to the restriction (12.3.33), in order to reduce the violation of crossing implied by equation (12.3.31).

This method has been thoroughly discussed by Tryon (1971b) who notes that equation (12.3.32) can be rewritten as

$$\rho(s)=[T_l^I(s)]^{-1}-[V_l^I(s)]^{-1}, \tag{12.3.34}$$

so that the method consists of constructing models for the difference between the inverse Veneziano amplitudes and the inverse physical amplitude—hence the alternative name for the method. Clearly, since $V_l^I(s)$ is real, equation (12.3.33) will be satisfied if $T_l^I(s)$ is unitary; also $\rho(s)$ will have a left-hand cut along the real axis $-\infty \leqslant s \leqslant 0$. It is also clear that any unitary $T_l^I(s)$ with a left-hand cut can be written in the form of equations (12.3.32) and (12.3.33), so that without further restrictions the result is completely arbitrary. Usually, however $\rho(s)$ is assumed to have no poles on the physical sheet, so that the zeros of $T_l^I(s)$ and $V_l^I(s)$ are constrained to occur at the same positions, with the same slopes. Hence the resonance spectrum of such models will tend to be close to that of the original Veneziano model. Furthermore, the values of $T_l^I(s)$ near threshold will be similar to those of $V_l^I(s)$ if the sub-threshold zeros suggested by current algebra are incorporated into the model $V_l^I(s)$, as is the case in, for example, the Lovelace–Veneziano model. However, despite these predetermined features, it is clear that there is considerable freedom left in the method, corresponding to the arbitrariness allowed in the function $\rho(s)$, (which need not even be the same for different partial waves), although it is, of course, possible to restrict the arbitrariness somewhat by demanding that other conditions be satisfied; for example the BNR relations (cf. Chapter 7) (Lipinski, 1970), or the absence of "ghosts" (Tryon, 1971b).

We turn now from the general discussion of this method (Tryon, 1971b) to the simple *ad hoc* prescription proposed by Lovelace (1969). In this model $\rho(s)$ is assumed to be the same for all partial waves, and is given a left-hand cut with

$$\operatorname{Im} \rho(s) = -\left(\frac{\nu+1}{\nu}\right)^{\frac{1}{2}}, \qquad s \leqslant 0. \tag{12.3.35}$$

The assumption is that the fine details of the near left-hand cut of $\rho(s)$ are unimportant, and that the essential requirement is to match the large-s discontinuity on the right-hand cut. Given equation (12.3.35), the real part can then be determined from an unsubtracted dispersion relation to give (cf. equation (9.2.8))

$$\rho(s) = -\frac{2}{\pi} \frac{1}{[\nu(\nu+1)]^{\frac{1}{2}}} \ln\left[\sqrt{\nu} + \sqrt{\nu+1}\right]. \tag{12.3.36}$$

The apparent singular branch points at $\nu = 0$ and -1 cancel, and as $s \to \infty$ $\rho(s) \to 0$, so that $T_l^I(s) \to V_l^I(s)$. Thus, in addition to retaining approximately the same resonance spectrum and threshold region amplitudes of the un-unitarized Lovelace–Veneziano amplitude, it also retains, to leading order in s (resumming the partial wave series) the same Regge asymptotic behaviour, incorporating the constraints of crossing symmetry (the phase-energy relations). Unitarity has been introduced at the cost of some violation of crossing symmetry at low energies. The resulting S-wave phase shifts are shown in Fig. 12.3.6. As can be seen, the narrow-resonance poles are shifted on to the second sheet to give normal resonances, and in the case of the S-wave daughter of the ρ (765) the result is a very broad resonance. The scattering lengths, in such models are, as we have noted, similar to those of the un-unitarized model, provided the normalization is not altered. However, the rho-width will, in general, be different, so that the normalization

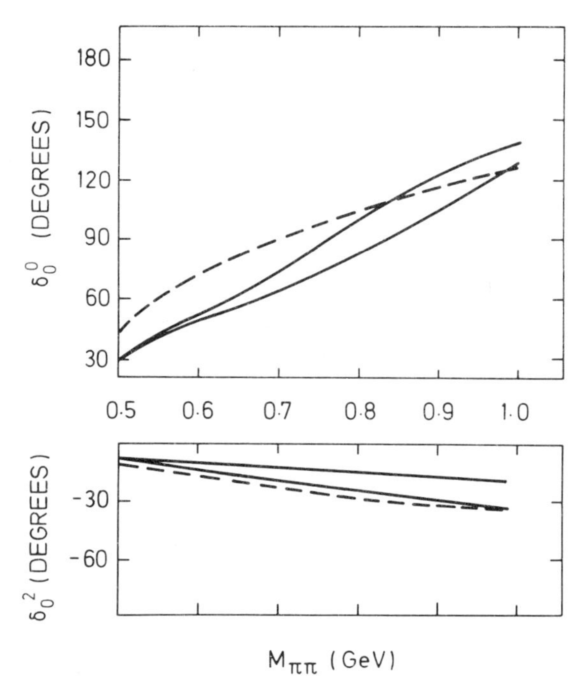

FIG. 12.3.6. The predicted S-wave phase shifts from unitarized Veneziano models. The dashed line is the prediction of Lovelace (1969) and the solid lines define the band allowed by Kang *et al.* (1971).

must be adjusted to retain the same rho-width. With the choice of $\rho(s)$ given by equation (13.3.36) the rho-width is in fact changed rather little so that the scattering lengths are still somewhat larger than those predicted by current algebra (cf. the discussion of Section 14.1).

The Lovelace unitarization procedure sacrifices low-energy crossing for unitarity. This defect may be remedied (at a cost) by using different parameterizations for $\rho_l^I(s)$, which are now allowed to depend on the particular partial wave, and adjusting the extra parameters which are introduced so that various low-energy crossing conditions (e.g. the BNR relations and Martin inequalities discussed in Chapter 7) are satisfied. An example of such a model is that of Kang *et al.* (1971), who have also taken care to ensure that the partial wave absorptive parts have the correct crossed threshold behaviour at $s = 0$. Their results are compared with those obtained by the Lovelace method in Fig. 12.3.6. It is interesting that the scattering lengths obtained by Kang *et al.*,

$$a_0^0 = 0{\cdot}186, \qquad a_1^1 = 0{\cdot}035, \qquad a_0^2 = -0{\cdot}045,$$

are now in excellent agreement with the current algebra predictions (cf. Chapter 14). The low-energy crossing properties of this model are much more satisfactory than those of the Lovelace model; however, this has been achieved at the cost of losing the simple high-energy behaviour preserved by Lovelace's method. We shall return to the Lovelace unitarization scheme for the rest of this discussion.

This model has been used quite extensively in phenomenological applications; for example, it has been extended to the coupled $\pi\pi$, $K\bar{K}$ system (Lovelace, 1969) and used in a study of K_{e4} decay (Roberts and Wagner, 1969). One interesting point that has been discussed is the off-shell dependence of the unitarized form. The off-shell K-matrix in potential scattering is given by

$$T_l^{I,\text{off}}(s) = \frac{K_l^{I,\text{off}}(s)}{1 - i(\nu/\nu+1)^{\frac{1}{2}} K_l^{I,\text{on}}(s)},$$

and it has been suggested (Wagner, 1969) that the analogous formula

$$T_l^{I,\text{off}}(s) = \frac{V_l^{I,\text{off}}(s)}{1 + \rho(s) V_l^{I,\text{on}}(s)}, \tag{12.3.37}$$

holds in the present case, with $\rho(s)$ the quantity introduced in equation (12.3.36). This is a plausible extension of the original approximation, in that the on-shell phase is retained, as required by unitarity, and for $l > 0$ threshold factors are correctly incorporated. There could be further explicit dependence on the off-shell mass. However, the prescription (12.3.37) does have the good property that zeros of the original $V_l^I(s)$, in particular those of the S-waves $V_0^I(s)$, are retained. Thus, if the off-shell zeros of current algebra are imposed on the original $V^I(s, t)$, which will induce nearby zeros of the associated $V_0^I(s)$, this property will survive unitarization and lead to a full amplitude still possessing the off-shell zeros (apart from very slight changes owing to contributions from D- and higher waves). It is in fact the S-waves which receive the most striking off-shell behaviour by equation (12.3.37). Wagner (1969) made the bold suggestion that the above off-shell dependence of $\pi\pi$ scattering amplitudes should furnish a form factor description of dipion production, assuming this to proceed dominantly by pion exchange, and a very similar approach suggested by current algebra formulae (cf. Chapter 14) was proposed by Gutay *et al.* (1969) (cf. Section 3.1(e)). No strong confirmatory evidence has been forthcoming.

(iii) *Tryon's Model*

After showing that the K-matrix unitarization methods could be regarded as a class of models for the differences between the inverse partial wave amplitudes and the corresponding Veneziano amplitudes, it was natural to consider constructing models for the difference of the partial wave amplitudes themselves, i.e. for the quantities

$$\Delta T_l^I(s) \equiv T_l^I(s) - V_l^I(s). \tag{12.3.38}$$

This Tryon proceeded to do (Tryon, 1971a,d) basing his model, which we now describe, on partial wave dispersion relations for the quantities $\Delta T_l^I(s)$. Since the quantities $V_l^I(s)$ are thought to have an essential singularity at infinity (at least for $I=0$ and 2) (Tryon, 1971c) this involves as a basic assumption that the essential singularities of T_l^I and V_l^I are identical. This replaces the assumption about zeros in the previous method. Tryon's aim is to unitarize all waves with l less than some value L, assuming that

$$\operatorname{Im} \Delta T_l^I(s) = 0, \qquad s \geq 4, \qquad l > L. \tag{12.3.39}$$

This is a phenomenologically sensible approach because elastic unitarity is only a good approximation for $W_s \leq 1$ GeV, over which range only a few partial waves have appreciable values; taking the small higher partial waves to be real will not therefore give any appreciable violation of unitarity. Tryon's method is for unitarizing low-energy amplitudes. It has the great virtue that with a finite number of unitary partial waves, important properties like exact crossing, and positivity for the un-unitarized waves, are retained, so that all the rigorous constraints of Chapter 7 which follow from these will be automatically satisfied. Let us consider crossing symmetry in this approximation. From (12.3.38), $\Delta T_l^I(s)$ has a left-hand cut $-\infty < s \leq 0$, and crossing symmetry requires that the discontinuity across this cut be given in terms of the discontinuity across the right-hand cut by [cf. equation (9.1.12)]

$$\operatorname{Im}[\Delta T_l^I(\nu)] = \frac{1}{\nu}\int_0^{-\nu-1} d\nu'\, P_l\left(1+\frac{2(\nu'+1)}{\nu}\right)$$
$$\times 2\sum_{I',l'} C_{II'}^{(st)}(2l'+1)\operatorname{Im}[\Delta T_{l'}^{I'}(\nu')]P_{l'}\left(1+\frac{2(\nu+1)}{\nu'}\right), \tag{12.3.40}$$

Where $C_{II'}^{(st)}$ is the crossing matrix of equation (2.21). If we impose condition (12.3.39), then the partial wave series for the absorptive parts along the right-hand cut terminates at finite $l=L$, and so the series in equation (12.3.40) converges for all l and ν. Thus, if the above equation is used in partial wave dispersion relations (for all l), and the subtractions made in a crossing symmetric manner, the resulting $\Delta T_l^I(s)$, and hence also the $T_l^I(s)$, will satisfy exactly all the requirements of crossing symmetry. The requirements of positivity are satisfied by construction, since the waves will be unitarized for $l \leq L$, and for $l > L$ will be given by the poles of the Veneziano formula, whose residues all have the correct sign for the trajectory parameter chosen ($a = 0{\cdot}483$, $b = 0{\cdot}017$, where $\alpha(s) = a + bs$) (Tryon, 1971b). Thus, the model satisfies automatically all the rigorous requirements of crossing and positivity.

Specializing to the case $L = 1$, as Tryon did in his practical application, there is only need for one subtraction, which can be made at $\nu_0 = \frac{2}{3}$. The subtraction parameter is then the quantity λ of equation (2.25). Very flexible trial functions (of up to 39 parameters) were used for Im T_l^I ($l \leq 1$) in the low-energy region, and were made to go over smoothly to V_l^I at higher energies. By varying the parameters, unitarity, and the dispersion relations for $l=0$ and 1, were imposed at a fine mesh of points in the low-energy region.‡ It was found that five of the parameters could be varied independently; the subtraction constant, the mass and width of the rho, and two parameters associated with the $I=0$ S-wave. The latter can, for example, be taken to be the energies at which the δ_0^0 phase passes through 90° and 135°. We note that in general no specific requirements on the resonance spectrum below 1 GeV (i.e. the ρ and ε) need be made, although, of course, one is at liberty to impose as much similarity between the parameters of these resonances§ and those of the narrow resonance models as one wishes. Tryon, however, preferred to fix just the rho-parameters at $m_\rho = 762$ MeV, $\Gamma_\rho = 120$ MeV, and to explore the solution space as the other parameters were varied.

The general features of the results were as follows. Firstly, once λ is fixed the dependence of the S-wave scattering lengths on the other parameters is very slight, so that, as λ is varied, a band of solutions is obtained in the (a_0^0, a_0^2) plane, reproducing the "universal curve" found earlier (Morgan and Shaw, 1970). Secondly, given m_ρ, Γ_ρ and λ, δ_0^2 is predicted, stable to within a few degrees below 1 GeV, for quite appreciable variations in the $I=0$ S-wave parameters. Furthermore, the predicted form is consistent with experiment. Finally, it is important to note that the model possesses

‡ In practice these conditions were enforced to 2%, but it is unlikely that enforcing them more accurately would change the results appreciably.

§ In the present discussion we are identifying resonances with phase shifts passing through 90°. In this sense, the ε (900) effect *is* a resonance. However, see discussion in Section 3.3(f).

almost exact crossing symmetry, unitarity‡ for the waves with $l \leq L$, and positivity for all higher partial waves, so that any consequences of these conditions are automatically satisfied. The model also has the appropriate values of m_ρ and Γ_ρ. It thus demonstrates that these conditions, and the existence of the rho with the appropriate mass and width, are by no means sufficient to achieve a tightly constrained range of possibilities for the solution. For example, the claim of Le Guillou *et al.* (1971) that they lead to a zero in $T_0^0(s)$ in the interval $1{\cdot}1 \leq s \leq 1{\cdot}7 m_\pi^2$ is not upheld (Tryon, 1971a). Even with the additional assumptions of the model, including a prescribed high-energy behaviour, a one-parameter ambiguity (the universal curve) remains in the S-wave scattering lengths; furthermore, even with these fixed there remains a range of possibilities for the functional form of δ_0^0. Roughly speaking, three parameters (in addition to m_ρ and Γ_ρ) are still needed to specify a unique solution, although, of course, this is still a considerable restriction on the multiplicity of possible solutions.

(f) Extension to other 0^-0^- Scattering Processes

For processes with an exotic channel like πK, KK scattering, the Veneziano model can be applied in a way closely analogous to that used in the $\pi\pi$ case (Lovelace, 1968; Kawarabayashi *et al.*, 1969; see also Lovelace (1969), and references therein). Using the isospin crossing relations of Chapter 2, all the amplitudes can be expressed in terms of one for which the u-channel is exotic by equations analogous to (12.3.5). One then writes Veneziano formulae for these amplitudes analogous to equation (12.3.6). Thus, if the s- and t-channels are governed by trajectories x, y respectively, then the leading Veneziano term is proportional to

$$V_{xy} = \frac{\Gamma[1-\alpha_x(s)]\Gamma(1-\alpha_y(t)]}{\Gamma[1-\alpha_x(s)-\alpha_y(t)]}. \tag{12.3.41}$$

If this is carried out, the resulting equations for the amplitudes of equations (2.27)–(2.33) are

$$\pi\pi \quad T_s^2 = 2f_{\pi\pi}^2 V_{\rho\rho}(t, u), \tag{12.3.42}$$

$$\pi K \quad T^{\pm} = \tfrac{1}{2} f_{\pi K}^2 [V_{K^*\rho}(s, t) \pm V_{K^*\rho}(u, t)], \tag{12.3.43}$$

$$KK \quad T_s^I = f_{KK}^2 [V_{\phi\rho}(t, u) - (-1)^I V_{\phi\rho}(u, t)]. \tag{12.3.44}$$

Equation (12.3.42) is, of course, just equation (12.3.6) with the constant renamed for later convenience, and the results for $K\bar{K}$ scattering follow from the above and equation (2.32). Since we have a dual model with simple Regge pole asymptotics, and an exotic channel in each case, the trajectories

‡ See footnote on previous page.

referred to are actually groups of exchange degenerate partners, as discussed in Section 12.2(a). Thus the subscript ρ refers to the exchange degenerate $\rho - P' - \omega - A_2$ trajectory, K^* to the exchange degenerate K^* (890) $- K^{**}$ (1420) trajectory, and ϕ to the exchange degenerate ϕ-f' trajectory which must for consistency decouple from the $\pi\pi$ channel. (See equations (12.2.8a,b,c)). That all these trajectories have the same slope, i.e.

$$\alpha_x(s) = 1 + \alpha'(s - m_x^2), \qquad (12.3.45)$$

is a result that follows generally in this type of dual model (Igi, 1968; Mandelstam, 1968b), and agrees well with experiment (see, e.g. Fig. 13.1). Finally, the contributions of resonances on the leading $\rho - P'$ trajectory factorize provided that

$$f_{K\pi}^2 = f_{\pi\pi} f_{KK}, \qquad (12.3.46)$$

a particular realization of which is the assumption of universal rho coupling

$$f_{\pi\pi}^2 = f_{KK}^2 = f_{\pi K}^2. \qquad (12.3.47)$$

This factorization property does not in general extend to any of the daughter trajectories.

As for the $\pi\pi$ case the above amplitudes can be modified by the addition of an arbitrary number of satellite terms. However, particularly interesting results are again obtained if we restrict ourselves to the single-term Veneziano formulae given above, and impose the Adler zeros as in the Lovelace–Veneziano model of Section 12.3(c). Therefore, when an external pion or kaon four-momentum goes to zero, we demand that a zero of the amplitude occurs, arising from the denominator gamma function of equation (12.3.41). This gives the conditions

$$\alpha_\rho(m_\pi^2) = \alpha_{K^*}(m_K^2) = \alpha_\phi(2m_K^2 - m_\pi^2) = \tfrac{1}{2}, \qquad (12.3.48)$$

which immediately lead to the mass formulae

$$m_\phi^2 - m_{K^*}^2 = m_{K^*}^2 - m_\rho^2 = m_K^2 - m_\pi^2. \qquad (12.3.49)$$

These striking results had been derived from SU(6) and agree rather well with experiment. Equation (12.3.48) is also the condition for the contributions of the states on the first daughter trajectory, $\alpha_\rho - 1$, to factorize (Takagi, 1968; see also Lovelace, (1969)).

We thus have a generalization of the Lovelace–Veneziano model to πK and KK as well as $\pi\pi$ scattering. As in the $\pi\pi$ case, the scattering lengths for πK and KK agree well with the predictions of current algebra‡ (cf. Section 14.4(a)) provided universal rho-coupling, equation (12.3.47) is adopted

‡ This is not so for $K\bar{K}$, since in this case the existence of resonances close to threshold will cause the smooth extrapolation assumptions used in current algebra calculations to break down.

together with a suitable overall normalization (Kawarabayashi *et al.*, 1969). There are also striking similarities in the predicted resonance patterns. The ϕ, K^* are, like the rho, accompanied by S-wave daughter resonances, and the f', K^{**} are, like the f, accompanied by P-wave daughters; all these are predicted to couple strongly to the two-body channels. Detailed predictions for the masses and widths have been given by Lovelace (1969), who has unitarized the model using a coupled channel K-matrix method which is a direct extension of the single-channel case described in Section 12.3(e). This also gives predictions for the non-resonant phase shifts, and again a close parallel is found between the $\pi\pi$ and $K\pi$ cases. For example, the exotic $I=2\,\pi\pi$ S-wave and $I=\frac{3}{2}\,K\pi$ S-wave are both predicted to be small and negative. While this latter prediction is borne out by experiment, the details of the daughter spectrum do not accord well with observation, as we have already noted in the $\pi\pi$ case (see also Chapter 13).

Despite these problems with the details of the daughter spectrum, it is clear that the Lovelace–Veneziano model has some striking successes in accounting for the main features of the $\pi\pi$, $K\pi$, KK and $K\bar{K}$ systems. However, this success does not extend to the case of processes with no exotic channel (Canning, 1969; Osborn, 1970). For example, for $\pi\eta$ scattering, the imposition of the Adler zeros via the trajectory functions, as above, leads to the result $m_\pi = m_\eta$ in obvious disagreement with experiment. Furthermore the close link between the Veneziano model (with or without satellites) with the Adler zero imposed, and conventional current algebra calculations, breaks down in the absence of an exotic channel. These points are discussed further in Section 14.4.

12.4. Concluding Remarks

In this chapter we have traced the development from finite energy sum rules (FESR) through general ideas of duality, to highly specific dual models; in particular, the Veneziano model. At each stage the assumptions, and hence the predictions, became more specific and also more controversial. Thus, in Section 12.1 we showed how FESR enabled inferences to be made about high-energy scattering from low-energy data without the necessity of making very specific assumptions. From this grew the ideas of semi-local, and local, two-component duality discussed in Section 12.2. These give their most striking results when applied to processes with an exotic channel, especially when combined with the assumption of simple Regge pole dominance of high-energy scattering. Many of these predictions are brilliantly confirmed by experiment; perhaps the most important are the consequences of weak exchange degeneracy, equation (12.2.8). Unlike most results, which refer to a specific process, these refer to trajectory functions, and so apply to all processes where the exchange is present.

Perhaps the other most exciting result of Section 12.2 is the suggestion of the existence of large numbers of resonances lying on daughter trajectories, which opens up the possibility of making models with Regge pole behaviour at high energies and resonance dominance at all energies. This possibility is realized, in the narrow resonance approximation, by the Veneziano model discussed in Section 12.3. Although this model embodies the very specific assumptions of resonance dominance and simple Regge pole behaviour at high energies (all in the framework of strict local duality) there is still appreciable freedom because of the possibility of satellites. These allow for considerable variation in the properties of the daughter resonances. Thus, while in general the daughter trajectories are unit spaced, as shown in Fig. 12.3.2, it is also possible for them to be spaced by two units as in Fig. 12.2.3; the latter is, in fact, the case suggested by Schmid circle analysis of simple Regge pole terms (cf. Section 12.2(b)). For a time, it was hoped that this ambiguity would be resolved by retaining only the leading Veneziano term, since if the Adler zero is imposed on this (the Lovelace–Veneziano model) a close correlation with the results of current algebra is obtained, at least for processes with exotic channels. However, this success does not extend to processes without exotic channels, and it is in any case clear that the details of the daughter spectrum predicted by this model are in conflict with experiment.

The other obvious shortcoming of the Veneziano formula as a dynamical model is its violation of unitarity; it is a narrow-resonance model. Some attempts to alleviate this while retaining the main features of the scheme have been described in Section 12.3(e). While these models are interesting and useful, no completely satisfactory, unique prescription has been found. More satisfactory in principle than these methods is the attempt to construct an exact dual field theory. However, with such a tightly knit scheme, the scope to vary parameters is extremely circumscribed. Since present schemes achieve highly unrealistic parameters, they have little bearing on phenomenology. We again refer to the reviews of Schwarz (1973), Veneziano (1974), Mandelstam (1974) and Frampton (1974) for further details.

Finally, although in Section 12.3 we have devoted our discussion entirely to the Veneziano model, and some of its developments, it should be noted that this is not the only attempt to construct crossing symmetric models with asymptotic Regge behaviour. An alternative dual, narrow-resonance model is that of Virasoro (1969), which has resonances in all three channels, and so is not applicable to processes with exotic channels. Interesting models outside the narrow resonance framework are those of Cohen-Tannoudji *et al.* (1971), and Moffatt and co-workers (Moffatt, 1971; Curry *et al.*, 1971).

However, these models have not yet received the wide application or detailed study accorded to the Veneziano model, on which our discussion has focussed, and for details of these alternative approaches we refer the reader to the original literature.

Chapter 13

Resonance Spectra

The topic of this chapter forms an interface between detailed questions of $\pi\pi$ and πK dynamics, and the whole developing field of classification physics for hadrons (for a general review see Rosner (1974a). If we plot the accepted meson resonances (Particle Data Group, 1974) on a J versus M^2 diagram (Fig. 13.1), certain regularities—i.e. groupings of states into families of related quantum numbers—are immediately seen. It is the attempt to systematize these family groupings with which we are concerned here. (We do not discuss the recently discovered ψ family of resonances—see footnote, p. 350.)

We shall concentrate on two alternative types of scheme: those stemming from some variant of the quark model, and those inspired by notions of duality. In each case, the classification of hadrons and their decays is only one area of application, and we shall be focussing on just one portion of this, that relating to mesons, with special emphasis on the natural parity sector accessible via $\pi\pi$ and πK states. The quark model has had impressive success in describing the observed baryon spectrum (see e.g. Dalitz (1967) and Litchfield (1974)); it is of great interest to see whether this success is maintained in the simpler and more striking case of the meson spectrum. Wider aspects of dual models have been summarized in Chapter 12.

13.1. Meson States in the Quark Model

Before passing to details it is worthwhile calling to mind the general status and scope of the quark model (for a general review, and references to earlier

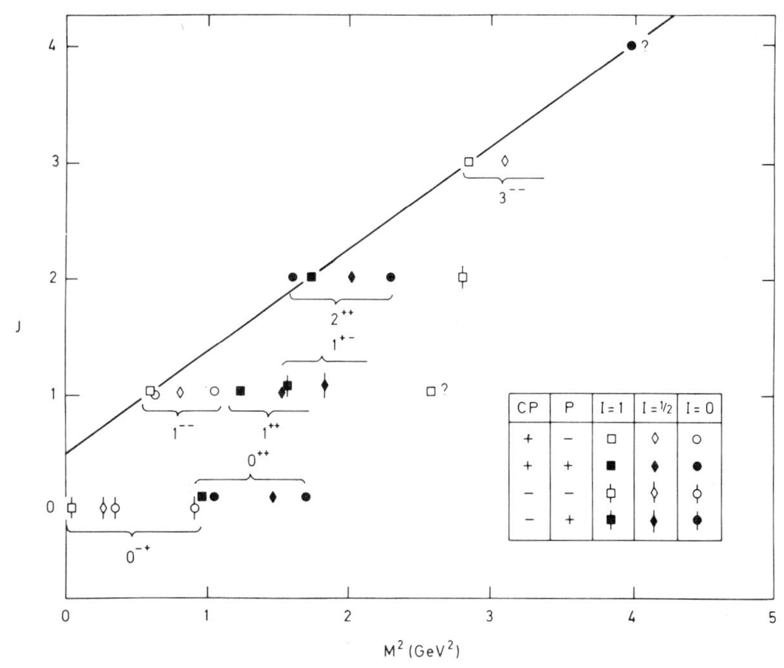

FIG. 13.1. Plot of meson states, spin J versus squared mass M^2. Of the possible candidates given in Table 13.1.3 only those with established spins are shown.

work, see Lipkin (1973)). The notion of hadrons being built of more "elementary" spin-half constituents, usually assumed to have fractional charge, has had impressive explanatory success in a number of areas, for example deep inelastic lepton scattering and high energy hadron scattering, as well as the systematics of resonance spectra and decays with which we are concerned here. As a result, there is a widespread belief that quark ideas represent a significant approximation to the truth, despite the unsatisfactory and even paradoxical features of present formulations. The most obvious examples of these are the use of overall symmetric wave functions for the spin-half quarks in describing the baryon spectrum (Dalitz, 1967), and the failure to observe free quarks. At present, the suggested resolutions of these questions, and hence interpretations of the model, are many and varied. They range from dynamical pictures in which the quarks are regarded as material particles moving in smooth potentials, to algebraic approaches in which only the symmetry aspects of the quark model are regarded as physically significant, the quarks themselves being relegated to the status of a mathematical convenience. This latter viewpoint is associated particularly with Gell-Mann, who has also emphasized (Gell-Mann, 1972) that one should distinguish the two concepts of *constituent* quarks and *current* quarks. The former are the elementary building blocks of the hadrons, the observed states being identified with simple combinations of them. The latter entities correspond to the quark fields out of which weak and electromagnetic currents of hadrons are built, and which appear in the heuristic derivation of the commutation relations of current algebra (see Appendix D). This distinction plays an important role in recent developments in the phenomenology of meson decays, which we discuss below. Firstly however, we discuss the classification of states in the context of simple (constituent) quark models.‡

(a) The L-Excitation Quark Model: General Features

Many of the features of Fig. 13.1 are simply explained by a model in which meson states are pictured as quark–antiquark ($q\bar{q}$) bound states with constituents moving slowly (non-relativistically) within a deep potential well. Likewise, baryons are viewed as (qqq) bound states. In the simplest version of the model, a triplet of spin-half quarks $q \equiv (p, n, \lambda)$ is postulated with the quantum number shown in Table 13.1.1. Approximate SU(3) symmetry is

TABLE 13.1.1. Quantum Numbers of the Quarks. (Here S denotes the strangeness)

	B	I	I_3	Y	S	Q
p	$\frac{1}{3}$	$\frac{1}{2}$	$\frac{1}{2}$	$\frac{1}{3}$	0	$\frac{2}{3}$
n	$\frac{1}{3}$	$\frac{1}{2}$	$-\frac{1}{2}$	$\frac{1}{3}$	0	$-\frac{1}{3}$
λ	$\frac{1}{3}$	0	0	$-\frac{2}{3}$	-1	$-\frac{1}{3}$

imposed on the model by requiring the quarks to be approximately degenerate in mass, and SU(3) to be an approximate symmetry of the overall Hamiltonian. The SU(3) classification of the observed mesons then follows from the decomposition§

$$\{3\}_q \otimes \{\bar{3}\}_{\bar{q}} = \{1\} \oplus \{8\}. \tag{13.1.1}$$

Similarly, the baryons are required to fall into {1}, {8} and {10}, representations of SU(3). Moreover, if the ($\bar{q}q$) bound states (and likewise the qqq states for baryons) are assumed to occur in an ascending series of angular momentum states starting from an $L = 0$ ground state, further systematic features are accounted for.

‡ The possible existence of charmed quarks c is not discussed. If they do exist, the implied $I = 0$ $c\bar{c}$ states can mix with the known $I = 0$ states, altering the predictions discussed here. However, to the extent that such new states are heavy and the mixing small, this effect will be small.
§ We loosely refer to the singlet and octet together as a "nonet", irrespective of whether they are or are not assigned to a single SU(6) representation (see below).

The quark and anti-quark spins combine to form either the singlet or triplet state $S = 0$ or 1. Thus for the ground state ($L = 0$) one has the states ${}^3S_1(J^{PC} = 1^{--})$ and ${}^1S_0$ (0^{-+}), corresponding to the two lowest-lying nonets. For non-zero L, the spin and orbital angular momentum combine to give a succession of angular momentum levels with ${}^{2S+1}L_J = {}^3L_{L+1}$, ${}^3L_L$, ${}^3L_{L-1}$ and ${}^1L_L$. The corresponding J^{PC} values are listed in Table 13.1.2.‡ Successive L

TABLE 13.1.2. Spin, Parity and Charge Conjugation (J^{PC}) Assignments in the Quark Model L-bands. (Here S denotes the total quark spin.)

	$S=1$			$S=0$
J	$L+1$	L	$L-1$	L
$P \equiv (-)^{L+1}$	$(-)^J$	$(-)^{J+1}$	$(-)^J$	$(-)^{J+1}$
C	$(-)^J$	$(-)^{J+1}$	$(-)^J$	$(-)^J$
$\tau \equiv P(-)^J$	+	−	+	−
CP	+	+	+	−

values yield bands of particles of alternating parity (cf. Fig. 13.1) and a similar effect is observed in the baryon spectrum. If, furthermore, we assume that M^2 increases linearly with L, then the states of a given family coincide with those on a linear, exchange-degenerate Regge trajectory—for example the ${}^3L_{L+1}(I = 1, Y = 0)$ family would correspond to the states on the $\rho - A_2$ trajectory. This sort of spectrum would arise naturally in, for example, a model with a simple harmonic oscillator potential and a relativistic wave equation (Dalitz, 1966). It follows that the spacing between the bands is given by the "universal" slope of the meson trajectories—about $0{\cdot}9$ GeV^{-2}. The splittings within a given band, corresponding to different spin orientations, are much smaller than this, being typically of order $\Delta M^2 \approx 0{\cdot}4\Delta J$.

Finally, a global feature of the model which should be stressed before passing on to details is the predicted absence of "exotics". *Exotic particles of the first kind* are those with internal quantum numbers which cannot be formed from $q\bar{q}$ (mesons) or qqq (baryons), for example an $I = \frac{3}{2}$, $Y = \pm 1$ meson.§ *Exotic particles of the second kind* refers to mesons with forbidden J^{PC} quantum numbers, i.e. natural parity mesons with odd CP (cf. Table 13.1.2). No exotic mesons have so far been seen, and, although two possible states have been suggested in the more extensively studied baryon case, the evidence is inconclusive.

‡ The charge conjugation quantum number C properly relates only to the neutral, non-strange members of the multiplets.

§ We indicate the hypercharge Y rather than strangeness $S(\equiv Y$ for mesons) to avoid confusion with the total quark spin quantum number, also denoted S.

(b) SU(3) and its Breaking: Ideal Mixing

According to equation (13.1.1) the mesons should occur in nonets, comprising an SU(3) singlet and octet. The octet consists of a $Y=1$, $I=\frac{1}{2}$ state K and its anti-particle $\bar{K}$, a $Y=0$ $I=1$ state V, and a $Y=0$ isoscalar state S_8. The explicit quark content of the various states is

$$K^+=p\bar{\lambda}, \qquad K^0=n\bar{\lambda},$$

$$V^+=p\bar{n}, \qquad V^0=\frac{1}{\sqrt{2}}(p\bar{p}-n\bar{n}), \qquad V^-=n\bar{p}, \tag{13.1.2}$$

$$S_8=\frac{1}{\sqrt{6}}(p\bar{p}+n\bar{n}-2\lambda\bar{\lambda}).$$

The SU(3) singlet, which will in general be of different mass, has the same I, Y values as S_8, and explicit quark content.

$$S_1=\frac{1}{\sqrt{3}}(p\bar{p}+n\bar{n}+\lambda\bar{\lambda}). \tag{13.1.3}$$

Because of SU(3) breaking the states S_1 and S_8 will in general mix to give physical states, which we denote by

$$S_0=S_8\cos\theta+S_1\sin\theta,$$
$$S_0'=-S_8\sin\theta+S_1\cos\theta, \tag{13.1.4}$$

where θ is the "mixing angle" for the supermultiplet in question. Measurability is ascribed to this quantity in two contexts. Firstly, if the SU(3) breaking interaction is assumed to transform as the $I=Y=0$ member of an SU(3) octet‡—the simplest possibility conserving I and Y—the masses within an octet obey the Gell-Mann–Okubo mass formula

$$4m_K^n=m_V^n+3m_8^n. \tag{13.1.5}$$

This, together with equation (13.1.4), allows us to determine the mixing angle from the expression

$$4m_K^n=m_V^n+3(m_0^n\cos^2\theta+m_0'^n\sin^2\theta). \tag{13.1.6}$$

These formulas are sometimes given for the masses ($n=1$), sometimes for the squared masses ($n=2$). The latter is suggested for mesons by the form of

‡ This assumption works well in describing the mass splittings within the $\frac{1}{2}^+$ octet and $\frac{3}{2}^+$ decouplet of baryons.

the relevant relativistic wave equations, and we shall follow it here.‡ The difference is only crucial for the 0^- octet, since in other cases the mass splittings are small compared with the mean mass of the octet, and θ is only weakly dependent on the choice $n = 1$ or 2.

The angle θ can also be determined from the SU(3) relations among reduced hadronic decay widths $\tilde{\Gamma}\,(a \to bc)$, where a, b and c are members of SU(3) supermultiplets A, B and C. The reduced width $\tilde{\Gamma}$ is a measure of the coupling strength and is obtained from the width Γ by dividing by a phase space factor Φ. The choice of Φ is not unambiguous, and in an explicit dynamical model the relations between the reduced widths would themselves be subject to SU(3) breaking effects. Aside from this, one has an independent measure of the mixing angle, and thus, in principle, a check on the assumed description of the mixing.

A particularly simple mixing pattern occurs if we attribute the mass splittings within a nonet solely to a mass difference between the non-strange quarks ($m_p^2 = m_n^2 = m^2$) and the strange quark ($m_\lambda^2 = m^2 + \Delta$), which is assumed to be heavier. With this hypothesis, the mass operator for the states S_1 and S_8 (equations (13.1.2), (13.1.3)) is not diagonal. If we do diagonalize it to find the physical states S_0, S_0', we obtain

$$S_0 = -\lambda\bar{\lambda}, \qquad S_0' = \frac{1}{\sqrt{2}}(p\bar{p} + n\bar{n}), \tag{13.1.7}$$

leading to the mass relations

$$m_0'^2 = m_V^2, \qquad 2m_K^2 = m_0^2 + m_0'^2. \tag{13.1.8}$$

The corresponding value of the mixing angle is (equation (13.1.4))

$$\cos\theta = (\tfrac{2}{3})^{\frac{1}{2}}, \qquad \text{i.e. } \theta \approx 35{\cdot}3°. \tag{13.1.9}$$

Nonets satisfying equation (13.1.8) (and hence with mixing angle given by equation (13.1.9)), are referred to as "ideal". It should be stressed that in this derivation we have assumed that, in the absence of SU(3) breaking, the singlet and octet masses are degenerate, i.e. $m_1 = m_8$. This is not an SU(3) result, but holds for example in SU(6); even then only for those nonets which belong to a single representation. We shall return to this point in more detail in paragraph (c) below.

Since the above scheme specifies the mixing angle, it has, as a consequence of our earlier discussion, implications for decay widths as well as for masses. Implicit in this is the assumption that one can factor out mass splitting effects through external phase space factors with the resulting "reduced widths" governed by the appropriate exact SU(3) relations. The meson families

‡ Our motive for this choice will become clear in paragraph (d) below.

appear as mixed nonets, and the various reduced widths should be expressible through SU(3) Clebsch–Gordon coefficients in terms of singlet and octet coupling constants g_1 and g_8 and the mixing angle θ.‡ The *pattern* of decays is governed by the ratio $\lambda \equiv g_1/g_8$ and the value of the mixing angle. Ideal nonets have $\cot\theta = \sqrt{2}$. Additional restrictions follow (Zweig, 1964) if two-body decays $A \to B + C$ are assumed to proceed as in Fig. 13.1.1 with the quark and anti-quark of M_1 separating to form M_2 and M_3 through the

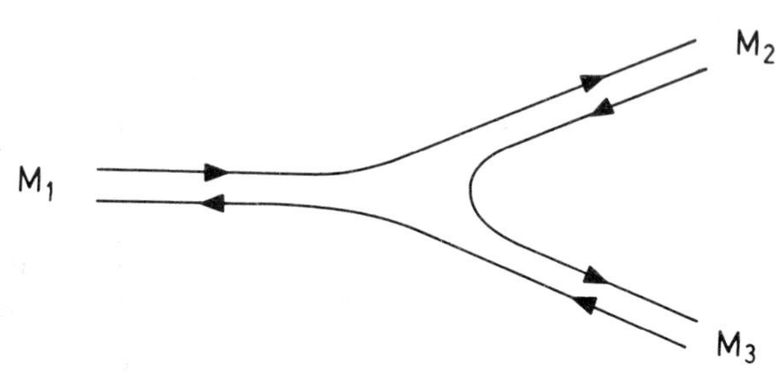

FIG. 13.1.1. Quark annihilation diagram for the decay $M_1 \to M_2 + M_3$.

creation of an additional $q\bar{q}$ pair, i.e. if they can be associated with a planar connected quark diagram. It follows immediately that the $\lambda\bar{\lambda}$ state, in general a linear combination of S_0 and S_0', cannot couple to states with no λ quark content. This requirement, which we term *Zweig's Rule*, determines the ratio $\lambda \equiv g_1/g_8$, referred to above, to be $\sqrt{2}$, thus imposing further restrictions on the decay patterns. They take a particularly simple form for ideal nonets, since the $\lambda\bar{\lambda}$ state then coincides with the physical state S_0, which hence is forbidden to decay to $\pi\pi$, $\rho\pi$ etc. If the mixing is not ideal, the quark selection rules for the decays of the S_0 and S_0' configurations could still apply, but they would not correspond to the physically observed states. When we turn to the larger groupings suggested by the quark model, it is interesting to study how the SU(3) parameters λ and θ vary from supermultiplet to supermultiplet within a given band. For example, within the $L = 1$ band one groups together the 2^{++}, 1^{++} and 0^{++} (triplet) levels. For the natural parity levels (2^+ and 0^+), upon which there are data, there are indications that, although the mixing angles are different, the λ-ratios are rather similar.

(c) Phenomenology of the Low-lying States

A possible identification of the observed resonances with the states predicted by the simple L-excitation quark model is indicated in Table 13.1.3.

‡ e.g. in our convention

$$\frac{\tilde{\Gamma}(f \to \pi\pi)}{\tilde{\Gamma}(f' \to \pi\pi)} = \left(\frac{g_8 \sin\theta + g_1 \cos\theta}{-g_8 \cos\theta + g_1 \sin\theta}\right)^2.$$

TABLE 13.1.3. Possible Assignments of Meson States for $L \leq 2$ within the L-excitation Quark Model. The states underlined are those which can couple to the $\pi\pi$ or πK systems. (S denotes the total quark spin.) The symbol ? indicates uncertainty either in status or quantum number assignments (Particle Data Group, 1974)[a]

L	S	J^{PC}	$I=1$	$I=\frac{1}{2}$	$I=0$	$I=0$
0	0	0^{-+}	π (140)	K (494)	η (549)	η' (958)
	1	1^{--}	ρ (770)	K^* (892)	ω (783)	ϕ (1020)
1	1	0^{++}	δ (976)	κ (1250)?	S^* (1000)	ε (1250)?
	1	1^{++}	A_1 (1100)?	K_A (1250)[b]?	D (1258)?	E (1416)?
	1	2^{++}	A_2 (1310)	K_N (1420)	f (1270)	f' (1514)
	0	1^{+-}	B (1235)	K_A (1340)[b]?		
2	1	1^{--}	ρ' (1600)		ω (1675)?	
	1	2^{--}		L (1765)[b]?		
	1	3^{--}	g (1686)	K_N (1800)		
	0	2^{-+}	A_3 (1640)?	L (1765)[b]?		

[a] Notional average values are used where the masses are uncertain.
[b] States with $Y = \pm 1$ and $C = \pm 1$ can mix.

Since the experimental information on meson systems is somewhat fragmentary, it is not surprising that many of the states in the $L = 2$ band have yet to be identified. An exception is the $\pi\pi$ channel, for which phase shift analyses exist over the whole energy region of the $L = 0$, 1 and 2 bands. It is a striking success of the quark model that all the predicted states with quantum numbers accessible from a $\pi\pi$ system have been identified, and that no additional states have yet been observed.‡ It will be of great interest to see if this success is repeated when πK phases become available over a similar range.

We now examine the phenomenology of the $L = 0$ and 1 bands in more detail.

(i) *The L = 0 Band*

All members of both the predicted nonets are well established, apart from a slight doubt over the spin of the η' (958) for which both $J = 0$ and 2 are consistent with present experimental information (cf. Butterworth, 1973). In what follows we adopt the conventional $J = 0$ assignment.

For the 0^{-+} nonet, the mass formula, equation (13.1.6), indicates relatively weak mixing. This conclusion is independent of whether the linear

‡ A possible exception to this would be the $I = J = 1$ ρ' state with $m \approx 1200$ MeV and $\omega\pi$ as dominant decay mode, which has been suggested on the basis of pole interpretations of electromagnetic form factors. If this state existed, it would, presumably, be classified within the quark model as an $L = 0$ radial excitation, analogous to the $P11$ (1470) state in πN scattering. There is, however, no clearcut evidence in favour of this ρ' state at present (Ballam *et al.*, 1974).

TABLE 13.1.4. SU(3) Comparisons of Vector Meson Decay Widths ($1^- \to 0^- 0^-$)

Decay	Width[a] (MeV)	$\tilde{\Gamma} \equiv \Gamma/\Phi'$ [d]	SU(3) Factor f	$g^2 \equiv f\tilde{\Gamma}$
$\rho \to \pi\pi$	150 ± 10	1·92	$\frac{3}{2}$	$2{\cdot}88 \pm 0{\cdot}20$
$K^* \to K\pi$	50	1·67	2	3·33
$\phi \to K\bar{K}$	4·2	2·26	$\sec^2\theta = 1{\cdot}67$[b]	3·78
			$= 1{\cdot}5$[c]	3·40

[a] Particle Data Group (1974).
[b] From the Gell-Mann–Okubo mass formula, equation (13.1.6).
[c] Assuming ideal mixing.
[d] The phase space factor Φ' is given in equation (13.1.18).

($n = 1$) or quadratic ($n = 2$) version is used. Confirmation, and discrimination, between the two cases is to be sought in the 2γ decay modes. At present, although the branching ratio for the η' (958) is known, the total width, and hence the partial width is not.

In contrast, the 1^{--} nonet approximately conforms to the ideal mixing scheme described above. Thus equations (13.1.8) are satisfied to within a few percent, and equation (13.1.6) yields $\cos^2\theta = 0{\cdot}60$ compared to 0·67 for exact ideal mixing. This is confirmed by examining the various predictions for decay widths. For example, SU(3) predicts the following relations for decays into pseudoscalar mesons (e.g. see Gourdin (1967))

$$\tilde{\Gamma}(\rho \to \pi\pi) : \tilde{\Gamma}(K^* \to K\pi) : \tilde{\Gamma}(\phi \to K\bar{K}) = \tfrac{2}{3} : \tfrac{1}{2} : \cos^2\theta. \quad (13.1.10)$$

This is in reasonably good agreement with present data as is shown in Table 13.1.4. Similarly, if we consider the direct couplings to photons measured in the decays $V \to l^+ l^-$, or in colliding beam experiments, we have (Gourdin, 1967)

$$\gamma_\rho^{-2}/(\gamma_\omega^{-2} + \gamma_\phi^{-2}) = 3; \qquad \gamma_\omega^{-2}/\gamma_\phi^{-2} = \cot^2\theta. \quad (13.1.11)$$

The experimental values are (Le Francois, 1971) $3{\cdot}0 \pm 0{\cdot}6$, $1{\cdot}5 \pm 0{\cdot}3$, again confirming the almost ideal mixing angle indicated by the mass formula.

(ii) *The* $L = 1$ *Band*

This band is more problematic, and of the four expected levels ($2^{++} 1^{++}$, 0^{++}, 1^{+-}) only the 2^{++} tensor nonet is well established. Like the 1^{--} nonet, this approximately conforms to the pattern of an ideal nonet. The mass formulas, equations (13.1.8), are approximately satisfied, and equation (13.1.6) gives $\cos^2\theta = 0{\cdot}73$, again fairly close to the value 0·67 required by exact ideal mixing.

The f' (1514) is then an (almost) pure $\lambda\bar{\lambda}$ state, so that by Zweig's rule the 2π decay should be strongly suppressed relative to $K\bar{K}$. The 2π decay has not yet been seen, $K\bar{K}$ being the only decay mode detected. SU(3) analysis of the decays to $(1^-)+(0^-)$ mesons yields a mixing angle of $34{\cdot}5°\pm 9°$, again consistent with the mass formula, and with ideal mixing (Samios *et al.*, 1974).

By comparison, the natural parity nonet of 0^{++} scalar is of more doubtful standing. There is no shortage of potential members—besides the δ (970) $I=1$ resonance (decaying to $\pi\eta$ and probably also to $K\bar{K}$), and the κ (1250) $I=\frac{1}{2}$ state.‡ Three $I=0$ candidates have at various times been considered (ε (700), S^* (997) and ε' (1250)). One suggestion (Rosner, 1974a) is to accept these three states and interpret them as originating from two pristine nonet members and an additional singlet (a "dilaton") with mixing among the physically observed states. However, there is no compelling evidence for a resonance interpretation of the δ_0^0 $\pi\pi$ phase shift in the region of the rho-mass, and an alternative interpretation of the phenomena in the $I=0$ channel has been given by Morgan (1974b). The ε (900) ($\delta_0^0\approx 90°$) and ε' (1240) ($\delta_0^0\approx 270°$) effects are regarded as aspects of one and the same broad elastic resonance, ε (1100–1300), interrupted by coupling to the S^* (997).§ It is further suggested that this latter effect (which lies close to the $K\bar{K}$ threshold and couples to $\pi\pi$ and $K\bar{K}$) corresponds to resonance poles on both the near-lying unphysical sheets. This enables the S^* to be associated with a conventional Breit–Wigner formula, and appropriate reduced widths (or coupling constants) for $S^*\to\pi\pi$ and $K\bar{K}$ to be extracted.‖ Taken with estimates for the reduced widths of the other $0^+\to 0^-0^-$ decays, these are shown to admit an SU(3) solution, with non-ideal mixing, cot $\theta_s\approx 0{\cdot}4$ instead of $\sqrt{2}$.¶ Numerical details are tentative. Further experimental work is needed both on the $I=0$ reactions, especially $\pi\pi\to K\bar{K}$, and also on the other decay modes $\kappa\to K\pi$ and $\delta\to\pi\eta$ and $K\bar{K}$, which in principle give a direct measure of the octet-coupling constant.‡‡

This leaves the unnatural parity nonets ($J^{PC}=1^{++}$, 1^{+-}) to be considered. Apart from the B (1237) ($I=1$, $J^{PC}=1^{+-}$) information on these states is

‡ Evidence for the κ has been mainly concentrated on the low-mass side where the πK $I=\frac{1}{2}$ S-wave phase shift is observed to rise monotonically towards 90°; however Lauscher *et al.* (1974) (*ABCLV* collaboration) analysing $K^-\pi^+$ production at 10 and 16 GeV/c with $M_{\pi K}$ extending up to 1·68 GeV find evidence for a 0^+ resonance with mass 1245 ± 30 MeV and width 485 ± 80 MeV. The resulting phase shift joins on well to the previous determinations at lower masses, (cf. the discussion of Section 3.3(g)).

§ This offers an escape from the long-noted difficulty with the Gell-Mann–Okubo mass formula for this nonet.

‖ For the 0^+ states, and in particular for the S^* (1000), the question of how kinematic factors should be included in effecting symmetry comparisons is of overriding importance (see Section 3.3(f)).

¶ The other parameter determining the decay pattern, $\lambda_S\equiv(g_1/g_8)_S$ comes out similar to that determined in the tensor decays ($\lambda_S\approx\lambda_T\approx 1{\cdot}2$), and thus close to the value indicated by Zweig's rule. If upheld, this would have interesting consequences for higher symmetries (Morgan, 1974d) (cf. Table 13.1.5).

‡‡ An alternative scheme of assignments for the 0^+ has been proposed by Conforto (1974) calling for a very narrow κ-effect at 900 MeV. This is not excluded by present data.

rudimentary and ambiguous. In particular, there is the notorious problem of the status, resonance or otherwise, of the A_1 ($\approx$1100)-effect ($I=1$, $J^{PC}=1^{++}$), and its strange particle analogues comprising the so-called Q-bump.‡ The A_1 is seen as an enhancement in the $\rho\pi$ production cross-section in the reaction $\pi N \rightarrow \rho\pi N$. Unfortunately, such a bump could also arise from diffraction dissociation $\pi \rightarrow \rho$"π", followed by rescattering of the "pion" from the target (Deck, 1964). Furthermore, "phase shift analysis" of the produced 3π final states using the isobar model (Ascoli *et al.* (1973) and earlier references therein) does not reveal a resonance-like movement of the 1^+ production phase through the A_1 region, unlike the analogous 2^+ phase over the A_2 mass region, which does. This would seem to cast considerable doubt on the resonance pedigree of the A_1. However, it has been pointed out (Goradia *et al.*, 1974) that the failure of the isobar model ansatz to satisfy unitarity in the final state two-body sub-energies is particularly serious for the 1^+ state at the A_1 mass, and probably vitiates the above negative conclusion. One would look to see progress on the status of the A_1 and Q effects in the near future.§

Of the $I=0$ 1^+ states wanted by the quark model, the only respectable candidate is D (1285)$\rightarrow K\bar{K}\pi$, $\eta\pi\pi$ which could be one of the $I=0$ counterparts to the A_1. Earlier reports of an H (990)$\rightarrow 3\pi$ as a possible $I=0$ companion for the B (1235) lack confirmation (see Particle Data Group, 1974).

This leaves B (1235)$\rightarrow \omega\pi$ as the only well-attested 1^+ resonance decay. This can give ω-mesons of helicity $\lambda=0$ or ± 1, and the relative admixture of the two (or equivalently the ratio of D- to S-wave decays) is an important parameter when considering higher symmetry schemes (Rosner, 1974a). The current experimental value (Particle Data Group, 1974) for the ratio $\Gamma\,(l=2)/[\Gamma\,(l=2)+\Gamma\,(l=0)]$ is approximately (6 ± 3)%.

(d) The Quark Model and SU(6)

The possibility of grouping particles according to the L-excitation quark model outlined above implies some degree of approximate symmetry with regard to both internal SU(3) quantum numbers and ordinary spin. Once intrinsic spin is included the scope of proposed symmetries is restricted by the requirements of Lorentz invariance. However, if the mesons are pictured as built of $q\bar{q}$ bound states with the quarks moving non-relativistically, one has the possibility of classifying mesons at rest according to enlarged

‡ Since C is not a good quantum number for states of non-zero strangeness, except in the limit of SU(3) symmetry, mixing can occur between the strange particle counterparts of the A_1 and the B (Goldhaber, 1967).

§ A resolution of these problems has been proposed by Bowler and Game (1974).

symmetries such as SU(6), with intrinsic spin and SU(3) treated on the same footing. The mesons are pictured as $q\bar{q}$ combinations with q and $\bar{q}$ each having six possible "alignments" (i.e. $q \equiv p^{\uparrow}, p^{\downarrow}, n^{\uparrow}, n^{\downarrow}, \lambda^{\uparrow}, \lambda^{\downarrow}$, where $\uparrow$ and $\downarrow$ denote intrinsic spin projections $\pm\frac{1}{2}$ along some reference direction). The quarks form a six-dimensional representation of this larger group, and the ground state mesons are associated with the group $U(6)_q \otimes U(6)_{\bar{q}}$.‡ Conjoining the relative orbital angular momentum one is lead to the group $U(6)_q \otimes U(6)_{\bar{q}} \otimes O(3)_{\text{orbital}}$ as a classification scheme for the meson spectrum (Weyers (1973) calls this the "rest symmetry"). In this scheme, the decomposition into SU(3) supermultiplets of the $L=0$ ground state is

$$(6_q, 6_{\bar{q}});\, L=0 \begin{cases} S=0, \{1\} \oplus \{8\}, J^{PC}=0^{-+} \\ S=1, \{1\} \oplus \{8\}, J^{PC}=1^{--}, \end{cases}$$

and the next ($L=1$) level is

$$(6_q, 6_{\bar{q}});\, L=1 \begin{cases} S=0, \{1\} \oplus \{8\}, J^{PC}=1^{+-} \\ S=1, \{1\} \oplus \{8\}, J^{PC}=2^{++} \\ S=1, \{1\} \oplus \{8\}, J^{PC}=1^{++} \\ S=1, \{1\} \oplus \{8\}, J^{PC}=0^{++}. \end{cases}$$

With such a large group the scope for symmetry breaking is considerable, and the aim of classification theory is to identify hierarchies of increasingly broken subsymmetries. One can imagine generalized spin and SU(3) couplings breaking the rest symmetry into irreducible representations of the sub-group $SU(6)_S \otimes O(3)$,

$$(6_q, 6_{\bar{q}});\, L \to (6_q \otimes 6_{\bar{q}});\, L = (1 \oplus 35);\, L.$$

In this way one is led to a separation of the original [36] representation of $(6_q, 6_{\bar{q}})$ into a singlet [1] and a [35] under $SU(6)_S$. (The subscript S for spin is introduced to distinguish this scheme from the analogous one, employing W-spin, to be discussed below.) If we write the six-dimensional representation of the quarks as $[6] \equiv [\{3\}, \frac{1}{2}]$, where the SU(3) and spin content are shown explicitly, then we have for the $L=0$ case

$$[6] \otimes [6] = [1] \oplus [35], \tag{13.1.12}$$

‡ The notation $U(6)_q \otimes U(6)_{\bar{q}}$ means one is classifying meson states according to their quark and anti-quark content *independently*. This is different from the approach adopted in the analogous case of SU(3) (equation (13.1.1)), where it is assumed from the start that, as regards the internal quantum numbers, the q and $\bar{q}$ belong to adjoint representations of SU(3)—in other words, one operates on the q and $\bar{q}$ indices simultaneously. The analogous distinction, in the case where one just has ordinary spin, is between spin-independence and rotational invariance. The former excludes and the latter admits $\boldsymbol{\sigma}^{(1)} \cdot \boldsymbol{\sigma}^{(2)}$ terms in the Hamiltonian leading to a splitting between singlet and triplet levels ($\{2\} \otimes \{2\} = \{1\} \oplus \{3\}$).

with

$$[1]=[\{1\},0],$$
$$[35]=[\{1\},1]\oplus[\{8\},1]\oplus[\{8\},0]. \qquad (13.1.13)$$

Hence the singlet and octet 1^- states would be degenerate in mass in exact $SU(6)_S$, whereas the 0^- singlet and octet belong to different SU(6) representations. Furthermore, if we assume that the symmetry breaking interaction transforms in a simple way under SU(6)—as a member of a [35] representation—then the ideal mixing relations, equation (13.1.8), are predicted, together with the result‡ (Gursey and Radicati, 1964)

$$m^2_{K^*}-m^2_\rho=m^2_K-m^2_\pi. \qquad (13.1.14)$$

The success of these predictions, together with the approximate ideal mixing pattern of the $L=1$ 2^{++} mesons, is evidence for $SU(6)_S\otimes O(3)$ as an approximate classification symmetry for the mesons. As such it is important to keep in mind its considerable success in the baryon spectrum, where for example the two lowest lying bands of resonances fit beautifully into the [56] $L=0^+$ and [70], $L=1^-$ representations of the same group. (For a recent review, see Rosner (1974a).

When we turn from classifying states at rest to the consideration of decay processes, $SU(6)_S$ fails badly. For exact $SU(6)_S$, quark spin is conserved, so that the familiar decay processes $\rho\to\pi\pi$ and $\Delta\to N\pi$ are forbidden $[(S=1)\not\to(S=0)+(S=0),\ (S=\frac{3}{2})\not\to(S=\frac{1}{2})+(S=0)]$. Clearly, in seeking to maintain the de-coupling of intrinsic and orbital spin for moving states, one is contravening Lorentz invariance. A way out of this dilemma and its undesirable consequences was pointed out by Lipkin and Meshkov (1965). They showed that the reflection and rotation invariance properties of collinear processes (such as two-body decays) leads to invariance under W-spin (or $SU(2)_W$) rather than S-spin (or $SU(2)_S$).§ The distinction concerns the occurrence of the intrinsic parity P_{int} in the expression for the generators operating on the quark and anti-quark states.

$$W_x=P_{\text{int}}S_x,\qquad W_y=P_{\text{int}}S_y,\qquad W_z=S_z. \qquad (13.1.15)$$

The W-spin operators satisfy angular momentum selection rules. Furthermore, the concept of W-spin can be extended to arbitrary systems of quarks and anti-quarks by assuming additivity (again see the discussion in Lipkin

‡ This result would not be satisfied if we had chosen to use linear rather than quadratic mass relations. Hence our choice.

§ Strictly, it is the little W-group of all W-spin rotations about the z-axis and all 180° rotations about axes perpendicular to the z-axis. However, for one-body quark model operators which only have components with $W=0$ and $W=1$ it comes to the same thing (Lipkin, 1973).

(1973)). Thus the W-spin generators acting on any system of quarks and anti-quarks can be expressed in terms of generators of $U(6)_q \otimes U(6)_{\bar{q}}$,

$$W_z = S_z q + S_z \bar{q}, \qquad W_x = S_x q - S_x \bar{q}, \qquad W_y = S_y q - S_y \bar{q}. \quad (13.1.16)$$

Reclassifying in terms of W-spin instead of S-spin produces no change for baryons. Mesons experience the phenomenon of "W–S flip", i.e.

$$S = 1,\ S_z = \pm 1 \rightarrow W = 1,\ W_z = \pm 1$$
$$S = 1,\ S_z = 0 \rightarrow W = 0,\ W_z = 0$$
$$S = 0,\ S_z = 0 \rightarrow W = 1,\ W_z = 0, \qquad (13.1.17)$$

so that the decays $\rho \rightarrow \pi\pi$, $\Delta \rightarrow N\pi$ are allowed by W-spin conservation. It was therefore proposed (Lipkin and Meshkov, 1965) that $SU(6)_W$‡ should be adopted as an approximate symmetry for two-body decays, suggesting $SU(6)_W \otimes O(3)$ as an alternative classification group for moving states.§ We would, however, expect to find symmetry violations, since the notion of W-spin conservation is based on the study of collinear processes. For example, in an explicit quark model, a two-body decay will not be collinear because of the internal motions of the quarks. Violations are observed, the most notable being that the ratio of D- to S-waves in the decay $B \rightarrow \pi\omega$ is predicted to be 2 : 1, whereas the experimental ratio is approximately 6% (see Section 13.1(c) above).

(e) Higher Symmetries and Decay Widths

The above discussion suggests that any attempt to predict relations between decay widths based on symmetries higher than SU(3) must incorporate a model, or prescription, for broken $SU(6)_W$. There are three main approaches to the problem.

(i) *Explicit Quark Models*

In these models $SU(6)_W$ is automatically broken by the internal motions of the quarks, which are regarded as particles bound in a smooth potential. These models not only give an account of $SU(6)_W$ breaking, but go beyond $SU(6)_W$ in relating decays belonging to different representations, thus unconnected by the symmetry. For example, the decays $(L = 2) \rightarrow (L = 0) + (L = 0)$ will be related to those with $(L = 1) \rightarrow (L = 0) + (L = 0)$.

‡ The subscript W indicates that W-spin operators have replaced the usual S-spin operators. The *algebraic* structure of $SU(6)_W$ and $SU(6)_S$ is identical.
§ Strictly, the relevant group is $SU(6)_W \otimes O(2)_{L_z}$ (Weyers, 1973).

Many calculations have been performed in this framework, usually based upon some variant of the non-relativistic quark model with oscillator wave function. (Greenberg, 1964; Dalitz, 1967; Faiman and Hendry, 1968, 1969; Copley *et al.*, 1969). Following this work, an extension of the model, with several improvements over earlier versions, has been studied by Feynman *et al.* (1971), and used to calculate mesonic decay widths. In particular, the PCAC condition (see Appendix D) is invoked to relate the emission of pseudo-scalar mesons to matrix elements of the appropriate axial vector current. There are serious problems arising from the use of PCAC for K and η emission both as regards consistency with the results obtained using the presumably more reliable pion PCAC hypothesis, and comparison with experiment. However, the predicted partial widths for the well-established states with $L \leqslant 1$ are within 50% of the experimental values in those cases where PCAC is used for the pion. An alternative model, based on using the Bethe–Salpeter equation, thereby avoiding the difficulties associated with the PCAC Condition, has been given by Böhm *et al.* (1974).

(ii) *l-Broken* $SU(6)_W$

This approach (Colglazier and Rosner, 1971; Faiman and Plane, 1972) focusses on the systematics of the relative orbital angular momentum L_{orb} of the two final hadrons. For decays of the form $(L = 1) \to (L = 0) + (L = 0)$, L_{orb} has to be 0 or 2 from angular momentum and parity conservation, and the choices are similarly restricted in analogous decays. It turns out that the most flagrant violations of $SU(6)_W$ concern comparisons between S- and D-wave decays—for example the previously cited failure to predict the S-, D-wave breakdown of the $B \to \omega\pi$ decay. It was therefore proposed that this link be broken, and only the $SU(6)_W$ relations among the S-wave contributions and the D-wave contributions separately be retained. As a result, one has two parameters instead of one to describe all the decays of the form $L = 1 \to$ pairs of $L = 0$ mesons (Colglazier and Rosner, 1971). In order to effect the comparisons, one needs two further ingredients—a doctrine about the effects of mixing on the decays for each nonet (ideal mixing together with Zweig's Rule is the usual assumption), and a decision as to what phase space factor one should use. Colglazier and Rosner used

$$\Phi^l \equiv p^{2l+1}/m_A^2, \tag{13.1.18}$$

for the decay $A \to BC$, where p is the centre-of-mass final state momentum. This choice of Φ turns out to be the major discriminant between this and the approach based on the Melosh transformation. We therefore postpone the comparison with experiment until after our discussion of this latter approach.

(iii) *The Melosh Transformation: Constituent and Current Quarks*

The preceding discussion implies that hadrons can be classified into representations of $SU(6)_{W;\text{constituents}}$, so called because of its natural occurrence as a classification symmetry in simple "constituent" quark models. The same *algebra* also occurs when we consider the properties of currents in simple quark models. Thus, if we assume the current commutation relations suggested by free quark models, and restrict ourselves to "good" operators,‡ the same algebraic structure of $SU(6)_W$ is again obtained. These operators therefore generate a symmetry group $SU(6)_{W;\text{currents}}$ which can also be used to classify the hadrons. However, while the hadron states transform in a simple way under $SU(6)_{W;\text{constituents}}$, their classification under $SU(6)_{W;\text{current}}$ is very complicated, each state having components from many representations.§ It is on the conjectured form of the relationship between these two distinct classifications that recent progress has been made.||

The new approach is thus restricted to "transition" symmetries, only applicable to matrix elements of the form ⟨Hadron|current|Hadron⟩ rather than the formerly conjectured "vertex" symmetries. The above ideas can therefore be applied to the prediction of radiative decays, $A \to B + \gamma$ or pionic decays $A \to B + \pi$. In the latter case (Gilman, Kugler and Meshkov 1973, 1974; Hey and Weyers, 1973; Gilman, 1974), the first step is to relate the pionic decay matrix elements to the matrix elements of the axial charge, Q^5, using PCAC (cf. Appendix D). Following Fubini and Furlan (1965), the identification is made in the infinite momentum frame, when it leads to the formula

$$\Gamma^{(0)}(A \to B\pi) = \frac{C}{2J_A + 1} p_\pi^{(0)} \frac{(m_A^2 - m_B^2)^2}{m_A^2} \sum_\lambda |\langle A, \lambda | Q_\alpha^5 | B, \lambda \rangle|^2, \tag{13.1.19}$$

where p_π is the centre-of-mass pion momentum, and the superscript (0) implies the limit $m_\pi \to 0$. Apart from the $(2J_A + 1)^{-1}$ term and the constant C, which depends on the isospins, the kinematic factors preceding the spin sum are a specific consequence of using the PCAC approximation in the infinite momentum frame (Horn, 1966). This phase space factor

$$\Phi^{\text{PCAC}} \equiv p_\pi^{(0)}(m_A^2 - m_B^2)^2 / m_A^2, \tag{13.1.20}$$

‡ The "good" operators are the ones whose commutation relations lead to well-defined, convergent, sum rules when saturated by a complete set of states in the infinite momentum frame.

§ Assuming identity of the two $SU(6)_W$'s leads to immediate contradictions. For example, the Adler–Weisberger relation (see Chapter 14) would be saturated by contributions from N (938) and Δ (1236) alone, giving $G_A/G_V = -\frac{5}{3}$, in disagreement with experiment.

|| In what follows our treatment will of necessity be brief. Fuller reviews have been given by Weyers (1973), Hey, Rosner and Weyers (1973) and Rosner (1974a). A useful elementary discussion is that of Hey (1974).

then replaces the assumed form Φ^{2l+1} of equation (13.1.18), since the symmetry predictions will be obtained for the matrix elements of $\langle A|Q^5_\alpha|B\rangle$.‡ It is in the final step of evaluating these terms that appeal is made to the conjectured Melosh transformation. The hadronic states A and B are taken to lie in representations of $SU(6)_{W;\text{constituents}}$, the appropriate classification symmetry for hadrons moving along a given direction. The axial charge Q^5 is assumed to lie in a representation of the distinct $SU(6)_{W;\text{currents}}$. In order to relate different decay matrix elements, one needs the form of Q^5, $V^{-1}Q^5V$, transformed into the $SU(6)_{W;\text{constituents}}$ basis. Melosh (1973) worked out such a transformation for free quarks. The transformed axial charge then comprises two terms, one with the quantum numbers of the π, the other with those of the A_1. It is assumed that this simple algebraic structure extends to systems of interacting quarks, but with arbitrary coefficients for the two contributions. This yields a two-parameter description of all pionic decays $A \to B + \pi$ of members of one L-band to those of another. Although the Melosh approach and l-broken $SU(6)_W$ in general lead to different answers,§ there is in the present case a one-to-one correspondence between the two, and the difference in the outcome arises solely from the phase space factor employed (Kugler, 1973). Changing from Φ^l to Φ^{PCAC} modifies the SU(3) predictions discussed earlier; these can be used to extend the predictions for pionic decays to include other possible decay modes.

(iv) *Results for* $L = 1$ *Meson Decays*

In this section we will restrict ourselves to brief comments on the decays $(L=1) \to (L=0)+(L=0)$. There are analogous predictions for other decays $[(L=1) \to (L=1)+(L=0)$; $(L=2) \to (L=0)+(L=0)$, etc.], for which we refer the reader to Gilman, Kugler and Meshkov (1974) and Rosner (1974a).

In Table 13.1.5 the experimental widths, where available, are compared with the predictions of the three approaches discussed above. The quark model results cited are taken from Böhm *et al.* (1974). Those for both versions of "broken $SU(6)_W$" are adaptations of Rosner (1974a), and neglect η–η' mixing in the pseudoscalar meson nonet. These also assume ideal mixing and Zweig's Rule for the 2^{++} tensor nonet, although the predictions for the $I = \frac{1}{2}$ and 1 states are independent of these assumptions. In addition, we have given predictions for the 0^{++} nonet, using masses,

‡ One might think to correct for the fact that Φ^{PCAC} is independent of the orbital angular momentum of the decay (Hey and Weyers, 1973). However, the numerical calculations discussed below use the unmodified form of equation (13.1.20).

§ An important example is furnished by the radiative couplings of baryon resonances, measured in photoproduction experiments. In this case the Melosh transformation leads to less specific predictions than l-broken $SU(6)_W$, which can be regained as a special case.

TABLE 13.1.5. Width Comparisons for Decays of $L = 1$ Mesons to $L = 0$ Mesons
(a) S-Waves

Decay	Γ_{Expt} (MeV)	Γ_{theory}(MeV) "Broken SU(6)$_W$" Φ^{PCAC}	Φ^{l}	$\tilde{\Gamma}$
$B\,(1237) \to (\omega\pi)_S$	113 ± 19	113 (input)		$S^2/96$
$A_1\,(1100) \to (\rho\pi)_S$	≈ 300	196	411	$S^2/24$
$\delta\,(976) \to \eta\pi$	50 ± 20	83	163	$S^2/96$
$\kappa\,(1250) \to K\pi$	450–550	734	365	$3S^2/128$
$\varepsilon\,(1250) \to \pi\pi$	[635][b]		552	$0{\cdot}64\,(3S^2/64)$[b]
$S^*\,(1000) \to \pi\pi$	[182][b]		216	$0{\cdot}61\,(S^2/64)$[b]

(b) D-Waves

Decay	Γ_{Expt} (MeV)	Γ_{theory}(MeV) Quark model[a]	"Broken SU(6)$_W$" Φ^{PCAC}	Φ^{l}	$\tilde{\Gamma}$
$A_2\,(1310) \to \rho\pi$	72 ± 2	82–110	72 (input)	50	$3D^2/80$
$\to \eta\pi$	15 ± 1	14–18	16	19	$D^2/240$
$\to \eta'\pi$	$\leqslant 1$		6	1·3	$D^2/120$
$\to K\bar{K}$	≈ 5	7–9		15	$D^2/160$
$K^*\,(1420) \to K\pi$	55 ± 3	55–74	56	76	$3D^2/320$
$\to K^*\pi$	30 ± 3	3–38	27	16	$9D^2/640$
$\to K\eta$	2 ± 2	2		2·5	$D^2/960$
$f\,(1270) \to \pi\pi$	141 ± 25	140–180	113	175	$3D^2/160$
$\to K\bar{K}$	7 ± 6	8–10		7	$D^2/160$
$f'\,(1514) \to K\bar{K}$	40 ± 10	45–60		65	$D^2/80$
$B\,(1237) \to (\omega\pi)_D$	7 ± 2		26	14	$D^2/48$
$A_1\,(1100) \to (\rho\pi)_D$	?		11	2	$D^2/48$

[a] From Böhm *et al.* (1974), who simultaneously predict various decay widths for the $L = 0$ and 2 bands; the S-wave decays are unfortunately not given.
[b] The estimates for Γ_{expt} and the explicit coefficients appearing in $\tilde{\Gamma}$ are taken from the "S^*–ε" interference model fit of Morgan (1974b).

mixing angle and relative singlet–octet coupling taken from the phenomenological analysis of Morgan (1974b).‡ Since this fit assumed the Φ^{l} phase space factor, we have given results for this form only, and they are very tentative. With the exception of the δ (970), very broad widths are

‡ In general, the reduced widths for $S_0(f', S^*$ etc.) and $S'_0(f, \varepsilon$ etc.)$\to \pi\pi$, $K\bar{K}$ and $\eta\eta$, and the corresponding $K^* \to K\pi$ decay are predicted to lie in the ratios $S_0^{\pi\pi} : S_0^{K\bar{K}} : S_0^{\eta\eta} : S_0'^{\pi\pi} : S_0'^{K\bar{K}} : S_0'^{\eta\eta} : K^{*K\pi} = 3(\lambda - c)^2 : (2\lambda + c)^2 : (\lambda + c)^2 : 3(\lambda c + 1)^2 : (2c\lambda - 1)^2 : (-\lambda c + 1)^2 : \frac{9}{2}(1 + c^2)$, where $c = \cot\theta$, and $\lambda \equiv (g_1/g_8)$.

predicted: one therefore anticipates that the phenomenological characterization of this multiplet may be especially ambiguous. For the S-wave decays the overall scale has been set from $\Gamma[B \to (\omega\pi)_S]$,‡ for the D-waves from $\Gamma(A_2 \to \rho\pi)$ when Φ^{PCAC} is used, and from a compromise between $\Gamma(A_2 \to \rho\pi)$ and $\Gamma(f \to \pi\pi)$ when Φ' is used. As can be seen from the table, the situation for D-wave decays is encouraging. Although many of the successful relations between different widths follow from SU(3), this does not apply to the $A_2 \to \rho\pi$, $A_2 \to \eta\pi$ and $B \to (\omega\pi)_D$ decays. On the other hand, very little is known empirically about the S-wave decays of $L = 1$ Mesons.

Generally, until the spectrum and decay parameters are better known, judgement must be suspended as to the overall success of applying broken $SU(6)_W$ ideas to meson decays.

13.2. Regge Poles and Duality

Having seen how particles can be grouped according to their internal quantum numbers, we now turn to the complementary classification into families of ascending angular momentum via Regge trajectories. The evidence from the baryon spectrum suggests that all particles lie on approximately linear Regge trajectories $\alpha(t)$, with slopes clustering about a universal value $\alpha'(t) \approx 0{\cdot}9\ \text{GeV}^{-2}$. The meson spectrum (Fig. 13.1) conforms to this pattern, and, furthermore, affords the most striking instance of the identification of Regge poles in exchange ($t < 0$) with their manifestation ($t > 0$) as ascending families of physical resonances. The best authenticated of all the exchanged Regge poles is the ρ-trajectory which fits very satisfactorily to the ρ–g sequence (Fig. 13.1).

Relations between Regge trajectories follow from considerations of duality. A striking empirical feature is the approximate degeneracy of those meson trajectories which differ from each other by signature alone (cf. Appendix C). This effect—weak exchange degeneracy—is summarized in equation (12.2.8) and Fig. 12.2.1 and is explained naturally by the combined ideas of "absence of exotics" and "two-component semi-local duality". These topics have been discussed in some detail in Section 12.2. The predictions for the resonance spectrum become more specific when one turns to models incorporating local duality, such as the Veneziano model described in Section 12.3. The most striking result is the prediction of an infinite sequence of "daughter" trajectories lying parallel to, but below, the leading ρ–f trajectory (we specialize to the case of $\pi\pi$ scattering) and (as stressed in Section 12.3(b)) there are essentially two alternatives for the

‡ This is taken from $\Gamma_S/(\Gamma_S + \Gamma_D)$ times the B-width, assuming that $B \to \pi\omega$ is the only decay mode ($B \not\to \delta\pi$ etc.). It could be that S-wave decays are thereby substantially overestimated (Rosner, 1974b).

resonance pattern within this framework. The first is that given by the single-term Veneziano formula equation (12.3.6) which predicts the singly-spaced spectrum shown in Fig. 12.3.2, in which each particle on the leading trajectory with spin $J = N$ is accompanied by a tower of resonances with spins $J = N-1,\ N-2, \ldots, 0$. Within the single-term model the partial widths into $\pi\pi$ are also predicted. The values are given in Table 12.3.1. We note in particular that the states along the first daughter trajectory couple strongly to $\pi\pi$. They are by contrast predicted to decouple from the $\pi\omega$ channel. Finally, each $I = 0$ state in Fig. 12.3.2 is, in general, accompanied by a second $I = 0$ state within its SU(3) nonet; this will in turn be accompanied by its own daughter spectrum. Simple duality arguments lead directly to the decoupling of these states from the $\pi\pi$ system (cf. the argument following equations (12.2.8)).

All these remarks, except the last, apply to the single-term Veneziano model. When satellite terms (equation (12.3.10)) are added, the widths predicted for all the daughter states will in general be altered. Moreover, by a suitable choice of coefficients (cf. (12.3.14)) alternate daughters may be eliminated completely, so that the spectrum of Fig. 12.2.3 results. As discussed in Section 12.3(b), this is the same spectrum as results from Schmid circle analysis of the simplest Regge pole forms for spin-flip amplitudes. Such amplitudes are the ones for which simple Regge pole ideas (implicit in the Veneziano model) are most successful, at least for rho- and omega-exchange.

We thus have two simple possibilities for the daughter spectrum, single or double spacing. It remains to consider which, if either, of these two patterns is consistent with experiment. Before doing this, it is interesting to contrast the predictions with those of the L-excitation quark model discussed in the previous section. In addition to the leading ρ–f trajectory with $J = L+1$ this predicts a second trajectory with $J = L-1$ lying two units below. This prediction is identical to that obtained by retaining the first two trajectories of the doubly-spaced spectrum of Fig. 12.2.3. In neither case is the simple prediction for the "daughters" expected to be exact. The forces responsible for mass splittings in the quark model, and unitarity effects in the dual model, lead one to expect mass shifts. Nonetheless, the empirical success of the quark model for the $\pi\pi$ spectrum, in particular the observation of appropriate resonances ε (1250) and ρ' (1600)‡ approximately coincident with the f and g mesons, can also be taken as evidence for the two leading trajectories of the doubly-spaced dual model. Whether there are further lower lying daughter trajectories is still unknown.

‡ Evidence for the existence of the ρ' (1600) is obtained in the 4π channel from electromagnetic interactions. A small coupling to $\pi\pi$ has also been established. (See Chapter 6, and discussions in Sections 3.3, 5.3 and 5.4.)

Turning to the singly spaced spectrum of Fig. 12.3.2 we would have in addition a trajectory one unit below the leading trajectory, giving S-, P- and D-wave daughters of the ρ, f, g mesons respectively. Furthermore, in the single-term Veneziano model, which suggests this spectrum, these resonances couple strongly to $\pi\pi$ and decouple from $\pi\omega$. The large experimental $\pi\pi$ S-wave in the rho-region has often been taken as evidence for the S-wave daughter of the rho predicted by the single-term Veneziano formula. However, the resonance interpretation of this wave is, as we have seen, highly ambiguous (Section 3.3). Whatever position one takes on this it is clear from $\pi\pi$ phase shift analyses that the predicted P-wave and D-wave resonances, respectively in the f^0 and g-regions, coupling strongly to $\pi\pi$, do not exist (cf. Chapter 3). It would seem, therefore, that the resonance pattern for the $\pi\pi$ system predicted by the single-term Veneziano model (in particular the Lovelace–Veneziano model of Section 12.3(c)) is in definite conflict with experiment. However, if one abandons the predicted couplings of the single-term model, assuming say strong coupling to $\omega\pi$, weak to $\pi\pi$,‡ the evidence is much less decisive. Nonetheless, there is no decisive evidence for these extra states, so that the doubly spaced spectrum looks more promising. Whether lower lying daughter trajectories exist—which might be interpreted within the quark model as radial excitations—is an open question.

‡ A P-wave daughter of the f, with just these decay characteristics, has been suggested on the basis of a theoretical interpretation of electromagnetic form factor data (Shaw, 1972) and there exist data which could support such an interpretation (Ballam *et al.*, 1974).

Chapter 14

Predictions from Current Algebra and PCAC

Current algebra, especially when combined with the hypothesis of PCAC, brought fresh ideas and concepts to bear on the study of hadrons and their interactions. Its genesis was in the study of weak interactions, and in attempts to understand the observed regularities among decay coupling constants for hadrons and leptons. The end product is well known—the identification of the vector and axial vector weak currents of hadrons as physically observable quantities and the assumption that these manifest broken $SU(2)\otimes SU(2)$ symmetry. The symmetry aspect is expressed via the equal time commutators of the associated charges and currents; the breaking of the symmetry is reflected in the non-conservation of the axial current. The notion that the divergence of this non-conserved current is dominated by the pion-pole contribution to the appropriate matrix elements (which is the essential idea of PCAC) allows the approximate translation of many statements concerning matrix elements of currents into statements involving scattering amplitudes containing pions.

The application of these ideas, culminating in Weinberg's famous linear model for low-energy $\pi\pi$ scattering (Weinberg, 1966), has had a very great impact on our understanding of $\pi\pi$ interactions. Viewed operationally, Weinberg's achievement in predicting $\pi\pi$ scattering lengths, with the ratio of weak decay rates $\Gamma(\pi \to \mu\nu)/\Gamma(\mu \to e\nu\bar{\nu})$ as the only experimental input, is a remarkable one. Furthermore, the resulting picture was unanticipated from other theoretical viewpoints. The novel feature was the prediction of small S-wave scattering lengths of opposite sign ($a_0^0 > 0$, $a_0^2 < 0$), but comparatively large slopes for the scattering amplitudes, leading to a large and rapidly rising $I=0$ S-wave phase shift, despite the small value of the scattering length. This type of behaviour is closely linked to the occurrence of sub-threshold zeros in the S-wave amplitudes. Previous calculations, often based on partial wave dispersion relations, had emphasized solutions which tended to satisfy the relation $2a_0^0 \approx 5a_0^2$, obtained by a smooth continuation to threshold of the analogous exact relation at the symmetry point $s=t=u=\frac{4}{3}$.

In this chapter, we shall be concerned mainly with drawing conclusions *from* current algebra and PCAC *about* the $\pi\pi$ system, but it is also of interest to run the argument in reverse, since the truth, or otherwise, of the input assumptions is of fundamental interest for our understanding of hadron symmetry. The $\pi\pi$ system is one of particular importance in this respect, because it is entirely composed of the particles which are singled out in PCAC (at least for hypercharge-conserving currents); also because the complete crossing symmetry leads to severe restrictions on the form of the off-shell extrapolations which occur in many formulations of these ideas, and which constitute the main source of ambiguity in their application. Furthermore, any difficulties which manifest themselves in $\pi\pi$ scattering will have ramifications for all applications involving two soft pions, that is, for many of the most important applications of current algebra and PCAC.

The basic assumptions which are made in this chapter—the notion of PCAC and the current commutators used—are doubtless familiar to many readers. They are summarized in Appendix D, partly for completeness, and to specify notations, and partly to indicate which of the several alternative approaches to PCAC we follow. In this chapter, we assume this background material and proceed directly to a description of Weinberg's calculation of $\pi\pi$ scattering, which is the central part of the chapter. Following this, we discuss the relation between this calculation and some other important topics; specifically, the use of current algebra sum rules, the KSFR relation, and some dual models. In the penultimate section, we explore the question of corrections to Weinberg's technique for proceeding from the off-shell current algebra constraints to his on-shell predictions. This introduces no new qualitative results, but is important if a precise vindication of his basic assumptions is to be sought from future experimental results on low-energy scattering. Finally, we make some brief remarks on the application of similar ideas to other closely related processes.

14.1. Weinberg's Calculation of Low-Energy $\pi\pi$ Scattering

This calculation, like the derivation of the Goldberger–Treiman relation given in Appendix D, falls into two parts. Firstly the use of the commutators and the definition of the pion field to derive formally exact results for the off-shell $\pi\pi$ scattering amplitude at special points; secondly, the passage to on-shell predictions by implementing PCAC using a simple linear model for the low-energy scattering amplitude both on- and off-shell.

(a) Predictions for Off-Shell $\pi\pi$ Scattering

The first step is to use the definition of the pion field (equation (D.10) of Appendix D)

$$\partial_\lambda A^{\lambda\alpha}(x)=\frac{m_\pi^2 f_\pi}{\sqrt{2}}\phi^\alpha(x), \qquad (14.1.1)$$

together with the reduction formalism, to define the off-shell continuation of the $\pi\pi$ scattering amplitude. For the process

$$\pi^\alpha(p_1)+\pi^\beta(p_2)\to\pi^\gamma(p_3)+\pi^\delta(p_4),$$

we consider the matrix element

$$\langle\gamma,\delta|T|\beta,\alpha\rangle=A(s,t,u)\delta_{\alpha\beta}\delta_{\gamma\delta}+B(s,t,u)\delta_{\alpha\gamma}\delta_{\beta\delta}+C(s,t,u)\delta_{\alpha\delta}\delta_{\beta\gamma}, \qquad (14.1.2)$$

where A, B and C are the Chew–Mandelstam amplitudes, and T is defined in terms of the S-matrix by equation (2.9a). From the PCAC definition, equation (14.1.1), and the reduction formalism, we can immediately define an off-shell continuation of T in p_2 and p_4, i.e.

$$\begin{aligned}\int d^4x\,d^4y\langle p_3,\gamma|T\{\partial_\mu A^{\mu\delta}(x),\partial_\nu A^{\nu\beta}(y)\}|p_1,\alpha\rangle\,\mathrm{e}^{-ip_4x}\,\mathrm{e}^{ip_2y}\\ =\frac{f_\pi^2m_\pi^4}{2}\int d^4x\,d^4y\langle p_3,\gamma|T\{\phi^\delta(x),\phi^\beta(y)\}|p_1,\alpha\rangle\,\mathrm{e}^{-ip_4x}\,\mathrm{e}^{ip_2y}\\ =-\frac{i(2\pi)^4\delta^{(4)}(p_3+p_4-p_1-p_2)f_\pi^2m_\pi^4(16\pi)}{(p_4^2+m_\pi^2)(p_2^2+m_\pi^2)(2\pi)^3(4p_1^0p_3^0)^{\frac{1}{2}}}\langle\gamma,\delta|T|\alpha,\beta\rangle. \qquad (14.1.3)\end{aligned}$$

Likewise, for one pion (β) off-shell one has

$$\begin{aligned}\int d^4y\langle\gamma,\delta|\partial_\mu A^{\mu\beta}(y)|\alpha\rangle\,\mathrm{e}^{ip_2y}\\ =\frac{-\sqrt{2}(16\pi)(2\pi)^4\delta^{(4)}(p_3+p_4-p_1-p_2)f_\pi m_\pi^2}{(p_2^2+m_\pi^2)(2\pi)^{\frac{9}{2}}(8p_1^0p_3^0p_4^0)^{\frac{1}{2}}}\langle\gamma,\delta|T|\alpha,\beta\rangle. \qquad (14.1.4)\end{aligned}$$

It is now comparatively straightforward to obtain some exact consequences of these definitions and the usually assumed commutation relations, in particular that between the axial charges

$$\delta(x^0-y^0)[Q_\alpha^5(x^0),Q_\beta^5(y^0)]=i\varepsilon_{\alpha\beta\gamma}Q_\gamma(y^0)\delta(x^0-y^0), \qquad (14.1.5)$$

and the more model-dependent σ-commutator

$$\delta(x^0-y^0)\left[Q^5_\alpha(x^0), \frac{\partial Q^5_\beta(y^0)}{\partial y^0}\right] = \left(\frac{f_\pi m_\pi^2}{\sqrt{2}}\right)\sigma(x^0)\delta_{\alpha\beta}\delta(x^0-y^0). \quad (14.1.6)$$

The reader unfamiliar with these relations should refer to Appendix D.

The first consequence of these equations is a particular case of Adler's theorem for the emission of one soft pion (Adler, 1965a,b). Partial integration of the left-hand side of equation (14.1.4) yields immediately

$$\langle\gamma, \delta|T|\alpha, \beta\rangle = D(p_2^2+m_\pi^2)p_{2\mu}\int d^4y\langle\gamma, \delta|A^{\mu\beta}(y)|\alpha\rangle\, e^{ip_2y},$$

where D is a factor which is regular at $p_{2\mu}=0$. Thus, in the limit $p_{2\mu}\to 0$ (i.e. $s=t=u=m_\pi^2=1$) the amplitude vanishes, since the matrix element of the axial current between a single-pion and a two-pion state has no pole in this limit. The resultant null condition

$$A(s=t=u=1)=0, \quad (14.1.7)$$

(which implies $B(s=t=u=1)=C(s=t=u=1)=0$ also), is usually called the *Adler zero* (note that with just one pion off-shell the independent specification of s, t and u is sufficient to fix the off-shell point $(s+t+u=\Sigma m_i^2)$; off-shell masses will sometimes not be shown explicitly even when two pions are off-shell).

The other results come from applying standard methods for the emission of two soft pions. The left-hand side of (14.1.3) is transformed twice by partial integration to yield

$$\begin{aligned}\int d^4x\, d^4y\, e^{ip_2y}\, e^{-ip_4x}\{&\delta(x^0-y^0)\langle\gamma|[A^{0\delta}(x), \partial_\nu A^{\nu\beta}(y)]|\alpha\rangle\\ &-ip_{4\mu}\delta(x^0-y^0)\langle\gamma|[A^{0\beta}(y), A^{\mu\delta}(x)]|\alpha\rangle\\ &+p_{2\nu}p_{4\mu}\langle\gamma|T\{A^{\mu\delta}(x), A^{\nu\beta}(y)\}|\alpha\rangle\}.\end{aligned} \quad (14.1.8)$$

The delta functions arise from the operators $\partial/\partial x^0$ and $\partial/\partial y^0$ acting on θ-functions implicit in the time-ordered products. Thus it is that the equal-time commutators come into play. The next stage is to take the limit $p_{2\mu}=p_{4\mu}=k_\mu\to 0$ so that $p_{1\mu}=p_{3\mu}\equiv p_\mu$ and hence $t=0$ and $s=m_\pi^2+2p\cdot k$, $u=m_\pi^2-2p\cdot k$. In this limit the third term is of order $k_\mu k_\nu$, and so vanishes faster as $k\to 0$ than the other terms. It can therefore be neglected in this limit. Furthermore, since the first two terms contribute to different amplitudes (the first can be shown to be symmetric, and the second antisymmetric, in β, δ) we can treat them separately.

Let us consider the first (symmetric) term. On integrating over the spacelike variable, we obtain ($k_\mu = 0$)

$$\int dx_0\, dy_0\, \delta(x^0 - y^0)\left\langle \gamma \left\| \left[Q_\delta^5(x^0), \frac{\partial Q_\beta^5(y^0)}{\partial y^0} \right] \right\| \alpha \right\rangle. \tag{14.1.9}$$

This is just the σ-commutator specified in equation (14.1.6), where we have assumed it is proportional to $\delta_{\beta\delta}$. More generally, this term can be shown to be symmetric in β, δ (as demonstrated in Appendix D, cf. section (b)) so that from equation (14.1.2) it can contribute to the terms

$$\delta_{\alpha\gamma}\delta_{\beta\delta}B + \tfrac{1}{2}\{\delta_{\alpha\beta}\delta_{\gamma\delta} + \delta_{\alpha\delta}\delta_{\beta\gamma}\}(A + C).$$

Thus the assumption that it is proportional to $\delta_{\beta\delta}$ gives the result

$$A(s = u = 1, t = 0) + C(s = u = 1, t = 0) = 0. \tag{14.1.10}$$

We shall call this the *σ-condition.*

Finally, there is a third condition relating to the second term of (14.1.8). Going to the special frame $\mathbf{k} = 0$, and performing the spacelike integration, gives

$$-ik_0\delta^{(3)}(\mathbf{p}_2 - \mathbf{p}_3 + \mathbf{p}_1 - \mathbf{p}_4) \int dx_0\, dy_0\, \delta(x^0 - y^0)$$
$$\times \langle \gamma \| [Q_\beta^5(y^0), Q_\delta^5(x^0)] \| \alpha \rangle \exp(-ip_2^0 y_0 + ip_4^0 x_0).$$

This is specified by the $\{Q^5, Q^5\}$ commutator of equation (14.1.5). From this relation one derives for the $\delta\beta$ anti-symmetric part of T

$$\langle \gamma\delta | T | \alpha\beta \rangle = \frac{1}{8\pi f_\pi^2}(p \cdot k)\{\delta_{\beta\gamma}\delta_{\alpha\delta} - \delta_{\gamma\delta}\delta_{\alpha\beta}\},$$

so that by equation (14.1.2) (with A and C written as functions of $s-u$, $s+u$ and t)

$$\left[\frac{A(s-u, s+u=2, t=0) - C(s-u, s+u=2, t=0)}{s-u}\right]_{s=u=1} = \frac{1}{16\pi f_\pi^2} = \frac{L}{4}, \tag{14.1.11}$$

follows, where we have introduced the characteristic length

$$L = \frac{1}{4\pi f_\pi^2} \approx 0{\cdot}094 \tag{14.1.12}$$

We shall refer to equation (14.1.11) as the *Adler–Weisberger condition.* Since this is the only one of our three conditions that is not null, it is the quantity L which will fix the normalization of our result (relating it to the weak pion decay constant), as we shall see below.

(b) On-Shell Results: The Linear Form

The on-shell predictions result from using the three conditions, equations (14.1.7), (14.1.10) and (14.1.11) to determine the three parameters of a simple linear form in the energy variables. The idea is as follows. The invariant amplitude A is in general a function of the kinematic variables s, t and u (symmetric in the latter two), and also of the external masses $m_i^2 = -p_i^2$, with the standard relation

$$s + t + u = \sum_{i=1}^{4} m_i^2. \tag{14.1.13}$$

In the regions $0 < s, t, u < 4$ (and $0 \leq m_i^2 \leq 1 \equiv m_\pi^2$), there are no singularities in these variables, so that it is reasonable to use a polynomial interpolation. Furthermore, if the unitarity branch cuts starting at $s = 4$ etc. are assumed to be weak, then they can be ignored without appreciable error (we return to this point later). Since, (a) the range to be interpolated over ($0 \leq s, t, u \leq 4$) is small compared to the typical range of variation of the strong interactions (which is of the order of $m_\rho^2 \approx 30$), and (b) the deviations from constancy on extrapolating from 0 to 1 in the pion mass are expected to be roughly of order $\frac{1}{9}$ (as discussed in Appendix D), an approximation which is linear in the kinematic invariants should be adequate, although (a) assumes that there are no nearby resonances. It is an important property of the linear approximation that no explicit dependence on the external masses p_i^2 need appear. (For the amplitude A, any term linear in p_1^2 must, by Bose statistics, be accompanied by a term in p_2^2 with the same coefficient, and, by T-invariance, the same holds for p_3^2 and p_4^2; but $\Sigma(-p_i^2) = s + t + u$.) The conclusion of this argument is that $A(s, t, u)$ is to be approximated by

$$A(s, t, u) = \alpha + \beta s + \gamma(t + u). \tag{14.1.14}$$

The three parameters are then fixed by the three off-shell constraints.

Before applying the off-shell constraints, one should note the relations between the low-energy parameters already implied just from assuming linearity (i.e. by crossing symmetry within the context of this simple form). On-shell there are just two parameters

$$\begin{aligned} A(s, t, u) &= (\alpha + 4\gamma) + (\beta - \gamma)s + \gamma(s + t + u - 4), \\ &\equiv \bar{\alpha} + \bar{\beta} s. \end{aligned} \tag{14.1.15}$$

The scattering amplitudes for eigenstates of isospin are given by

$$\begin{aligned} T^0 &= 5\bar{\alpha} + 4\bar{\beta} + 2\bar{\beta}s = a_0^0 + b_0^0(s/4 - 1), \\ T^1 &= \bar{\beta}(t - u) \qquad\quad = \tfrac{3}{4}a_1^1(t - u), \\ T^2 &= 2\bar{\alpha} + 4\bar{\beta} - \bar{\beta}s \ = a_0^2 + b_0^2(s/4 - 1), \end{aligned} \tag{14.1.16}$$

where the definitions of the S- and P-wave scattering lengths a_0^0 and a_0^2, and a_1^1, and of the slope parameters b_0^0 and b_0^2 are recalled in the right-hand equations. From these relations, it follows that

$$2a_0^0 - 5a_0^2 = 18a_1^1 = 24\bar{\beta}, \tag{14.1.17}$$

and

$$b_0^0 = -2b_0^2 = 8\bar{\beta}. \tag{14.1.18}$$

Furthermore, the zeros, z_0 and z_2 of the isospin even amplitudes $[T^I(s = z_I) = 0]$ are related by

$$4z_0 + 5z_2 = 12, \tag{14.1.19}$$

(as emphasized by Pennington and Pond (1971)), and since the linear approximation is purely S- and P-wave, the zeros of the S-wave amplitudes are similarly related.

This completes the relations obtained by using crossing symmetry and the assumption of a linear form. It remains to apply the off-shell constraints. Firstly, from the Adler–Weisberger condition, equation (14.1.11), we have the immediate result

$$2a_0^0 - 5a_0^2 = 18a_1^1 = 24\bar{\beta} = 6L \approx 0{\cdot}54. \tag{14.1.20}$$

(For consistency we will use the value $L = 0{\cdot}090$ rather than the slightly more precise value $0{\cdot}094$, since later we will compare with more complex calculations which use the former value. For any reasonable estimate of the errors in the calculation, this difference is insignificant.)

To fix the other on-shell parameter, it is convenient to express the Adler zero (equation (14.1.7)), and the sigma condition (14.1.10), in terms of the $I_s = 2$ amplitude $T^2(s, t, u)$. To do this we express the invariant amplitudes A, B and C in terms of T^2. The two conditions then give

$$\text{Adler condition:} \quad T^2(s = 1, t = u = 1) = 0, \tag{14.1.21}$$

$$\text{Sigma condition:} \quad T^2(s = 0, t = u = 1) = 0. \tag{14.1.22}$$

For the assumed linear form in s, t and u, this obviously implies the on-shell condition

$$T^2(s = 2, t = u = 1) = 0. \tag{14.1.23}$$

We shall argue, when we go on to discuss corrections to the linear approximation, that this version of Weinberg's second on-shell prediction is likely to prove the most stable against the inclusion of higher order terms (rather than the result $2a_0^0 + 7a_0^2 = 0$ derived below). Equation (14.1.23) immediately gives $z_2 = 2$, which together with equation (14.1.19) gives

$$z_2 = 2; \qquad z_0 = 0{\cdot}5. \tag{14.1.24}$$

Everything is now fixed. One has

$$\bar{\alpha} = -\bar{\beta} = -L/4, \tag{14.1.25}$$

so that

$$\begin{aligned} T^0 &= \tfrac{1}{4}L(2s-1), \\ T^1 &= \tfrac{1}{4}L(t-u), \\ T^2 &= -\tfrac{1}{4}L(s-2). \end{aligned} \tag{14.1.26}$$

The scattering lengths and slope parameters are given by

$$\begin{aligned} &a_0^0 = \tfrac{7}{4}L = 0{\cdot}157; \qquad b_0^0 = 2L = 0{\cdot}180 \\ &a_1^1 = \tfrac{1}{3}L = 0{\cdot}030; \\ &a_0^2 = -\tfrac{1}{2}L = -0{\cdot}045; \quad b_0^2 = -L = -0{\cdot}090. \end{aligned} \tag{14.1.27}$$

The S-wave scattering length ratio is given by

$$2a_0^0 + 7a_2^0 = 0. \tag{14.1.28}$$

$\pi\pi$ amplitudes calculated from partial wave dispersion relations have often been classified in terms of the parameters of a linear expansion (of the on-shell amplitude) about the symmetry point (cf. equation (2.25))

$$A(s, t, u) = -\lambda + \tfrac{1}{4}\lambda_1(s - \tfrac{4}{3}).$$

Matching to the form (14.1.15) in terms of $\bar{\alpha}$ and $\bar{\beta}$ and inserting Weinberg's values for these parameters (14.1.25) gives

$$\lambda = -\tfrac{1}{12}\lambda_1 = -\tfrac{1}{12}L.$$

This exceptionally small value of λ gives Weinberg's solution its special character relative to earlier calculations.

The most striking feature of the above results is that the S-wave scattering lengths are small (and hence difficult to determine from experiment), and the slope parameters b_0^I play a very important role in determining the behaviour of the S-wave phase shifts in the low-energy physical region. The $I = 0$ phase (estimated from Re $T_0^0(s)$ as given by equation (14.1.26)) grows quite rapidly, whereas the $I = 2$ phase falls, and at a slower rate.

This predicted behaviour of the S-waves was the real novelty of Weinberg's picture, since his P-wave scattering length is of roughly the same magnitude as one infers from extrapolating an energy-dependent Breit–Wigner fit to the rho-meson. This S-wave behaviour, which has subsequently been at least qualitatively confirmed by experiment (cf. Chapter 3) is closely connected with the S-wave zeros which occur at the positions given in equation (14.1.24), rather close to threshold on the natural scale of energy variation. We stress that the positions of these zeros are not determined, as is sometimes implied, by the Adler zero at $s=t=u=1$. By itself this condition implies nothing about the *positions* of the on-shell zeros without the assumption (in addition to "smoothness") of the σ-condition, or some condition equivalent to it, as for example in a wide class of dual models (cf. Section 14.2 below). On the other hand, either condition does suggest the *existence* of zeros, very likely somewhere near threshold.

Finally, a practical point concerning these sub-threshold S-wave zeros is that they imply poles in the quantities $\text{Re}\, f^{-1}(q^2) = q \cot \delta/\omega$, so that conventional effective range expansions are extremely inaccurate, and should not be used, for example, for extracting the S-wave scattering lengths from analyses of K_{l4} decays.

14.2. Relationship to Other Approaches

(a) Sum Rules and the KSFR Relation

Although we have emphasized the Weinberg calculation, an alternative, and earlier, approach was via sum rules. Again, the object is to proceed from formal off-shell results to useful on-shell ones via PCAC. For example, consider the Adler–Weisberger condition, equation (14.1.11). This can be recast in the form of a forward sum rule, using the fact that $A(s, t, u) - C(s, t, u)$ with $s+u=2$ and $t=0$ is the difference of the forward elastic scattering amplitudes for zero-mass π^- and π^+ mesons on a π^+ target. One can then write a forward sum rule for the left-hand side of equation (14.1.11), using the Optical Theorem, and derive the relation

$$\frac{2}{f_\pi^2} = \frac{1}{\pi} \int \frac{d(s-u)}{s-u} \left[\tfrac{1}{3}\sigma_{\text{tot}}^{I=0} + \tfrac{1}{2}\sigma_{\text{tot}}^{I=1} - \tfrac{5}{6}\sigma_{\text{tot}}^{I=2}\right]. \qquad (14.2.1)$$

This is called the Adler sum rule (Adler, 1965c), and is exactly analogous to the Adler–Weisberger relation for πN scattering (hence the appropriateness of our name for the condition from which it is derived).

Invoking PCAC, Adler proceeded to evaluate the sum rule using on-shell cross-sections, rather than those for zero mass scattering, by analogy with the πN case where the procedure met with great success. He used the approximation of resonance saturation, and, on finding that the rho- and f^0-resonance gave only 42% and 14%, respectively, of the required contribution, concluded that a large low-energy $I=0$ S-wave contribution was necessary. (We have modified the precise numbers in Adler's paper, as we have not used the Goldberger–Treiman relation on the left-hand side of equation (14.2.1) and we have used the more recent width values $\Gamma_\rho = 153$ MeV and $\Gamma_{f^0} = 185$ MeV.) This conclusion has subsequently been confirmed by the large "experimental" δ_0^0 phase shift indicated throughout this review, and ties in with the rapidly rising phase shift suggested by the Weinberg form near threshold. (In his paper Adler parameterized the effect by a large scattering length, but stressed that this was just one possible way.)

We have been able to derive an interesting result from the Adler sum rule because it concerns the t-channel $I = 1$ amplitude, and so is expected to converge rapidly. If the leading Regge term for $I = 2$ exchange has $\alpha_2(0) < 0$, then the σ-condition equation (14.1.10) can also be used to give a sum rule in the same way, i.e.

$$0 = \int d(s-u)\{\tfrac{1}{3}\sigma_{\text{tot}}^{I=0} - \tfrac{1}{2}\sigma_{\text{tot}}^{I=1} + \tfrac{1}{6}\sigma_{\text{tot}}^{I=2}\}. \tag{14.2.2}$$

Interesting results follow if we consider the saturation of this sum rule and the Adler sum rule with just the rho- and epsilon-resonances in the narrow width approximation (Gilman and Harari, 1968). (The epsilon need not necessarily be a true resonance, but can alternatively be regarded as a pole approximation to a non-resonant continuum.) In this approximation, and taking $m_\varepsilon = m_\rho$, equation (14.2.2) gives

$$2\Gamma_\varepsilon = 9\Gamma_\rho, \tag{14.2.3}$$

and on substituting this into the Adler sum rule the relation

$$\Gamma_\rho = \frac{m_\rho^3}{48\pi f_\pi^2}, \tag{14.2.4}$$

is obtained. This is, apart from unimportant terms of order $(m_\pi/m_\rho)^2$, the celebrated KSFR relation (Kawarabayashi and Suzuki, 1966; Fayyazuddin and Riazuddin, 1966), and is in rough agreement with experiment. For example, with $m_\rho = 765$ MeV, and $f_\pi = 128$ MeV, the rho-width is predicted to be $\Gamma_\rho = 180$ MeV. Unfortunately, the derivation of equation (14.2.4) is uncompelling, because the sum rule equation (14.2.2) has similar convergence properties to a sum rule for the same amplitude evaluated at the symmetry point, and it has been rather convincingly shown that this latter sum rule, if it converges at all, does so much too slowly for such a drastically low cut-off to be a satisfactory approximation (Tryon 1969a).

An alternative method to derive (14.2.3) from the sum rule (14.2.2) is to assume, in addition to resonance dominance, an extreme form of local duality. Thus, if one assumes that the resonance contributions in $I = 0$, and 1 cancel locally, rather than merely on the average, then the rho-contribution must be cancelled by that of the epsilon, so that equation (14.2.3) again follows. However, this argument is also unconvincing, because it is known that the local duality assumption fails badly in the f^0-region (see Chapter 12 for a discussion of this point).

We conclude that the KSFR relation can only be obtained from the Adler sum rule by assuming the simple relation given in equation (14.2.3), for which we have so far been unable to furnish a convincing argument. It is therefore interesting to consider the original derivation, (Kawarabayashi and Suzuki, 1966; Fayyazuddin and Riazuddin, 1966). This followed from a consideration of matrix elements of time-ordered products of axial currents and their derivatives between the rho-state and the vacuum. By means of similar manipulations to those described above for matrix elements between pion states, these authors derived a formal relation for $\rho \to 2\pi$ decay in the limit of zero pion four momentum, when of course $p_\rho = 0$ also. If we write $t = -p_\rho^2$ this relation is

$$f_\pi^2 f_{\rho\pi\pi}(t=0) = f_\rho(t=0), \tag{14.2.5}$$

where $f_{\rho\pi\pi}$ is related to the rho-width by (cf. equation 3.1.25)

$$\frac{f_{\rho\pi\pi}^2(t=m_\rho^2)}{4\pi} = \frac{12\Gamma_\rho}{m_\rho}\left[1-\frac{4m_\pi^2}{m_\rho^2}\right]^{-\frac{3}{2}}, \tag{14.2.6}$$

and f_ρ is defined by

$$(2\pi)^{\frac{3}{2}}\sqrt{2q^0}\langle 0|V^\lambda(0)|\rho(q)\rangle \equiv f_\rho \varepsilon^\lambda.$$

Assuming rho-dominance, the pion form factor is

$$F_\pi(t) = \frac{f_\rho f_{\rho\pi\pi}}{m_\rho^2 - t}.$$

Imposing the requirement $F_\pi(0) = 1$, gives at $t = 0$

$$f_\rho = m_\rho^2/f_{\rho\pi\pi}.$$

Substituting into equation (14.2.5), we then have

$$f_{\rho\pi\pi}^2(t=0) = m_\rho^2/f_\pi^2.$$

Finally, if we assume that

$$f_{\rho\pi\pi}(t=0) = f_{\rho\pi\pi}(t=m_\rho^2), \tag{14.2.7}$$

and also neglect terms of order $(m_\pi/m_\rho)^2$,‡ we can use the definition of $f_{\rho\pi\pi}$ to re-derive equation (14.2.4).

The assumption of equation (14.2.7) of course involves a long extrapolation, and taking $f_{\rho\pi\pi}$ constant implies that the left-hand cut of the $I = J$

‡ By retaining these terms the predicted rho-width can be lowered from 180 MeV to ≈ 140 MeV, but in view of the crudeness of the approximations this is probably not significant.

$=1\,\pi\pi$ amplitude is negligible, a result that would not be borne out if, for example, it was evaluated from the crossed rho-term alone. However, it has been observed (Kamal, 1969) that a large S-wave contribution of a magnitude roughly described by equation (14.2.3) would largely cancel the crossed rho-term; once again the derivation of the KSFR relation relies on the existence of a large S-wave contribution, as did the derivation based on the Adler sum rule discussed above. This latter derivation can be recast in various ways, but all of them rest on the validity of equation (14.2.3). For example, one could use the fixed $t=0$ dispersion relation for the $I_s=1$ amplitude to evaluate the scattering length a_1^1, set at the current algebra value. In this case, equation (14.2.3) causes the u-channel contributions to cancel exactly, and the s-channel cut is just given by the rho-pole, which leads immediately to the desired result.

Two morals can be drawn from this discussion. The first is that one has no reason to expect the KSFR relation to be precise, since both methods of derivation involve equally crude resonance dominance approximations. Secondly, for a model satisfying dispersion relations automatically to produce the current algebra normalization of the scattering lengths, in particular to give the P-wave scattering length a_1^1, it is not sufficient for it to have the correct rho-width; it must also have the large $I=0$ S-wave contribution roughly indicated by equation (14.2.3).

(b) Veneziano Models

In Section 12.3 we discussed the Lovelace–Veneziano model and noted that the scattering length values obtained were similar to Weinberg's. In particular, the ratio of scattering lengths is very close, although only the Adler zero (but not the σ-condition) was explicitly imposed. It is interesting to understand how this comes about. As we shall see, in a wide class of dual models the imposition of the Adler zero actually entails the σ-condition being satisfied, a fact that we shall exploit in our discussion below of corrections to the Weinberg calculation.

As noted in Chapter 12, it is useful when considering dual models to emphasize the amplitudes corresponding to channels with exotic quantum numbers—that is, the amplitudes with $I=2$ in the s, t and u channels, respectively. We have already written the Adler zero and the σ-condition in terms of these amplitudes in equations (14.1.22) and (14.1.23), i.e.

$$\text{Adler condition} \quad T^2(s=1, t=u=1)=0, \tag{14.2.8}$$

$$\text{Sigma condition} \quad T^2(s=0, t=u=1)=0. \tag{14.2.9}$$

In the linear approximation in s, t and u, these give the on-shell result

$$T^2(s=2, t=u=1)=0, \tag{14.2.10}$$

and determine the amplitude up to an overall normalization constant. Now recall, that since T^2 is an exotic amplitude, the Veneziano formula for it (cf. equation (12.3.3)) has the property of containing no explicit s-dependence. It immediately follows from this that if the Adler zero, equation (14.2.8), is imposed, the σ-condition, equation (14.2.9), and the on-shell result, equation (14.2.10), will automatically be satisfied. If the model is normalized to the Adler–Weisberger relation, which in terms of the exotic amplitudes is (cf. equations (14.1.11) and (12.3.5))

$$\left[\frac{T^2(u, t, s) - T^2(s, t, u)}{s - u}\right]_{\substack{s=u=1\\ t=0}} = \frac{L}{4}, \tag{14.2.11}$$

all the current algebra conditions are satisfied. Thus, in dual models of this type, if the Adler zero is imposed, the σ-condition is redundant, being guaranteed by the much stronger assumption of no explicit s-dependence. We note that this will be so not only for the one-term Veneziano formula, but in a wide class of models. For example, the addition of satellites with constant coefficients will not affect the above arguments in any way. In fact, the argument is clearly valid for any narrow resonance model with no exotic states, and no explicit mass dependence.

Finally, we return to the specific example of the Lovelace–Veneziano model, which we suppose normalized to the Adler–Weisberger condition, equation (14.1.11), so that the Adler sum rule, equation (14.2.1), is automatically satisfied. In this case not only are the scattering lengths similar to the Weinberg values, but in addition the local cancellation property in the sum rule, equation (14.2.2), is also satisfied, the relation $2\Gamma_\varepsilon = 9\Gamma_\rho$ being exact for zero mass external pions. It thus follows from our previous discussion that the KSFR relation would be satisfied if the ρ- and ε-resonances alone dominated the Adler sum rule. In fact, the higher mass resonances are not negligible, and reduce the predicted rho-width by a factor of $2/\pi$ (for zero mass pions) below the KSFR value (Osborn, 1969).

14.3. Corrections to the Weinberg Form

In Section (14.1) we have shown how Weinberg obtained on-shell results from off-shell conditions by assuming a linear dependence on s, t, u and the squared external masses. Once linearity is assumed, and the Adler zero imposed, the ambiguities in principle present in off-shell continuation can be eliminated by general conditions. However, such a linear form has some obvious shortcomings, in particular the neglect of higher order terms, and of the threshold branch points at s, t, or $u = 4$. The effects of incorporating these are often referred to as "corrections to the Weinberg form" in the hope that the latter correctly describes the main characteristics of the amplitudes. It is important to investigate corrections to check that they are indeed small, since, if they are not, the conclusions based on the Weinberg form may not survive.

The ambiguities present in off-shell continuation referred to above stem from our ignorance of the matrix elements of the axial current invoked in the definitions, equations (14.1.3) and (14.1.4). For the general case,‡ they have been discussed in the context of integral representations by Basdevant and Hayot (1971). In the present context however, we are concerned, not with the ambiguities present in the most general off-shell continuation compatible with the Adler zero, but with the ambiguity remaining after some form of the PCAC assumption of "smooth" or "normal" variation has been imposed. It is due to this latter assumption that current algebra has implications for $\pi\pi$ scattering, equations (14.1.1), (14.1.3) and (14.1.4) being merely definitions.§

In what follows, we discuss the two types of correction in turn. As we shall see, while those due to higher order terms in s, t and u can only be evaluated in specific models, in a reasonable framework they turn out to be very small. In contrast, the effects of the threshold branch points are both more important, and, in the context of smoothly interpolating forms, can be evaluated in a less model dependent way. Their greater importance comes about because, while the scattering lengths predicted are expected to be small, the linear terms predicted are large, and lead, via unitarity, to a cut discontinuity which increases rapidly immediately above threshold. As we shall show, this "boost" from unitarity leads to an increase in the predicted $I=J=0$ parameters, especially b_0^0. The qualitative features, however, remain unchanged from Weinberg's model.

(a) Higher Order Terms

Here we discuss corrections arising from higher order terms in s, t and u in equation (14.1.14), still retaining the assumption that the elastic thresholds can be neglected. This latter point will be discussed in Section 14.3(b) below.

(i) *Hard Pions*

A popular method for calculating corrections to simple current algebra results is the so-called "hard pion" approach, useful reviews of which in this context has been given by Arnowitt (1969) and Nath (1974). A number of calculational techniques have been used in this field—Ward identities (Schnitzer and Weinberg, 1967; Gerstein and Schnitzer, 1968), effective Lagrangians (Arnowitt *et al.*, 1967, 1968), and dispersion relations (Brown

‡ The Adler zero can in general be imposed, since it follows directly from equation (14.1.4) as shown in Section 14.1.

§ See the discussion of PCAC in Appendix D.

TABLE 14.3.1. Values of the $\pi\pi$ Scattering Lengths a_J^I and Slope Parameters b_J^I obtained in the Weinberg Model, compared with those obtained in other models using real expansions. Bracketted quantities are input.

Model \ a_J^I	a_0^0	a_0^2	a_1^1	b_0^0	b_0^2	a_2^0	a_2^2	$-a_0^0/a_0^2$
Weinberg	0·157	−0·045	0·030	0·180	−0·090	—	—	3·5
Quadratic form (s, t, u)	0·181	−0·045	0·036	0·221	−0·082	(0·0015)	(0·0002)	4·0
Lovelace–Veneziano (Exact)	0·172	−0·045	0·035	0·202	−0·086	0·0009	0·0001	3·8
Lovelace–Veneziano (quadratic approx.)	0·170	−0·045	0·034	0·197	−0·085	(0·0009)	(0·0001)	3·8
Hard pion	0·171	−0·045	0·036					3·8

and West, 1967, 1968; Geffen, 1967; Das *et al.*, 1967). The essential features are the same in each case, and are quite simple. The basic idea is to augment the linear form of equation (14.1.14)—which, for example, appears as a seagull term in an effective Lagrangian approach, or as a subtraction term in a dispersion theory approach—by epsilon- and rho-meson resonance poles. The meson vertices of these terms are approximated by minimal order polynomials in the four-momenta (this fixes the off-shell behaviour), and the resulting corrections are expressed in terms of the rho- and epsilon-resonance masses and widths. These corrections turn out to be of order $(m_\pi/m_\rho)^2$ and $(m_\pi/m_\varepsilon)^2$, and are in practice small. In fact, if we take the results of Arnowitt (1969), and modify his parameter values to give $m_\varepsilon = m_\rho$, $\Gamma_\rho = 120$ MeV, $\Gamma_\varepsilon = 500$ MeV ($\lambda_A = 0{\cdot}3$, $\lambda^2 = 1{\cdot}5$, $\lambda_2 = 0$, in his notation) then the scattering lengths given in Table (14.3.1) are obtained, which are not very different from Weinberg's values.

(ii) *Duality and the Quadratic Form*

The most general quadratic form which can be written for $A(s, t, u)$ is (Khuri, 1968)

$$A(s, t, u) = \alpha + \beta s + \gamma(t+u) + \delta s^2 + \varepsilon(st + su - tu)$$
$$+ \omega\left(6 - \sum_{i>j} m_i^2 m_j^2\right) + ftu + g(t^2 + u^2). \qquad (14.3.1)$$

However, in contrast to Khuri's discussion, we shall base ours on the assumptions suggested by duality ideas, and discussed in Section 14.2. That is, we assume (i) the absence of explicit s-dependence in the exotic amplitude $T^2(s, t, u)$, (ii) the absence of explicit mass dependence ($\omega = 0$), and impose (iii) the Adler zero and (iv) the Adler–Weisberger condition. The sigma condition will (as discussed in Section 14.2) be automatically satisfied, and it is a matter of straightforward calculation to express everything in terms of the D-wave scattering lengths a_2^I, and the quantity L occurring in the Adler–Weisberger condition equation (14.1.11). The resulting equations for the S-wave scattering lengths are (Morgan and Shaw, 1972)

$$a_0^0 = \tfrac{7}{4}L + 15a_2^0 + \tfrac{195}{16}a_2^2, \qquad (14.3.2a)$$

$$a_0^2 = -\tfrac{1}{2}L + \tfrac{15}{8}a_2^2. \qquad (14.3.2b)$$

The P-wave scattering length a_1^1 and the slope coefficients b_0^I can then be estimated from the relations

$$2a_0^0 - 5a_0^2 = 18a_1^1 - 30(2a_2^0 - 5a_2^2), \qquad (14.3.3a)$$

$$b_0^0 = 6a_1^1 - 10a_2^0 + 100a_2^2, \qquad (14.3.3b)$$

$$b_0^2 = -3a_1^1 + 20a_2^0 - 20a_2^2, \qquad (14.3.3c)$$

which follow from the quadratic form equation (14.3.1). We thus see that for a wide class of models the corrections can be expressed in the quadratic approximation in terms of the D-wave scattering lengths. To estimate the corrections numerically, we will use the values of the D-wave scattering lengths obtained from forward dispersion relations, as discussed in Chapter 10, (Morgan and Shaw, 1969, 1970). These values, and the resulting threshold parameter values from equations (14.3.2) and (14.3.3) using $L = 0{\cdot}090$ are given in Table 14.3.1 labelled "Quadratic Form (s, t, u)".

Within the framework of the quadratic approximation we have established a relation between corrections to the linear approximation and the D-wave scattering lengths, valid in a wide class of dual models, including the Lovelace–Veneziano model. For this latter case, we can compare the results of the quadratic approximation obtained by inserting the predicted a_2^I in equations (14.3.2) and (14.3.3), with the results of evaluating the model exactly. As can be seen from Table 14.3.1, the results are very similar, showing that in the Lovelace–Veneziano model at least, the quadratic approximation is rather a good one.

We can thus conclude that in this class of models, the corrections to Weinberg are, like those in the "hard pion" calculations, rather small.

(b) Unitarity Corrections

It is clear from the foregoing discussion that crossing symmetry plays a key role in restricting the possible off-shell continuation of the scattering amplitude. Therefore, to retain predictive power one needs to retain this property when attempting to incorporate the effects of the elastic thresholds. The most straightforward way to do this is to interpolate the amplitude not by crossing symmetric polynomials in $s = 4 + 4q_s^2$ etc., but by crossing symmetric polynomials in terms of the momenta themselves, i.e. q_s, q_t and q_u. Elastic unitarity gives rise to a square-root branch point at threshold in the s or q_s^2 planes, which is mapped away on transforming to the q_s-plane. Further singularities are exposed, but these are less important. As a result, it is reasonable, when working in the momentum plane, to assume a smooth interpolating function right up to, and beyond, the elastic threshold. In practice, it is convenient not to work in terms of the momenta themselves, but in terms of

$$\begin{aligned} Q_s &\equiv \tfrac{1}{2}(4-s)^{\frac{1}{2}}, \qquad && s<4 \\ &= -\frac{i}{2}(s-4)^{\frac{1}{2}}, \qquad && s>4, \end{aligned} \tag{14.3.4}$$

and the analogous variables Q_t, Q_u, introduced originally by Iliopolous (1967). These are proportional to the momenta, but are real for $0 < s, t, u$

<4, so that, if the amplitude is represented by polynomials in these variables, all the coefficients are real. We require to interpolate between the points at which the formal off-shell results of Section 14.1(a) hold (i.e. $s=0, 1$; $Q_s=1, \sqrt{\frac{3}{2}}$), and threshold ($s=4$; $Q_s=0$). We do this using, instead of real polynomials in s, t and u, real polynomials in Q_s, Q_t and Q_u.

The first application of this method was by Iliopolous (1967, 1968) who considered the most general quadratic form in Q_s, Q_t and Q_u, i.e.

$$A(s,t,u)=x_1+x_2Q_s+x_3(Q_t+Q_u)+x_4Q_s^2+x_5(Q_t^2+Q_u^2). \quad (14.3.5)$$

Terms of the form Q_s^{2m+1}, Q_t^{2n+1} etc. are forbidden by analyticity. The terms in x_1, x_4 and x_5 alone would reproduce the Weinberg form of equation (14.1.14). Absorptive parts of the form

$$\text{Im}\, T_0^I(q^2)=\alpha_I q,$$

are introduced for the S-waves by the two extra terms in x_2 and x_3, and these two coefficients are determined by imposing the threshold unitarity constraints

$$\alpha_I=(a_0^I)^2. \quad (14.3.6)$$

The results are shown in Table 14.3.2 labelled "Iliopolous", and, as can be seen, the corrections are small. This result might have been expected, since the only contributions to the absorptive parts taken into account are those proportional to the squares of the scattering lengths, which are themselves small. However, the rapid rise of the $I=J=0$ phase shift, and hence of the associated absorptive part, arises from the linear term in the T_0^0 amplitude with coefficient b_0^0, so that the main effect of unitarity corrections can only be gauged when this is taken into account. This point was noted and used by Morgan and Shaw (1972), who extended the above calculations to include possible cubic terms. The on-shell absorptive part then has the form

$$\text{Im}\, T_0^I(q^2)=\alpha_I q+\beta_I q^3, \quad (14.3.7)$$

where, in addition to the condition (14.3.6), unitarity also requires

$$\beta_I=(a_0^I)^4+2b_0^I a_0^I-\tfrac{1}{2}(a_0^I)^2. \quad (14.3.8)$$

To determine all the extra coefficients in this case; it is also necessary to exploit unitarity for one external pion off-shell. For the details, and a full discussion of this approach to unitarity corrections we refer to the above paper. The results are shown in Table 14.3.2, labelled "cubic". The corrections are indeed larger than in the quadratic case investigated by Iliopolous, and in particular, a large increase occurs in the coefficient b_0^0. However, the solution still retains its qualitative similarity to that of Weinberg.

TABLE 14.3.2. Values of the $\pi\pi$ Scattering Lengths a_J^I and Slope Parameters b_J^I obtained in the Weinberg Model, compared with those obtained in models allowing for threshold effects. Bracketted quantities are input.

Model \ a_J^I	a_0^0	a_0^2	a_1^1	b_0^0	b_0^2	a_2^0	a_2^2	$-a_0^0/a_0^2$
Weinberg	0·157	−0·045	0·030	0·180	−0·090	—	—	3·5
Iliopolous	0·171	−0·044	0·030	0·165	−0·091	0·0001	0·0001	3·9
Cubic	0·236	−0·041	0·028	0·348	−0·078	0·0016	0·0013	5·8
Dual quartic (QD)	0·211	−0·043	0·036	0·254	−0·079	(0·0015)	(0·0002)	4·9
Grassberger	0·201	−0·041	0·035					4·9
FDR (a)	0·16 ± 0·04	−0·05 ± 0·01	0·035 ± 0·002	0·245	−0·059	0·0016 ± 0·0002	0·0002 ± 0·0002	(3·2 ± 1·1)
FDR (b)	0·20 ± 0·04	−0·04 ± 0·01	0·036 ± 0·002	0·262	−0·062	0·0016 ± 0·0002	0·0002 ± 0·0002	(5 ± 1)

If we look at the D-wave scattering lengths predicted by the cubic form (Table 14.3.2), we see that while that for a_2^0 is in accord with the dispersion relation estimates, that for a_2^2 is too large. The above authors corrected this by including quadratic terms under the same duality-inspired assumptions that led to equations (14.3.2), i.e. no explicit mass dependence, and no explicit s-dependence in the exotic amplitude $T^2(s, t, u)$. The two extra parameters over the cubic calculation are determined by demanding that the D-wave scattering length values agree with those suggested by the dispersion relation analysis. The results of this calculation are shown in Table 14.3.2 labelled "dual quartic". Qualitatively the same remarks apply regarding the magnitude of the corrections as in the cubic case. Although the way in which the quartic terms are introduced in the dual quartic form is clearly model dependent, the model is reasonable, and the resulting form exhibits some desirable features. As in all the models we have discussed, crossing is satisfied exactly. However, unitarity is implemented much more satisfactorily. Furthermore, the rigorous inequalities on the $\pi^0\pi^0$ S-wave, equations (7.31) to (7.36) are all found to be satisfied for both the cubic and dual quartic forms.

Finally we note that similar results to the above have been obtained by Grassberger (1972b). This author uses an explicitly crossing symmetric dispersion representation, rather than an expansion, to describe the on- and off-shell $\pi\pi$ amplitude. Unitarity can thus be incorporated exactly into the framework, but at the cost of assuming the behaviour of the phase shifts from $500 \leq W \leq 900$ MeV from experiment. Those are joined smoothly to a linear expansion for $\tan \delta_l^I$ in the threshold region, and the scattering lengths (which also occur in a subtraction term) are adjusted so that the off-shell conditions are again satisfied. The resulting values are extremely close to those of the dual quartic form, as seen in Table 14.3.2.

(c) Comparison with Dispersion Relation Results

To conclude this section, we compare the above results with those of phenomenological dispersion relation analyses. In Chapter 10, it was seen that analyses based both on forward dispersion relations and on the Roy equations, together with existing data at somewhat higher energies, led to the conclusion that, if one parameter (conveniently taken to be the scattering length ratio $-a_0^0/a_0^2$, or a_0^0 itself) was fixed, the values of all the other threshold parameters were essentially determined. In Table 14.3.2, the entries labelled FDR are forward dispersion relation results taken from Morgan and Shaw (1969, 1970), the Roy's relation results of Basdevant, *et al.* (1972a,b) being very similar. The ratio is chosen to agree with the dual quartic form, and the agreement for the other parameters shows that, in this model, the relations between the threshold parameters obtained from the dispersion relation analysis are reasonably well satisfied. This detailed tying together of these very different approaches—one based on current algebra with reasonable interpolations over the triangle region; the other based on "experimental" input in the rho-region, together with the application of the techniques of dispersion relations—is very satisfactory for the assumptions of current algebra.

Unfortunately, the single input parameter a_0^0 (or a_0^0/a_0^2) for the phenomenological analysis is not yet well determined (cf. Chapter 6), although recent K_{e4} results are encouraging (see Fig. 5.1.5). Until it is, the detailed correctness of the current algebra calculations remains open.‡ If a_0^0 does not have the predicted value, one could relax the sigma condition and try to test the other current algebra conditions. If the σ-condition is abandoned, the other conditions predict a relation between a_0^0 and a_0^2, which is exhibited for the various models in the paper of Morgan and Shaw (1972). For the dual quartic model and $0<a_0^0<0{\cdot}3$, the resulting curve lies close to the "universal curve" in the (a_0^0, a_0^2) plane obtained from the dispersion relation analyses of Chapter 10, so that if a_0^0 is in this range one would conclude that the Adler–Weisberger condition was reasonably well satisfied. For $a_0^0>0{\cdot}3$ the current algebra calculation becomes somewhat unstable against changes in the model used for interpolation over the low-energy region, and further investigation would be required before conclusions could be drawn.

14.4. Applications to Related Processes

(a) 0^-0^- Scattering Lengths

In his original paper on $\pi\pi$ scattering, Weinberg (1966) also derived a formula for the S-wave scattering lengths for pion scattering from an arbitrary hadron target in the "heavy target" approximation, that is, in the approximation that the pion mass is negligible compared to the target mass. The formula is

$$a_0^I = -\frac{Lm_t}{m_t+1}[I(I+1)-I_t(I_t+1)-2], \tag{14.4.1}$$

where I is the total isospin, I_t and m_t the target isospin and mass, and L the parameter defined in equation (14.1.12). If one wishes to abandon the heavy target approximation, and implement PCAC using linear forms as in the

‡ There are in addition problems in reconciling estimates of the P-wave scattering length a_1^1 obtained from analyses of dipion production data with those from dispersion relation analyses. We refer to Chapter 6 and Section 3.3 for further details; see also Basdevant *et al.* (1975).

case of $\pi\pi$, then the $SU(2)\otimes SU(2)$ scheme must be generalized to $SU(3)\otimes SU(3)$. That is, the strangeness-changing weak currents are also considered, the enlarged set of charges obeying the commutation relations of the larger algebra of $SU(3)\otimes SU(3)$. Similarly, the PCAC equation (14.1.1) is generalized to the whole octet, for example the kaon field being related to the strangeness-changing axial vector current by the equation

$$\partial_\lambda A^{\lambda\Delta S=1}(x) = m_K^2 f_K \phi_K^+(x). \tag{14.4.2}$$

Similarly to the $\pi\pi$ case, σ-terms involving commutators between axial charges and their derivatives occur, but now involving the strangeness-changing currents as well as the strangeness-conserving currents involved in the $SU(2)\otimes SU(2)$ algebra. The assumed absence of $I=2$ σ-terms in the $\pi\pi$ case is extended to the absence of exotic σ-terms in general (i.e. $I=2$ in $\pi\pi$, $I=\frac{3}{2}$ in $K\pi$ etc.). These assumptions, together with that of a linear form in s, t, u and p_i^2, where p_i are the external particle four momenta, are sufficient to determine the amplitude *in those cases where there is an exotic channel present* (Osborn, 1970). For example, for the πK case (Griffith, 1968) the general linear form is

$$\begin{aligned} T^+(s,t) &= \alpha + \beta t + \gamma(s+u) + \delta(p_1^2 + p_2^2), \\ T^-(s,t) &= \varepsilon(s-u), \end{aligned} \tag{14.4.3}$$

where α, β, γ, δ, ε are constants, and p_1, p_2 are the kaon four momenta. The $T^\pm(s,t)$ are the amplitudes of definite isospin in the t-channel ($\pi\pi\to K\bar{K}$) as defined in equations (2.28) and (2.29). On-shell, this already leads to the relation between the πK (s-channel) scattering lengths (a_l^I)

$$a_0^{\frac{1}{2}} - a_0^{\frac{3}{2}} = 6m_K(a_1^{\frac{1}{2}} - a_1^{\frac{3}{2}}). \tag{14.4.4}$$

The Adler zeros for one soft pion ($s=u=m_K^2$, $t=m_\pi^2$), and one soft kaon ($s=u=m_\pi^2$, $t=m_K^2$), eliminate two constants, and two more are eliminated by the requirement that the πK σ-terms should have no $I=\frac{3}{2}$ components. The amplitude can finally be normalized by an "Adler–Weisberger" condition for either two soft pions, or two soft kaons. Consistency requires

$$f_\pi = f_K, \tag{14.4.5}$$

and the result is

$$a_0^{\frac{1}{2}} - a_0^{\frac{3}{2}} = \frac{3m_K}{m_K+1}L. \tag{14.4.6}$$

The other scattering lengths are given by equation (14.4.4) and the relations

$$a_0^{\frac{1}{2}} + 2a_0^{\frac{3}{2}} = 0, \tag{14.4.7}$$

$$a_1^{\frac{1}{2}} = \frac{1}{2}\left(\frac{L}{m_K+1}\right); \qquad a_1^{\frac{3}{2}} = 0. \tag{14.4.8}$$

Numerically, $a_0^{\frac{1}{2}} = 0{\cdot}14$, $a_0^{\frac{3}{2}} = -0{\cdot}07$ and $a_1^{\frac{1}{2}} = 0{\cdot}01$. The πK S-wave parameters b_0^I, defined by

$$T_0^I(q^2) = a_0^I + b_0^I \nu + \cdots, \tag{14.4.9}$$

have the values

$$b_0^{\frac{1}{2}} = 0{\cdot}96L = 0{\cdot}09 m_\pi^{-2}; \qquad b_0^{\frac{3}{2}} = -0{\cdot}64L = -0{\cdot}06 \tag{14.4.10}$$

There are several parallels between these and the $\pi\pi$ results. For example, the S-wave scattering lengths are small and of opposite sign, in contrast to the S-wave dominance relation ($a_0^{\frac{1}{2}} = a_0^{\frac{3}{2}}$ in this case), and again this behaviour is associated with sub-threshold S-wave zeros. The $I = \frac{1}{2}$ S-wave behaviour is somewhat similar to the $I = 0$ $\pi\pi$ S-wave, as are the exotic $I = \frac{3}{2}$ $K\pi$ and $I = 2$ $\pi\pi$ waves, which is easily seen by comparing the above formulas with the $\pi\pi$ results.

Of course, in the above approach extrapolations to $m_K = 0$ have been made using kaon PCAC, and one suspects that this may not be a very good approximation, since on the natural scale of hadron masses, m_K is not very small. However, it is interesting to note that the S-wave scattering lengths are just those given by the heavy target approximation, equation (14.4.1), which uses only pion PCAC. Furthermore, corrections to the above linear approximation have been estimated by both the SU(3) extension of the "hard pion" approach (Pond, 1971), and by using an "Iliopolous" type expansion in the various channel momenta (Alder *et al.*, 1972) in a way similar to that discussed in the $\pi\pi$ case in Section 14.3. In both approaches the corrections turn out to be rather small, comparable to those for $\pi\pi$ scattering.‡

Turning to the 0^-0^- processes with no exotic channel—e.g. $\pi\eta$ scattering—the assumption of no exotic sigma terms is no longer useful, and has to be replaced by a more specific assumption about the nature of symmetry breaking. This problem has been discussed by Osborn (1970) assuming the symmetry breaking model of Gell-Mann, Oakes and Renner (1968), which is consistent with the values of the non-exotic σ-terms found in the previous calculations for processes with an exotic channel. The most interesting point concerning this work is not the scattering length values, but the insight into the application of the Adler zero to Veneziano models. As

‡ As an alternative, Fox and Griss (1974) adopt the somewhat *ad hoc* procedure of using the naïve current algebra prediction extrapolated on shell for the subtraction constants in a fixed-t dispersion relation at $t = 0$. This leads to a 50% increase in the estimate for $a_0^{\frac{1}{2}}$ over that of the uncorrected linear model.

discussed in Section 14.2, the imposition of the Adler zero automatically gave the necessary σ-condition in the case of $\pi\pi$, because of the absence of explicit s dependence in the exotic $I=2\pi\pi$ amplitude. A similar mechanism occurs in other cases with exotic channels. However, for a process without exotic channels, there is no s-independent amplitude in the Veneziano model, and imposition of the Adler zero does not automatically lead to the desired condition on the σ-terms. This perhaps explains why for processes with no exotic channel, e.g. $\pi\eta$, the Lovelace–Veneziano model is strikingly less successful than for those with an exotic channel, as noted in Section 12.3.

(b) Dipion Production

A very interesting topic in the present context is dipion production:

$$\pi + N \to \pi + \pi + N, \tag{14.4.11}$$

studied close to threshold. This was first calculated using conventional soft pion techniques by Chang (1967), and using the phenomenological Lagrangian method by Olsson and Turner (1968). The results are, of course, similar, but the latter authors also considered the effect of relaxing the sigma condition. In the phenomenological Lagrangian approach (with no hard pion corrections), pion–pion scattering is just given by a seagull term (Fig. 14.4.1), which gives the result of the Weinberg approximation described

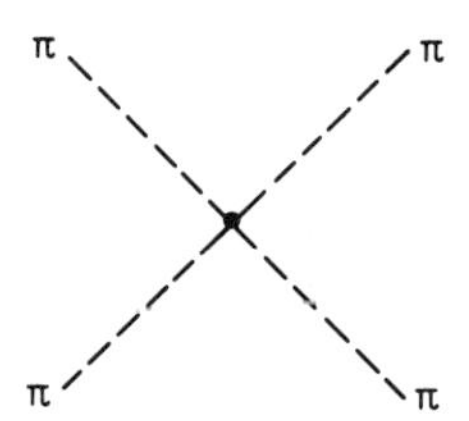

FIG. 14.4.1. Seagull diagram for $\pi\pi$ scattering.

above. As discussed earlier, the commutators of SU(2) ⊗ SU(2), together with PCAC, only specify the amplitude up to an arbitrary constant, which in this approximation is conveniently taken to be a_0^0/a_0^2. The value of this ratio is fixed at $-\frac{7}{2}$, if the sigma condition is imposed. For dipion production, the dominant diagrams are the ones shown in Fig. 14.4.2, the other possibilities giving only small contributions near threshold. The peripheral diagram clearly contains the same a_0^0/a_0^2 ambiguity if the sigma condition is not imposed. Leaving this quantity free in the calculations, the authors found on comparing with the available data on $\pi^+\pi^-$ production that values near $-\frac{7}{2}$ were favoured, the value $-\frac{3}{2}$, for example, lying well below the data. A subsequent analysis (Bunyatov, 1974), using more precise data on $\pi^\pm p \to \pi^+\pi^\pm n$, both confirms this result, and yields the more precise values

$$a_0^0 = 0{\cdot}18 \pm 0{\cdot}02, \qquad a_0^2 = -0{\cdot}07 \pm 0{\cdot}01. \tag{14.4.12}$$

Since information on the validity of the σ-condition (and hence of the scattering length ratio) is not easily obtained, this result is of considerable interest, although if precise conclusions are to be reliably established from such analyses, a detailed study of possible corrections will be needed (Arnowitt, 1969).

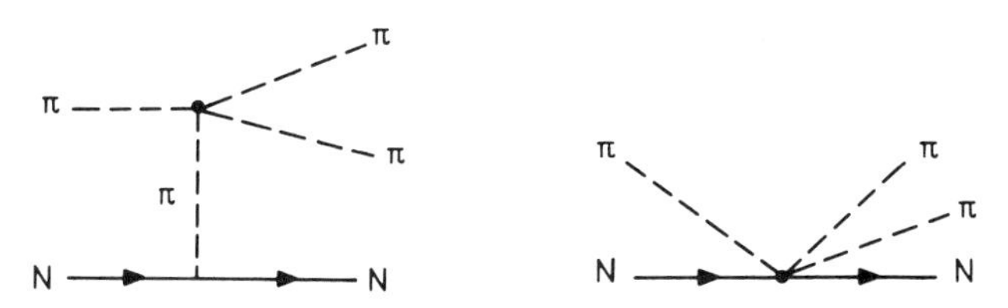

FIG. 14.4.2. Soft pion contributions to single-pion production.

(c) Decay Processes

There are also interesting current algebra predictions for some of the decay processes which have been discussed in other chapters—in particular for K_{l4}, $\eta \to 3\pi$ and $K \to 3\pi$ decays. However, these applications require extra assumptions on the nature of the weak and electromagnetic interactions, as well as use of the extended algebra of $SU(3) \otimes SU(3)$. The status of such calculations is very unclear. Thus, for example, the predictions for K_{l4} are closely related to those for K_{l3} decay, which have been a source of controversy for many years.‡ Similarly, there are difficulties with the applications to $K \to 3\pi$ and $\eta \to 3\pi$ decays. Whereas the simplest approach to $K \to 3\pi$ decay proves reasonably successful, precisely analogous arguments lead to the result that $\eta \to 3\pi$ decay is forbidden, a result which has been the subject of controversy since its derivation. The problems here can be circumvented, but not solved, by the use of the pion-pole model in which the decay amplitudes are somewhat arbitrarily assumed to be dominated by the diagram of Fig. 12.3.4. The Dalitz plot distribution is thus controlled by the off-shell continuation of the $\pi\pi$ scattering amplitude. This model has been

‡ A recent experiment (Donaldson *et al.*, 1974) suggests that the long-standing problems with K_{l3} decay may be purely experimental in origin. For a thorough discussion of earlier experiments, as well as the theory of both K_{l3} and K_{l4} decay, see Chounet *et al.* (1972).

discussed in the context of the Lovelace–Veneziano model in Chapter 12, which gives a similar $\pi\pi$ amplitude (up to a constant which is irrelevant for this application) to that predicted from current algebra, provided the σ-condition is imposed in the latter case. The success of this model in accounting for the slopes of the Dalitz plot distributions for these decays (cf. Chapter 12) could be regarded as further qualitative support for this assumption. Unfortunately, until the general difficulties referred to above are understood, the status of the model remains at best obscure. For a review of these questions, and detailed references, we refer to Sutherland (1971). We would conclude that, while these calculations are of great interest because of the serious problems in the application of $SU(3) \otimes SU(3)$ symmetry which they explore, until these problems are understood, they add little to our knowledge of the $\pi\pi$ interaction itself.

Part IV

Appendices

Appendix A

Notations and Conventions

(a) Units

Throughout this book, we have used so-called "natural units" in which $\hbar = c = 1$. In addition we have mostly used energy units such that $\mu \equiv m_\pi = 1$. If other units are used, e.g. GeV, they are explicitly written. A conversion table for lengths and areas between the system of natural units and other systems is given in Table A.1. Finally, although we have set $c = 1$, nevertheless we usually write lab. momenta as GeV/c (or MeV/c) to distinguish them from energies, which would be GeV.

(b) The Symbols $0 \sim \approx \rightarrow$ and $\equiv$

We have used these symbols in the usual mathematical way, but for completeness we define them below.

The expression $f(x) \rightarrow A$ means that the function $f(x)$ "tends to" the quantity A. If $f(x) \sim A$ as $x \rightarrow x_0$, then the leading term of $f(x)$ equals A in this limit, *including* any overall constant. If, however, the overall factor is *omitted* then we write; $f(x) = 0(A)$ as $x \rightarrow x_0$. Finally, if $f(x) \equiv A$ then $f(x)$ is *identically* equal to A, as, for example, in a definition, and if $f(x) \approx A$, then $f(x)$ is *approximately* equal to A, i.e. different from A but sufficiently close to it.

(c) Four-Vectors

The general contravariant four-vector is denoted V^μ with $\mu = 0, 1, 2, 3$, i.e. $V^\mu \equiv (V^0, \mathbf{V})$, and the general covariant four-vector V_μ is obtained by

TABLE A.1. Conversions between Natural Units and Other Systems for Lengths and Areas

(a) *Lengths*

	fm	n.u.	GeV^{-1}	mb GeV
1 fm	—	0·707	5·068	1·973
1 n.u.	1·414	—	7·165	2·790
1 GeV^{-1}	0·197	0·140	—	0·389
1 mb GeV	0·507	0·359	2·568	—

(b) *Areas*

	mb	n.u.	GeV^{-2}	$\text{mb}^{\frac{1}{2}}\,\text{GeV}^{-1}$
1 mb	—	0·0500	2·568	1·603
1 n.u.	19·99	—	51·33	32·03
1 GeV^{-2}	0·389	0·0195	—	0·624
1 $\text{mb}^{\frac{1}{2}}\,\text{GeV}^{-1}$	0·624	0·0312	1·603	—

contracting with the metric tensor $g_{\mu\nu}$, i.e.

$$V_\mu = g_{\mu\nu} V^\nu.$$

We use a metric $g_{00} = -1$, $g_{11} = g_{22} = g_{33} = 1$, and $g_{\mu\nu} = 0$ if $\mu \neq \nu$. So for the space-time four-vector $x^\mu = (t, \mathbf{x})$, $x_\mu = (-t, \mathbf{x})$. Scalar products are

$$A \,.\, B = A_\mu B^\mu = A^\mu B_\mu = \mathbf{A} \,.\, \mathbf{B} - A^0 B^0, \tag{A.1}$$

and, for example, if the four-momentum of a particle is $p^\mu \equiv (E, \mathbf{p})$, then

$$p^2 = p^\mu p_\mu = \mathbf{p}^2 - E^2 = -m^2.$$

Repeated indices are always summed over the values 0, 1, 2, 3 as above. We also use

$$\partial_\mu \equiv \frac{\partial}{\partial x^\mu} = g_{\mu\nu} \partial^\nu,$$

and

$$\Box_x^2 \equiv \sum_\mu \frac{\partial^2}{\partial x_\mu \, \partial x^\mu},$$

for the single and double differential operators in four dimensions.

Pauli metric. Our notation can be converted into the standard Pauli metric notation by writing $V_4 = V^4 = iV^0 = -iV_0$, and summing over 1, 2, 3, 4 instead of 0, 1, 2, 3. With this slight change our notation coincides with that of Marshak, Riazzuddin and Ryan (1969) in metric, normalization and spinor conventions, which are given below. A "dictionary" to convert to the alternative metric $g^{00} = 1$, $g^{ii} = -1$ is given by Adler and Dashen (1968).

(d) Gamma Matrices and the Dirac Equation

The Dirac γ-matrices that we use satisfy the anti-commutation relations

$$\{\gamma_\mu, \gamma_\nu\} = 2g_{\mu\nu}, \qquad (\mu, \nu = 0, 1, 2, 3), \tag{A.2}$$

and in terms of them

$$\gamma_5 \equiv i\gamma_0\gamma_1\gamma_2\gamma_3, \tag{A.3}$$

and

$$\sigma_{\mu\nu} \equiv \frac{1}{2i}[\gamma_\mu, \gamma_\nu]. \tag{A.4}$$

An explicit representation is

$$\gamma_0 = i\begin{pmatrix} \mathbf{1} & 0 \\ 0 & -\mathbf{1} \end{pmatrix}, \qquad \gamma_5 = \begin{pmatrix} 0 & -\mathbf{1} \\ -\mathbf{1} & 0 \end{pmatrix},$$

and

$$\gamma_k = \begin{pmatrix} 0 & -i\boldsymbol{\sigma}_k \\ i\boldsymbol{\sigma}_k & 0 \end{pmatrix}, \tag{A.5}$$

where $\boldsymbol{\sigma}_k$ $(k = 1, 2, 3)$ are the Pauli matrices,

$$\sigma_1 = \begin{pmatrix} 0 & 1 \\ 1 & 0 \end{pmatrix}, \qquad \sigma_2 = \begin{pmatrix} 0 & -i \\ i & 0 \end{pmatrix}, \qquad \sigma_3 = \begin{pmatrix} 1 & 0 \\ 0 & -1 \end{pmatrix}.$$

The Dirac equations for the particle spinor $u_s(\mathbf{p})$ and the antiparticle spinor $v_s(\mathbf{p})$ are

$$(i\gamma \,.\, p + m)u_s(\mathbf{p}) = 0; \qquad (i\gamma \,.\, p - m)v_s(\mathbf{p}) = 0. \tag{A.6}$$

The conjugate spinors satisfy

$$\bar{u}_s(\mathbf{p})(i\gamma \,.\, p + m) = 0; \qquad \bar{v}_s(\mathbf{p})(i\gamma \,.\, p - m) = 0. \tag{A.7}$$

The spinors are normalized such that

$$\bar{u}_s(\mathbf{p})u_{s'}(\mathbf{p}) = -\bar{v}_s(\mathbf{p})v_{s'}(\mathbf{p}) = \delta_{ss'}, \tag{A.8}$$

or, equivalently

$$u_s^\dagger(\mathbf{p})u_{s'}(\mathbf{p}) = v_s^\dagger(\mathbf{p})v_{s'}(\mathbf{p}) = \left(\frac{p^0}{m}\right)\delta_{ss'}. \tag{A.9}$$

(e) Field Theory

Most of this book uses the methods of S-matrix theory without specific reference to fields, but there are instances, notably in Section 11.2 and

Chapter 14, which call for manipulations with pion field operators. We have adopted a normalization such that the non-interacting fields $\phi^{\text{in,out}}(x)$ (which are the asymptotic, $|t|\to\infty$, forms of the interacting fields) have expansions into creation and annihilation operators of the form

$$\phi^{\text{in,out}}(x)=\frac{1}{(2\pi)^{\frac{3}{2}}}\int\frac{d^3k}{(2k^0)^{\frac{1}{2}}}[a(\mathbf{k})\,\mathrm{e}^{ikx}+a^\dagger(\mathbf{k})\,\mathrm{e}^{-ikx}], \tag{A.10}$$

with

$$[a(\mathbf{k}),\,a^\dagger(\mathbf{k}')]=\delta^{(3)}(\mathbf{k}-\mathbf{k}'). \tag{A.11}$$

Single-particle states are given by

$$|\mathbf{k}\rangle=a^\dagger(\mathbf{k})|0\rangle,$$

implying the normalization of states given in Chapter 2. This entails the following form of the *reduction formula* for the interacting fields

$$\begin{aligned}&\langle p_3\,p_4|S|\,p_1\,p_2\rangle\\&=I_{fi}-\frac{1}{(2\pi)^3}\frac{(p_3^2+\mu^2)(p_4^2+\mu^2)}{(4p_2^0p_4^0)^{\frac{1}{2}}}\iint d^4x\,d^4y\langle p_3|T(\phi(x)\phi(y))|\,p_1\rangle\;e^{-ip_4x}\,e^{ip_2y}\end{aligned} \tag{A.12}$$

where

$$T(\phi(x)\phi(y))\equiv\theta(x^0-y^0)\phi(x)\phi(y)+\theta(y^0-x^0)\phi(y)\phi(x).$$

The term I_{fi} is discussed in detail in Chapter 2.

The normalization is also illustrated by considering a specific model of an interacting field theory. For pions, the simplest is the so-called "ϕ^4 theory" (see e.g. the review of Matthews and Salam (1954)) in which the Lagrangian density is

$$\begin{aligned}\mathscr{L}=\sum_\alpha\Big\{&\tfrac{1}{2}\sum_\nu[(\partial_\nu\phi^\alpha(x))(\partial^\nu\phi^\alpha(x))\\&-\mu^2\phi^\alpha(x)\phi^\alpha(x)]-4\pi\lambda(\phi^\alpha(x)\phi^\alpha(x))^2\Big\},\end{aligned} \tag{A.13}$$

where we have introduced the isospin index α ($\alpha=1,2,3$) for pions. Physical pion states are

$$\begin{aligned}|\pi^\pm,k\rangle&=\frac{1}{\sqrt{2}}[a^{(1)\dagger}(\mathbf{k})\pm ia^{(2)\dagger}(\mathbf{k})]|0\rangle,\\|\pi^0,k\rangle&=a^{(3)\dagger}(\mathbf{k})|0\rangle.\end{aligned} \tag{A.14}$$

To first order, the Lagrangian (A.13) gives the $\pi\pi$ scattering amplitude

$$\langle p_3, \gamma; p_4, \delta | T | p_1, \alpha; p_2, \beta \rangle$$
$$= -\lambda [\delta_{\alpha\beta}\delta_{\gamma\delta} + \delta_{\alpha\gamma}\delta_{\beta\delta} + \delta_{\alpha\delta}\delta_{\beta\gamma}]. \quad \text{(A.15)}$$

Thus the normalization is such that, in first-order perturbation theory, the parameter λ is the same as the quantity conventionally introduced in the expansion of the $\pi\pi$ scattering amplitude about the symmetry point $s = t = u = 4m_\pi^2/3$ (see equation (2.25)).

(f) Isospin Conventions

Isospin amplitudes for meson–meson scattering are written down in Chapter 2. They are based on the following conventions for the single-particle isospin states:

$$|\pi, 1\rangle = -|\pi^+\rangle, |\pi, 0\rangle = |\pi^0\rangle, |\pi, -1\rangle = |\pi^-\rangle,$$
$$|K, \tfrac{1}{2}\rangle = |K^+\rangle, |K, -\tfrac{1}{2}\rangle = |K^0\rangle, \quad \text{(A.16)}$$
$$|\bar{K}, \tfrac{1}{2}\rangle = |\bar{K}^0\rangle, |\bar{K}, -\tfrac{1}{2}\rangle = -|K^-\rangle.$$

In addition, we also use the "covariant" pion states $|\pi_i\rangle$ $(i = 1, 2, 3)$ which are defined by

$$|\pi_i\rangle \equiv a^{(i)\dagger}(\mathbf{k})|0\rangle.$$

From (A.14), the physical charge states are given by

$$|\pi^\pm\rangle = \frac{1}{\sqrt{2}}[|\pi_1\rangle \pm i|\pi_2\rangle],$$
$$|\pi^0\rangle = |\pi_3\rangle. \quad \text{(A.17)}$$

Appendix B

Spin Analysis of Di-meson Production

In this appendix we outline the spin formalism for discussing di-meson production processes such as $\pi N \to \pi\pi N$, $\pi\pi\Delta$, $K\bar{K}N$ and the analogous KN processes (see Chapter 3). Topics discussed are: kinematics, and helicity and transversity amplitudes in the direct and exchange channels (Section (a)); the crossing matrix connecting them (Section (b)); intensity formulas and their relation to density matrix elements for di-meson production (Section (c)); the extension of the formalism to encompass experiments with information on baryon polarization (Section (d)); and finally, some brief remarks on the form of specific exchange contributions (Section (e)).

(a) Kinematics and Amplitudes

Consider the di-meson process illustrated in Fig. B.1. This depicts the reaction

$$M_2(0^-) + B_1(\tfrac{1}{2}^+) \to M_4(0^-) + M_5(0^-) + B_3(\tfrac{1}{2}^+ \text{ or } \tfrac{3}{2}^+ \text{ etc.}),$$

e.g. $\pi N \to \pi\pi N$. For the theoretical description, a "baryon-first" convention has been adopted. Furthermore, in making a spin analysis, it is convenient to picture the above process as going via two stages, first the production of a di-meson state, D_4, then the decay $D_4 \to M_4 M_5$. Since all possibilities are considered for the mass and spin of D_4, this does not entail any dynamical assumption.

The primary s-channel process for which spin formalism is to be provided is therefore

$$B_1 M_2 \to B_3 D_4 \qquad (s\text{-channel}).$$

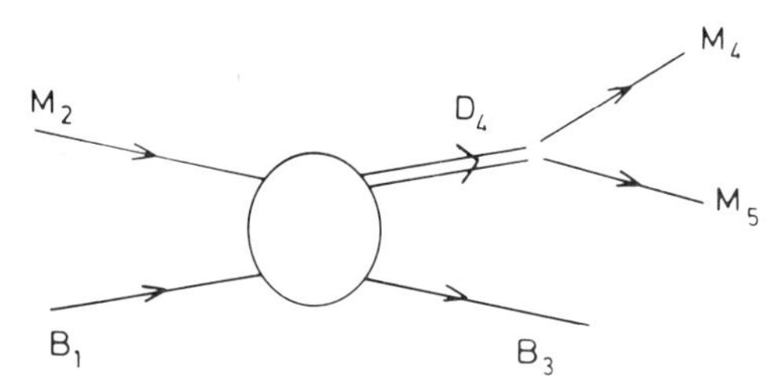

FIG. B.1. Di-meson production $M_2(0^-)+B_1(\frac{1}{2}^+)\to M_4(0^-)+M_5(0^-)+B_3(\frac{1}{2}^+, \frac{3}{2}^+ \text{ etc.})$, as a sum of contributions from "intermediate states".

It is also useful for classifying exchanges to consider the related t-channel process

$$\bar{D}_4 M_2 \to B_3 \bar{B}_1 \qquad (t\text{-channel}).$$

Convenient kinematic variables to describe the process of Fig. B.1 are s, the total CM energy squared; t, the (squared) momentum transfer to the baryon vertex; M, the di-meson mass ($M \equiv M_{45}$, e.g. $M_{\pi\pi}$); and a pair of polar angles to describe the di-meson decay. The (squared) momentum transfer, t, for the process $12 \to 34$ is given in terms of the s-channel CM scattering angle, θ_s, by the relation‡

$$4st = (m_1^2 - m_2^2 - m_3^2 + m_4^2)^2 - \mathcal{S}_{12}^2 + 2\mathcal{S}_{12}\mathcal{S}_{34}\cos\theta_s - \mathcal{S}_{34}^2, \tag{B.1}$$

where $\mathcal{S}_{12}(\mathcal{S}_{34})$ determines the initial (final) CM momentum,

$$\begin{pmatrix}\mathcal{S}_{12}\\ \mathcal{S}_{34}\end{pmatrix} = 2\sqrt{s}\begin{pmatrix}q_i\\ q_f\end{pmatrix}, \tag{B.2}$$

and is given by the formula

$$\mathcal{S}_{ij} = [(s-(m_i+m_j)^2)(s-(m_i-m_j)^2)]^{\frac{1}{2}}. \tag{B.3}$$

The boundaries of the physical region are given by the vanishing of the function (Kibble, 1960),

$$\begin{aligned}\Phi = {} & st(m_1^2+m_2^2+m_3^2+m_4^2-s-t) - s(m_2^2-m_4^2)(m_1^2-m_3^2)\\ & - t(m_1^2-m_2^2)(m_3^2-m_4^2)\\ & -(m_1^2+m_4^2-m_2^2-m_3^2)(m_1^2m_4^2-m_2^2m_3^2)\\ = {} & s(t_{\min}-t)(t-t_{\max}).\end{aligned} \tag{B.4}$$

‡ Note the distinction between the production angle, θ_s, and the decay angles θ, ϕ, for the quasi-two body system D_4.

The limiting values t_{min} and t_{max} for the s-channel physical region are given by

$$4s\binom{t_{min}}{t_{max}} = (m_1^2 - m_2^2 - m_3^2 + m_4^2)^2 - (\mathscr{S}_{12} \mp \mathscr{S}_{34})^2, \qquad \text{(B.5)}$$

corresponding to $\cos\theta_s = \pm 1$. Trigonometric functions of θ_s may be re-expressed in terms of t via the formulae

$$\sin(\theta_s/2) = \left\{\frac{t_{min} - t}{t_{min} - t_{max}}\right\}^{\frac{1}{2}}, \qquad \cos(\theta_s/2) = \left\{\frac{t - t_{max}}{t_{min} - t_{max}}\right\}^{\frac{1}{2}}. \qquad \text{(B.6)}$$

At large s, $t_{max} \sim -s$ and

$$\begin{aligned} t_{min} \sim & -(m_1^2 - m_3^2)(m_2^2 - m_4^2)/s \\ & -(m_1^2 + m_2^2 - m_3^2 - m_4^2)(m_1^2 m_2^2 - m_3^2 m_4^2)/s^2 \\ & + 0(1/s^3), \end{aligned} \qquad \text{(B.7)}$$

so that, for the general unequal mass case, $t_{min} = 0(1/s)$, but for the special case of $m_1 = m_3$ (or $m_2 = m_4$), e.g. $\pi N \to \pi\pi N$, it is $0(1/s^2)$.

Spin alignments are specified by assigning helicity values (components of each particle's spin along its direction of motion) λ_1 and λ_3 for the initial and final baryon states, and m ($m \equiv \lambda_4$) for the intermediate di-meson state. For some purposes, it is more convenient to use transversity components (Kotanski, 1966), i.e. to specify spin components along the normal to the production plane for $B_1 M_2 \to B_3 D_4$.‡

One must specify the co-ordinate frames in which helicities or transversities are being analysed.§ The most commonly used are the s- and t-channel frames associated with the reactions $12 \to 34$ and $\bar{4}2 \to 3\bar{1}$ respectively. Particle decay modes such as $D_4 \to M_4 M_5$ should then be described in the D_4 rest system with respect to the same co-ordinate frames in order to achieve simple expressions for the contributions to the decay angular distributions from specific helicity or transversity states (see below). Specifically, in the helicity description, the polar angles (θ, ϕ) for the decay are referred to axes $Oxyz$ with Oy normal to the production plane and Oz along the di-meson direction of motion; for the s-channel, this is opposite to the direction of the recoil final baryon, B_4; for the t-channel it is opposite to the direction of the incoming meson, M_2. The two frames are connected by a rotation about Oy (see formulae (B.13), (B.14), and (B.15) below). The corresponding trans-

‡ In general, when there are more than two particles in the final state, the normal to other planes defined by the production process, for example the plane defined by the final particles B_3, M_4 and M_5 in the overall CM, can also be used.

§ A particle's helicity and the associated helicity eigenvectors are not invariant against changes of reference frame (Lorentz transformations) except for boosts along the particle's direction.

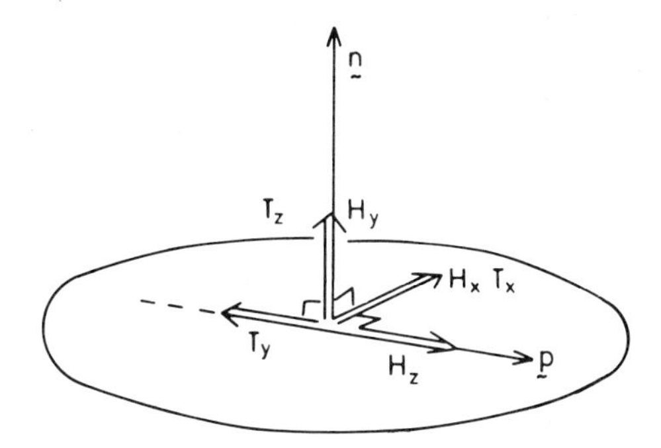

FIG. B.2. Reference axes (H_x, H_y, H_z) and (T_x, T_y, T_z) for the helicity and transversity frames, respectively, defined in terms of **p**, the particle's direction of motion, and **n**, the normal to the production plane, in the overall centre-of-mass.

versity axes have Oz normal to the production plane and, in the convention which we adopt, Oy opposite to the particle's direction of motion

$$(x_T, y_T, z_T) = (x_H, -z_H, y_H), \tag{B.8}$$

(see Fig. B.2), so that the transversity frame is obtained from the helicity frame by a rotation of $-\pi/2$ about the x-axis (Euler angles $\pi/2$, $\pi/2$, $-\pi/2$) leading to the rotations (B.9) and (B.10) below. Decay angular distributions can be expressed in terms of the transversity amplitudes, using the polar angles (α, β) of the transversity co-ordinate frame, related to the previous (θ, ϕ) via (B.8) ($\sin\alpha\cos\beta = \sin\theta\cos\phi$ etc.).

The general two-body process $12 \to 34$, and its related t-channel process $\bar{4}2 \to 3\bar{1}$, are described by helicity amplitudes $H^{(s)}_{\lambda_3\lambda_4;\lambda_1\lambda_2}$, $H^{(t)}_{\lambda_3\lambda_1;\lambda_4\lambda_2}$, or transversity amplitudes $T^{(s)}_{\tau_3\tau_4;\tau_1\tau_2}$, $T^{(t)}_{\tau_3\tau_1;\tau_4\tau_2}$, related by the equations

$$T^{(s)}_{\tau_3\tau_4;\tau_1\tau_2} = \sum_{\lambda_1\lambda_2\lambda_3\lambda_4} e^{-i\pi/2(\tau_1+\tau_2-\tau_3-\tau_4)} d^{s_1}_{\tau_1\lambda_1}(\pi/2)\, d^{s_2}_{\tau_2\lambda_2}(\pi/2) \times d^{s_3}_{\tau_3\lambda_3}(-\pi/2)\, d^{s_4}_{\tau_4\lambda_4}(-\pi/2)\, e^{+i\pi/2(\lambda_1+\lambda_2-\lambda_3-\lambda_4)} H^{(s)}_{\lambda_3\lambda_4;\lambda_1\lambda_2}, \tag{B.9}$$

$$T^{(t)}_{\tau_3\tau_1;\tau_4\tau_2} = \sum_{\lambda_1\lambda_2\lambda_3\lambda_4} e^{-i\pi/2(\tau_4+\tau_2-\tau_3-\tau_1)} d^{s_4}_{\tau_4\lambda_4}(\pi/2)\, d^{s_2}_{\tau_2\lambda_2}(\pi/2) \times d^{s_3}_{\tau_3\lambda_3}(-\pi/2)\, d^{s_1}_{\tau_1\lambda_1}(-\pi/2)\, e^{i\pi/2(\lambda_4+\lambda_2-\lambda_1-\lambda_3)} H^{(t)}_{\lambda_3\lambda_1;\lambda_4\lambda_2}. \tag{B.10}$$

The normalization of these amplitudes is defined by their use in the intensity formulae discussed in Section (c) below. Parity conservation imposes a relation between a given helicity amplitude and the corresponding amplitude with all the helicities reversed.

$$H_{\lambda_3\lambda_4;\lambda_1\lambda_2} = \eta(-)^{\Sigma(s_i+\lambda_i)} H_{-\lambda_3-\lambda_4;-\lambda_1-\lambda_2}, \tag{B.11}$$

where the s_i are the spins of the participating particles and η the product of their intrinsic parities. This halves the number of independent helicity amplitudes. For the transversity amplitudes, the corresponding rule

$$T_{\tau_3\tau_4;\tau_1\tau_2} = \eta(-)^{\tau_1+\tau_2-\tau_3-\tau_4} T_{\tau_3\tau_4;\tau_1\tau_2}, \tag{B.12}$$

simply states that half the transversity amplitudes are zero.

(b) Crossing

The crossing relation between the s- and t-channels is particularly simple for the transversity amplitudes‡

$$T^{(t)}_{-\tau_3-\tau_1;-\tau_4-\tau_2} = (-)^{s_1+s_2+s_3+s_4} e^{-i\pi(\tau_2-\tau_3)} \times e^{i(\tau_1\chi_1-\tau_2\chi_2+\tau_3\chi_3-\tau_4\chi_4)} T^{(s)}_{\tau_3\tau_4;\tau_1\tau_2}, \tag{B.13}$$

where the rotation angles χ_i are explicit functions of s and t and the particle masses (Cohen-Tannoudji *et al.*, 1968). Relevant formulae are given below, (B.18), (B.19) and (B.20). The corresponding relations for the helicity amplitudes‡ are

$$H^{(s)}_{\lambda_3\lambda_4;\lambda_1\lambda_2} = \sum_{\mu_i} e^{i\pi(\lambda_2-\lambda_3)} \prod_i d^{s_i}_{\mu_i\lambda_i}(\chi_i) H^{(t)}_{\mu_3\mu_1;\mu_4\mu_2}, \tag{B.14}$$

$$H^{(t)}_{\mu_3\mu_1;\mu_4\mu_2} = \sum_{\lambda_i} e^{-i\pi(\lambda_2-\lambda_3)} \prod_i d^{s_i}_{\lambda_i\mu_i}(-\chi_i) H^{(s)}_{\lambda_3\lambda_4;\lambda_1\lambda_2}, \tag{B.15}$$

which are implied by (B.9), (B.10) and (B.13) and the general properties of the rotation matrices.

We now adapt the notation to fit the situation in which we are interested where particle 2 has spin zero and particle 4 has variable spin, j, which must be indicated, and natural parity. We therefore write the helicity amplitudes in the form $H^j_{\lambda_3\lambda_1;m}$ with m now standing for di-meson helicity, formerly λ_4.

From the parity relation (B.11), we only need refer to half the helicity amplitudes, and we choose for the independent set those with λ_3 positive. For the important case where not only particle 1 but also particle 3 has spin $\frac{1}{2}$, we just write $H^j_{++;m}$, $H^j_{+-;m}$ for the non-flip and flip amplitudes, respectively. For the corresponding transversity amplitudes we write $T^j_{\pm\pm;\tau}$, $T^j_{\mp\pm;\tau}$ where in the former case $\tau+j$ must be odd, and in the latter case $\tau+j$ must be even.

Following Ader *et al.* (1968), it is useful to introduce the following combinations of helicity amplitudes corresponding to the exchange of given dominant parity in the t-channel, natural or unnatural. For the present case,

‡ We follow the phase conventions of Cohen-Tannoudji *et al.* (1968). For the case, not relevant to the present discussion, where particles 1 and 4 are fermions there is an additional minus sign.

where D_4 has natural parity the appropriate combinations are

$$H^{j\pm}_{\lambda_3\lambda_1;m} \equiv \frac{1}{\sqrt{2}}[H^j_{\lambda_3\lambda_1;m} \mp (-1)^m H^j_{\lambda_3\lambda_1;-m}]. \tag{B.16}$$

For $m = 0$, the normalization is unmodified,

$$H^{j-}_{\lambda_3\lambda_1;0} \equiv H^j_{\lambda_3\lambda_1;0}.$$

As $s \to \infty$, these amplitudes are dominated respectively by natural (+) and unnatural (−) exchange.‡ The transversity amplitudes already accomplish this separation, being associated with predominantly natural and unnatural exchange according as the sum of the di-meson's transversity and spin, $j+\tau$, is odd or even. Since parity conservation requires $j+\tau_1-\tau_3-\tau$ to be odd, this implies that $\tau_1-\tau_3$ must be even (odd) for predominantly natural (unnatural) exchanges.

From this it follows that the crossing relation for transversity amplitudes (B.13), and consequently those for helicity amplitudes (B.14) and (B.15), preserves the separation into states corresponding to the exchange of predominant naturality (±). Furthermore, one sees from (B.13) that the baryon crossing angles χ_1 and χ_3 always appear in the combinations $\tau_1\chi_1 + \tau_3\chi_3$ with $\tau_1-\tau_3$ even or odd. In particular, for the important case when $s_3 = \frac{1}{2}$ ($\pi N \to \pi\pi N$), the combinations are $\frac{1}{2}(\chi_1 \pm \chi_3)$ respectively. In the more general case when particle 3 has spin s_3, one obtains, after some algebra, the following crossing relation for the (±) helicity amplitude combinations:

$$H^{j(s)\pm}_{\lambda_3\lambda_1;m} = e^{-i\pi\lambda_3} \sum_{\mu_1(\mu_3>0)} [d^{\frac{1}{2}}_{\mu_1\lambda_1}(\chi_1)\, d^{s_3}_{\mu_3\lambda_3}(\chi_3) \mp \varepsilon(-)^{\mu_1-\mu_3}\, d^{\frac{1}{2}}_{-\mu_1\lambda_1}(\chi_1)\, d^{s_3}_{-\mu_3\lambda_3}(\chi_3)]$$

$$\times \sum_{m'\geq 0} 2^{-(p+p')/2}[d^j_{m'm}(\chi_4) \mp (-)^{m'}\, d^j_{-m'm}(\chi_4)] H^{j(t)\pm}_{\mu_3\mu_1;m'}, \tag{B.17}$$

where $p = 0$ or 1 according as $m \neq 0$ or $m = 0$, and likewise for p' with m', and where $\varepsilon = \eta(-)^{\Sigma s_i}$ is +1 for the case presently under consideration when particle 3 is $\frac{1}{2}^+$. The helicity crossing transformation is thus seen to remain a direct product of a transformation on the baryon spins and a transformation on the di-meson spin, even after forming the (±) combinations. For the particular case when particle 3 has spin $\frac{1}{2}$ (e.g. $\pi N \to \pi\pi N$), the resulting

‡ An exact separation into contributions respectively from natural and unnatural exchange can be made for the *partial wave* helicity amplitudes for the production process $B_1M_2 \to B_3D_4$ (Gell-Mann *et al.*, 1964). On expanding the above $H^{j\pm}_{\lambda_3\lambda_1;m}$ in terms of these, there are in general contributions from both natural and unnatural exchanges, but one or other possibility dominates as $s(\cos\theta_t) \to \infty$. The corrections to the dominant exchange arise because the parity operation applied to say the final state changes $\theta_t \to \pi - \theta_t$ and $d^J_{\lambda\mu}(\theta_t)$ in general contains even and odd powers of $\cos\theta_t$.

transformation matrix operating on the baryon spin takes the form

$$C^{st(\pm)}_{B_{\lambda_3\lambda_1:\mu_3\mu_1}}=\begin{bmatrix}\cos\chi^{(\pm)} & \sin\chi^{(\pm)}\\ -\sin\chi^{(\pm)} & \cos\chi^{(\pm)}\end{bmatrix},$$

with, as anticipated above,

$$\chi^{(\pm)}=\tfrac{1}{2}[\chi_1\pm\chi_3].$$

For the case of equal baryon masses ($m_1=m_3=m_B$), the transformation coefficients take the special form

$$\cos\chi^{(-)}=\left(\frac{t_{\min}}{t}\right)^{\frac{1}{2}}\cos(\theta_s/2);\quad \sin\chi^{(-)}=\left(\frac{t_{\max}}{t}\right)^{\frac{1}{2}}\sin(\theta_s/2),$$

$$\cos\chi^{(+)}=\left(\frac{4m_B^2-t_{\min}}{4m_B^2-t}\right)^{\frac{1}{2}}\cos(\theta_s/2),\tag{B.18}$$

$$\sin\chi^{(+)}=-\left(\frac{4m_B^2-t_{\max}}{4m_B^2-t}\right)^{\frac{1}{2}}\sin(\theta_s/2).$$

The square root singularity at $t=0$ corresponds to the t-channel pseudo-threshold (which, in general, is located at $t=(m_1-m_3)^2$). As a result, t-channel helicity amplitudes can contain singular contributions at $t=0$. On substitution into the spin-averaged intensity formula to be discussed below (B.28), all singular contributions cancel, as is obvious from the direct product form of the helicity crossing transformation (B.17).‡ It is sometimes urged that one should never work with t-channel components, on account of the above singularity. In fact, the t-channel density matrix elements (c.f. equation (B.32)) are, like their s-channel counterparts, smooth functions of t at $t=0$; and have the special virtue, for which their use was originally proposed (Gottfried and Jackson, 1964), of isolating specific t-channel exchanges.

The spin rotation angle with which we are chiefly concerned is χ_4, which relates to the di-meson D_4, and which is given by

$$\cos\chi_4=-\frac{[\alpha(t-t_{\min})+\mathcal{T}_{24}(t_{\min})]}{\mathcal{T}_{24}(t)}\underset{s\to\infty}{\sim}-\frac{[t+m_4^2-m_2^2]}{\mathcal{T}_{24}(t)},\tag{B.19}$$

$$\sin\chi_4=\frac{-2m_4[s(t_{\min}-t)(t-t_{\max})]^{\frac{1}{2}}}{\mathcal{T}_{24}(t)\mathcal{S}_{34}}\underset{s\to\infty}{\sim}\frac{-2m_4\sqrt{-t}}{\mathcal{T}_{24}(t)}.\tag{B.20}$$

‡ One could even work in a hybrid helicity representation with the di-meson helicity analysed in the t-channel and the baryon helicities in the s-channel. However, one would not then have a simple expression for the production angular distribution in terms of partial wave helicity amplitudes.

Here $\mathscr{S}_{34}$ is the quantity defined in equation (B.3), $\mathscr{T}_{24}(t)$ is the analogous t-channel quantity

$$\mathscr{T}_{24}(t) = 2q_t^{(24)}(t)^{\frac{1}{2}} = [(t-(m_4+m_2)^2)(t-(m_4-m_2)^2)]^{\frac{1}{2}}, \tag{B.21}$$

and α denotes the ratio

$$\alpha = (s + m_4^2 - m_3^2)/\mathscr{S}_{34}. \tag{B.22}$$

The asymptotic forms of $\cos \chi_4$, $\sin \chi_4$ at large s are also exhibited in (B.19), from which it is seen that large s means compared to the squares of the participating masses. From (B.6), (B.7) and (B.18) the crossing angles governing the baryon spin for the equal mass ($\frac{1}{2}^+ \to \frac{1}{2}^+$) case are seen to take the following simple form for large s and fixed-t

$$\begin{aligned} &\cos \chi^{(-)} = 0; \qquad \sin \chi^{(-)} = 1, \\ &\cos \chi^{(+)} = \frac{2m_B}{(4m_B^2 - t)^{\frac{1}{2}}}; \qquad \sin \chi^{(+)} = \frac{-(-t)^{\frac{1}{2}}}{(4m_B^2 - t)^{\frac{1}{2}}}. \end{aligned} \tag{B.23}$$

For the unnatural exchanges, the crossing matrix is anti-diagonal, whilst for the natural exchanges it is predominantly diagonal. The former is a special property of the equal mass case; when the masses of m_1 and m_3 are unequal, the $C_B^{(-)}$ crossing matrix reverts to being predominantly diagonal.

An important property peculiar to the helicity amplitudes is that the helicity indices completely specify the components of spin Δ_i, Δ_f along the initial and final directions of motion ($\Delta_i = \lambda_1 - \lambda_2$; $\Delta_f = \lambda_3 - \lambda_4$, for $12 \to 34$) leading‡ to characteristic power zeros as the scattering angle θ_s tends to 0 and π

$$\begin{aligned} H^{(s)}_{\lambda_3\lambda_1;m} &\propto [\sin \theta_s/2]^{|\Delta_i - \Delta_f|}[\cos \theta_s/2]^{|\Delta_i + \Delta_f|} \\ &= [\sin \theta_s/2]^{|\lambda_1 - \lambda_3 + m|}[\cos \theta_s/2]^{|\lambda_1 + \lambda_3 - m|} \\ &\underset{s\to\infty}{\sim} \left[\frac{t_{\min} - t}{s}\right]^{|\lambda_1 - \lambda_3 + m|/2} \end{aligned} \tag{B.24}$$

In the last line, we have passed to the large s, fixed-t limit using (B.6). This has an important consequence for exchanges of given parity coupling to amplitudes involving baryon flip ($\lambda_1 - \lambda_3 \neq 0$) and with the di-meson helicity, m, non-zero. As discussed previously, such exchanges predominantly couple to di-meson states of helicity m and $-m$ with equal, or equal and opposite, amplitudes. The kinematic zero $[t_{\min} - t]^{|\lambda_3 - \lambda_1 + m|/2}$ required by (B.24) for the $-m$ amplitude is thus transmitted to the $+m$ amplitude which,

‡ The full helicity amplitude $H^{(s)}_{\lambda_3\lambda_4;\lambda_1\lambda_2}$ can be expanded as a sum of partial wave helicity amplitudes multiplied by the appropriate rotation matrices, all of which contain the required factors in cos (sin) $\theta_s/2$.

kinematically, only requires $[t_{\min}-t]^{|\lambda_3-\lambda_1-m|/2}$.‡ In the framework of the absorption model, such amplitudes are expected to be considerably modified by absorptive cuts, especially the no-net flip amplitude when $\lambda_3-\lambda_1 = m$ (see discussion in Section 3.1).

There are analogous kinematic zeros for the t-channel amplitudes

$$H^{(t)}_{\lambda_3\lambda_1;m} \propto [\cos\theta_t/2]^{|m+\lambda_3-\lambda_1|}[\sin\theta_t/2]^{|m-\lambda_3+\lambda_1|}. \tag{B.25}$$

For large s and small fixed-t, the above functions of the t-channel scattering angle, θ_t, are given by

$$\cos\theta_t/2 \approx \left(\frac{t_{\min}-t}{t_{\min}}\right)^{\frac{1}{2}}; \qquad \sin\theta_{t/2} \approx \left(\frac{t}{t_{\min}}\right)^{\frac{1}{2}}, \qquad m_1 \neq m_3, \tag{B.26}$$

and

$$\begin{pmatrix}\cos\theta_t/2\\ \sin\theta_t/2\end{pmatrix} \approx \left[\frac{1}{2}\left(1 \pm \left(\frac{t}{t_{\min}}\right)^{\frac{1}{2}}\right)\right]^{\frac{1}{2}}, \qquad m_1 = m_3. \tag{B.27}$$

The resulting small-t behaviours of the $H^{(t)}$'s are consistent with those of the $H^{(s)}$'s and the crossing relations.

(c) Intensity Formulae and the Spin Density Matrix for Di-Meson Production

The unpolarized intensity (target unpolarized and no measurement of baryon recoil polarization) is given by

$$I(s, t, M, \theta, \phi) \equiv \frac{\partial^4\sigma}{\partial t\,\partial M\,\partial\cos\theta\,\partial\phi} = \frac{1}{2}\sum_{\lambda_1\lambda_3} |H_{\lambda_3\lambda_1}|^2, \tag{B.28}$$

which fixes our normalization for the helicity amplitudes. The full helicity amplitude $H_{\lambda_3\lambda_1}(s, t, M, \theta, \phi)$ may be decomposed into a sum of contributions corresponding to intermediate di-meson states of spin j and helicity m such as we have been discussing. Thus, one writes§

$$H_{\lambda_3\lambda_1}(\theta, \phi) = \sum_{j=0}^{\infty}\sum_{m=-j}^{j} (2j+1)^{\frac{1}{2}} H^j_{\lambda_3\lambda_1;m}\, d^j_{m0}(\theta)\, e^{im\phi}. \tag{B.29}$$

On substituting into the above intensity formula (B.28), this gives

$$I(\theta, \phi) = \tfrac{1}{2} \sum_{\lambda_3\lambda_1}\sum_{j_1m_1}\sum_{j_2m_2} (2j_1+1)^{\frac{1}{2}}(2j_2+1)^{\frac{1}{2}}\, e^{i(m_1-m_2)\phi} \times H^{*j_1}_{\lambda_3\lambda_1;m_1} H^{j_2}_{\lambda_3\lambda_1;m_2}\, d^{j_1}_{m_10}(\theta)\, d^{j_2}_{m_20}(\theta). \tag{B.30}$$

‡ In our convention, one only refers to amplitudes with $\lambda_3-\lambda_1 \geq 0$.

§ The choice $(2j+1)^{\frac{1}{2}}$ for the spin-weight factor causes the H^j_m's (modulus squared) to contribute with equal coefficients in the integrated intensity.

The bilinear expressions in the helicity amplitudes appearing in this equation are often re-expressed as (production) *density matrix elements*,‡ so that

$$I(\theta, \phi)=N \sum_{j_1 m_1 j_2 m_2} (2j_1+1)^{\frac{1}{2}}(2j_2+1)^{\frac{1}{2}} \rho^{j_1 j_2}_{m_1 m_2}\, d^{j_1}_{m_1 0}(\theta)\, d^{j_2}_{m_2 0}(\theta)\, \mathrm{e}^{i(m_1-m_2)\phi}, \tag{B.31}$$

with

$$\rho^{j_1 j_2}_{m_1 m_2}=\frac{1}{2N} \sum_{\lambda_3 \lambda_1} H^{*j_1}_{\lambda_3 \lambda_1; m_1} H^{j_2}_{\lambda_3 \lambda_1; m_2}, \tag{B.32}$$

and with N taken either as the integrated intensity $(4\pi\langle \bar{I}\rangle \equiv \partial^2\sigma/\partial M\, \partial t)$ or as unity. In the former case, one has the *normalized density matrix elements* and integration of (B.31) gives the trace condition

$$\sum_{jm} \rho^{jj}_{mm}=1. \tag{B.33}$$

The density matrix elements are not all independent,

$$\rho^{j_1 j_2}_{m_1 m_2}=(\rho^{j_2 j_1}_{m_2 m_1})^*=(-)^{m_1+m_2} \rho^{j_1 j_2}_{-m_1 -m_2}, \tag{B.34}$$

as follows from the defining formula and parity conservation (B.11). Furthermore, the independent elements are not all observable. Imaginary parts, where these are allowed by the above symmetry relations, are never observable (for experiments where the baryon polarization is undetected), and, except for the special case of pure spin j, the real parts give non-orthogonal contributions to the intensity.

A compact way of expressing the independent information in the intensity distribution is to expand in spherical harmonics (only the real parts are needed)

$$I(\theta, \phi)=4\pi\langle \bar{I}\rangle \sum_{l=0}^{\infty} \sum_{m=0}^{l} a_{lm} \operatorname{Re} Y^m_l(\theta, \phi), \qquad (a_{00}=(4\pi)^{-\frac{1}{2}}). \tag{B.35}$$

This is often re-written as

$$I(\theta, \phi)=4\pi\langle \bar{I}\rangle \sum_{l} \left\{\langle \operatorname{Re} Y^0_l\rangle \operatorname{Re} Y^0_l(\theta, \phi)+2 \sum_{m=1}^{\infty} \langle \operatorname{Re} Y^m_l\rangle \operatorname{Re} Y^m_l(\theta, \phi)\right\}, \tag{B.36}$$

‡ Where there is information on the final baryon spin alignment, for example from its decay, the concept of *joint density matrix elements*

$$\rho^{j_1 j_2}_{\lambda_3 \lambda'_3 m_1 m_2}=\frac{1}{2N} \sum_{\lambda_1} H^{*j_1}_{\lambda_3 \lambda_1; m_1} H^{j_2}_{\lambda'_3 \lambda_1; m_2}$$

is sometimes used—e.g. for $\pi N \to \pi\pi\Delta$.

and $\langle \text{Re } Y_l^m \rangle$ is often written as $\langle Y_l^m \rangle$ for short. The coefficients $\langle Y_l^m \rangle$ can be expressed as sums of Re $\rho_{m_1m_2}^{j_1j_2}$ using the standard formulae for expressing products of rotation matrices (Rose, 1957) via the relation (Ochs, 1972, cf. also Jacob and Wick (1959))

$$a_{lm} \equiv \langle \text{Re } Y_l^m \rangle$$

$$= (4\pi)^{-\frac{1}{2}} \sum_{j_1j_2m_1m_2} C(l, j_1, j_2; m, m_1, m_2) \text{ Re } \rho_{m_1m_2}^{j_1j_2}, \tag{B.37}$$

where

$$C(l, j_1, j_2; m, m_1, m_2) = (-)^m[(2l+1)(2j_1+1)(2j_2+1)]^{\frac{1}{2}} \times \begin{pmatrix} l & j_1 & j_2 \\ m & -m_1 & m_2 \end{pmatrix} \begin{pmatrix} l & j_1 & j_2 \\ 0 & 0 & 0 \end{pmatrix}. \tag{B.38}$$

As an example, the a_{lm} coefficients for the case where just S-, P- and D-waves contribute ($j_{\max} = 2$) are listed in Table B.1 (Sekulin, 1973). This illustrates the principle that only combinations of density matrix elements are directly observable.

Further information on density matrix elements can be extracted from experiment in the form of inequalities. The starting point is the set of Schwarz inequalities

$$|\rho_{m_1m_2}^{j_1j_2}| \leqslant (\rho_{m_1m_1}^{j_1j_1} \rho_{m_2m_2}^{j_2j_2}), \tag{B.39}$$

which follow immediately from the definition (B.32). ($\rho_{mm}^{jj} \geqslant 0$ also follows directly.) These inequalities can be exploited to infer bounds on combinations of density matrix elements of particular interest (for example see Grayer *et al.* (1972b) and Sekulin (1973)). Their existence also suggest that ways should be sought of exhibiting all the density matrix elements for a process within a framework which embodies the inequalities. A scheme for doing this has been proposed by Doncel *et al.* (1972) and applied to data on $\pi\pi$ production by Laurens (1971).

Combinations of density matrix elements which are of especial interest are those corresponding to given dominant parity in the t-channel. We have already (following Ader *et al.* (1968)), introduced the associated helicity amplitudes, $H_{\lambda_3\lambda_1;m}^{j\pm}$ in (B.16). The associated density matrix elements for di-meson production can likewise be decomposed into the combinations

$$\rho_{m_1m_2}^{j_1j_2\pm} \equiv \tfrac{1}{2}(\rho_{m_1m_2}^{j_1j_2} \mp (-)^{m_2} \rho_{m_1-m_2}^{j_1j_2}), \tag{B.40}$$

where

$$\rho_{m_1m_2}^{j_1j_2\pm} = \frac{1}{8N} \sum_{\lambda_3\lambda_1} H_{\lambda_3\lambda_1;m_1}^{*j_1\pm} H_{\lambda_3\lambda_1;m_2}^{j_2\pm}. \tag{B.41}$$

TABLE B.1. Legendre Coefficients for Di-meson Production in Terms of Production Density Matrix Elements (from Sekulin (1973))

$$a_{00}=\sqrt{\frac{1}{4\pi}}$$

$$a_{10}=\sqrt{\frac{1}{4\pi}}\{4\sqrt{\tfrac{3}{5}}\rho_{11}^{21}+4\sqrt{\tfrac{1}{5}}\rho_{00}^{21}+2\rho_{00}^{10}\}$$

$$a_{11}=\sqrt{\frac{1}{4\pi}}\{2\sqrt{\tfrac{6}{5}}\rho_{21}^{21}+2\sqrt{\tfrac{3}{5}}\rho_{10}^{21}+2\sqrt{\tfrac{1}{5}}\rho_{0-1}^{21}+2\rho_{10}^{10}\}$$

$$a_{20}=\sqrt{\frac{1}{4\pi}}\{-\tfrac{4}{7}\sqrt{5}\rho_{22}^{22}+\tfrac{2}{7}\sqrt{5}\rho_{11}^{22}+\tfrac{2}{7}\sqrt{5}\rho_{00}^{22}+2\rho_{00}^{20}-2\sqrt{\tfrac{1}{5}}\rho_{11}^{11}+2\sqrt{\tfrac{1}{5}}\rho_{00}^{11}\}$$

$$a_{21}=\sqrt{\frac{1}{4\pi}}\{\tfrac{2}{7}\sqrt{30}\rho_{21}^{22}+\tfrac{2}{7}\sqrt{5}\rho_{10}^{22}+2\rho_{10}^{20}+2\sqrt{\tfrac{3}{5}}\rho_{10}^{11}\}$$

$$a_{22}=\sqrt{\frac{1}{4\pi}}\{-\tfrac{4}{7}\sqrt{5}\rho_{20}^{22}-\tfrac{1}{7}\sqrt{30}\rho_{1-1}^{22}+2\rho_{20}^{20}-\sqrt{\tfrac{6}{5}}\rho_{1-1}^{11}\}$$

$$a_{30}=\sqrt{\frac{1}{4\pi}}\{-12\sqrt{\tfrac{1}{35}}\rho_{11}^{21}+6\sqrt{\tfrac{3}{35}}\rho_{00}^{21}\}$$

$$a_{31}=\sqrt{\frac{1}{4\pi}}\{-2\sqrt{\tfrac{3}{35}}\rho_{21}^{21}+4\sqrt{\tfrac{6}{35}}\rho_{10}^{21}-6\sqrt{\tfrac{2}{35}}\rho_{0-1}^{21}\}$$

$$a_{32}=\sqrt{\frac{1}{4\pi}}\{2\sqrt{\tfrac{3}{7}}\rho_{20}^{21}-2\sqrt{\tfrac{6}{7}}\rho_{1-1}^{21}\}$$

$$a_{33}=-\sqrt{\frac{1}{4\pi}}\{6\sqrt{\tfrac{1}{7}}\rho_{2-1}^{21}\}$$

$$a_{40}=\sqrt{\frac{1}{4\pi}}\{\tfrac{2}{7}\rho_{22}^{22}-\tfrac{8}{7}\rho_{11}^{22}+\tfrac{6}{7}\rho_{00}^{22}\}$$

$$a_{41}=\sqrt{\frac{1}{4\pi}}\{-\tfrac{2}{7}\sqrt{5}\rho_{21}^{22}+\tfrac{2}{7}\sqrt{30}\rho_{10}^{22}\}$$

$$a_{42}=-\sqrt{\frac{1}{4\pi}}\{\tfrac{2}{7}\sqrt{15}\rho_{20}^{22}-\tfrac{2}{7}\sqrt{10}\rho_{1-1}^{22}\}$$

$$a_{43}=-\sqrt{\frac{1}{4\pi}}\{2\sqrt{\tfrac{5}{7}}\rho_{2-1}^{22}\}$$

$$a_{44}=\sqrt{\frac{1}{4\pi}}\{\sqrt{\tfrac{10}{7}}\rho_{2-2}^{22}\}$$

By parity conservation, there are no cross terms between H^+ and H^- amplitudes, as can be verified algebraically using (B.11), (B.16) and (B.32). If $m_2 = 0$, only the unnatural parity combination is admitted, the exchange parity diagonalization being in this case exact even for finite s. Equation (B.41) brings out that the representation (B.40) is really a separation of the density matrix into two blocks (+ and −), as is obvious if one changes the basis of states from $|m_2\rangle$ to $|m_2\pm\rangle = (1/\sqrt{2})(|m_2\rangle \mp (-)^{m_2}|-m_2\rangle)(m_2 \neq 0)$, and likewise for m_1. As an explicit example, consider the case with just S- and P-wave di-mesons. The untransformed density matrix (using the constraints of (B.34)) has the form

$$\begin{bmatrix} \rho_{ss} & \rho_{1s}^* & \rho_{0s}^* & -\rho_{1s}^* \\ \rho_{1s} & \rho_{11} & \rho_{10} & \rho_{1-1} \\ \rho_{0s} & \rho_{10}^* & \rho_{00} & -\rho_{10}^* \\ -\rho_{1s} & \rho_{1-1} & -\rho_{10} & \rho_{11} \end{bmatrix}, \tag{B.42}$$

in an obvious simplified notation. On transforming to the (±) representation one has, instead.

$$\begin{array}{c} \\ (-) \\ \\ (+) \end{array}\left[\begin{array}{ccc|c} \multicolumn{3}{c|}{(-)} & (+) \\ \rho_{ss} & \sqrt{2}\rho_{1s}^* & \rho_{0s}^* & 0 \\ \sqrt{2}\rho_{1s} & \rho_{11}-\rho_{1-1} & \sqrt{2}\rho_{10} & 0 \\ \rho_{0s} & \sqrt{2}\rho_{10}^* & \rho_{00} & 0 \\ \hline 0 & 0 & 0 & \rho_{11}+\rho_{1-1} \end{array}\right]. \tag{B.43}$$

In general, with the di-meson spin running up to j, there are two Hermitian blocks of order $[(j+1)(j+2)]/2$ and $j(j+1)/2$ respectively. As was emphasized above, not all the density matrix elements are observable (in the present example, Im ρ_{10}, Im ρ_{0s} and one combination of the diagonal elements, say ρ_{11}, are unobservable). However, the unobserved quantities can be bounded by use of the Schwarz inequalities referred to above. This information can be converted into bounds on the eigen-values of the density-matrix (Grayer *et al.*, 1972b); it can thus afford indications on the *rank* of the density matrix. For the case where the final baryon, B_3, is $\frac{1}{2}^+$, only two independent combinations of the unobserved baryon helicity enter because of parity conservation (B.11). The density matrix elements are thus of the form

$$\rho_{m_1m_2}^{j_1j_2\pm} = \frac{1}{4N}[H_{++;m_1}^{*j_1\pm} H_{++;m_2}^{j_2\pm} + H_{+-;m_1}^{j_1\pm} H_{+-;m_2}^{j_2\pm}]. \tag{B.44}$$

The ranks of ρ^+ and ρ^-, whose sum gives the rank of ρ, can thus never exceed two. In the above S- and P-wave example, this implies that one of the eigen-values of ρ^- is necessarily zero. If the production amplitudes for the three di-meson spin states corresponding to (predominantly) unnatural exchange are coherent in baryon spin or in phase, the rank of ρ^- drops to one. This situation is at least approximately realized in 17 GeV/c $\pi^+\pi^-$ production at small t and in the vicinity of the rho-mass (again see Grayer *et al.* (1972b)).

(d) Experiments on Polarized Targets

The production helicity amplitudes which we have been discussing are considerably underdetermined unless there are data with polarized targets. The scope for polarization experiments is much larger than for genuine two-body reactions, where measurable effects can only result from polarization normal to the plane of production, due to parity conservation.

It is convenient for discussing polarization possibilities to revert to the transversity amplitudes (see (B.9) and preceding discussion). If one refers the di-meson decay to the transversity co-ordinates with polar angles (α, β) introduced previously (see (B.8)), then the full transversity amplitude $T_{\tau_3\tau_1}(\alpha, \beta)$ can be expanded into a sum of contributions from specific di-meson partial waves and spin alignments

$$T_{\tau_3\tau_1}(\alpha, \beta) = \sum_{j=0}^{\infty} \sum_{\tau=-j}^{j}{}' (2j+1)^{\frac{1}{2}} T^{j}_{\tau_3\tau_1;\tau}\, d^{j}_{\tau 0}(\alpha)\, e^{i\tau\beta}, \tag{B.45}$$

where the prime on the τ summation indicates that only contributions with $\tau_1 - \tau_3 - \tau + j$ odd are to be included. The unpolarized intensity is then given by

$$I_U(\alpha, \beta) = \tfrac{1}{2} \sum_{\tau_3\tau_1} |T_{\tau_3\tau_1}(\alpha, \beta)|^2, \tag{B.46}$$

an incoherent sum of contributions from $\tau_1 = \pm\frac{1}{2}$, τ_3 ranging from $-s_3$ to s_3. The corresponding intensity formula, I_P, for an experiment using a polarized target is

$$I_P(\alpha, \beta) = \mathrm{Tr}\, \tfrac{1}{2}[\mathbf{1} + \mathbf{P} \,.\, \boldsymbol{\sigma}]\Theta, \tag{B.47}$$

where Θ is the spin density matrix for the initial nucleon,

$$\Theta_{\tau_1\tau_1'} = \sum_{\tau_3} T^*_{\tau_3\tau_1} T_{\tau_3\tau_1'}, \tag{B.48}$$

and $\boldsymbol{\sigma} \equiv (\sigma_x, \sigma_y, \sigma_z)$ are the Pauli matrices. Measurements with the target polarized normal to the plane of production, P_z (Fig. B.2), allow the contributions from $\tau_1 = \frac{1}{2}$ and $-\frac{1}{2}$ to be separated. Imposing the other component of transverse polarization (P_x) or longitudinal polarization (P_y)‡ gives information on cross-terms between $\tau_1 = \frac{1}{2}$ and $\tau_1 = -\frac{1}{2}$ components. For the important example where the final baryon has spin $\frac{1}{2}$, the scope of the various measurements is as follows:

$$\begin{aligned} &U: (|T_{++}|^2 + |T_{-+}|^2) + (|T_{+-}|^2 + |T_{--}|^2), \\ &P_z: (|T_{++}|^2 + |T_{-+}|^2) - (|T_{+-}|^2 + |T_{--}|^2), \\ &P_x: 2\,\mathrm{Re}\,(T^*_{++}T_{+-} + T^*_{-+}T_{--}), \\ &P_y: 2\,\mathrm{Im}\,(T^*_{++}T_{+-} + T^*_{-+}T_{--}). \end{aligned} \tag{B.49}$$

If the di-meson spin runs through all values up to j, then there are $2(j+1)^2$ independent complex helicity components. An experiment on an unpolarized target (U) yields $(2j+1)(j+1)$ pieces of information in the form of $\langle Y_l^m \rangle$ coefficients. Likewise, an experiment with the target polarized normal to the plane of production (P_z) yields a further $(2j+1)(j+1)$ items of information. Transverse polarization in the plane of production (P_x) and longitudinal polarization (P_y) each allow a further $(2j+1)j$ coefficients to be measured. Evidently, not all these are independent. It is clear from equation (B.49) that no observable quantity changes if each of the pairs (T_{++}, T_{+-}) and T_{-+}, T_{--}) is multiplied by an overall phase. However, for the case where $j_{\max} = 1$, one can show that to within discrete ambiguities, measurements of U, P_z and P_x are sufficient to fix all the amplitudes, apart from the two multiplicative phases referred to above. It is likely that this result extends to all j, with increasing scope for discrete ambiguities. Measurement of longitudinal polarization would eliminate some of these. To go further and also to measure the relative phase of the amplitudes corresponding to $\tau_f = +$ and $\tau_f = -$, one would need to perform polarization correlation experiments with measurement of the final baryon polarization (R and A measurements). In practice, for OPE dominated reactions, the realistic prospect is of experiments with a transverse polarized target, but, as we have seen, this will give a very large increase of information, enabling amplitude analyses to be made with far fewer theoretical assumptions.

(e) Examples of Specific Final States and Exchanges

(i) $N\pi \to N\varepsilon$ ($\frac{1}{2}^+0^- \to \frac{1}{2}^+0^+$)

This is a good example to illustrate the relation between helicity and transversity components, and "st" crossing. The two components of trans-

‡ A particular experimental arrangement with the target polarized normal to the beam will yield information on both P_z and P_x, the relative sensitivities being a function of the design. P_y measurements are technically much more difficult.

versity, T_{+-} and T_{-+} which are allowed by (B.12) are related to the helicity amplitudes H_{++}, H_{+-} by (B.9) and (B.10)

$$T_{\pm\mp}^{(s,t)} = H_{+-}^{(s,t)} \pm iH_{++}^{(s,t)}, \tag{B.50}$$

and the s, t crossing relations for transversity amplitudes (B.13) are

$$T_{\pm\mp}^{(t)} = \pm i \exp\left[\pm\frac{i}{2}(\chi_1-\chi_3)\right] T_{\mp\pm}^{(s)}. \tag{B.51}$$

Solving the corresponding relation for the H's, one recovers the relation (cf. B.17)

$$\begin{bmatrix} H_{++}^{(s)} \\ H_{+-}^{(s)} \end{bmatrix} = -i \begin{bmatrix} \cos\chi^{(-)} & \sin\chi^{(-)} \\ -\sin\chi^{(-)} & \cos\chi^{(-)} \end{bmatrix} \begin{bmatrix} H_{++}^{(t)} \\ H_{+-}^{(t)} \end{bmatrix}. \tag{B.52}$$

For this reaction, only unnatural parity exchanges are admitted, π, A_1 etc. The most general expression for the helicity amplitudes is obtained by forming appropriate spinor expectation values of the invariant amplitude‡ $\gamma_5[A+\frac{i}{2}B\gamma\,.\,(p_2+p_4)]$ where p_2 and p_4 are the four-momenta of the π and ε respectively (Diu and Le Bellac, 1968). The resulting t-channel helicity amplitudes have the form (we write $m_1 = m_3 = m_B$)

$$\begin{aligned} H_{++}^{(t)} &= \frac{iA\sqrt{-t}}{2m_B} - \frac{iB}{2\sqrt{-t}}(m_4^2 - m_2^2), \\ H_{+-}^{(t)} &= \frac{-iB\sqrt{\Phi}}{2m_B\sqrt{-t}}, \end{aligned} \tag{B.53}$$

where Φ is defined in equation (B.4).

The related s-channel amplitudes are given by

$$\begin{aligned} H_{++}^{(s)} &= \left(\frac{A}{2m_B}(-t_{\min})^{\frac{1}{2}} - \frac{B(m_4^2-m_2^2)}{2(-t_{\min})^{\frac{1}{2}}}\right)\cos\theta_s/2 \underset{s\to\infty}{\sim} \frac{Bs}{2m_B}, \\ H_{+-}^{(s)} &= \left(\frac{-A}{2m_B}(-t_{\max})^{\frac{1}{2}} + \frac{B(m_4^2-m_2^2)}{2(-t_{\max})^{\frac{1}{2}}}\right)\sin\theta_s/2 \underset{s\to\infty}{\sim} \frac{-A\sqrt{-t}}{2m_B}, \end{aligned} \tag{B.54}$$

as follows from (B.6), (B.18) and (B.52). These formulae illustrate a number of general points. π-exchange contributions (both elementary and Reggeized) are all contained in the function A. This couples solely to non-flip in the t-channel and predominantly (as $s\to\infty$) to flip in the s-channel. A_1 contributions enter in the term B; however this corresponds

‡ A and B are the analogues of the familiar invariant amplitudes for πN elastic scattering.

to the coupling of an axial-vector field—thus with 0^- contributions as well as 1^+. The true 1^+-exchange contributions only enter in the t-channel flip amplitude corresponding to the triplet-S $N\bar{N}$ state. At finite energies the B terms give singular $(-t)^{-\frac{1}{2}}$ contributions to $H^{(t)}_{++}$, $H^{(t)}_{+-}$, which however conspire to cancel in the intensity (Froggatt and Morgan, 1969). At large s, the B contribution is predominantly flip in the t-channel and non-flip in the s-channel. It is non-vanishing at $t=0$, in contrast to the π-exchange contribution to the flip amplitude which contains the factor $(-t)^{\frac{1}{2}}$. For large but finite s and very small t, the π-exchange term reverts to being predominantly non-flip, the ratio of non-flip to flip being given by

$$\frac{H^{(s)}_{++}}{H^{(s)}_{+-}} \sim \left(\frac{-t_{\min}}{t_{\min}-t}\right)^{\frac{1}{2}}. \tag{B.55}$$

The energy dependencies indicated in (B.54) are for elementary exchanges. Reggeized A_1-exchange should be rather similar to Reggeized π-exchange, assuming parallel trajectories and normal slopes.

(ii) $N\pi \to N\rho$ $(\frac{1}{2}^+0^- \to \frac{1}{2}^+1^-)$

There are now six helicity and transversity components, related by

$$T^{(s,t)}_{\tau_3\tau_4;\tau_1\tau_2} = \frac{1}{\sqrt{2}}\begin{bmatrix} 1 & i \\ i & 1 \end{bmatrix}_{\tau_1\lambda_1} \otimes \frac{1}{\sqrt{2}}\begin{bmatrix} 1 & -i \\ -i & 1 \end{bmatrix}_{\tau_3\lambda_3}$$

$$\otimes \begin{bmatrix} 1 & -i\sqrt{2} & -1 \\ -i\sqrt{2} & 0 & -i\sqrt{2} \\ -1 & -i\sqrt{2} & 1 \end{bmatrix}_{\tau_4\lambda_4} H^{(s,t)}_{\lambda_3\lambda_4;\lambda_1\lambda_2}. \tag{B.56}$$

For the non-vanishing transversity components crossing is given by (B.13)

$$\begin{aligned} T^{(t)}_{\pm\pm;0} &= \mp i \exp(\mp i\chi^{(+)}) T^{(s)}_{\mp\mp;0}, \\ T^{(t)}_{\pm\mp;\tau} &= \mp i \exp(\pm i\chi^{(-)} + i\tau\chi_4) T^{(s)}_{\mp\pm;-\tau}, \qquad (\tau = \pm 1). \end{aligned} \tag{B.57}$$

The corresponding crossing relations for the $(-)$ combinations of helicity amplitudes (B.16) is (cf. B.17)

$$\begin{bmatrix} H^{(s)-}_{++;0} \\ H^{(s)-}_{+-;0} \\ H^{(s)-}_{++;1} \\ H^{(s)-}_{+-;1} \end{bmatrix} = -i \begin{bmatrix} \cos\chi^{(-)} & \sin\chi^{(-)} \\ -\sin\chi^{(-)} & \cos\chi^{(-)} \end{bmatrix} \otimes \begin{bmatrix} \cos\chi_4 & -\sin\chi_4/\sqrt{2} \\ \sqrt{2}\sin\chi_4 & \cos\chi_4 \end{bmatrix} \begin{bmatrix} H^{(t)-}_{++;0} \\ H^{(t)-}_{+-;0} \\ H^{(t)-}_{++;1} \\ H^{(t)-}_{+-;1} \end{bmatrix}, \tag{B.58}$$

where direct product notation implies that the matrix in $\chi^{(-)}$ operates on the baryon spin indices and the matrix in χ_4 operates on the di-meson helicity index. The $(+)$ combinations transform according to

$$\begin{bmatrix} H^{(s)+}_{++;1} \\ H^{(s)+}_{+-;1} \end{bmatrix} = -i \begin{bmatrix} \cos\chi^{(+)} & \sin\chi^{(+)} \\ -\sin\chi^{(+)} & \cos\chi^{(+)} \end{bmatrix} \begin{bmatrix} H^{(t)+}_{++;1} \\ H^{(t)+}_{+-;1} \end{bmatrix}. \tag{B.59}$$

The relevant crossing angles are defined in (B.18), (B.19) and (B.20).

Both unnatural (π, A_1 etc.) and natural (ω, A_2 etc.) contributions may now enter. Elementary π-exchange couples to the unique t-channel amplitude $H^{(t)-}_{++;0}$ and this of course generalizes to all di-meson spins (only $m=0$ di-mesons can contribute in the t-channel). The unique t-channel amplitude has the form

$$H^{(t)-}_{++;0} = \frac{ic(t)\mathcal{T}_{24}\sqrt{-t}}{2m_B m_4}, \tag{B.60}$$

where $c(t)$ contains the pion pole factor and possibly additional vertex factors, for example from Reggeization.‡ The corresponding s-channel amplitudes are

$$\begin{aligned}
H^{(s)-}_{++;0} &= \frac{c\mathcal{T}_{24}}{2m_B m_4}\cos\chi_4(-t_{\min})^{\frac{1}{2}}\cos\theta_s/2 \underset{s\to\infty}{\sim} 0, \\
H^{(s)-}_{+-;0} &= \frac{-c\mathcal{T}_{24}}{2m_B m_4}\cos\chi_4(-t_{\max})^{\frac{1}{2}}\sin\theta_s/2 \underset{s\to\infty}{\sim} \frac{c\sqrt{-t}\,(t+m_4^2-m_2^2)}{2m_B m_4}, \\
H^{(s)-}_{++;1} &= \frac{\sqrt{2}c\mathcal{T}_{24}}{2m_B m_4}\sin\chi_4(-t_{\min})^{\frac{1}{2}}\cos\theta_s/2 \underset{s\to\infty}{\sim} 0, \\
H^{(s)-}_{+-;1} &= \frac{-\sqrt{2}c\mathcal{T}_{24}}{2m_B m_4}\sin\chi_4(-t_{\max})^{\frac{1}{2}}\sin\theta_s/2 \underset{s\to\infty}{\sim} \frac{\sqrt{2}c}{m_B}(-t).
\end{aligned} \tag{B.61}$$

The two $|m|=1$ amplitudes are seen to have the correct kinematic factors as $\theta_s \to 0$ and π (B.24) from the relation (derivable from (B.5), (B.6), (B.19) and (B.20))

$$\frac{\mathcal{T}_{24}}{m_4}\sin\chi_4 = \frac{\mathcal{S}_{12}}{\sqrt{s}}\sin\theta_s.$$

The above formulae (B.61) brings out a number of aspects of the π-exchange contribution, some analogous to those discussed previously for $N\pi \to N\varepsilon$, and some new ones. As for $N\pi \to N\varepsilon$, the s-channel couplings are

‡ A Reggeized π may have additional couplings (not containing the π-pole) to $H^{(t)-}_{++;1}$.

predominantly flip at high energies and the $m = 0$ coupling $H^{(s)-}_{+-;0}$ vanishes as $\sqrt{-t}$ near the forward direction. At very small t, the non-flip coupling re-asserts itself as discussed above for $N\varepsilon$ production (B.55). π-exchange couples to $H^{(s)-}_{+-;1}$ not to $H^{(s)+}_{+-;1}$; consequently the kinematic factor $(-t)$ has to appear (cf. (B.24)). However, as discussed following that equation, this situation is readily disrupted by absorptive cuts which do not correspond to the exchange of definite parity.

Another novel feature, as compared to $N\pi \to N\varepsilon$, is the explicit appearance of the di-meson crossing angle χ_4. In equation (B.61) the $m = 0$ s-channel flip amplitude is seen to vanish when $\cos\chi_4 = 0$; at high energies this occurs for $-t = m_4^2 - m_2^2 \approx 0{\cdot}5\ \mathrm{GeV}^2$ for rho-production. The di-meson crossing matrix also imposes a characteristic di-meson mass dependence, with the $m = 1$ coupling becoming more important relative to $m = 0$ as $m_4 \to 0$.

(iii) $N\pi \to \Delta\rho$ $(\frac{1}{2}^+0^- \to \frac{3}{2}^+1^-)$

Pion exchange contributes to the single t-channel amplitude according to the formula (see for example Jackson and Pilkuhn (1964))

$$H^{(t)}_{\frac{1}{2}\frac{1}{2};0} = \frac{c(t)\mathcal{T}_{24}\mathcal{T}_{13}[(m_1+m_3)^2 - t]^{\frac{1}{2}}}{4m_3m_4}, \tag{B.62}$$

where $\mathcal{T}_{13}$ is defined by analogy with (B.21). The st crossing relation may be cast into the factorized form (B.17) but now with $\varepsilon = \eta(-)^{\Sigma s_i} = -1$. In the high-energy limit, the formula for the crossing angles χ_1 and χ_3 becomes

$$\begin{aligned} \cos\chi_1 &= \frac{-(t+m_1^2-m_3^2)}{\mathcal{T}_{13}}, \qquad & \sin\chi_1 &= \frac{2m_1\sqrt{-t}}{\mathcal{T}_{13}}, \\ \cos\chi_3 &= \frac{(t+m_3^2-m_1^2)}{\mathcal{T}_{13}}, \qquad & \sin\chi_3 &= \frac{-2m_3\sqrt{-t}}{\mathcal{T}_{13}}, \end{aligned} \tag{B.63}$$

leading to s-channel amplitudes of the form

$$H^{(s)-}_{\lambda_3\lambda_1;m} = \frac{B_{\lambda_3\lambda_1}M_m\, c(t)}{4m_3m_4}, \tag{B.64}$$

where

$$\begin{aligned} B_{3+} &= \sqrt{3}\sqrt{-t}\,m_3(m_1+m_3), \\ B_{1+} &= t(2m_3+m_1) + (m_3^2 - m_1^2)(m_1+m_3), \\ B_1 &= \sqrt{-t}\,(t + 2m_3^2 + m_1m_3 - m_1^2), \\ B_{3-} &= -\sqrt{3}\,m_3 t, \end{aligned} \tag{B.65}$$

and with M_m given as before by $\mathcal{T}_{24}(\cos \chi_4, \sqrt{2} \sin \chi_4)$ (cf. (B.58), (B.19) and (B.20))

$$M_0 = -(t + m_4^2 - m_2^2); \qquad M_1 = -2\sqrt{2} m_4 \sqrt{-t}. \tag{B.66}$$

At large but finite energies, the kinematic factors $(-t)^{|\lambda_1 - \lambda_3 + m|/2}$ should be replaced by the corresponding powers of $(t_{\min} - t)$. The main s-channel contribution is now non-flip so that, unlike the case of $N\pi \to N\rho$, the OPE signal is not suppressed relative to the background in the forward direction.

For S- and P-wave di-meson production there are 16 helicity amplitudes and 30 measurables for experiments off unpolarized targets. Formulae are given in Irving (1973).

Appendix C

High-energy Scattering and Regge Poles

As elsewhere in hadron physics, the concept of Regge poles recurs throughout the present review as a key unifying idea, linking together families of resonances, and relating them to characteristics of high-energy exchange reactions. In the latter context, the contributions of Regge cuts, arising presumably from the exchange of two or more Regge poles, are also important. In this Appendix, we will give a brief outline of the particular results relevant to our theme. For fuller discussions, we refer the reader to the book by Collins and Squires (1968), and to subsequent reviews (e.g. those of Collins (1971) and Phillips and Ringland (1972)).

A Regge pole is a pole of the scattering amplitude in the complex angular momentum plane, and is characterized by a *trajectory* $\alpha(t)$ (real for $t \leq 0$), a *signature* $\tau = \pm 1$, and a *coupling strength* (*residue*). The two former quantities are intrinsic properties of the Regge pole, the latter defines its contribution to a specific reaction. The standard formula for such a (boson) Regge pole contribution to the *t-channel* scattering amplitude is

$$T_R(t, s) = \left(\frac{\pi}{2}\right) b(t)[2\alpha(t)+1]\frac{[1+\tau\, e^{-i\pi\alpha(t)}]}{\sin \pi\alpha(t)} P_\alpha(-z_t), \tag{C.1}$$

where, for the case of $\pi\pi$ scattering (cf. equation (2.8))

$$z_t \equiv \cos \theta_t = \frac{s-u}{t-4\mu^2}.$$

Continued to negative t and large positive s, the above formula implies the following asymptotic form (to leading order in s at fixed-t) for the contribution of a t-channel boson Regge pole to a high-energy s-*channel* scattering amplitude

$$T_R(t, s) \underset{s\to\infty}{\sim} R_s(s, t) \equiv \beta(t)\frac{[1+\tau\, e^{-i\pi\alpha(t)}]}{\sin \pi\alpha(t)}\left(\frac{s}{s_0}\right)^{\alpha(t)}, \tag{C.2}$$

where, following convention, we have introduced a convenient scale factor, s_0. The residue function $\beta(t)$ is related to the $b(t)$ of equation (C.1) above by

$$\beta(t) = \frac{\pi^{\frac{1}{2}}[2\alpha+1]\Gamma[\alpha+\frac{1}{2}]}{2\Gamma[\alpha+1]\sin \pi\alpha}\left(\frac{s_0}{q_t^2}\right)^{\alpha(t)} b(t). \tag{C.3}$$

The analogous high-energy u-channel amplitude is given by

$$R_u(u, t) \equiv \tau\beta(t)\frac{[1+\tau\, e^{-i\pi\alpha(t)}]}{\sin \pi\alpha(t)}\left(\frac{u}{s_0}\right)^{\alpha(t)}. \tag{C.4}$$

The need for the signature quantum number arises because the existence of exchange forces gives rise to characteristic $(-1)^l$ factors in the representation of the partial wave amplitude (cf. equations (7.16)–(7.18)) which render its continuation to complex values of angular momentum non-unique by Carlson's Theorem (Titchmarsh, 1939). To restore the uniqueness of this continuation, even and odd combinations in l must be formed in order to remove this factor (for a full discussion of this point see the references cited above).

For $t > 4\mu^2$ (the t-channel physical region) the Regge pole generates t-channel bound states and resonances with spin l as Re $\alpha(t)$ passes through even (odd) integers l, depending on $\tau = +1(-1)$. The contribution to an individual integral partial wave amplitude of the expression $T_R(t, s)$ of equation (C.1) is

$$T_l^R = \frac{b[2\alpha+1]}{(\alpha-l)(\alpha+l+1)}\,\frac{[1+\tau\, e^{-i\pi\alpha}]}{2}. \tag{C.5}$$

By expanding in the energy, t, near a right signature point (i.e. l even (odd) for $\tau = +1(-1)$), the partial wave amplitude can be expressed in Breit–Wigner form with

$$M_R\Gamma_{\rm tot} = {\rm Im}\,\alpha/\alpha'; \qquad M_R\Gamma_{\rm el} = -b/\alpha', \qquad (\alpha' \equiv d\,{\rm Re}\,\alpha/dt). \tag{C.6}$$

A further connection with particles is obtained from equation (C.2). Provided that $\alpha(t) \leq 1$ for $t \leq 0$ (the s-channel scattering region), R_s is bounded as $s \to \infty$; thus Regge pole exchange generalizes the idea of single-particle exchange without the usually attendant s^l catastrophe. Phenomenologically it is found that bosons lie on approximately linear trajectories with $\alpha' \approx 1\ \text{GeV}^{-2}$, and $\alpha(0) < 1$, e.g. $\alpha_\rho(0) \approx 0{\cdot}5$. An exception is the so-called Pomeron (P) trajectory with $\alpha_P(0) = 1$, which apparently contains no physical particles, but is introduced to explain diffraction scattering, and the constancy of many total cross-sections at intermediate energies.

The contribution $T_R(t, s)$ of (C.1) corresponds to a signatured amplitude

$$T_R^{(\tau=\pm 1)}(t, s) = T_R^{\text{RHC}}[t, s(z_t, t)] \pm T_R^{\text{LHC}}[t, s(-z_t, t)]. \tag{C.7}$$

The portion with just a right-hand cut is

$$T_R^{\text{RHC}} = \frac{\pi b(t)[2\alpha(t)+1]}{2 \sin \pi\alpha(t)} P_\alpha(-z_t). \tag{C.8}$$

This has an s-discontinuity

$$\begin{aligned} D_s^R &\equiv \frac{1}{2i} \text{Disc}_s \, [T_R^{\text{RHC}}] \\ &= -(\pi/2) b(t)[2\alpha(t)+1] P_\alpha(z_t)\theta(z_t - 1), \end{aligned} \tag{C.9}$$

i.e. extending from $s = 0$ to infinity.‡ The double spectral function associated with T_R^{RHC} is given by

$$\begin{aligned} \rho^R &\equiv \frac{1}{2i} \text{Disc}_t \, [D_s^R] \\ &= -\frac{1}{2i} \text{Disc}_t \, [b(t)(2\alpha+1)](\pi/2) P_\alpha(z_t)\theta(z_t - 1). \end{aligned} \tag{C.10}$$

A number of other properties are associated with the form (C.2), and we will mention just a few of them.

(i) *Factorization*

If the t-channel process is $ab \to cd$, then near the pole the residue $\beta(t)$ factorizes into the form

$$\beta(t) = \beta^{ab}\beta^{cd},$$

where β^{ab} and β^{cd} are the couplings of the Reggeon to ab and cd, respectively. This leads to many predictions, most of which are unfortunately difficult to test. For example, if we assume that at high energies the t-channel Pomeron dominates $\pi\pi$, πN and NN scattering, then there results the relation

$$\sigma(\pi\pi)\sigma(NN) = [\sigma(\pi N)]^2, \tag{C.11}$$

between total cross-sections. This relation is also obtained in simple additive

‡ In some applications (cf. Section 11.1) it is convenient to redefine T_R to have correct s-channel thresholds but with the same asymptotic ($s \to \infty$) behaviour.

quark models (see e.g. Kokkedee (1969)) which further yield the specific ratios

$$\sigma(\pi\pi) : \sigma(\pi N) : \sigma(NN) = 4 : 6 : 9.$$

The pp and $\pi^+\pi^+$ total cross-sections are believed to be dominated by Pomeron exchange even at intermediate energies (say 5–30 GeV). Over this range, the pp total cross-section is essentially constant at 39 ± 1 mb, suggesting a value for the $\pi^+\pi^+$ total cross-section of about 17 mb. Similar values are obtained from Regge pole fits to πN and NN data and using equation (C.11) (Barger and Phillips, 1971).

(ii) *Exchange Degeneracy*

In general, trajectories with the same quantum numbers, but opposite signatures, are physically distinct, and give different contributions to the scattering amplitude. If, however, exchange forces are absent for some reason, then this is no longer the case, and $\alpha_+(t) = \alpha_-(t)$, $\beta_+(t) = \beta_-(t)$. This result is known as Exchange Degeneracy (EXD), and the weaker result $\alpha_+(t) = \alpha_-(t)$, but $\beta_+(t) \neq \beta_-(t)$, is called Weak EXD. Experimentally, it is found that Weak EXD is satisfied approximately for several pairs of trajectories, e.g. the $\rho - A_2$ boson trajectories. This finds a simple interpretation in the framework of duality, as discussed in Section 12.2(a).

(iii) *Nonsense Zeros*

In equation (C.2), zeros of sin $\pi\alpha(t)$ will give rise to poles of R_s and must be removed if "nonsense" states are not to result (i.e. states with $l < 0$, or imaginary masses). The zeros of the factor $[1 + \tau \exp(-i\pi\alpha)]$ (the *signature factor*) will do this for poles occurring at wrong signature points, but not at right signature points. In the latter case, zeros must be inserted in the residue function $\beta_\pm(t)$. If, for some particular reaction, EXD holds, then this implies zeros in the residue function $\beta_\mp(t)$ of the partner trajectory, at what are now wrong signature points. These latter zeros are called "nonsense wrong signature zeros" (NWSZ). By factorization, such zeros will propagate to other reactions, and, if present, will show up as dips in differential cross-sections. This is discussed more fully in Section 12.2(a).

(iv) *Conspiracy and Evasion*

It can be shown that t-channel amplitudes must satisfy linear relations (constraints) at certain values of t associated with t-channel thresholds and pseudo-thresholds (crossed-channel thresholds), in order to satisfy analyticity and crossing symmetry. This may be achieved in two distinct ways: either each Regge pole parameter separately satisfies the constraint, which is known as "evasion"; or only the sum does, which is known as a "conspiracy". The latter solution implies the existence of correlations between the parameters of different Regge poles.

(v) *Daughters*

Another general concept introduced in many models and theories is that of "daughter trajectories". The idea is that, associated with a given "leading trajectory" $\alpha \equiv \alpha_{\text{leading}}(s)$ (i.e. the one with the largest value of $\alpha(0)$), there may exist a number of lower-lying (daughter) trajectories for which $\alpha \equiv \alpha_{\text{daughter}} = \alpha_{\text{leading}} - n$ (either exactly or approximately) with $n = 1, 2, 3, \ldots$.

Applications of the foregoing ideas occur at a number of places in the main text. Chapter 3 introduces the exchanged pion-pole in production processes, the concepts of evasive coupling, nonsense wrong signature zeros, and the phenomenological status of daughter trajectories. Regge exchanges (P, ρ etc.) as a model for high-energy contributions to sum rules feature in Chapter 10. Section 11.1 is concerned with the strip approximation, the attempt to achieve a dynamical bootstrap based on Regge poles. The duality notions which form the topic of Chapter 12 are grounded in the ideas of Regge poles, stressing in particular the approximate equivalence of descriptions in terms of direct-channel effects and exchanged Regge poles. This involves the concept of families of Regge trajectories—the leading trajectory (e.g. ρ–f–g) and its daughter trajectories. This idea is discussed further in connection with the resonance spectrum in Chapter 13.

Appendix D

Current Algebra and PCAC

(a) Weak Interaction Currents

In this appendix we briefly review those basic ideas of current algebra and PCAC used in the application to $\pi\pi$ scattering discussed in Chapter 14. However, before turning to these specific topics, it is useful to recall some generalities about weak interaction currents, since this whole area of physics grew out of insights into the structure of weak interactions.

In the current–current theory of weak interactions,‡ the weak interaction Lagrangian is written as the product of the total weak interaction current $C^\lambda(x)$ and its conjugate. More precisely

$$\mathscr{L}_w(x) = \frac{G}{\sqrt{2}} C_\lambda(x) C^{\lambda\dagger}(x) + h.c,$$

where G is the Fermi coupling constant. The current $C^\lambda(x)$ is the sum of the hadronic weak current $J^\lambda(x)$ and the leptonic weak current $L^\lambda(x)$, i.e.

$$C^\lambda(x) = J^\lambda(x) + L^\lambda(x),$$

and, for example, the explicit structure of $L^\lambda(x)$ is displayed by§

$$L^\lambda(x) = i \sum_{l=e,\mu} \bar{\nu}_l(x) \gamma^\lambda (1+\gamma^5) l(x). \tag{D.1}$$

‡ For a simple introduction to the topics we shall discuss see Martin (1967); an exhaustive treatment is given by Marshak, Riazuddin and Ryan (1969).

§ We only explicitly discuss the charged current. There will be analogous terms to represent the neutral current.

The hadronic current contains both $\Delta Y = 0$, and 1 components, and, in the modification due to Cabbibo, $J^\lambda(x)$ is written as

$$J^\lambda(x) = \cos\theta_c J^{\lambda,0}(x) + \sin\theta_c J^{\lambda,1}(x),$$

where experimentally the Cabbibo angle $\theta_c \approx 0{\cdot}24$ (Ebel *et al.*, 1971). However, since we shall be concerned only with the hypercharge-conserving current we will drop the superscript and just write this as $J^\lambda(x)$.

The hadronic current has a part $V^\lambda(x)$ which transforms like a vector, and a part $A^\lambda(x)$ which transforms like an axial vector, and it is of particular interest to consider the form of the matrix elements of these currents between single-nucleon states, as they are the matrix elements occurring in the basic nuclear β-decay,

$$n(p) \to p(p') + e^- + \bar{\nu}_e.$$

If we assume that the G-parity of V^λ and A^λ is $+1$ and -1, respectively, then Lorentz invariance enables us to write

$$(2\pi)^3\left(\frac{p^{0\prime}p^0}{m^2}\right)^{\frac{1}{2}}\langle p|V^\lambda(0)|n\rangle = i\bar{u}(p')[g_V(q^2)\gamma^\lambda + f_V(q^2)\sigma^{\lambda\mu}q_\mu]u(p), \qquad \text{(D.2)}$$

$$(2\pi)^3\left(\frac{p^{0\prime}p^0}{m^2}\right)^{\frac{1}{2}}\langle p|A^\lambda(0)|n\rangle = i\bar{u}(p')[g_A(q^2)\gamma^\lambda\gamma^5 - if_A(q^2)q^\lambda\gamma^5]u(p), \qquad \text{(D.3)}$$

where $q^\lambda = (p - p')^\lambda$, and g_V, f_V, g_A and f_A are form factors.

The conserved vector current (CVC) hypothesis is to identify V^λ and $V^{\lambda\dagger}$ with the $(1+i2)$ and $(1-i2)$ components of the divergenceless isospin current of strong interactions. We then have $\partial_\lambda V^\lambda = 0$, and this condition can be shown to imply (Ward, 1950) that there is no renormalization of g_V i.e. $g_V(0) = 1$. The contrasting lack of a similar condition for A^λ is most simply demonstrated from the existence of π-decay (Taylor, 1958). Lorentz invariance enables us to write the matrix element for π^--decay as

$$(2\pi)^{\frac{3}{2}}(2p^0)^{\frac{1}{2}}\langle 0|A^\lambda(0)|\pi^-(p)\rangle = if_\pi p^\lambda, \qquad \text{(D.4)}$$

where f_π is a constant, related to the decay width by

$$\Gamma(\pi \to \mu\nu) = \frac{G^2 f_\pi^2}{8\pi}\frac{m_\mu^2}{m_\pi^3}(m_\pi^2 - m_\mu^2)^2,$$

which gives

$$f_\pi = 128{\cdot}4\ \text{MeV} = 0{\cdot}920\ m_{\pi^-}. \qquad \text{(D.5)}$$

Applying translational invariance to (D.4) we have

$$\langle 0|A^\lambda(x)|\pi^-(p)\rangle = e^{ipx}\langle 0|A^\lambda(0)|\pi^-(p)\rangle,$$

and hence

$$\langle 0|\partial_\lambda A^\lambda(x)|\pi^-(p)\rangle = \frac{f_\pi m_\pi^2 \, e^{ipx}}{(2p^0)^{\frac{1}{2}}(2\pi)^{\frac{3}{2}}} \neq 0.$$

Thus g_A has renormalization effects, and in fact we know from data on nuclear β-decays that

$$g_A = (1{\cdot}23 \pm 0{\cdot}01)g_V.$$

However, the fact that g_A/g_V does not differ very much from unity suggests that the divergence of A^λ while not zero, may still be relatively small, and we shall see below that this notion may be made more precise by the hypothesis of a "partially" conserved axial current, i.e. a current which is conserved in the limit of zero pion mass.

(b) Current Commutators

From the CVC hypothesis the three components of $V^{\lambda,\alpha}(x)$, $(\alpha = 1, 2, 3)$, are divergenceless, and the generators of isospin rotations are given by

$$Q_\alpha(x^0) = \int d^3x \, V^{0,\alpha}(x), \qquad (\alpha = 1, 2, 3),$$

and satisfy the equal-time commutation relations

$$[Q_\alpha(x^0), Q_\beta(x^0)] = i\varepsilon_{\alpha\beta\gamma}Q_\gamma(x^0). \tag{D.6}$$

Analogous charges may be defined for the axial currents, i.e.

$$Q_\alpha^5(x^0) = \int d^3x \, A^{0,\alpha}(x), \qquad (\alpha = 1, 2, 3),$$

and these obey

$$[Q_\alpha(x^0), Q_\beta^5(x^0)] = i\varepsilon_{\alpha\beta\gamma}Q_\gamma^5(x^0). \tag{D.7}$$

The set of commutators (D.6) and (D.7), may now be closed (i.e. form an *algebra*) by postulating the result

$$[Q_\alpha^5(x^0), Q_\beta^5(x^0)] = i\varepsilon_{\alpha\beta\gamma}Q_\gamma(x^0), \tag{D.8}$$

which can be motivated by appealing to models in which the observed structure of the leptonic current also extends to the "bare" hadronic currents with the same "bare" coupling constant (e.g. the current quark model). The set of commutators (D.6)–(D.8) then constitutes the algebra of the symmetry SU(2) $\otimes$ SU(2), the role of the $[Q^5, Q^5]$ commutator being to set a scale for the axial current, as exemplified in the famous Adler–Weisberger relation (Adler, 1965c; Weisberger, 1965). The *symmetry* is not exact because the pion mass is not zero, and the axial current is not conserved, but it is assumed that the commutators still hold, i.e. that the *algebra* is still valid. These ideas can be extended to the larger symmetry of $SU(3) \otimes SU(3)$.

The commutators (D.6)–(D.8) express the idea of $SU(2) \otimes SU(2)$ as a broken symmetry of hadronic interactions. We shall also require one further commutator which incorporates assumptions on how the symmetry is broken. This is the commutator between the axial charge and its derivative, which is assumed to form an isotopic singlet, i.e.

$$\delta(x^0 - y^0)\left[Q^5_\alpha(x^0), \frac{\partial Q^5_\beta(y^0)}{\partial y^0}\right] = \left(\frac{f_\pi m_\pi^2}{\sqrt{2}}\right)\sigma(x^0)\, \delta_{\alpha\beta}\, \delta(x^0 - y^0), \qquad \text{(D.9)}$$

and which we shall refer to as the "σ-term". This commutator cannot be deduced from the others we have considered because it involves time derivatives, so that commutators with the Hamiltonian (including the symmetry breaking part) appear. In fact we shall be dealing with terms of the form

$$\int dy^0 \left\langle f \left| \left[Q^5_\alpha(y^0), \frac{\partial Q^5_\beta(y^0)}{\partial y^0}\right] \right| i \right\rangle$$

$$= \int dy^0 \left\langle f \left| \left[Q^5_\beta(y^0), \frac{\partial Q^5_\alpha(y^0)}{\partial y^0}\right] + \frac{\partial}{\partial y^0}[Q^5_\alpha(y^0), Q^5_\beta(y^0)] \right| i \right\rangle.$$

Since the second term vanishes from (D.8) and the conservation of the vector current, the above is necessarily symmetric in α and β and will in general be a mixture of $I = 0$ and 2 components. By (D.9) we have required it to be pure $I = 0$. This is to make an assumption about the symmetry breaking, the assumption having been originally suggested by the σ-model (Gell-Mann and Levy, 1960). We discuss this commutator in Chapter 14.

(c) PCAC Hypothesis

In order to use the above results to make predictions about pion–pion scattering, or any other strong interaction process involving pions, it is necessary to relate matrix elements of the axial current to those of the corresponding pionic processes. This is achieved by the use of the partially conserved axial current (PCAC) hypothesis alluded to above, and which we will now discuss in more detail. We will adopt the field-theoretic approach to PCAC, which is a convenient one for deriving the low-energy theorems which are required. For a fuller discussion of this approach, and its relation to other definitions of PCAC we refer to the previously cited references (Martin, 1967; Marshak, Riazuddin and Ryan, 1969), and the excellent review by Coleman (1968).

Consider the isovector, pseudo-scalar field ϕ^α defined in terms of the observable A^λ by

$$\partial_\lambda A^{\lambda\alpha}(x) \equiv \frac{m_\pi^2 f_\pi}{\sqrt{2}} \phi^\alpha(x). \tag{D.10}$$

By forming the matrix element $\langle 0|\phi^\alpha|\pi^\beta, p\rangle$, and using the definition of f_π (D.4), we see immediately that ϕ^α is correctly normalized to serve as an interpolating field, in the reduction formalism, for the creation of a pion with isospin index α. However, (Borchers, 1960) one can devise an infinite number of alternatives to (D.10) which also produce correctly normalized states on-shell. The particular choice of (D.10) defines a specific off-shell dependence of the matrix elements of the pion field in terms of the matrix elements of the divergence of the axial current $A^\lambda(x)$. The PCAC hypothesis asserts that the matrix elements of ϕ^α (i.e. of $\partial_\lambda A^{\lambda,\alpha}$) are smoothly, or better (following Coleman) *normally* varying with the extrapolated off-shell pion mass. By "normally" varying we mean that the scale of variation is given by the nearest singularities. If these are located at a distance m_3^2, then in extrapolating an amplitude from zero mass to the physical pion mass, one should only make an error of order $(m_\pi/m_3)^2 \equiv m_3^{-2}$. The usefulness of this procedure depends critically on the lightness of the pion, which guarantees that m_3^{-2} is in general reasonably small. When implementing the necessary extrapolation, nearby singularities should, as far as possible, be explicitly taken into account.

As an illustration of the usefulness of the PCAC hypothesis we consider the Goldberger–Treiman relation (Goldberger and Treiman, 1958) which is obtained by taking the matrix elements of the pion field between single-nucleon states. It is instructive to recall the well-known derivation of the result, since it clearly illustrates the two-step way in which the analogous $\pi\pi$ results are derived. The steps are: firstly to derive formally exact, but physically empty, off-mass shell results which are a consequence of our definition of ϕ; and secondly, to convert these results to physically useful ones by exploiting the PCAC normal variation assumption, referred to above. It is this second step which, as discussed in Chapter 14, demands most attention in the more intricate case of $\pi\pi$ scattering, since it is here that the possibility of ambiguities arises.

Rewriting (D.10) for the charge current $A^{\lambda+} = A^{\lambda 1} + iA^{\lambda 2}$ gives

$$\partial_\lambda A^{\lambda+}(x) = f_\pi m_\pi^2 \phi^+(x), \tag{D.11}$$

where ϕ^+ creates a π^+. Taking the matrix element of (D.11) between states of one neutron and one proton gives

$$\langle p|\partial_\lambda A^{\lambda+}(x)|n\rangle - f_\pi m_\pi^2 \langle p|\phi^+(x)|n\rangle. \tag{D.12}$$

The r.h.s. of this equation may be written

$$\frac{f_\pi m_\pi^2}{(2\pi)^3}\left(\frac{m^2}{p^{0\prime}p^0}\right)^{\frac{1}{2}}\frac{i\sqrt{2}g_{\pi NN}}{q^2+m_\pi^2}K_{\pi NN}(q^2)\bar{u}(p')\gamma^5 u(p), \tag{D.13}$$

where $g_{\pi NN}$ is the πNN coupling constant ($g_{\pi NN}^2/4\pi \approx 14{\cdot}6$), and $K_{\pi NN}(q^2)$ is the pionic nucleon form factor ($K_{\pi NN}(q^2=-m_\pi^2)=1$). The l.h.s. of (D.12) may be evaluated directly from equation (D.3). After using the Dirac equation we have

$$\langle p|\partial_\lambda A^{\lambda+}(0)|n\rangle$$
$$=\frac{i}{(2\pi)^3}\left(\frac{m^2}{p^{0\prime}p^0}\right)^{\frac{1}{2}}[2mg_A(q^2)+q^2 f_A(q^2)]\bar{u}(p')\gamma^5 u(p). \tag{D.14}$$

Equating (D.13) and (D.14) at $q^2=0$ gives

$$f_\pi g_{\pi NN}K_{\pi NN}(0)=\sqrt{2}mg_A(0).$$

Finally, if we assume that $K_{\pi NN}$ is a slowly varying function of q^2 such that

$$K_{\pi NN}(0)\approx K_{\pi NN}(q^2=-m_\pi^2)=1,$$

we have the Goldberger–Treiman relation

$$f_\pi=\frac{\sqrt{2}mg_A}{g_{\pi NN}}. \tag{D.15}$$

Since the nearest singularity of $K_{\pi NN}(q^2)$ is at $q^2=-9m_\pi^2$ we expect the result to be valid to an accuracy of $m_\pi^2/9m_\pi^2\approx 11\%$. The accepted values of the two sides of (D.15) are

$$f_\pi = 128\ \text{MeV}; \qquad \frac{\sqrt{2}mg_A}{g_{\pi NN}}\approx 122\ \text{MeV},$$

and so the discrepancy is actually a little less than expected. Earlier estimates of the r.h.s. of (D.15) were considerably smaller because the accepted experimental value for g_A was lower, 1·18 instead of the new value of 1·23, and led to a discrepancy of $\approx 10\%$.

References

Abarbanel, H. D. and Goldberger, M. L. (1968). *Phys. Rev.* **165,** 1594.
Abbud, F., Lee, B. W. and Yang, C. N. (1967). *Phys. Rev. Lett.* **18,** 980.
Abrams, G. S. *et al.* (1970). *Phys. Rev. Lett.* **25,** 617.
Abrams, R. J. *et al.* (1967). *Phys. Rev. Lett.* **18,** 1209.
Abrams, R. J. *et al.* (1970). *Phys. Rev.* **D1,** 1917.
Ademollo, M. *et al.* (1967). *Phys. Rev. Lett.* **19,** 1402.
Ader, J. P., Bonnier, B. and Meyers, C. (1972). *Nucl. Phys.* **B45,** 554.
Ader, J. P., Bonnier, B. and Meyers, C. (1973). *Phys. Lett.* **46B,** 403.
Ader, J. P. *et al.* (1968). *Nuovo Cim.* **56A,** 952.
Adler, S. L. (1965a). *Phys. Rev.* **137,** B1022.
Adler, S. L. (1965b). *Phys. Rev.* **139,** B1638.
Adler, S. L. (1965c). *Phys. Rev.* **140,** B736.
Adler, S. L. and Dashen, R. F. (1968). "Current Algebras and Applications to Particle Physics", Benjamin, N.Y.
Adylov, G. T. *et al.* (1974). *Phys. Lett.* **51B,** 402.
Aguilar-Benitez, *et al.* (1973). *Phys. Rev. Lett.* **30,** 672.
Albrecht, W. *et al.* (1971). *Nucl. Phys.* **B27,** 615.
Alder, J. C. *et al.* (1972). *Nucl. Phys.* **B46,** 573.
Alexanian, M. and Wellner, M. (1965). *Phys. Rev.* **137B,** 155.
Almahed, S. and Lovelace, C. (1972). *Nucl. Phys.* **B40,** 157.
Alston-Garnjost, M. *et al.* (1971). *Phys. Lett.* **36B,** 152.
Altarelli, G. and Rubinstein, H. R. (1969). *Phys. Rev.* **183,** 1469.
Alvensleben, W. *et al.* (1971). *Phys. Rev. Lett.* **26,** 273.
Amati, D., Fubini, S. and Stanghellini, A. (1962). *Phys. Lett.* **1,** 29.
Amati, D., Leader, E. and Vitale, B. (1960a). *Nuovo Cim.* **17,** 68
Amati, D., Leader, E. and Vitale, B. (1960b). *Nuovo Cim.* **18,** 409.
Amati, D., Leader, E. and Vitale, B. (1960c). *Nuovo Cim.* **18,** 458.
Amati, D., Leader, E. and Vitale, B. (1963). *Phys. Rev.* **130,** 750.
Amati, D. and Vitale, B. (1955). *Nuovo Cim.* **2,** 719.

Anderson, J. C. *et al.* (1973). *Phys. Rev. Lett.* **31,** 562.
Anisovich, V. V. (1963). *Soviet Phys. JETP* **17,** 1072.
Anisovich, V. V. and Ansel'm, A. A. (1966). *Soviet Phys. Usp.* **9,** 117 (Russian Volume **88,** 287).
Anninos, P. *et al.* (1968). *Phys. Rev. Lett.* **20,** 402.
Apel, W. D. *et al.* (1972). *Phys. Lett.* **41B,** 542.
Apel, W. D. *et al.* (1975). Paper contributed to the Palermo Conference, June, 1975.
Arbab, F. and Donohue, J. T. (1970). *Phys. Rev.* **D1,** 217.
Arnowitt, R. (1969). Proceedings of the Conference on $\pi\pi$ and $K\pi$ Interactions, Argonne National Laboratory, p. 619.
Arnowitt, R., Friedman, M. H. and Nath, P. (1967). *Phys. Rev. Lett.* **19,** 812.
Arnowitt, R., Friedman, M. H. and Nath, P. (1968). *Phys. Rev.* **174,** 1999, 2008; **175,** 1802.
Ascoli, G. *et al.* (1970). *Phys. Rev. Lett.* **25,** 962.
Ascoli, G. *et al.* (1971). *Phys. Rev. Lett.* **26,** 929.
Ascoli, G. *et al.* (1973). *Phys. Rev.* **D7,** 669.
Atkinson, D. (1962). *Phys. Rev.* **128,** 1908.
Atkinson, D. (1968a). *Nucl. Phys.* **B7,** 375.
Atkinson, D. (1968b). *Nucl. Phys.* **B8,** 377.
Atkinson, D. (1969). *Nucl. Phys.* **B13,** 415.
Atkinson, D. (1970). *Acta phys. austriaca,* Supp. VII, 32.
Atkinson, D. (1971). *Springer Tracts in Modern Physics,* **57,** 248.
Atkinson, D. and Contogouris, A. P. (1965). *Nuovo Cim.* **39,** 1082.
Atkinson, D., Dietz, K. and Morgan, D. (1966). *Ann. Phys. (N.Y.)* **37,** 77.
Atkinson, D. and Morgan D. (1966). *Nuovo Cim.* **41,** 559.
Auberson, G. (1970). *Nuovo Cim.* **68A,** 281.
Auberson, G. *et al.* (1970). *Nuovo Cim.* **65A,** 743.
Auberson, G. and Epele, L. (1974). *Nuovo Cim.* **25A**, 453.
Auberson, G., Piguet, O. and Wanders, G. (1968). *Phys. Lett.* **28B,** 41.
Auberson, G. and Wanders, G. (1965). *Phys. Lett.* **15,** 61.
Auerbach, S. P., Pennington, M. R. and Rosenzweig, C. (1973). *Phys. Lett.* **45B,** 275.
Ausländer, V. L. *et al.* (1969). Preprint 248, Novosibirsk.
Ayres, D. S. *et al.* (1972). Proceedings of the XVIth International Conference on High Energy Physics, Chicago-Batavia (A.I.P.).
Ayres, D. S. *et al.* (1973a). "π-π Scattering—1973" (A.I.P.), p. 302.
Ayres. D. S. *et al.* (1973b). "π-π Scattering—1973" (A.I.P.), p. 284.

Baillon, P. *et al.* (1972). *Phys. Lett.* **38B,** 555.
Baker, G. A. (1965). *Adv. Theor. Phys.* **1,** 1.
Baker, S. L. (1972). Proceedings of the XVIth International Conference on High Energy Physics, Chicago-Batavia (A.I.P.).
Baker, S. L. *et al.* (1973). Imperial College London preprint IC/HEP/73/12.
Bakker, A. M. (1971). "Meson Resonances and Related Electromagnetic Phenomena", ed. by R. H. Dalitz and A. Zichichi, International Physics Series.
Balachandran, A. P. and Nuyts, J. (1968). *Phys. Rev.* **172,** 1821.
Balakin, V. E. *et al.* (1972). *Phys. Lett.* **41B,** 205.
Bali, N. F. and Chiu, S. Y. (1967). *Phys. Rev.* **153,** 1579.
Ball, J. S., Lee, P. S. and Shaw, G. L. (1973). *Phys. Rev.* **D7,** 2789.
Ballam, J. *et al.* (1974). *Nucl. Phys.* **B76,** 375.
Bander, M., Coulter, P. W. and Shaw, G. L. (1965). *Phys. Rev. Lett.* **14,** 230.

Bander, M. and Shaw, G. L. (1965). *Phys. Rev.* **139,** B956.
Barbarino, G. *et al.* (1972). *Nuovo Cim. Lett.* **3,** 689.
Barbaro-Galtieri, A. *et al.* (1973). Conference on $\pi\pi$ Scattering and Related Topics, Tallahassee (A.I.P.); see also Matison *et al.* 1974.
Barbiellini, G. *et al.* (1973). *Nuovo Cim. Lett.* **6,** 557.
Barbour, I. M. and Schult, R. L. (1967a). *Phys. Rev.* **155,** 1712.
Barbour, I. M. and Schult, R. L. (1967b). *Phys. Rev.* **164,** 1791.
Bardadin-Otwinowska, M. *et al.* (1974). *Nucl. Phys.* **B72,** 1.
Bardeen, W. and Lee, B. W. (1969). *Phys. Rev.* **177,** 2389.
Barger, V. and Phillips, R. J. N. (1971). *Nucl. Phys.* **B32,** 93.
Barnham, K. W. J. *et al.* (1973). *Phys. Rev.* **D7,** 1884.
Barrelet, F. (1972). *Nuovo Cim.* **8A,** 331.
Bartel, W. *et al.* (1968). *Phys. Lett.* **28B,** 148.
Bartoli, B. *et al.* (1972). *Phys. Rev.* **D6,** 2374.
Barton, G. and Rosen, S. P. (1962). *Phys. Rev. Lett.* **8,** 414.
Bartsch, J. *et al.* (1970). *Nucl. Phys.* **B22,** 109.
Bartsch, J. *et al.* (1972). *Nucl. Phys.* **B46,** 46.
Basdevant, J. L. (1970). Padé Approximants, *in* "Methods of Subnuclear Physics", Vol. IV, ed. M. Nikolic, Gordon and Breach.
Basdevant, J. L. (1972). *Fortschr. der Phys.* **20,** 283.
Basdevant, J. L. (1975). University of Paris report PAR-LPTHE 75–11.
Basdevant, J. L. and Hayot, F. (1971). *Phys. Lett.* **36B,** 375.
Basdevant, J. L. and Lee, B. W. (1969a). *Nucl. Phys.* **B13,** 182.
Basdevant, J. L. and Lee, B. W. (1969b). *Phys. Lett.* **29B,** 437.
Basdevant, J. L. and Lee, B. W. (1970). *Phys. Rev.* **D2,** 1680.
Basdevant, J. L. and Schomblond, C. (1973). *Phys. Lett.* **45B,** 48.
Basdevant, J. L. and Zinn-Justin, J. (1971). *Phys. Rev.* **D3,** 1865.
Basdevant, J. L., Bessis, D. and Zinn-Justin, J. (1968). *Phys. Lett.* **27B,** 230.
Basdevant, J. L., Bessis, D. and Zinn-Justin, J. (1969a). *Nuovo Cim.* **60A,** 185.
Basdevant, J. L. *et al.* (1969*b*). *Nuovo Cim.* **64A,** 585.
Basdevant, J. L., Le Guillou, J. C. and Navelet, H. (1972). *Nuovo Cim.* **7A,** 363.
Basdevant, J. L., Froggatt, C. D. and Petersen, J. L. (1972a). *Phys. Lett.* **41B,** 173.
Basdevant, J. L., Froggatt, C. D. and Petersen, J. L. (1972b). *Phys. Lett.* **41B,** 178.
Basdevant, J. L. *et al.* (1973). Submitted to IInd Aix-en-Provence International Conference on Elementary Particles, 1973, University of Paris preprint PAR-LPTHE 73–8.
Basdevant, J. L., Froggatt, C. D. and Petersen, J. L. (1974). *Nucl. Phys.* **B72,** 413.
Basile, P. *et al.* (1971). *Phys. Lett.* **36B,** 619.
Baton, J. P., Laurens, G. and Reignier, J. (1967a). *Nucl. Phys.* **B3,** 349.
Baton, J. P., Laurens, G. and Reignier, J. (1967b). *Phys. Lett.* **25B,** 419.
Baton, J. P., Laurens, G. and Reignier, J. (1970a). *Phys. Lett.* **33B,** 525.
Baton, J. P., Laurens, G. and Reignier, J. (1970b). *Phys. Lett.* **33B,** 528.
Baubillier, M. *et al.* (1972). Proceedings of the XVIth International Conference on High Energy Physics, Chicago-Batavia (A.I.P.) (see Durosoy *et al.* (1973)).
Bauer, T. H. (1970). Cornell University Thesis, and *Phys. Rev. Lett.* **25,** 485.
Bég, M.A.B. (1962). *Phys. Rev. Lett.* **9,** 67.
Bég, M. A. B. and De Celles, P. C. (1962). *Phys. Rev. Lett.* **8,** 46.
Beier, E. W. *et al.* (1972). *Phys. Rev. Lett.* **29,** 511.
Beier, E. W. *et al.* (1973). *Phys. Rev. Lett.* **30,** 399.
Belavin, A. A. and Narodetsky, I. M. (1968). *Phys. Lett.* **26B,** 668.

Bell, J. S. and Steinberger, J. (1965). Proceedings of the Oxford International Conference on Elementary Particles, p. 198.
Benaksas, D. *et al.* (1972). *Phys. Lett.* **39B,** 289.
Bensinger, J. R. *et al.* (1971). *Phys. Lett.* **36B,** 134.
Berends, F. A., Donnachie, A. and Oades, G. C. (1968). *Phys. Rev.* **171,** 1457.
Berger, E. L. (1971). "Phenomenology in Particle Physics, 1971", Proceedings of Conference held at Caltech, ed. C. B. Chiu, G. C. Fox and A. J. G. Hey, p. 83.
Bernardini, M. *et al.* (1973). Paper submitted to the Bonn Conference.
Bessis, D. and Pusterla, M. (1967). *Phys. Lett.* **25B,** 279.
Bessis, D. and Pusterla, M. (1968). *Nuovo Cim.* **54A,** 243.
Bettini, A. *et al.* (1971). *Nuovo Cim.* **1A,** 333.
Beusch, W. (1970). "Experimental Meson Spectroscopy", Columbia University Press, N.Y., p. 185.
Bialas, A. and Zalewski, K. (1968). *Nucl. Phys.* **B6,** 465.
Bingham, H. H. *et al.* (1972a). *Nucl. Phys.* **B41,** 1.
Bingham, H. H. *et al.* (1972b). *Phys. Lett.* **41B,** 635.
Binnie, D. M. *et al.* (1973). *Phys. Rev. Lett.* **31,** 1534.
Birge, R. W. *et al.* (1965). *Phys. Rev.* **139,** B1600.
Biswas, N. *et al.* (1968). *Phys. Lett.* **27B,** 513.
Bizzarri, R. *et al.* (1969). *Nucl. Phys.* **B14,** 169.
Bizzarri, R. (1972). "Symposium on Nucleon–Antinucleon Annihilations, Chexbres, Switzerland, 1972", ed. by L. Montanet, CERN Publication 72–10.
Bizzarri, R. (1973). University of Rome preprint.
Bjorken, J. D. (1960). *Phys. Rev. Lett.* **4,** 473.
Blankenbecler, R. *et al.* (1960). *Ann. Phys. (N.Y.)* **10,** 62.
Blum, W. *et al.* (1975). Paper contributed to the Palermo Conference, June, 1975.
Bogoliubov, N. N., Medvedev, B. V. and Polivanov, M. K. (1958). "Problems in the Theory of Dispersion Relations", Fizmatgiz, Moscow, 1958.
Böhm, M. Joos, H. and Krammer, M. (1974). *Nucl. Phys.* **B69,** 345.
Bolotov, V. N. *et al.* (1974). *Nucl. Phys.* **73,** 365.
Bonnier, B. (1974). *Nucl. Phys.* **B75,** 333.
Bonnier, B. (1975). CERN preprint. TH2008.
Bonnier, B. and Gauron, P. (1970). *Nucl. Phys.* **B21,** 465.
Bonnier, B. and Gauron, P. (1972). *Nucl. Phys.* **B36,** 11.
Bonnier, B. and Gauron, P. (1973). *Nucl. Phys.* **B52,** 506.
Bonnier, B. and Procureur, J. (1973). *Nucl. Phys.* **B61,** 155.
Bonnier, B. and Vinh Mau, R. (1968). *Phys. Rev.* **165,** 1923.
Borchers, H. (1960). *Nuovo Cim.* **15,** 784.
Bourquin, M. *et al.* (1971). *Phys. Lett.* **36B,** 615.
Bowcock, J. E. and Kannelopoulos, T. (1968). *Nucl. Phys.* **B3,** 417.
Bowcock, J. E. and John, G. (1969). *Nucl. Phys.* **B11,** 659.
Bowen, D. *et al.* (1972). *Phys. Rev. Lett.* **29,** 890.
Bowler, M. G. (1973). CERN preprint/DPhII/PHYS 73–40.
Bowler, M. G. and Game, M. A. V. (1974). Oxford University preprint 38/74.
Bramon, A. (1973). *Nuovo. Cim. Lett.* **8,** 639.
Bramon, A. and Greco, M. (1973). *Nuovo Cim.* **14A,** 323.
Brandenburg, G. W. *et al.* (1974). Contributed paper to the XVIIth International Conference on High Energy Physics, London.
Bremermann, H. R., Oehme, R. and Taylor, J. G. (1958). *Phys. Rev.* **109,** 2178.
Brockway, D. V. (1970). University of Illinois COO-1195-197 (unpublished).

Brodsky, S. J. (1971). Proceedings of the 5th International Symposium on Electron and Photon Interactions at High Energy, Cornell University.
Brodsky, S. J., Kinoshita, T. and Terazawa, H. (1970). *Phys. Rev. Lett.* **25,** 972.
Bros, J., Epstein, H. and Glaser, V. (1965). *Communs. Math. Phys.* **1,** 240.
Brown, G. E. (1970). *Com. Nucl. Particle Physics* **4,** 140.
Brown, G. E. and Durso, J. W. (1971). *Phys. Lett.* **35B,** 120.
Brown, L. M. and Singer, P. (1964). *Phys. Rev.* **133,** B812.
Brown, S. G. and West, G. B. (1967). *Phys. Rev. Lett.* **19,** 812.
Brown, S. G., and West, G. B. (1968). *Phys. Rev.* **168,** 1605; **174,** 1777, 1786.
Bugg, D. V. (1974). *Phys. Lett.* **52B,** 102.
Bugg, D. V., *et al.* (1971). *Nucl. Phys.* **B26,** 445.
Bulos, F. *et al.* (1971a). *Phys. Rev. Lett.* **26,** 149.
Bulos, F. *et al.* (1971b). *Phys. Rev. Lett.* **26,** 1453.
Bulos, F. *et al.* (1971c). *Phys. Rev. Lett.* **26,** 1457.
Bunaytov, S. A., Gulkanian, H. R. and Kurbatov, V. S. (1973). Dubna preprint.
Bunyatov, S. A. (1974). Contribution to the IVth International Conference on Meson Spectroscopy, Boston, Massachusetts.
Butterworth, I. (1973). Proceedings of the IInd Conf. Int. sur les Particules Elementaires, Aix-en-Provence, p. 173.

Cabibbo, N. and Maksymowicz, A. (1965). *Phys. Rev.* **137,** B438 (Errata **168,** 1926 (1968)).
Canning, G. P. (1969). *Nucl. Phys.* **B14,** 437.
Carlson, C. E. and Tung, W. K. (1972). *Phys. Rev.* **D4,** 2873.
Carmony, D. D. *et al.* (1971). *Phys. Rev. Lett.* **27,** 1160.
Carroll, A. S. *et al.* (1974). *Phys. Rev. Lett.* **32,** 247.
Carroll, J. T. *et al.* (1972). *Phys. Rev. Lett.* **28,** 318.
Caser, M., Piguet, C. and Vermeulen, J. L. (1969). *Nucl. Phys.* **B14,** 119.
Cashmore, R. J. (1973). Proceedings of the 14th Scottish Universities Summer School in Physics, p. 611.
Castillejo, L., Dalitz, R. H. and Dyson, F. J. (1956). *Phys. Rev.* **101,** 453.
Chan, L. H. *et al.* (1966). *Phys. Rev.* **141,** 1298; **147,** 1194.
Chan, L., Hagopian, V. and Williams, P. K. (1970). *Phys. Rev.* **D2,** 583.
Chan Hong-Mo (1972). RHEL preprints RPP/T/21, and RL–73–062.
Chang, L. N. (1967). *Phys. Rev.* **162,** 1497.
Charap, J. M. and Fubini, S. P. (1959). *Nuovo Cim.* **14,** 540.
Charap, J. M. and Fubini, S. P. (1960). *Nuovo Cim.* **15,** 73.
Charap, J. M. and Tausner, M. J. (1960). *Nuovo Cim.* **18,** 316.
Charlesworth, J. A. *et al.* (1973). "$\pi\pi$ Scattering—1973" (A.I.P.), p. 317.
Chemtob, M. and Riska, D. O. (1971). *Phys. Lett.* **35B,** 115.
Chemtob, M., Durso, J. W. and Riska, D. O. (1972). *Nucl. Phys.* **B38,** 141.
Chew, G. F. (1966). "The Analytic S-matrix", W. A. Benjamin, New York.
Chew, G. F. and Frautschi, S. C. (1961a). *Phys. Rev.* **124,** 264.
Chew, G. F. and Frautschi, S. C. (1961b). *Phys. Rev. Lett.* **7,** 394.
Chew, G. F. *et al.* (1957a). *Phys. Rev.* **106,** 1337.
Chew, G. F. *et al.* (1957b). *Phys. Rev.* **106,** 1345.
Chew, G. F. and Jones, C. E. (1964). *Phys. Rev.* **135,** B208.
Chew, G. F. and Low, F. E. (1959). *Phys. Rev.* **113,** 1640.
Chew, G. F. and Mandelstam, S. (1960). *Phys. Rev.* **119,** 467.
Chew, G. F. and Mandelstam, S. (1961). *Nuovo Cim.* **19,** 752.

Chew, G. F. Mandelstam, S. and Noyes, H. P. (1960). *Phys. Rev.* **119,** 478.
Chikovani, G. *et al.* (1966). *Phys. Lett.* **22,** 233.
Chisholm, J. S. R. (1963). *J. math. Phys.* **4,** 1506.
Chiu, C. B. and Kotanski, A. (1968). *Nucl. Phys.* **B7,** 615.
Chiu, C. B. Phillips, R. J. N., Riddell, R. J. and Rarita, W. (1968). *Phys. Rev.* **165,** 1615.
Cho, C. F. and Sakurai, J. J. (1969). *Phys. Lett.* **30B,** 119.
Cho, C. F. and Sakurai, J. J. (1970). *Phys. Rev.* **D2,** 517.
Chounet, L. M., Gaillard, J. M. and Gaillard, M. K. (1972). *Phys. Reports* **4C,** 199.
Christenson, J. H. *et al.* (1964). *Phys. Rev. Lett.* **13,** 138.
Cini, M. and Fubini, S. (1960). *Ann. Phys. (N.Y.)* **10,** 352.
Ciulli, S. and Fischer, J. (1961). *Nucl. Phys.* **24,** 456.
Ciulli, S. (1969a). *Nuovo Cim.* **61A,** 787.
Ciulli, S. (1969b). *Nuovo Cim.* **62A,** 301.
Cline, D. *et al.* (1968). *Phys. Rev. Lett.* **21,** 1268.
Cline, D. B., Braun, K. J. and Scherev, V. (1970). *Nucl. Phys.* **B18,** 77.
Cnops, A. M. *et al.* (1968). *Phys. Lett.* **27B,** 113.
Cohen, D. *et al.* (1973). *Phys. Rev.* **D7,** 661.
Cohen, K. J. *et al.* (1972). Proceedings of the 3rd Philadelphia Conference on Experimental Meson Spectroscopy, Philadelphia (A.I.P.), p. 242.
Cohen-Tannoudji *et al.* (1971). *Phys. Rev. Lett.* **26,** 112.
Cohen-Tannoudji, G., Morel, A. and Navelet, H. (1968). *Ann. Phys. (N.Y.)* **46,** 239.
Coleman, S. (1968). *In*, "Hadrons and their Interactions", editor A. Zichichi, Academic Press.
Colglazier, E. W. and Rosner, J. L. (1971). *Nucl. Phys.* **B27,** 349.
Collins, P. D. B. (1966). *Phys. Rev.* **142,** 1163.
Collins, P. D. B. (1971). *Phys. Reports* **1C,** 105.
Collins, P. D. B. and Johnson, R. C. (1969a). *Phys. Rev.* **177,** 2472.
Collins, P. D. B. and Johnson, R. C. (1969b). *Phys. Rev.* **182,** 1755.
Collins, P. D. B. and Johnson, R. C. (1969c). *Phys. Rev.* **185,** 2020.
Collins, P. D. B. and Squires, E. J. (1968). Regge Poles in Particle Physics, *Springer Tracts in Modern Physics* **45,** Springer-Verlag, Berlin.
Collins, P. D. B. and Teplitz, V. L. (1965). *Phys. Rev.* **140,** B663.
Colton, E. *et al.* (1971). *Phys. Rev.* **D3,** 2028.
Common, A. K. (1968). *Nuovo Cim.* **53A,** 946.
Common, A. K. (1969). *Nuovo Cim.* **63A,** 451.
Common, A. K. (1970). *Nuovo Cim.* **69A,** 115.
Common, A. K. and Pidcock, M. K. (1972). *Nucl. Phys.* **B42,** 194.
Common, A. K., Hodgkinson, D. and Pidcock, M. K. (1974). *Nucl. Phys.* **B74,** 429.
Common, A. K. and Wit, R. (1971). *Nuovo Cim.* **3A,** 179.
Conforto, G. (1973). Private communication.
Conforto, G. (1974). University of Rome preprint.
Contogouris, A. P., Tran Thanh Van, J. and Lubatti, H. J. (1967). *Phys. Rev. Lett.* **19,** 1352.
Conway, P. D. (1970). *Nuovo Cim.* **67A,** 497.
Copley, L. A., Karl, G. and Obryk, E. (1969). *Phys. Lett.* **29B,** 117.
Copley, L. A. and Masson, D. (1967). *Phys. Rev.* **164,** 2059.
Cords, D. *et al.* (1974). Contributed paper, Proceedings of the XVIIth International Conference on High Energy Physics, London.

Cottingham, W. N. and Vinh Mau, R. (1963). *Phys. Rev.* **130,** 735.
Cottingham, W. N. *et al.* (1973a). *Phys. Rev.* **D8,** 800.
Cottingham, W. N. *et al.* (1973b). *Phys. Lett.* **44B**, 1.
Crennell, D. J. *et al.* (1971). *Phys. Rev. Lett.* **27,** 1674.
Curry, P. *et al.* (1971). *Phys. Rev.* **D3,** 1233.
Cutkosky, R. and Deo, B.B. (1968). *Phys. Rev.* **174,** 1859.

Dalitz, R. H. (1953). *Phil. Mag.* **44,** 1068.
Dalitz, R. H. (1954). *Phys. Rev.* **94,** 1046.
Dalitz, R. H. (1966). Proceedings of the XIIIth International Conference on High Energy Physics, Berkeley, (University of California Press, 1967).
Dalitz, R. H. (1967). Lectures at the 2nd Hawaii Topical Conference on Particle Physics (University of Hawaii Press, Honolulu, Hawaii, 1967).
Das, T., Mathur, V. and Okubo, S. (1967). *Phys. Rev. Lett.* **19,** 900.
Dashen, R. and Kane, G. L. (1974). *Phys. Rev.* **D11,** 136.
Davier, M. *et al.* (1969). SLAC-PUB-666 (unpublished).
Davier, M. *et al.* (1973). *Nucl. Phys.* **B58,** 31.
Davies, A. T. (1970). *Nucl. Phys.* **B21,** 359.
Day, T. B., Snow, G. A. and Sucher, J. (1960). *Phys. Rev.* **118,** 864.
Deck, R. T. (1964). *Phys. Rev. Lett.* **13,** 169.
Deinet, W. *et al.* (1969). *Phys. Lett.* **30B,** 359.
Devenish, R. and Lyth, D. H. (1972). *Phys. Rev.* **D5,** 47.
Devons, S. *et al.* (1971). *Phys. Rev. Lett.* **27,** 1614.
Devons, S. *et al.* (1973). *Phys. Lett.* **47B,** 271.
Diaz, J. *et al.* (1970). *Nucl. Phys.* **B16,** 239.
Diebold, R. (1972). Proceedings of the XVIth International Conference of High Energy Physics, Chicago-Batavia (A.I.P.).
Dietz, K. and Domokos, G. (1964). *Phys. Lett.* **11,** 167.
Dilley, J. (1971). *Nucl. Phys.* **B25,** 227.
Diu, B. and Le Bellac, M. (1968). *Nuovo Cim.* **53A,** 158.
Dolen, R., Horn, D. and Schmid, C. (1968). *Phys. Rev. Lett.* **19,** 402.
Donaldson, G. *et al.* (1974). *Phys. Rev.* **D9,** 2960.
Doncel, M. G. *et al.* (1972). *Nucl. Phys.* **B38,** 477.
Donnachie, A. and Gabathuler, E. (1970). Vector Meson Production and Omega–Rho Interference: Proceedings of Daresbury Study Weekend June 1970 —DNPL/R7.
Donnachie, A., Hamilton, J. and Lea, A. T. (1964). *Phys. Rev.* **135,** B515.
Donnachie, A. and Hamilton, J. (1965). *Ann. Phys.* (*N.Y.*) **31,** 410.
Donnachie, A., Heinz, R. M. and Lovelace, C. (1966). *Phys. Lett.* **22,** 322.
Donnachie, A. Kirsopp, R. G. and Lovelace, C. (1968). *Phys. Lett.* **26B,** 161.
Donnachie, A. and Thomas, P. R. (1973). *Nuovo Cim. Lett.* **7,** 285.
Donnachie, A. and Thomas, P. R. (1974). Private communication. Also P. R. Thomas, Ph.D. thesis, University of Manchester, 1974.
Dowell, J. D. *et al.* (1974). Contributed paper, Proceedings of the XVIIth International Conference on High Energy Physics, London.
Durand, L. and Chiu, Y. T. (1965). *Phys. Rev.* **139,** B646.
Durusoy, N. B. *et al.* (1973). *Phys. Lett.* **45B,** 517.
Dürr, H. P. and Pilkuhn, H. (1965). *Nuovo Cim.* **40,** 899.

Ebel, G. *et al.* (1971). *Nucl. Phys.* **B33,** 317.
Ecker, G. and Honnerkamp, J. (1973). *Nucl. Phys.* **B52,** 211.
Eden, R. J. (1967). "High Energy Collisions of Elementary Particles", Cambridge University Press.
Eden, R. J. *et al.* (1966). "The Analytic S-Matrix", Cambridge University Press.
Eguchi, T. and Igi, K. (1971). *Phys. Rev. Lett.* **27,** 1319.
Eguchi, T. and Igi, K. (1972). *Phys. Lett.* **40B,** 345.
Eguchi, T., Shimada, T. and Fukugita, M. (1974). *Nucl. Phys.* **B74,** 102.
Ehrlich, R. D. *et al.* (1972). *Phys. Rev. Lett.* **28,** 1147.
Eisenhandler, E. *et al.* (1973). *Phys. Lett.* **47B,** 536 and 541.
Ely, R. P. *et al.* (1969). *Phys. Rev.* **180,** 1319.
Elvekjaer, F. (1972). *Nucl. Phys.* **B43,** 445.
Elvekjaer, F. and Nielsen, H. (1971). RHEL Preprint RPP/C/22 (unpublished).
Elvekjaer, F. and Pietarinen, E. (1972). *Nucl. Phys.* **B45,** 621.
Epstein, G. N. and McKellar, B. H. J. (1972). *Nuovo Cim. Lett.* **5,** 807.
Erdélyi, A. *et al.* (1953). "Higher Transcendental Functions" (Batemann Manuscript Project), McGraw-Hill.
Estabrooks, P. and Martin, A. D. (1972a). *Phys. Lett.* **41B,** 350.
Estabrooks, P. and Martin, A. D. (1972b). *Phys. Lett.* **42B,** 229.
Estabrooks, P. and Martin, A. D. (1973a). "$\pi\pi$-scattering" (A.I.P.), p. 357.
Estabrooks, P. and Martin, A. D. (1973b). Proceedings of the Pion Exchange Meeting, Daresbury Study Weekend No. 6, DNPL/R.30.
Estabrooks, P. and Martin, A. D. (1974a). *Nucl. Phys.* **B79,** 301.
Estabrooks, P. and Martin, A. D. (1974b). *Phys. Lett.* **B53**, 253.
Estabrooks, P. *et al.* (1973). "$\pi\pi$-scattering" (A.I.P.), p. 37.
Estabrooks, P., Martin, A. D. and Michael, C. (1974a). *Nucl. Phys.* **B72,** 454.
Estabrooks, P. *et al.* (1974b). *Nucl. Phys.* **B81,** 70.
Extermann, P. *et al.* (1975). Paper submitted to the Palermo Conference, July 1975.

Faiman, D. and Hendry, A. W. (1968). *Phys. Rev.* **173,** 1720.
Faiman, D. and Hendry, A. W. (1969). *Phys. Rev.* **180,** 1572.
Faiman, D. and Plane, D. (1972). *Nucl. Phys.* **B50,** 379.
Fayyazuddin and Riazuddin (1966). *Phys. Rev.* **147,** 1071.
Felst, R. (1973). Desy report 73/56.
Feynman, R. P., Kislinger, M. and Ravndal, F. (1971). *Phys. Rev.* **D3,** 2706.
Finkelstein, J. (1969). *Phys. Rev. Lett.* **22,** 362.
Firestone, A. *et al.* (1971). *Phys. Rev. Lett.* **26,** 1460.
Firestone, A. *et al.* (1972). *Phys. Rev.* **D5,** 2188.
Flatté, S. M. *et al.* (1972). *Phys. Lett.* **38B,** 232.
Fong, D. (1968). Caltech Thesis (unpublished).
Ford, W. T. *et al.* (1972). *Phys. Lett.* **38B,** 335.
Fox, G. C. and Quigg, C. (1973). *A. Rev. nucl. Sci.* **23,** 219.
Fox, G. C. and Griss, M. L. (1974). *Nucl. Phys.* **B80,** 403.
Frampton, P. (1970). *Phys. Rev.* **D1,** 3141.
Frampton, P. (1974). "Dual Models", Benjamin, N.Y.
Froissart, M. (1961). *Phys. Rev.* **123,** 1053.
Frautschi, S. C. (1963). "Regge Poles and S-matrix Theory", Benjamin, New York.

Frautschi, S. C. and Walecka, J. D. (1960). *Phys. Rev.* **120,** 1486.
Frazer, W. R. (1961). *Phys. Rev.* **123,** 2180.
Frazer, W. R. and Fulco, J. R. (1959). *Phys. Rev. Lett.* **2,** 365.
Frazer, W. R. and Fulco, J. R. (1960a). *Phys. Rev.* **117,** 1603.
Frazer, W. R. and Fulco, J. R. (1960b). *Phys. Rev.* **119,** 1420.
French, B. (1968). Proceedings of the 14th International Conference on High Energy Physics, Vienna, 1968, p. 91.
Freund, P. G. O. (1968). *Phys. Rev. Lett.* **20,** 235.
Froggatt, C. D. and Morgan, D. (1969). *Phys. Rev.* **187,** 2044.
Froggatt, C. D. and Morgan, D. (1970). *Phys. Lett.* **33B,** 582.
Froggatt, C. D. and Morgan, D. (1972). *Phys. Lett.* **40B,** 655.
Froggatt, C. D. and Petersen, J. L. (1975). *Nucl. Phys.* **B91,** 454.
Frye, G. and Warnock, R. L. (1963). *Phys. Rev.* **130,** 478.
Fubini, S. and Furlan, G. (1965). *Physics* **1,** 229.
Fujii, Y. and Fukugita, M. (1974). *Nucl. Phys.* **B85,** 179.
Fulco, J. G., Shaw, G. L. and Wong, D. (1965). *Phys. Rev.* **137B,** 1242.
Furuichi, S. (1967). *Suppl. Prog. Theor. Phys.* **39,** 190.

Galster, S. *et al.* (1971). *Nucl. Phys.* **32B,** 221.
Gammel, J. L. and McDonald, F. A. (1966). *Phys. Rev.* **142,** 1245.
Garibotti, C. R. and Villani, M. (1969). *Nuovo Cim.* **61A,** 747.
Geffen, D. (1967). *Phys. Rev. Lett.* **19,** 770.
Gell-Mann, M. (1972). *Acta phys. austriaca Suppl.* IX, p. 733.
Gell-Mann, M. and Levy, M. (1960). *Nuovo Cim.* **16,** 705.
Gell-Mann, M. and Pais, A. (1955). *Phys. Rev.* **97,** 1387.
Gell-Mann, *et al.* (1964). *Phys. Rev.* **133,** B145.
Gell-Mann, M., Oakes, R. J. and Renner, B. (1968). *Phys. Rev.* **175,** 2195.
Gell-Mann, M. and Zachariasen, F. (1961). *Phys. Rev.* **124,** 953.
Gerstein, I. S. and Schnitzer, H. J. (1968). *Phys. Rev.* **170,** 1638.
Gilbert, W. (1957). *Phys. Rev.* **108,** 1078.
Gilman, F. J. (1974). Proceedings of the IVth International Conference on Experimental Meson Spectroscopy, Boston, Massachusetts, U.S.A.
Gilman, F. J. and Harari, H. (1968). *Phys. Rev.* **165,** 1803.
Gilman, F. J. Kugler, M. and Meshkov, S. (1973). *Phys. Lett.* **45B,** 481.
Gilman, F. J. Kugler, M. and Meshkov, S. (1974). *Phys. Rev.* **D9,** 715.
Glashow, S. (1961). *Phys. Rev. Lett.* **7,** 469.
Gleeson, A. M., Meggs, W. J. and Parkinson, M. (1970). *Phys. Rev. Lett.* **25,** 74.
Goble, R. L. and Rosner, H. L. (1972). *Phys. Rev.* **D5,** 2345.
Goebel, C. J. (1956). *Phys. Rev.* **103,** 258.
Goebel, C. J. (1958). *Phys. Rev. Lett.* **1,** 337.
Goebel, C. J., Blackmon, M. L. and Wali, K. C. (1969). *Phys. Rev.* **182,** 1487.
Goebel, C. J. and Shaw, G. (1968). *Phys. Lett.* **27B,** 291.
Goldberger, M. L. and Treiman, S. B. (1958). *Phys. Rev.* **110,** 1178.
Goldhaber, A. S. *et al.* (1969). *Phys. Lett.* **30B,** 249.
Goldhaber, G. (1967). *Phys. Rev. Lett.* **19,** 976.
Goldhaber, G. (1970). "Experimental Meson Spectroscopy", Columbia University Press, N.Y., p. 59.
Good, M. L. (1957). *Phys. Rev.* **106,** 591.
Gopal, G. P., Migneron, R. and Rothery, A. (1971). *Phys. Rev.* **D3,** 2262.

Goradia, Y., Lasinski, T., Tabak, M. and Smadja, G. (1974). Berkeley report LBL-3011.
Gore, B. F. (1972). *Phys. Rev.* **D6,** 666.
Gottfried, K. and Jackson, J. D. (1964). *Nuovo Cim.* **34,** 735.
Gounaris, G. J. (1971). *Phys. Rev.* **D4,** 2788.
Gounaris, G. J. and Sakurai, J. J. (1968). *Phys. Rev. Lett.* **21,** 244.
Gourdin, M. (1967). "Unitary Symmetry", North Holland.
Gourdin, M. (1974). *Phys. Reports* **11C,** 30.
Grassberger, P. (1972a). *Nucl. Phys.* **B42,** 641.
Grassberger, P. (1972b). Bonn preprint P12-125.
Grassberger, P. and Kuhnelt, H. (1973). *Nucl. Phys.* **B56,** 536.
Graves-Morris, P. R. (1971). *Nuovo Cim. Lett.* **2,** 1231.
Gray, L. *et al.* (1973). *Phys. Rev. Lett.* **30,** 1091.
Grayer, G. *et al.* (1972a). Proceedings of the 4th International Conference on High Energy Collisions, Oxford.
Grayer, G. *et al.* (1972b). *Nucl. Phys.* **B50,** 29.
Grayer, G. *et al.* (1972c). Proceedings of the XVIth International Conference on High Energy Physics, Chicago-Batavia (A.I.P.).
Grayer, G. *et al.* (1972d). *Phys. Lett.* **39B,** 563.
Grayer, G. *et al.* (1973). "$\pi\pi$ Scattering—1973" (A.I.P.), p. 117.
Grayer, G. *et al.* (1974). *Nucl. Phys.* **B75,** 189.
Greenberg, O. W. (1964). *Phys. Rev. Lett.* **13,** 598.
Gribov, V. N. (1958). *Nucl. Phys.* **5,** 653.
Griffith, R. W. (1968). *Phys. Rev.* **176,** 1705.
Gross, D. J. (1967). *Phys. Rev. Lett.* **19,** 1303.
Gunion, J. F., Brodsky, S. J. Blankenbecler, R. (1972). *Phys. Lett.* **39B,** 649.
Gursey, F. and Radicati, L. (1964). *Phys. Rev. Lett.* **13,** 299.
Gutay, L. J. Meiere, F. T. and Scharenguivel, J. H. (1969). *Phys. Rev. Lett.* **23,** 431.

Hamilton, J. (1970). Lectures given at the Heidelberg–Karlsruhe International Summer Institute in Theoretical Physics, Heidelberg.
Hamilton, J. and Spearman, T. D. (1961). *Ann. Phys. (N.Y.)* **12,** 172.
Hamilton, J. *et al.* (1962a). *Nuovo Cim.* **20,** 519.
Hamilton, J., Spearman, T. D. and Woolcock, W. S. (1962b). *Ann. Phys. (N.Y.)* **17,** 1.
Hamilton, J. *et al.* (1962c). *Phys. Rev.* **128,** 1881.
Harari, H. (1968). *Phys. Rev. Lett.* **20,** 1395.
Harari, H. (1969). *Phys. Rev. Lett.* **22,** 562.
Harari, H. (1971a). *Phys. Rev. Lett.* **26,** 1400.
Harari, H. (1971b). *Ann. Phys. (N.Y.)* **63,** 432.
Harari, H. (1971c). Proceedings of the International Symposium on Electron and Photon Interactions at High Energies, Cornell University.
Harari, H. and Zarmi, Y. (1969). *Phys. Rev.* **187,** 2230.
Hedegaard-Jensen, N. (1974). *Nucl. Phys.* **B77,** 173.
Hedegaard-Jensen, H., Nielsen, H. and Oades, G. C. (1973). *Phys. Lett.* **46B,** 385.
Hey, A. J. G. (1974). Daresbury Lecture Series No. 13, Daresbury Report DL/R33.
Hey, A. J. G., Rosner, J. L. and Weyers, J. (1973). *Nucl. Phys.* **B61,** 205.
Hey, A. J. G. and Weyers, J. (1973). *Phys. Lett.* **44B,** 263.
Höhler, G. and Pietarinen, E. (1975). *Phys. Lett.* **53B,** 471.
Höhler, J., Strauss, R. and Wunder, H. (1968). Karlsruhe report (unpublished).

t'Hooft, G. (1971a). *Nucl. Phys.* **B33,** 173.
t'Hooft, G. (1971b). *Nucl. Phys.* **B85,** 167.
Hoogland, W. *et al.* (1974). *Nucl. Phys.* **B69,** 266.
Horn, D. (1966). *Phys. Rev. Lett.* **17,** 778.
Hoyer, P. (1973). (Private communication).
Hoyer, P., Estabrooks, P., and Martin A. D. (1974). *Phys. Rev.* **D10,** 80.
Hoyer, P., Roberts, R. G. and Roy, D. P. (1973). *Nucl. Phys.* **B56,** 173.
Hyams, B. *et al.* (1973). *Nucl. Phys.* **B64,** 134.
Hyams, B. *et al.* (1974). *Nucl. Phys.* **B73,** 202.

Iagolnitzer, D., Zinn-Justin, J. and Zuber, J. B. (1973). *Nucl. Phys.* **B60,** 233.
Igi, K. (1962). *Phys. Rev. Lett.* **9,** 76.
Igi, K. (1968). *Phys. Lett.* **28B,** 330.
Igi, K. and Matsuda, S. (1967). *Phys. Rev. Lett.* **18,** 625, 822.
Iliopolous, J. (1967). *Nuovo Cim.* **52A,** 192.
Iliopolous, J. (1968). *Nuovo Cim.* **53A,** 552.
Irving, A. C. (1973). *Nucl. Phys.* **B63,** 499.

Jackson, J. D. and Pilkuhn, H. (1964). *Nuovo Cim.* **33,** 906.
Jacob, M. and Wick, G. C. (1959). *Ann. Phys.* (*N.Y.*) **7,** 404.
Jacobs, L. D. (1972). *Phys. Rev.* **D6,** 1291.
Jengo, R. and Remiddi, E. (1969). *Nuovo Cim. Lett.* **1,** 637.
Jin, Y. S. and Martin, A. (1964). *Phys. Rev.* **135,** B1369, B1375.
Johannesson, N. O. and Petersen, J. L. (1974). *Nucl. Phys.* **B68,** 397.
Johnson, R. C. (1971). New Theory of Padé Approx. Durham preprint.
Johnson, R. C. (1972). Durham preprint.
Jones, J. A., Allison, W. M. and Saxon, D. H. (1974). *Nucl. Phys.* **B83,** 93.
Jongejans, B. and Voorthuis, H. (1971). "Meson Resonances and Related Electromagnetic Phenomena", ed. by R. H. Dalitz and A. Zichichi, International Physics Series, Bologna.

Kacser, C. (1963). *Phys. Rev.* **130,** 355.
Kamal, A. N. (1969). *Phys. Rev.* **180,** 1454.
Kane, G. L. (1970). "Experimental Meson Spectroscopy", ed. by C. Balcay and A. H. Rosenfeld, Columbia University Press, 1970, p. 1.
Kane, G. L. and Ross, M. (1969). *Phys. Rev.* **177,** 2353.
Kang, J. S. and Lee, B. W. (1971). *Phys. Rev.* **D3,** 2814.
Kang, K. S., Lacombe, M. and Vinh Mau, R. (1971). *Phys. Rev.* **D4,** 3005.
Kapadia, P. D. (1967). *Nucl. Phys.* **B3,** 291.
Katz, W. M. *et al.* (1969). Proceedings of the Conference on $\pi\pi$ and πK Interactions, Argonne National Laboratory.
Kawarabayashi, K., Kitakado, S. and Yabuki, H. (1969). *Phys. Lett.* **28B,** 432.
Kawarabayashi, K. and Suzuki, M. (1966). *Phys. Rev. Lett.* **16,** 255.
Kellett, B. H. (1971). *Nucl. Phys.* **B26,** 237.
Kennedy, J. and Spearman, T. D. (1962). *Phys. Rev.* **126,** 1596.
Khuri, N. N. (1968). *Phys. Rev.* **168,** 1884.
Khuri, N. N. and Treiman, S. B. (1960). *Phys. Rev.* **119,** 1115.
Kibble, T. W. B. (1960). *Phys. Rev.* **117,** 1159.
Kimel, J. D. (1970). *Phys. Rev.* **D2,** 862.

Knudsen, C. P. and Martin, B. R. (1973). *Nucl. Phys.* **B61,** 307.
Kokkedee, J. J. J. (1969). "The Quark Model", Benjamin.
Komen, G. J. (1973). *Nuovo Cim.* **19A,** 265.
Kotanski, A. (1966). *Acta phys. pol.* **29,** 699; **30,** 629.
Kramer, G. (1970). *Springer Tracts in Modern Physics* **55,** 152.
Kugler, M. (1973). "Particles and Fields, 1973", eds. H. Bingham, M. Davier and G. Lynch, American Institute of Physics, New York.
Kupsch, P. (1969a). *Nucl. Phys.* **B11,** 573.
Kupsch, P. (1969b). *Nucl. Phys.* **B12,** 155.
Kupsch, P. (1970a). *Nuovo Cim.* **66A,** 202.
Kupsch, P. (1970b). *Communs. Math. Phys.* **19,** 65.

Laurens, G. (1971). Saclay preprint (thesis) (unpublished).
Lauscher, P. *et al.* (1974). *Nucl. Phys.* **B86,** 189.
Lee, B. W. (1969). *Nucl. Phys.* **B9,** 649.
Lee, T. D. and Wu, C. S. (1966). *A. Rev. nucl. Sci.* **16,** 471.
Lee, T. D. and Yang, C. N. (1956). *Nuovo Cim.* **3,** 749.
Le Francois, J. (1971). In Proceedings of the 5th International Symposium on Electron and Photon Interactions at High Energies, Cornell University.
Le Guillou, J. C., Morel, A. and Navelet, H. (1971). *Nuovo Cim.* **5A,** 659.
Lehman, H. (1958). *Nuovo Cim.* **10,** 579.
Lehman, H. (1972). *Phys. Lett.* **41B,** 529.
Lemoigne, Y. *et al.* (1974). Contributed paper, Proceedings of the XVIIth International Conference on High Energy Physics, London.
Levinson, N. (1949). *K. dansk. Vidensk. Selsk. Mat.-Phys. Mod.* **25,** No. 9.
Linglin, D. *et al.* (1973). *Nucl. Phys.* **B57,** 64.
Lipinski, H. M. (1970). University of Wisconsin report (unpublished).
Lipkin, H. J. and Meshkov, S. (1965). *Phys. Rev. Lett.* **14,** 670.
Lipkin, H. J. (1973). *Phys. Reports* **8C,** 173.
Litchfield, P. J. (1974). Proceedings of the XVIIth International Conference on High Energy Physics, London.
Liu, Y. C. and Okubo, S. (1967). *Phys. Rev. Lett.* **19,** 190.
Logunov, A. A., Soloviev, L. D. and Tarkhelidze, A. N. (1967). *Phys. Lett.* **24B,** 181.
Lopez, C. (1973). *Nuovo Cim.* **18A,** 794.
Lopez, C. and Mennessier, G. (1975). *Phys. Lett.* **58B,** 437.
Losty, M. J. *et al.* (1974). *Nucl. Phys.* **B69,** 185.
Louie, J. *et al.* (1974). *Phys. Lett.* **48B,** 385.
Lovelace, C. (1961). *Nuovo Cim.* **21,** 305.
Lovelace, C. (1968). *Phys. Lett.* **28B,** 264.
Lovelace, C. (1969). Proceedings of the Conference on $\pi\pi$ and $K\pi$ Interactions, Argonne National Laboratory, p. 562.
Lukaszuk, L. and Martin, A. (1967). *Nuovo Cim.* **52,** 122.
Lyth, D. H. (1965). *Rev. mod. Phys.* **37,** 709.
Lyth, D. H. (1971a). *Nucl. Phys.* **B30,** 173.
Lyth, D. H. (1971b). *Nucl. Phys.* **B30,** 195.
Lyth, D. H. (1971c). *Phys. Rev.* **D3,** 1991.
Lyth, D. H. (1972). *Nucl. Phys.* **B48,** 537.

MacGregor, M. H., Arndt, R. A. and Wright, R. M. (1969). *Phys. Rev.* **182,** 1714.
Mahoux, G., Roy, S. M. and Wanders, G. (1974). *Nucl. Phys.* **B70,** 297.

Malamud, E. and Schlein, P. E. (1967). *Phys. Rev. Lett.* **19,** 1056.
Malamud, E. (1969). Proceedings of the Conference on $\pi\pi$ and Kπ Interactions, Argonne.
Mandelstam, S. (1958a). *Phys. Rev.* **112,** 1344.
Mandelstam, S. (1958b). *Phys. Rev.* **115,** 1741.
Mandelstam, S. (1958c). *Phys. Rev.* **115,** 1752.
Mandelstam, S. (1960). *Phys. Rev. Lett.* **4,** 84.
Mandelstam, S. (1963). *Nuovo Cim.* **30,** 1127, 1148.
Mandelstam, S. (1968a). *Phys. Rev.* **166,** 1539.
Mandelstam, S. (1968b). *Phys. Rev. Lett.* **21,** 1724.
Mandelstam, S. (1974). *Phys. Reports* **13C,** 259.
Männer, W. (1974). Contribution to the IVth International Conference on Experimental Meson Spectroscopy, Boston, Massachusetts, U.S.A., April 1974, and CERN preprint.
Marshak, R. E., Riazuddin and Ryan, C. P. (1969). "Theory of Weak Interactions in Particle Physics", Wiley, 1969.
Martin, A. (1966a). *Nuovo Cim.* **42,** 930.
Martin, A. (1966b). *Nuovo Cim.* **44,** 1219.
Martin, A. (1967). *Nuovo Cim.* **47A,** 265.
Martin, A. (1968). *Nuovo Cim.* **58A,** 303.
Martin, A. (1969). *Nuovo Cim.* **63A,** 167.
Martin, A. and Cheung, F. (1970). "Analytic Properties and Bounds of the Scattering Amplitude", Gordon and Breach.
Martin, A. D. (1973). Invited Talk at the IVth International Symposium on Multiparticle Hadrodynamics, Pavia, and CERN preprint TH 1741.
Martin, A. D. (1974). Private Communication.
Martin, B. R. (1967). *Fortschr. Phys.* **15,** 357.
Martin, B. R. and Miller, C. E. (1972). *Nucl. Phys.* **B43,** 157.
Martin, B. R. and Miller, C. E. (1973). *Nucl. Phys.* **B54,** 227.
Martin, B. R. and de Rafael, E. (1967). *Phys. Rev.* **162,** 1453.
Martin, B. R. and Shaw, G. (1973). Unpublished calculations (see Pennington (1974b)).
Masson, D. (1967). *J. math. Phys.* **8,** 512.
Masson, D. (1971). *In* "The Padé Approximant in Theoretical Physics", editors G. Baker and J. Gammel, Academic Press.
Mast, T. S. *et al.* (1969). *Phys. Rev.* **183,** 1200.
Matison, M. J. *et al.* (1974). *Phys. Rev.* **D9,** 1872.
Matthews, P. T. (1950). *Phys. Rev.* **80,** 292.
Matthews, P. T. (1951a). *Phil. Mag.* **42,** 221.
Matthews, P. T. (1951b). *Phys. Rev.* **81,** 936.
Matthews, P. T. and Salam, A. (1954). *Phys. Rev.* **79,** 910.
Mathews, R. D. (1971). Berkeley Report UCRL-20247 (unpublished).
Matsuda, S. (1968). *Phys. Rev.* **169,** 1169.
Meiere, F. T. and Sugawara, M. (1967). *Phys. Rev.* **153,** 1702.
Melosh, H. J. (IV) (1973). Ph.D. thesis, California Institute of Technology (unpublished).
Mercer, R. *et al.* (1971). *Nucl. Phys.* **B32,** 381.
Michael, C. (1973a). *Nucl. Phys.* **B57,** 292.
Michael, C. (1973b). *Nucl. Phys.* **B63,** 431.
Michael, C. (1973c). *Nucl. Phys.* **B61,** 199.

Michel, L. (1953). *Nuovo Cim.* **10,** 319.
Mitra, A. M. and Roy, S. (1964). *Phys. Rev.* **135,** B146.
Moffatt, J. W. (1961). *Phys. Rev.* **121,** 926.
Moffatt, J. W. (1971). *Phys. Rev.* **D3,** 1222.
Montanet, L. (1972). Proceedings of the Symposium on Nucleon–Antinucleon Annihilations, Chexbres, CERN Publication 72-10.
Montanet, L. (1974). Proceedings of the Symposium on Nucleon-Antinucleon Annihilations, Liblice-Prague, CERN Publication 74-18.
Morgan, D. (1965). *Nuovo Cim.* **36,** 813.
Morgan, D. (1971). Proceedings of the XIth Cracow School of Theoretical Physics.
Morgan, D. (1972). Proceedings of the Seventh Finnish Summer School.
Morgan, D. (1974a). *Phys. Rev.* **D9,** 3210.
Morgan, D. (1974b). *Phys. Lett.* **51B,** 71.
Morgan, D. (1974c). Proceedings of the XVIIth International Conference on High Energy Physics, London.
Morgan, D. (1974d). Contributed paper, Proceedings of the XVIIth International Conference on High Energy Physics, London.
Morgan, D. (1975). *In* "New Directions in Hadron Spectroscopy", ed. S. L. Kramer (ANL), and Rutherford Lab. report RL-75-133.
Morgan, D. and Pennington, M. R. (1975). *Phys. Rev.* (to be published).
Morgan, D. and Pišút, J. (1970). *Springer Tracts in Modern Phys.* **55,** 1.
Morgan, D. and Shaw, G. (1969). *Nucl. Phys.* **B10,** 1387.
Morgan, D. and Shaw, G. (1970). *Phys. Rev.* **D2,** 520.
Morgan, D. and Shaw, G. (1972). *Nucl. Phys.* **B43,** 365.
Mueller, A. H. (1970). *Phys. Rev.* **D2,** 2963.

Nachtmann, O. and de Rafael, E. (1969). CERN Preprint, TH. 1031 (unpublished).
Nath, P. (1974). Proceedings of the IVth International Conference on Experimental Meson Spectroscopy, Boston, Massachusetts, U.S.A., April 1974.
Neveu, A. and Scherk, J. (1968). *Phys. Lett.* **27B,** 384.
Neveu, A. and Scherk, J. (1970). *Ann. Phys. (N.Y.)* **57,** 39.
Nicholson, H. *et al.* (1969). *Phys. Rev. Lett.* **23,** 603.
Nicholson, H. *et al.* (1973). *Phys. Rev.* **D7,** 2572.
Nielsen, H. and Oades, G. C. (1972a). *Nucl. Phys.* **B49,** 586.
Nielsen, H. and Oades, G. C. (1972b). *Nucl. Phys.* **B49,** 573.
Nielsen, H. and Oades, G. C. (1973). *Nucl. Phys.* **B55,** 301.
Nielsen, H. and Oades, G. C. (1974). *Nucl. Phys.* **B72,** 321.
Nielsen, H., Petersen, J. L. and Pietarinen, E. *Nucl. Phys.* **B22,** 525 (1970).

Ochs, W. (1972). *Nuovo Cim.* **12A,** 724.
Ochs, W. and Wagner, F. (1973). *Phys. Lett.* **44B,** 271.
Oades, G. C. (1966). *Suppl. Nuovo Cim.* **4,** 217.
Odorico, R. (1969). *Nuovo Cim. Lett.* **11,** 655.
Odorico, R. (1970). *Phys. Lett.* **33B,** 489.
Odorico, R. (1971). *Phys. Lett.* **34B,** 65.
Odorico, R. (1972). *Phys. Lett.* **38B,** 411.
Olsson, M. G. (1967). *Phys. Rev.* **162,** 1338.
Olsson, M. G. and Turner, L. (1968). *Phys. Rev. Lett.* **20,** 1127.
Omnès, R. (1958). *Nuovo Cim.* **8,** 316.

Osborn, H. (1969). *Nuovo Cim. Lett.* **1,** 513.
Osborn, H. (1970). *Nucl. Phys.* **B15,** 501.

Pais, A. and Treiman, S. B. (1968). *Phys. Rev.* **168,** 1858.
Particle Data Group (1974). *Phys. Lett.* **50B,** 1.
Partovi, M. H. and Lomon, E. L. (1970). *Phys. Rev.* **2D,** 1999.
Partovi, M. H. and Lomon, E. L. (1972). *Phys. Rev.* **5D,** 1192.
Pennington, M. R. (1973). American Institute of Physics Conference Proceedings, No. 13, p. 89.
Pennington, M. R. (1974a). *Nuovo Cim.* **25A,** 149.
Pennington, M. R. (1974b). *Ann. Phys. (N.Y.)* **92,** 164.
Pennington, M. R. and Pond, P. (1971). *Nuovo Cim.* **3A,** 548.
Pennington, M. R. and Protopopescu, S. D. (1972). *Phys. Lett.* **40B,** 105.
Pennington, M. R. and Protopopescu, S. D. (1973). *Phys. Rev.* **D7,** 1429.
Pervushin, V. N. and Volkov, M. K. (1974). J.I.N.R., Dubna preprint E2-7661.
Petersen, J. L. (1971). *Phys. Reports* **2C,** 155.
Phillips, R. J. N. and Ringland, G. A. (1972). "High Energy Physics", Vol. 5, p. 187. Ed. E. H. S. Burhop, Academic Press.
Pidcock, M. K. (1974). *Nucl. Phys.* **B83,** 253.
Piguet, O. and Wanders, G. (1969). *Phys. Lett.* **30B,** 418.
Piguet, O. and Wanders, G. (1972). *Nucl. Phys.* **B46,** 295.
Pilkuhn, H. *et al.* (1973). *Nucl. Phys.* **B65,** 460.
Pišút, J. (1970). "Springer Tracts in Modern Physics", Vol. 55, p. 8 Springer-Springer Verlag, 1970.
Pišút, J., Prešnajder, P. and Fischer, J. (1969). *Nucl. Phys.* **B12,** 586.
Pokorski, S., Raitio, R. O. and Thomas, G. H. (1972). *Nuovo Cim.* **7A,** 828.
Pomeranchuk, I. Ya (1958). *Soviet Phys. JETP* **7,** 499.
Pond, P. (1971). *Phys. Rev.* **D3,** 2210.
Prešnajder, P. and Pišút, J. (1969). *Nucl. Phys.* **B14,** 489.
Protopopescu, S. D. *et al.* (1973). *Phys. Rev.* **D7,** 1279.
Prukop, J. P. *et al.* (1974). Contributed paper, Proceedings of the XVIIth International Conference on High Energy Physics, London.

Roberts, R. G. and Wagner, F. (1969). Proceedings of the Conference on Weak Interactions, CERN.
Robertson, W. J., Walker, W. D. and Davis, J. L. (1973). *Phys. Rev.* **D7,** 2554.
Rose, M. E. (1957). "Elementary Theory of Angular Momentum", Wiley, New York.
Rosenfeld, A. H. *et al.* (1975). *Phys. Lett.* **55B,** 486.
Roskies, R. (1969). *Phys. Lett.* **30B,** 42.
Roskies, R. (1970a). *Nuovo Cim.* **65A,** 467.
Roskies, R. (1970b). *J. math. Phys.* **11,** 2913.
Roskies, R. (1970c). *Phys. Rev.* **D2,** 1649.
Rosner, J. L. (1969). *Phys. Rev. Lett.* **22,** 689.
Rosner, J. L. (1974a). *Phys. Reports* **11C,** 189.
Rosner, J. L. (1974b). Private communication.
Ross, M., Henyey, F. S. and Kane, G. L. (1970). *Nucl. Phys.* **B23,** 269.
Roy, S. M. (1971). *Phys. Lett.* **36B,** 353.
Roy, S. M. and Singh, V. (1974). Private communication, and to be published.
Rubinstein, H. R. *et al.* (1968). *Phys. Rev. Lett.* **21,** 491.

Sakurai, J. J. (1961). *Ann. Phys. (N.Y.)* **11,** 1.
Salam, A. (1950). *Phys. Rev.* **79,** 910.
Salam, A. and Ward, J. C. (1964). *Phys. Lett.* **13,** 168.
Samios, N. P., Goldberg, M. and Meadows, B. T. (1974). *Rev. mod. Phys.* **46,** 49.
Sander, O. R. *et al.* (1972). Proceedings of the XVIth International Conference on High Energy Physics, Chicago-Batavia (A.I.P.).
Sawyer, R. F. and Wali, K. C. (1960). *Phys. Rev.* **119,** 1429.
Saxon, D. H., Mulvey, J. H. and Chinowsky, W. (1971). *Phys. Rev.* **D2,** 1790.
Scharenguivel, J. H. *et al.* (1969). *Phys. Rev.* **186,** 1397.
Scharenguivel, J. H. *et al.* (1970). *Phys. Rev. Lett.* **24,** 332.
Schierholz, G. and Sundermeyer, K. (1972). *Nucl. Phys.* **B40,** 125.
Schlein, P. E. (1967). *Phys. Rev. Lett.* **19,** 1052.
Schmid, C. (1968a). *Phys. Rev. Lett.* **20,** 689.
Schmid, C. (1968b). *Phys. Rev. Lett.* **20,** 628.
Schmid, C. (1969). *Nuovo Cim.* **61A,** 289.
Schnitzer, H. J. and Weinberg, S. (1967). *Phys. Rev.* **164,** 1828.
Schwarz, J. H. (1973). *Phys. Reports* **8C,** 270.
Schweinberger, W. *et al.* (1971). *Phys. Lett.* **36B,** 246.
Sekulin, R. L. (1973). *Nucl. Phys.* **B56,** 227.
Shabalin, E. P. (1963). *Zh. éksp. teor. Fiz.* **44,** 765. English Translation *Soviet Phys. JETP* **17,** 517 (1963).
Shapiro, J. (1969). *Phys. Rev.* **179,** 1345.
Shapiro, J. and Yellin, J. (1968). LRL report UCRL-18500 (unpublished).
Shaw, G. (1968). *Phys. Lett.* **28B,** 44.
Shaw, G. (1972). *Phys. Lett.* **39B,** 255.
Shibata, E. I., Frisch, D. H. and Wahlig, M. A. (1970). *Phys. Rev. Lett.* **25,** 1227.
Siddle, R. *et al.* (1971). *Nucl. Phys.* **B35,** 93.
Skuja, A. *et al.* (1973). *Phys. Rev. Lett.* **31,** 653.
Sonderegger, P. and Bonamy, P. (1969). Lund Conference, paper No. 372, unpublished.
Stodolsky, L. and Sakurai, J. J. (1963). *Phys. Rev. Lett.* **11,** 90.
Sutherland, D. G. (1971). Proceedings of the 2nd Daresbury Study Weekend, Daresbury Nuclear Physics Laboratory, 1971.
Symanzik, K. (1954). *Phys. Rev.* **105,** 743.
Symanzik, K. (1969). *Nuovo Cim. Lett.* **11,** 1.
Symanzik, K. (1970). *Communs, Math. Phys.* **16,** 48.

Takagi, F. (1968). Tohoku preprint TU/69/40. See Lovelace (1969).
Taylor, J. C. (1958). *Phys. Rev.* **110,** 1216.
Thompson, G. *et al.* (1973). *Nucl. Phys.* **B69,** 220.
Tiktopoulos, G. (1970). *Phys. Rev. Lett.* **25,** 1463.
Titchmarsh, E. C. (1939). "Theory of Functions" (Second Edition), Oxford University Press.
Treiman, S. B. and Sachs, R. G. (1956). *Phys. Rev.* **103,** 1545.
Treiman, S. B. and Yang, C. N. (1962). *Phys. Rev. Lett.* **8,** 140.
Trippe, T. G. (1971). Recent Experimental Studies of the Kπ Interactions, *in* "Zero Gradient Synchrotron Workshops, Summer 1971", ANL/HEP 7208, Vol. 1, p. 6.
Truong, T. N., Vinh Mau R. and Yem, P. X. (1968). *Phys. Rev.* **172,** 1645.
Tryon, E. P. (1968). *Phys. Rev. Lett.* **20,** 769.

Tryon, E. P. (1969a). *Phys. Rev. Lett.* **22,** 110.
Tryon, E. P. (1969b). Proceedings of the Conference on $\pi\pi$ and $K\pi$ Interactions, Argonne National Laboratory.
Tryon, E. P. (1971a). *Phys. Lett.* **36B,** 470.
Tryon, E. P. (1971b). *Phys. Rev.* **D4,** 1202.
Tryon, E. P. (1971c). *Phys. Rev.* **D4,** 1216.
Tryon, E. P. (1971d). *Phys. Rev.* **D4,** 1221.
Tryon, E. P. (1972a). *Phys. Lett.* **38B,** 527.
Tryon, E. P. (1972b). *Phys. Rev.* **D5,** 1039.
Tryon, E. P. (1973). *Phys. Rev.* **D8,** 1586.
Tryon, E. P. (1974a). *Phys. Rev.* **D10,** 1595.
Tryon, E. P. (1974b). Hunter College Preprint.
Tryon, E. P. (1975). *Phys. Rev.* **D11,** 698.

di Vecchia, P. and Drago, F. (1969). *Nuovo Cim. Lett.* **1,** 917.
Veltman, M. (1968). *Nucl. Phys.* **B7,** 637.
Veneziano, G. (1968). *Nuovo Cim.* **57A,** 190.
Veneziano, G. (1974). *Phys. Reports* **9C,** 200.
Villet, G. *et al.* (1973). Preprint D.Ph.E. CEN Saclay (1973)—presented—to IInd Aix-en-Provence International Conference on Elementary Particles.
Virasoro, M. (1969). *Phys. Rev.* **177,** 2309.

Wagner, F. (1969). *Nuovo Cim.* **64A,** 189.
Wagner, F. (1973). *Nucl. Phys.* **B58,** 494.
Wagner, F. (1974). Proceedings of the XVIIth International Conference on High Energy Physics, London.
Wali, K. C. (1962). *Phys. Rev. Lett.* **9,** 120.
Walker, W. D. *et al.* (1967). *Phys. Rev. Lett.* **18,** 630.
Walker, W. D. (1973). "$\pi\pi$ Scattering—1973" (A.I.P.), p. 80.
Wanders, G. (1966). *Helv. Phys. Acta* **39,** 228.
Wanders, G. (1969). *Nuovo Cim.* **63A,** 108.
Ward, J. C. (1950). *Phys. Rev.* **78,** 182.
Ward, J. C. (1951). *Phys. Rev.* **84,** 897.
Watson, K. M. (1952). *Phys. Rev.* **88,** 1163.
Webber, B. R. (1971). *Phys. Rev.* **D3,** 1971.
Weinberg, S. (1966). *Phys. Rev. Lett.* **17,** 616.
Weinberg, S. (1971). *Phys. Rev. Lett.* **27,** 1688.
Weisberger, W. I. (1965). *Phys. Rev. Lett.* **14,** 1047.
Weyers, J. (1973). CERN report TH1743 (unpublished).
Wightman, A. S. (1956). *Phys. Rev.* **101,** 860.
Williams, P. K. (1969). *Phys. Rev.* **181,** 1963.
Williams, P. K. (1970). *Phys. Rev.* **D1,** 1312.
Wilson, R. (1971). Proceedings of the 5th International Symposium on Electron and Photon Interactions at High Energies, Cornell.
Wolf, G. (1969). *Phys. Rev.* **182,** 1538.
Wolfenstein, L. (1966). *Nuovo Cim.* **42,** 17.
Worden, R. P. (1971). Private communication.
Worden, R. P. (1972). *Nucl. Phys.* **B37,** 253.

Worden, R. P. (1973). Proceedings of the "Pion Exchange Meeting", Daresbury Laboratory 1973, ed. G. A. Winbow, Daresbury preprint DNPL R/30.
Wu, T. T. and Yang, C. N. (1964). *Phys. Rev. Lett.* **13,** 384.

Yang, C. N. and Mills, R. L. (1954). *Phys. Rev.* **96,** 191.
Yennie, D. R. (1971). "Hadronic Interactions of Photons and Electrons", editors J. Cumming and H. Osborn, Academic Press, p. 321.
Yndurain, F. (1970). *Phys. Lett.* **31B,** 368.
Yndurain, F. (1972). *Rev. mod. Phys.* **44,** 645.
Yuta, H. *et al.* (1971). *Phys. Rev. Lett.* **26,** 1502.
Yuta, H. *et al.* (1973). *Nucl. Phys.* **B52,** 70.

Zachariasen, F. (1961). *Phys. Rev. Lett.* **7,** 112, 268 (E).
Zinn-Justin, J. (1971a). *Springer Tracts in Modern Physics,* **57,** 248.
Zinn-Justin, J, (1971b). *Physics Rep.* **1C,** 57.
Zweig, G. (1964). CERN reports TH401, 412 (unpublished).
Zylbersztejn, A. *et al.* (1972). *Phys. Lett.* **38B,** 457.

Subject Index

A

Absorption model, 44–49
  absorption as a function of di-meson mass, 46–49
  dual absorption model, 317–318
  Williams' model, 45–47, 71, 80
ACU (see Analyticity, crossing and unitarity)
Adler sum rule, 377
Adler–Weisberger condition (relation), 373, 375, 377
  expressed in terms of exotic s-channel amplitudes, 381
Adler zero (condition), 372, 381
  expressed in terms of exotic s-channel amplitudes, 375
  in Veneziano models, 329, 380–381
A_1-effect, 355, 358, 365
A_1-exchange (in di-meson production), 40, 41, 51, 53, 417
A_2-exchange (in di-meson production), 40, 43, 52–59, 63, 64, 315, 425
Algebra of currents (see Current algebra)
A_2-meson, 120–121, 355, 365
Amplitude analysis of di-meson production
  Δ-reactions, 39, 83–84
  method of Estabrooks and Martin, 72–76, 80, 93
  method of Ochs and Wagner, 77–82, 93
  spin and phase coherence (SPC), 52–58, 73–75
Analytic continuation of πN scattering amplitudes, 140–144
Analyticity
  from rigorous field theory, 205–206
  Mandelstam analyticity, 16–19
  (see also, Dispersion relations)
Analyticity, crossing and unitarity, 248
  scope for low-energy $\pi\pi$ scattering, 248–274
Ancestors, 355
Atkinson's method, 217–218, 256, 295

B

Backward dispersion relations for πN scattering, 139–142, 144–145
Balachandran–Nuyts–Roskies (BNR) relations
  derivation, 212–214
  explicit form for S- and P-waves, 213
  used to construct partial-wave models, 256–258
Barrelet ambiguity (zeros), 87, 104, 334
Baryon vertex factor in di-meson production, 33–34
BNR relations (see Balachandran–Nuyts–Roskies (BNR) relations)
Bootstraps
  Reggeization, 285–288
  ρ-bootstrap, 244–247
  strip approximation, 280–291
  via finite-energy sum rules, 318–322
Bounds
  absolute low-energy bounds, 210–211
  Froissart bound, 207–208
  inverse amplitude bounds, 219–225

C

CDD poles (ambiguity), 239–241, 247
CERN–Munich experiment, analyses of, 93
Chew–Low extrapolation (see Goebel–Chew–Low extrapolation)
Chew and Mandelstam's (effective range) formula, 243
Chew–Mandelstam equations
  inconsistencies in P-wave solution, 246–247
  P-wave dominant solution, 244–247
  S-wave dominant solution, 241–244
Colliding beams (see e^+e^- colliding beams)
Conformal mapping technique for continuing πN amplitudes, 141
Conspiracy, 425
Conventions, 7–15, 397–401
Coulomb interference, 83
Coupling parameter λ
  current algebra value, 376
  definition, 13
  interpreted as 4π coupling constant, 400–401
  theoretical bounds, 211, 224–225
  (see also, $\lambda\phi^4$ theory)
Crossing, 11, 206, 250–253
Crossing conditions
  left-hand cut from crossing, 235
  partial-wave amplitudes, 236–237
  partial-wave models, 256–258
  (see also, Balachandran–Nuyts–Roskies (BNR) relations, Martin inequalities)
Crossing matrices
  baryon spin crossing matrix, 408
  di-meson density matrix elements, 31
  helicity amplitudes, 31, 406–410
  isospin crossing, 12, 14
  transversity amplitudes, 406–410
Crossing symmetric expansions, 13, 254–265
  use in current algebra, 374–377, 384–388
Cross-section formulas
  di-meson production, 27–29, 410–416
  meson-meson scattering, 10
Current algebra
  Adler sum rule, 377
  Adler–Weisberger condition, 373, 375, 377
  Adler zero, 372, 381
  compared to Veneziano model, 380–381, 391–392
  decay processes, 393
  dipion production, 392
  extrapolation on-shell via linear form, 374–377
  hard pion corrections to Weinberg form, 382–384
  higher order terms in Weinberg expansion, 384–385
  KSFR relation, 377–380, 381
  off-shell scattering, 371–373
  other 0^-0^- scattering lengths, 389–392
  scattering length predictions compared to dispersion relations, 388–389
  σ-commutator, 372, 430
  σ-condition, 373
  $SU(2) \otimes SU(2)$ algebra, 429–430
  threshold parameters, 376, 383, 387
  unitarity corrections to Weinberg's form, 385–388
  Weinberg's calculation of low-energy $\pi\pi$ scattering, 370–377
  (see also, PCAC)
Current commutators, 371–373, 429–430

D

Daughter trajectories
  compared with quark model, 367
  definition, 426
  dual model spectra, 321–322, 324–325, 366–368
Decay widths in higher symmetry schemes
  comparison with experiment, 364–366
  l-broken $SU(6)_W$, 362
  Melosh transformation, 363–364
  quark models, 361–362
  SU(3) predictions, 356–357
Deck mechanism, 358
δ-meson, 120–121, 355, 357, 365

Δ-reactions
definition, 25
formulas for $N\pi \to \Delta\rho$, 420–421
Density matrix for di-meson production
allowed range for unmeasured elements, 54–55
definition, 28, 411
exchange of dominant naturality, 412, 414
general properties, 411–414
(see also, Legendre moments)
D-function, 237–241
Di-meson production
comparison with charged pion photoproduction, 61–63
Coulomb interference, 83
effective trajectory functions 58–59
exchange contributions as a function of di-meson mass, 46–49, 63–64
exchange mechanisms other than OPE, 40–49
general formalism, 26, 402–421
intensity formulas, 27–28, 410–413
low-energy analyses (isobar model), 82–83
predictions for unobserved density matrix elements, 54–55
production amplitudes at very small t, 49–52
s- and t-channel frames, 27, 404–405
systematics of exchange contributions, 53
(see also, Amplitude analysis of di-meson production, Density matrix for di-meson production, Helicity amplitudes for di-meson production, Legendre moments, Methods for extracting di-meson phase shifts, OPE, Polarization formulas)
Discrepancy function for $\pi N \to \pi N$, 140–150
Disperson relations
(see Backward disperson relations for πN scattering, Fixed variable (t) disperson relations, Inverse amplitude disperson relations, Inverse partial-wave amplitude disperson relations, Partial-wave dispersion relations for πN scattering, Partial-wave dispersion relations for $\pi\pi$ scattering, Roy's equations)
Double spectral functions, 16–18, 151, 283–284, 424
Dual absorption model, 317–318
Duality
dual absorption model, 317–318
exchange couplings in di-meson production, 63–65
exotic channels, 313–317
global, 312
local, 319, 321, 323
semi-global, 312
semi-local, 312
two-component, 312–318
(see also, Daughter trajectories, Exchange degeneracy (EXD), Finite-energy sum rules (FESR), Schmid circles (loops), Veneziano model)
Duality diagrams, 316
Dual models, 334, 346–347
(see also, Veneziano model)
Dürr–Pilkuhn form factors, 41

E

Electromagnetic form factors
(see Nucleon form factors, Pion form factor)
e^+e^- colliding beams
evidence for ρ' (1600), 184
experimental results, 181–184
4π production 183–184
ω–ρ interference, 180–181
$\pi^+\pi^-$ production, 174–183, 197
two-photon annihilation, 184–186
(see also, Pion form factor)
ε-meson candidates
assignments in quark model, 355, 357
information from di-meson production, 104–105, 110, 117, 200
predictions from dual models, 330–331, 367
predictions from Lagrangian models, 302
symmetry predictions for widths, 365
Estabrooks and Martin's method, 72–76, 80, 93

η-meson
$\pi\pi\gamma$ decay, 164
3π decay, 169–173, 335–336, 393–394
(see also, $\pi\eta$ scattering)
Evasion, 425
Exchange degeneracy (EXD)
duality and EXD, 314–317
strong EXD, 425
weak EXD, 425
Exotic states
exchange degeneracy, 313–317
definition of first and second kind, 351

F

Factorization of Regge poles, 424
Factorization model, 67–69
FESR (see Finite-energy sum rules)
Field theory
conventions, 399
reduction formulas, 400
(see also, Lagrangian field theory models)
Final-state interactions
dipion production, 67, 82
$K_{\pi 2}$ decay, 165
$K_{\pi 3}$ decay, 169–173
K_{e4} decay, 158
Finite-energy sum rules (FESR)
applications to $\pi\pi$ scattering, 310–311
bootstraps, 318–322
derivation and general properties, 306–310
Fixed variable (t) dispersion relations
derivation, 17–19, 206
derivative relations at $t=0$, 227–228
inverse amplitudes, 219–226
phenomenological applications to πK scattering, 275–278
phenomenological applications to $\pi\pi$ scattering, 261–270
projection of partial waves, 226–231
region of validity, 19, 206
subtractions, 19, 206, 208
(see also, Forward sum rules, Roy's equations)
f-meson
assignment in quark model, 355
resonance parameters from di-meson production, 108–109, 200
symmetry predictions for width, 356, 365
width in Lagrangian models, 298–299, 302
width in Veneziano model, 330–331
f'-meson
assignment in quark model, 355
decouples from $\pi\pi$ by duality, 316
resonance parameters from di-meson production, 120, 200
symmetry predictions for width, 356, 365
width in Lagrangian models, 302
Form factors (see Dürr–Pilkuhn form factor, Nucleon form factors, Pion form factor)
Forward sum rules
current algebra, 377–378
dispersion theory, 262–264, 274–275
Froissart bound, 207–208
Froissart–Gribov representation, 209–210, 285

G

Gamma matrices, 399
Gell–Mann–Okubo mass formula, 352
g-meson
assignment in quark model, 355
resonance parameters from di-meson production, 110, 200
width in Veneziano model, 330–331
Goebel–Chew–Low extrapolation, 69–72, 93
Goldberger–Treiman relation, 431–432
Gottfried–Jackson (t-channel helicity) frame, 27, 404–405
Gounaris–Sakurai formula (see Pion form factor)

H

Hard pion approach to current algebra, 382–384
Helicity amplitudes for di-meson production, 405–411, 416, 421
examples for specific final states, 416–421

examples from data, 57, 74
exchange of naturality, 30, 406–407
kinematic zeros at small t, 409–410
relation to transversity amplitudes, 405
restrictions from parity, 405
s–t crossing relations, 31, 406–410
Helicity amplitudes for $\pi\pi \to N\bar{N}$, 139
High-energy behaviour
Froissart bound, 207–208
phase-energy relations, 209
Pomeranchuk theorems, 209
Regge poles, 422–426
h-meson, 111–112, 200, 331
Homographic transformation, 293

I

Ideal mixing
definition, 353
experimental evidence for, 355–357
Inclusive cross-section, 26, 63–65
Inelasticity in $\pi\pi$ scattering, 109, 112
Inverse amplitude dispersion relations, 219–226
Inverse partial-wave amplitude dispersion relations, 260–261
Isobar model, 82–83, 129–130, 132
Isospin
conventions, 11–14, 401
decompositions for physical channels, 12, 118
(see also, Crossing matrices)

K

$K\bar{K}$ production, 109, 112–117, 120–121
KK ($K\bar{K}$) scattering, 14, 120–121, 343
KN scattering, 153–154
$K\pi$ scattering
current algebra predictions, 390–391
dispersion relation phenomenology, 275–278
isospin formalism, 13–14, 118, 401
Lagrangian models, 300–305
phase shifts from $K\pi$ production, 117–120, 197, 200–201
scattering lengths, 120, 276–278, 303, 344, 390–391
Veneziano model, 343–345 (see also, Methods for extracting di-meson phase shifts)
κ-meson
assignment in quark model, 365
predictions from Lagrangian models, 302
resonance parameters from di-meson production, 118, 200
symmetry predictions for width, 365
$K_{\pi 2}$ decay, 164–169
$K_{\pi 3}$ decay
early models for spectrum, 169–171
Khuri–Treiman method, 171–173
Lovelace–Veneziano model, 334–336
predictions from current algebra, 393–394
K_{e4} decay
formalism, 156–159
Pais–Treiman method, 159
results for low-energy S-wave $\pi\pi$, 160–163, 197–199
K-function (see Mandelstam representation)
Khuri–Treiman method, 171–173
Kinematics
di-meson production, 26–32, 402–421
invariants, s, t and u, 7–8
$\pi\pi$ scattering, 7–11
K-matrix unitarization, 337–341
K^*-mesons
assignments in quark model, 355
exchange degeneracy, 315, 344
predictions from Lagrangian models, 302
resonance parameters from di-meson production, 118, 200
symmetry predictions for widths, 356, 365
(see also, κ-meson)
KSFR relation, 377–380, 381

L

Lagrangian field theory models (see $\lambda\phi^4$ theory, Padé approximants, σ-model)
λ (see Coupling parameter λ)

$\lambda\phi^4$ theory
parameter λ related to expansion about symmetry point, 400–401
solution via Padé approximants, 295–296, 300
Legendre moments
data for dipion production, 38, 78–79, 88–90
definition, 27, 411
explicit formulas for $J_{max} = 2$, 413
ratios compared to Williams' model, 46–47
relation to density matrix elements, 412
spin and phase consistency tests, 80–81
Lehman ellipse, 206 (see also, Martin–Lehman ellipse)
Levinson's theorem, 239, 240, 246
Linear expansion about symmetry point, 376
(see also, Crossing symmetric expansions)
Lovelace–Veneziano model, 328–331
(see also, Veneziano model)

M

Mandelstam analyticity, 16–19
Mandelstam iteration, 282, 284, 290
Mandelstam representation, 16–19
kernels *K*, *J*, 15, 17, 284
(see also, Double spectral functions)
Martin inequalities
derivation and general properties, 214–217
Roy's equations, 231, 271
used to construct partial-wave models, 256–258
Martin–Lehman ellipse, 206–207
Melosh transformation, 363–366
Methods for extracting di-meson phase shifts
amplitude analysis and extrapolation, 72–76, 80, 93
amplitude analysis of physical region *t*-averaged moments, 76–82
Estabrooks and Martin's method, 72–76, 80, 93
"evasive" extrapolation, 70
factorization model, 67–69
Goebel–Chew–Low extrapolation, 69–72, 93
Ochs and Wagner's method, 77–82, 93
Schlein's method, 69, 77
survey of principal methods, 197–198
(see also, e^+e^- colliding beams, K_{e4} decay)
Metric, 397–398
Mixing angles, 352–354
experimental evidence on, 355–357
Models based on analyticity, 248–279

N

N/D method
continuous *J*, 285–287
formulation, 237–241
pion form factor, 177–180
ρ-bootstrap, 244–247
solutions of Chew–Mandelstam equations, 243–247
Nonsense zeros, 425
evidence in ω-exchange, 43
exchange degeneracy, 314, 316, 317
Normalization conventions, 9, 10, 400
Notation, 7–15, 397–401
$N\bar{N}$ annihilation, 121–136
expected range of *J* values, 122–124
$N\bar{N}$ (at rest) $\rightarrow 3\pi$, 127–133, 135, 336–337
$N\bar{N} \rightarrow K\bar{K}$, 122, 126
$N\bar{N}$ (in flight) $\rightarrow 3\pi$, 133–134
$N\bar{N} \rightarrow$ other channels, 135–136
$N\bar{N} \rightarrow \pi^+\pi^-$, 122–126, 198
$N\bar{N} \rightarrow \pi^0\pi^0$, 132
question of *S*-wave capture, 127, 132
Nucleon form factors, 187–192
NN scattering, 150–153
N-reactions, 25

O

Ochs and Wagner's method, 77–82, 93
Odorico zeros (see Zero trajectories)
Olsson sum rule, 246, 262
ω-ρ mixing (see ρ–ω interference)

Omnès function, 146, 238–239
(see also, Phase representation of pion form factor)
OPE
backgrounds, 37, 40–49
compared to experiment, 36–40
form factor modifications, 41–43
formulas, 33–36
other trajectories, 43–44
Reggeised OPE, 58–59
very small t region, 49–52
(see also, Absorption model)
Optical theorem, 10

P

Padé approximants
definition and general properties, 293–295
$\lambda\phi^4$ theory, 295–296
multichannel calculations, 296, 300–305
ρ-exchange model, 298–299
σ–model, 296–298
SU(3) solution, 300–305
Pais–Treiman method, 159
Partial-wave amplitudes
analytic properties, 233
definition, 11
Froissart–Gribov representation, 209–210
parameterization in terms of δ and η, 15
projections from fixed-t dispersion relations, 226–231
unitarity, 15
Partial-wave dispersion relations for πN scattering, 143–150
Partial-wave dispersion relations for $\pi\pi$ scattering
Chew–Mandelstam S-wave dominant solution, 241–244
Chew–Mandelstam P-wave dominant solution, 244–247
derivation and crossing, 233–237
phenomenological applications, 258–260
solution by N/D method, 237–241
Partial-wave expansion, 15
convergence, 19, 206
PCAC
decay widths in the quark model, 362, 365
definition, 431
for kaons, 390–391
Goldberger–Treiman relation, 431–432
Lagrangian models, 291–292, 296–297
PCAC form factors, 42
soft-pion results, 371–373
Veneziano model and Adler zeros, 329, 344–345
(see also, Current algebra)
Phase-energy relations, 209
Phase representation of pion form factor, 176
(see also, Omnès function)
Photoproduction
Drell–Söding mechanism, 192–193
of π^+, π^- states, 61–63
vector mesons (ρ, ρ'), 192–195
Physical regions, 132, 403, 404
Pion charge radius, 189
Pion decay constant, 428, 432
$\pi\eta$ scattering, 120
problem for Veneziano model, 345, 392
Pion field
defined through PCAC, 431
reduction formula, 400
Pion form factor
definition, 175
dispersion relations, 175–180
Gounaris–Sakurai formula, 178–179, 181–183
ρ-dominance models, 177–179, 379
space-like region, 189–191
πK scattering (see $K\pi$ scattering)
$\pi N \to \pi\pi N$ (see Di-meson production)
πN scattering
analytic continuation to t-channel, 140–141, 143
backward dispersion relations, 139–142
consistency tests for $\pi\pi$ amplitudes, 146–150
partial-wave dispersion relations, 143–144
phase condition, 138
predictions for $\pi\pi$ phase shifts, 144–145

$\pi\pi$ coupling constant λ (see Coupling parameter λ)
$\pi\pi \rightarrow K\bar{K}$, 112–117, 120
$\pi\pi$ phase shift analysis
alternative analyses, 91–93
ambiguities, 85–87, 104
dipion production data from Saclay, LBL and CERN–Munich experiments, 87–91
$\pi^0\pi^0$ production, 101–102
(see also, Methods for extracting di-meson phase shifts)
$\pi\pi$ phase shifts
summary of principal experimental results, 198–200
summary of principal methods of extraction, 196–198
(see also, Methods for extracting di-meson phase shifts)
$\pi\pi \rightarrow \pi\omega$, 319–322, 326–328
$\pi\pi$ resonances, table, 200
$\pi\pi$ scattering lengths (see Scattering lengths ($\pi\pi$ and πK))
Polarization formulas, 415–416
Pomeranchuk theorems, 209
Poor man's absorption (PMA) model, 45–47, 71, 80
Positivity, 15, 206
Potential, use in strip approximation, 283–286

Q

Quark diagrams, 316, 354
Quark model
assignment of meson states, 355–358
comparison with dual model spectra, 367–368
decay widths, 361–366
exotics of first and second kind, 351
L-excitation quark model, 350–351
Melosh transformation, 363–366
SU(3), 352–354
SU(6), 358–361
Zweig's rule, 354
(see also, Quark diagrams, Quarks)
Quarks
constituent and current quarks, 363
quantum numbers, 350

R

Reduction formula, 400
Regge poles
basic formulas and definitions, 422–426
dispersion relations, 288
duality and exchange degeneracy, 314–319
duality and meson classification, 366–368
effective π-exchange trajectory, 58–59
from Mandelstam iteration, 285–290
non-OPE contributions to di-meson production, 43–49
phenomenology of inclusive reactions, 63–65
(see also, Daughter trajectories)
Resonance parameters, summary of experimental results, 200
ρ-dominance (see Vector meson dominance)
ρ-meson
coupling constants, 36, 276, 379
e^+e^- colliding beams, 182
di-meson production, 95–97
KSFR relation for width, 377–380, 381
Lagrangian models, 296, 298, 302
quark model, 355, 356
ρ-bootstrap, 244–247
strip approximation, 289, 290
summary of experimental results, 200
ρ-ω interference
di-meson production, 60–61
$e^+e^- \rightarrow \pi^+\pi^-$, 180–182
$\rho'(1250)$—possible daughter of f^0
absence of signal in dipion production, 104
existence predicted by form factor model, 190–192
existence predicted by Veneziano model, 327, 331
indications in $N\bar{N}$ annihilations, 131–132, 135–136
possible signal in $\pi\omega$ photoproduction, 368
$\rho'(1600)$
assignment in quark model, 355
possible evidence from dipion production, 111

evidence from e^+e^- annihilation, 184
evidence from photoproduction, 193–195
summary of experimental results, 200
symmetry predictions for width, 365
width in Veneziano model, 331
Rigorous results, 205–218
Roskies relations (see Balachandran–Nuyts–Roskies (BNR) relations)
Roy's equations
compared to Chew–Mandelstam equations, 246
derivation and general properties, 228–231
phenomenological applications, 270–274

S

Satellites, 326–328, 367
Scalar mesons, 117, 121, 357, 365
(see also, δ-meson, ε-meson candidates, κ-meson, S^*-meson)
Scattering amplitudes
for $\pi\pi$ states, 9–12
threshold expansion, 15
Scattering lengths ($\pi\pi$ and πK)
bounds, 210–211, 215, 219–226
current algebra predictions, 376, 383, 387, 389, 390–391
definition, 15
empirical estimate for πK, 120
fixed-t dispersion relation, 263, 277–278
Lagrangian field theory models, 299, 303
Lovelace–Veneziano model, 330, 340, 344
Roy's equations, 263, 269, 273
summary of empirical estimates for $\pi\pi$, 178–199
(see also, Universal curve)
Schlein's method, 69, 77
Schmid circles (loops), 319, 321–322, 328, 367
σ-commutator, 372, 430
σ-condition, 373
expressed in terms of exotic s-channel amplitudes, 375
Veneziano model, 380–381
σ-model, 296–298, 299, 430
Slope parameter, definition, 15
S^*-meson
assignment in quark model, 355, 357
coupling to $\pi\pi$ and $K\bar{K}$ channels, 116–117
resonance parameters from di-meson production, 112–117, 200
symmetry predictions for width, 365
two-channel model, 115–117, 276
width in Lagrangian models, 302–303, 305
Spin-analysis of di-meson production, 402–421
Spin and phase coherence (SPC), 52–58
Strip approximation
"new" strip approximation, 285–290
"old" strip approximation, 283–285, 290
Sum rules (see Finite-energy sum rules, Forward sum rules)
$SU(2) \otimes SU(2)$ symmetry, 429–430
SU(3) symmetry and its breaking, 352–354, 356, 357
SU(6) symmetry
mass formulas, 344, 360
predictions for decay widths, 364–366
SU(6) and the quark model, 358–361
$SU(6)_W$, 360–361
$SU(6)_W$ breaking, 361–364
Symmetry point for $\pi\pi$ scattering, 13
(see also, Crossing symmetric expansions)

T

$t_{\min}$, 32, 404
Threshold parameters (see Scattering lengths ($\pi\pi$ and πK), Universal curve)
Total cross-section ($\pi\pi$)
duality predictions, 313
Froissart bound, 207–208
phenomenological estimates, 424–425
Transversity amplitudes
examples for specific reactions, 417–418
general formulas, 404–410
polarization formulas, 415–416

U

Unitarity, 14
  for elastic scattering as convolution, 15
  optical theorem, 10, 14
  partial waves, 15
Unitary symmetry (see SU(3) symmetry, SU(6) symmetry)
Units, 8, 10, 397–398
Universal curve
  forward dispersion relations, 268–270
  inverse partial-wave dispersion relations, 261
  πK scattering, 277–278
  remarks on form of, 274–275
  Roy's equations, 273
  unitarized Veneziano model, 342–343
Up-Down ambiguity, 85–86, 99–101, 118–119, 266–267

V

Vector meson dominance
  nucleon form factors, 190–192
  peripheral di-pion production, 62
  pion form factor, 177–179, 379
  ρ-photoproduction, 192–193
Veneziano model
  $\bar{p}n$ annihilation, 336–337, 130–133
  comparison with current algebra, 345, 380–381, 391–392
  derivation and general properties, 322–326
  $K, \eta \to 3\pi$ decays, 334, 336
  Lovelace-Veneziano model, 328–331
  $\pi N \to \pi\pi N$, 42
  $\pi\pi \to \pi\omega$, 326–328
  other 0^-0^- processes, 343–345
  resonance spectra, 324, 325, 327, 328, 366–368
  resonance widths, 330–331
  satellites, 326–328, 367
  scattering lengths, 330, 340
  zero trajectories, 331–334

W

Watson's theorem, 67, 158
Weinberg's model, 370–377
Williams' model, 45–47, 71, 80
W-spin, 360

Z

Zero trajectories
  $\bar{N}N \to \pi\pi$, 125–126
  $\bar{N}N \to 3\pi$, 132–134
  Veneziano model, 331–334
  (see also, Barrelet ambiguity (zeros))
Zeros
  theorem for zeros, 220–221
  (see also, Adler zero, Zero trajectories)
Zweig's rule, 354